ECOLOGY, CONSERVATION, *and* MANAGEMENT *of* GROUSE

ECOLOGY, CONSERVATION, *and* MANAGEMENT *of* GROUSE

Brett K. Sandercock, Kathy Martin, and
Gernot Segelbacher, *Editors*

Studies in Avian Biology No. 39

A PUBLICATION OF THE COOPER ORNITHOLOGICAL SOCIETY

University of California Press
Berkeley   Los Angeles   London

University of California Press, one of the most distinguished university presses in the United States, enriches lives around the world by advancing scholarship in the humanities, social sciences, and natural sciences. Its activities are supported by the UC Press Foundation and by philanthropic contributions from individuals and institutions. For more information, visit www.ucpress.edu.

Studies in Avian Biology No. 39
For digital version, see the UC Press website.

University of California Press
Berkeley and Los Angeles, California

University of California Press, Ltd.
London, England

Library of Congress Cataloging-in-Publication Data

Ecology, conservation, and
management of grouse / Brett K. Sandercock,
Kathy Martin, and Gernot Segelbacher, editors.   p.   cm. —
(Studies in avian biology; no. 39)
"A publication of the Cooper Ornithological Society."
Includes bibliographical references and index.
ISBN 978-0-520-27006-0
(cloth : alk. paper)

1. Grouse—Ecology.   2. Grouse—Conservation.   I. Sandercock,
Brett K. (Brett Kevin), 1966-   II. Martin, Kathy (Katherine),
1949-   III. Segelbacher, Gernot.

QL696.G27E26   2011

598.6'3—dc23   2011020262

The paper used in this publication meets the minimum requirements of
ANSI/NISO Z39.48-1992 (R 1997)(Permanence of Paper).

Cover image: Male Willow Ptarmigan in Tatshenshini Valley, British Columbia.
Photo by Gilbert X. Ludwig.

PERMISSION TO COPY

DEDICATION

We dedicate this volume to the
memory of Professor Esa Ranta,
grouse biologist and friend.

CONTENTS

Part III • Population Biology

Part IV • Conservation and Management

CONTRIBUTORS

CAMERON L. ALDRIDGE
NREL, Colorado State University &
U.S. Geological Survey
Fort Collins Science Center
2150 Centre Avenue, Building C
Fort Collins, CO 80526
aldridgec@usgs.gov

ROGER D. APPLEGATE
Tennessee Wildlife Resource Agency
P.O. Box 40747
Nashville, TN 37204
roger.applegate@state.tn.us

MARC ARVIN-BEROD
Office National de la Chasse et de la Faune Sauvage
Service Départemental de la Haute Savoie BP 41
74320 Sevrier, France
sd74@oncfs.gouv.fr

MICHAEL T. ATAMIAN
Department of Natural Resources and
Environmental Science
University of Nevada
1000 Valley Road
Reno, NV 89512
(Current address: Washington Department
of Fish and Wildlife, 2315 North
Discovery Place
Spokane Valley, WA 99216,
michael.atamian@dfw.wa.gov)

JACQUELINE K. AUGUSTINE
Department of Evolution, Ecology and
Organismal Biology
The Ohio State University
4240 Campus Drive
Lima, OH 45804
augustine.63@osu.edu

VERNON C. BLEICH
Department of Biological Sciences
Idaho State University
921 South 8th Avenue, Stop 8007
Pocatello, ID 83209
vcbleich@gmail.com

ERIK J. BLOMBERG
Department of Natural Resources and
Environmental Science
University of Nevada
1000 Valley Road
Reno, NV 89512
eblomberg@cabnr.unr.edu

CLAIT E. BRAUN
Grouse Inc.
5572 North Ventana Vista Road
Tucson, AZ 85750
sg-wtp@juno.com

JEAN-FRANÇOIS BRENOT
Office National de la Chasse et de la Faune
Sauvage
Direction des Etudes et de la Recherche
Espace Alfred Sauvy
66500 Prades, France
prades@oncfs.gouv.fr

MICHAEL L. CASAZZA
U.S. Geological Survey
Western Ecological Research Center
Dixon Field Station
6924 Tremont Road
Dixon, CA 95620
mike_casazza@usgs.gov

PETER S. COATES
U.S. Geological Survey
Western Ecological Research Center
Dixon Field Station
6924 Tremont Road
Dixon, CA 95620
pcoates@usgs.gov

JOHN W. CONNELLY
Idaho Department of Fish and Game
1345 Barton Road
Pocatello, ID 83204
jcsagegrouse@aol.com

STEPHEN J. DEMASO
Gulf Coast Joint Venture
National Wetlands Research Center
700 Cajundome Boulevard
Lafayette, LA 70506
steve_demaso@fws.gov

JEAN-FRANÇOIS DESMET
Groupe de Recherches et d'Information
sur la Faune dans les Ecosystèmes de
Montagne
La Jaÿsinia
74340 Samoëns, France
jfdesmet@wanadoo.fr

SHAWN ESPINOSA
Nevada Department of Wildlife
1100 Valley Road
Reno, NV 89512
sespinosa@ndow.org

ROBERT M. GIBSON
School of Biological Sciences
University of Nebraska at Lincoln
348 Manter Hall
Lincoln, NE 68588
rgibson2@unl.edu

ANDREW J. GREGORY
Division of Biology
Kansas State University
116 Ackert Hall
Manhattan, KS 66506
grego1aj@k-state.edu

CHRISTIAN A. HAGEN
Oregon Department of Fish and Wildlife
61374 Parrell Road
Bend, OR 97702
christian.a.hagen@state.or.us

STEPHEN HAMILTON
Alberta Conservation Association
101-9 Chippewa Road
Sherwood Park, AB T8A 6J7
(Current address: Department of Biological
Sciences, University of Alberta,
Edmonton, AB T6G 2E9,
sgh1@ualberta.net)

SUSAN J. HANNON
Department of Biological Sciences
University of Alberta
CW 405, Biological Sciences Building
Edmonton, AB T6G 2E9
sue.hannon@ualberta.ca

KATIE M. HERMAN-BRUNSON
Department of Wildlife and Fisheries Sciences
South Dakota State University
Box 2140B
Brookings, SD 57007
(Current address: Meadowlark Elementary,
Gillette, WY 82717,
kbrunson@ccsd.k12.wy.us)

KENT C. JENSEN
Department of Wildlife and Fisheries Sciences
South Dakota State University
Box 2140B
Brookings, SD 57007
kent.jensen@sdstate.edu

RYAN S. JONES
Department of Wildlife and Fisheries Sciences
Texas A&M University
210 Nagle Hall
College Station, TX 77843
ryan.jones@mdc.mo.gov

NICHOLAS W. KACZOR
Department of Wildlife and Fisheries Sciences
South Dakota State University
Box 2140B
Brookings, SD 57007
(Current address: Division of Refuge Planning,
U.S. Fish and Wildlife Service,
Lakewood, CO 80228,
nick_kaczor@fws.gov)

ROBB S. A. KALER
Division of Biology
Kansas State University
116 Ackert Hall
Manhattan, KS 66506
rsakaler@yahoo.com

ROBERT W. KLAVER
Earth Resources Observation and Science Center
U.S. Geological Survey
47914 252nd Street
Sioux Falls, SD 57198
bklaver@usgs.gov

JOHN P. LEONARD
Department of Wildlife and Fisheries Sciences
Texas A&M University
210 Nagle Hall
College Station, TX 77843
jpleonard2000@yahoo.com

PATRICK LEONARD
Office National de la Chasse et de la Faune Sauvage
CNERA Faune de Montagne
La Bérardie
05000 Gap, France
patrick.leonard@oncfs.gouv.fr

ROEL R. LOPEZ
Department of Wildlife and Fisheries Sciences
Texas A&M University
210 Nagle Hall
College Station, TX 77843
roel@tamu.edu

THOMAS M. LOUGHIN
Department of Statistics and Actuarial Science
Simon Fraser University
250-13450 102nd Avenue
Surrey, BC V3T 0A3
tloughin@sfu.ca

JEFFERY J. LUSK
Nebraska Game and Parks Commission
Wildlife Division
2200 N 33rd
Lincoln, NE 68583
jeff.lusk@nebraska.gov

EDDIE K. LYONS
Harold and Pearl Dripps Department of Agriculture
McNeese State University
4380 Ryan Street
Lake Charles, LA 70609
elyons@mcneese.edu

DOUG MANZER
Alberta Conservation Association
Box 1139, Provincial Building
Blairmore, AB T0K 0E0
doug.manzer@ab-conservation.com

KATHY MARTIN
Department of Forest Sciences
University of British Columbia
2424 Main Mall
Vancouver, BC V6T 1Z4
kathy.martin@ubc.ca

TY W. MATTHEWS
University of Nebraska at Lincoln
School of Natural Resources
135 Hardin Hall
3310 Holdrege Street
Lincoln, NE 68583
tywmatthews@hotmail.com

CLINTON W. MCCARTHY
U.S. Forest Service
Intermountain Region (R4)
Federal Building
324 25th Street
Ogden, UT 84401
cmccarthy01@fs.fed.us

ROBERT A. MCCLEERY
Department of Wildlife Ecology and Conservation
314 Newins-Ziegler Hall
PO Box 110430
University of Florida
Gainesville, FL 32611
ramccleery@ufl.edu

LANCE B. MCNEW
Division of Biology
Kansas State University
116 Ackert Hall
Manhattan, KS 66506
lbmcnew@k-state.edu

JOSHUA J. MILLSPAUGH
Department of Fisheries and Wildlife Sciences
302 Anheuser-Busch Natural Resources Building
University of Missouri at Columbia
Columbia, MO 65211
millspaughj@missouri.edu

MARC MONTADERT
Office National de la Chasse et de la Faune Sauvage
Les Granges Michel
25300 Les Verrières-de-Joux, France
marc.montadert@yahoo.fr

BERTRAND MUFFAT-JOLY
Office National de la Chasse et de la Faune Sauvage
Service Départemental de la Haute Savoie BP 41
74320 Sevrier, France
sd74@oncfs.gouv.fr

DAVID D. MUSIL
Idaho Department of Fish and Game
324 South 417 East
Jerome, ID 83338
david.musil@idfg.idaho.gov

NEAL D. NIEMUTH
U.S. Fish and Wildlife Service
Habitat and Population Evaluation Team 3425
Miriam Avenue
Bismarck, ND 58501
neal_niemuth@fws.gov

CLAUDE NOVOA
Office National de la Chasse et de la Faune Sauvage
Direction des Etudes et de la Recherche
Espace Alfred Sauvy
66500 Prades, France
claude.novoa@oncfs.gouv.fr

MICHAEL S. O'DONNELL
ASRC Management Services
U.S. Geological Survey
Fort Collins Science Center
2150 Centre Avenue, Building C
Fort Collins, CO 80526
odonnellm@usgs.gov

CORY T. OVERTON
U.S. Geological Survey
Western Ecological Research Center
Dixon Field Station
6924 Tremont Road
Dixon, CA 95620
coverton@usgs.gov

SARA J. OYLER-MCCANCE
U.S. Geological Survey
Fort Collins Science Center
2150 Centre Avenue, Building C
Fort Collins, CO 80526
sara_oyler-mccance@usgs.gov

MICHAEL A. PATTEN
Oklahoma Biological Survey and Department
of Zoology
University of Oklahoma
111 E Chesapeake Street
Norman, OK 73019
mpatten@ou.edu

MARKUS J. PETERSON
Department of Wildlife and Fisheries Sciences
Texas A&M University
210 Nagle Hall
College Station, TX 77843
mpeterson@tamu.edu

JAMES C. PITMAN
Emporia Research and Survey Office
Kansas Department of Wildlife and Parks
1830 Merchant
Emporia, KS 66801
jim.pitman@ksoutdoors.com

LARKIN A. POWELL
School of Natural Resources
University of Nebraska at Lincoln
419 Hardin Hall
3310 Holdrege Street
Lincoln, NE 68583
lpowell3@unl.edu

THOMAS J. PREBYL
Division of Biology
Kansas State University
116 Ackert Hall
Manhattan, KS 66506
tpreb09@k-state.edu

CHRISTIN L. PRUETT
Department of Biological Sciences
Florida Institute of Technology
150 W University Boulevard
Melbourne, FL 32901
cpruett@fit.edu

KERRY P. REESE
Department of Fish and Wildlife Resources
University of Idaho
P.O. Box 441136
Moscow, ID 83844
kreese@uidaho.edu

JEAN RESSEGUIER
Office National de la Chasse et de la Faune Sauvage
Direction des Etudes et de la Recherche
Espace Alfred Sauvy
66500 Prades, France
prades@oncfs.gouv.fr

ROBERT J. ROBEL
Division of Biology
Kansas State University
116 Ackert Hall
Manhattan, KS 66506
rjrobel@k-state.edu

MARK A. RUMBLE
Forest and Grassland Research Laboratory
U.S. Forest Service Rocky Mountain
Research Station
8221 South Hwy 16
Rapid City, SD 57702
mrumble@fs.fed.us

TERRY L. RUSSI
USDI Bureau of Land Management
785 North Main Street, Suite E
Bishop, CA 93514
birdsong.123@gmail.com

BRETT K. SANDERCOCK
Division of Biology
Kansas State University
116 Ackert Hall
Manhattan, KS 66506
bsanderc@k-state.edu

ADAM C. SCHOLE
School of Natural Resources
University of Nebraska at Lincoln
135 Hardin Hall
3310 Holdrege Street
Lincoln, NE 68583
lvtcschole@yahoo.com

BENJAMIN S. SEDINGER
Department of Natural Resources and
Environmental Science
University of Nevada
1000 Valley Road
Reno, NV 89512
bsedinger@gmail.com

JAMES S. SEDINGER
Department of Natural Resources and
Environmental Science
University of Nevada
1000 Valley Road
Reno, NV 89512
jsedinger@cabnr.unr.edu

GERNOT SEGELBACHER
Department of Wildlife Ecology and Management
Albert-Ludwigs-University Freiburg
Tennenbacher Str 4
79106 Freiburg, Germany
gernot.segelbacher@wildlife.uni-freiburg.de

JAY F. SHEPHERD
Washington Department of Fish and Wildlife
16018 Mill Creek Boulevard
Mill Creek, WA 98012
jay.shepherd@dfw.wa.gov

NOVA J. SILVY
Department of Wildlife and Fisheries Sciences
Texas A&M University
210 Nagle Hall
College Station, TX 77843
n-silvy@tamu.edu

JUDY ST. JOHN
Rocky Mountain Center for Conservation Genetics
and Systematics
Department of Biological Sciences
University of Denver
Denver, CO 80208
stjohn.judy@gmail.com

CRAIG A. STRICKER
U.S. Geological Survey
Fort Collins Science Center
c/o Denver Federal Center
Denver, CO 80225
cstricker@usgs.gov

CHRISTOPHER C. SWANSON
Department of Wildlife and Fisheries Sciences
South Dakota State University
Box 2140B
Brookings, SD 57007
(Current address: Kulm Wetland Management
District, U.S. Fish and Wildlife Service, Kulm,
ND 58456,
chris_swanson@fws.gov)

J. SCOTT TAYLOR
Wildlife Division
Nebraska Game and Parks Commission
2200 N 33rd
Lincoln, NE 68583
scott.taylor@nebraska.gov

BENJAMIN E. TOOLE
Department of Wildlife and Fisheries Sciences
Texas A&M University
210 Nagle Hall
College Station, TX 77843
btoole@fs.fed.us

BASTIEN TRAN
Office National de la Chasse et de la Faune Sauvage
Direction des Etudes et de la Recherche
Espace Alfred Sauvy
66500 Prades, France
bastien.tran@gmail.com

ANDREW J. TYRE
School of Natural Resources
University of Nebraska at Lincoln
135 Hardin Hall
3310 Holdrege Street
Lincoln, NE 68583
atyre2@unl.edu

GREGORY T. WANN
Natural Resource Ecology Laboratory
Colorado State University
Fort Collins, CO 80526
gtw248@gmail.com

SCOTT WILSON
Smithsonian Migratory Bird Center,
National Zoological Park
3001 Connecticut Avenue, NW
Washington, DC 20008
wilsonsd@si.edu

SAMANTHA M. WISELY
Division of Biology
Kansas State University
116 Ackert Hall
Manhattan, KS 66506
wisely@k-state.edu

DONALD H. WOLFE
Sutton Avian Research Center
University of Oklahoma
P.O. Box 2007
Bartlesville, OK 74005
dwolfe@ou.edu

PREFACE

Grouse are an ecologically important group of birds with a holarctic distribution in the tundra, forest, and grassland ecosystems of Europe, Asia, and North America. Field studies of grouse have long played an important role in the ecological sciences, including population biology, behavioral ecology, evolutionary ecology, conservation biology, and wildlife management. In population biology, long-term studies of grouse have provided important insights into the ecological factors regulating and limiting the dynamics of cyclic populations under natural conditions. Grouse are key prey items for specialist predators such as gyrfalcons, but also for generalist predators such as coyotes, red fox, and other species. Population studies of grouse and their predators have provided insights into the trophic dynamics that structure populations of northern vertebrates. Grouse are also noteworthy because they display a remarkable diversity of mating systems, ranging from social monogamy to classic lek-mating systems. Studies of grouse mating systems, parental care, and their ecological correlates now provide illustrative examples in major textbooks in behavioral ecology. In recent years, development of molecular methods has provided new tools for exploring the evolutionary ecology of grouse, and there is growing recognition that life-history traits of grouse may covary with ecological gradients such as altitude, latitude, and levels of anthropogenic disturbance. Unfortunately, grouse populations in many areas have been negatively impacted by human activities, and understanding the effects of ongoing changes in land use and land cover change remain key issues. However, success of translocation efforts and other conservation actions provides encouragement for development of mitigation and recovery efforts. In fact, several species of grouse have been proposed to be umbrella species for sensitive ecosystems, including capercaillie in forests and prairie chickens in native grasslands, and conservation efforts for grouse could benefit entire communities. Last, grouse have had long-standing importance for recreational and subsistence harvest in northern regions, and extensive time series of bag data are available from many areas. Studies of hunted grouse made important early contributions to theoretical developments in sustainable harvest strategies for wildlife populations, which continue to be developed and refined with new experimental work.

Current issues in grouse biology are discussed every three years at regular meetings of the International Grouse Symposium. The symposium is international in scope, with previous meetings in Scotland (1978, 1981), England (1984), Germany (1987), Norway (1990), Italy (1993), the United States (1996), Finland (1999), China (2002), and France (2005). This book was launched as a project following the 11th International Grouse Symposium, which was held 11–15 September 2008, in Whitehorse, Yukon, and was the first grouse symposium to be held in Canada. The 11th symposium was organized with generous support from the Centre for Applied Conservation Research at the University of British Columbia, the Pacific and Yukon Region of Environment Canada, and

Yukon College. Following a successful meeting, participants and other wildlife biologists were invited to submit manuscripts for an edited volume on grouse biology. Authors submitted a total of 43 manuscripts, which were evaluated by two rounds of peer review. We extend our deep thanks to the reviewers who assisted this volume with careful and constructive evaluations of the manuscripts: Cameron Aldridge, Tony Apa, Todd Arnold, Erin Bayne, Jeff Beck, Hans-Heiner Bergman, Juan Bouzat, Clait Braun, Miran Cas, Isabella Cattadori, Jack Connelly, Kevin Doherty, Peter Dunn, Renata Duraes, Bradley Fedy, Sam Fuhlendorf, River Gates, Alicia Goddard, Michael Gregg, Christian Hagen, Susan Hannon, Rick Hoffman, Matt Holloran, Jodie Jawor, Kent Jensen, Robb Kaler, Siegfried Klaus, Michael Larson, Gilbert Ludwig, Tobias Ludwig, Jesus Martinez-Padilla, Conor McGowan, Lance McNew, Terry Messmer, Michael Morrow, Olafur Nielsen, Neal Niemuth, Claude Novoa, Michael Patten, Noah Perlut, Jim Pitman, Shin-Jae Rhim, Beth Ross, Michael Runge, Jonas Sahlsten, Stefan Schindler, Michael Schroeder, Jim Sedinger, Nova Silvy, Kjell Sjöberg, Mark Statham, Ilse Storch, Egbert Strauss, Dominik Thiel, Brett Walker, Chris Williams, Scott Wilson, Kimberly With, Don Wolfe, Andrew Yost, and Guthrie Zimmerman.

The final volume was jointly edited by a team of three associate editors, including Brett K. Sandercock, Kathy Martin, and Gernot Segelbacher. The editors thank the Norwegian Institute for Nature Research in Trondheim, Norway, and Kansas State University (BKS), Environment Canada and the University of British Columbia (KM), and the University of Freiburg (GS) for institutional support during preparation of this volume. The *Studies in Avian Biology* series has been published by the Cooper Ornithological Society since 1978 and we are proud to be part of this long-standing series. The former series editor, Carl Marti, provided great encouragement during initial development of the project, and we deeply regret that he was unable to see this work completed before his death. We thank Chuck Crumly for his efforts as Science Publisher for helping us to manage a successful transition to the University of California Press as the new publisher for the series.

Ecology, Conservation, and Management of Grouse contains 25 chapters authored by a total of 80 contributors. The global distribution of grouse spans a large number of ecological and political units, and the diversity of topics in the contributed chapters reflects the broad range of questions that can be investigated with this important group of birds. Our volume has been organized into four main sections. In Spatial Ecology, seven chapters include investigations of new landscape tools for modeling grouse occurrence, as well as intensive studies of radio-marked birds to understand movements in response to seasonal changes, disturbance from anthropogenic structures, and landscape fragmentation. In Habitat Relationships, six chapters examine the links between habitat use and reproductive performance in two sensitive species of grouse, Greater Sage-Grouse and Greater Prairie-Chickens. In Population Biology, the seven contributed chapters include an experimental manipulation of testosterone under field conditions, population studies of grouse demography under different ecological conditions, and intriguing evidence of life-history changes in response to human activities and breeding in alpine habitats. The final section, Conservation and Management, includes five contributed chapters investigating the troubling impacts of global change on alpine populations, successful use of translocations to reestablish an island population, and the impacts of harvest on the mortality rates of grouse in hunted populations. Much of the new data on the biology of grouse has been collected with an impressive set of tools that characterize modern ecology: biogeochemistry, molecular genetics, endocrinology, radiotelemetry, and remote sensing. Conservation and management of grouse is increasingly a global concern, but the exciting new data in this volume will be useful in identifying conservation priorities and enabling action documents for conservation agencies, wildlife biologists, and managers. Please enjoy learning about current themes in grouse research, and we hope that the new information in this volume of *Studies in Avian Biology* will aid conservation efforts for sensitive populations of grouse, their critical habitats, and associated species of wildlife.

BRETT K. SANDERCOCK
Manhattan, Kansas

KATHY MARTIN
Vancouver, British Columbia

GERNOT SEGELBACHER
Freiburg, Germany

30 August 2010

PART ONE

Spatial Ecology

CHAPTER ONE

Spatially Explicit Habitat Models for Prairie Grouse

Neal D. Niemuth

Abstract. Loss, fragmentation, and isolation of grassland habitat have greatly reduced the range and numbers of prairie grouse (*Tympanuchus* spp.) across North America. Because prairie grouse are resident, area-sensitive species with relatively limited dispersal abilities, landscape characteristics such as the amount, types, and configuration of habitat influence the presence, abundance, and persistence of prairie grouse populations. Therefore, a landscape approach that uses spatially explicit models to guide prairie grouse conservation is both appropriate and necessary. To be effective for conservation, landscape models must incorporate prairie grouse biology, be developed at appropriate scales, and use accurate data with spatial and thematic resolution that are sufficiently fine to target sites for specific conservation actions. Uncertainties regarding the ecology of prairie grouse need to be addressed, including the form of relationships between the amount of habitat and the presence, density, and persistence of prairie grouse; and how landscape characteristics influence local movements, dispersal, and gene flow. Because many spatially explicit landscape models are developed using lek data, additional information is needed as to what lek counts represent to local prairie grouse populations. Adoption and implementation of a landscape approach to prairie grouse conservation will require that management perspectives be broadened to explicitly include landscapes and that development of landscape models shifts, at least in part, from the realm of research to that of management. Successful conservation of prairie grouse will require resolution of substantial socioeconomic and political obstacles, as well as an increased commitment from the conservation community to broad-scale habitat conservation.

Key Words: conservation, Greater Prairie-Chicken, landscape ecology, Lesser Prairie-Chicken, scale, Sharp-tailed Grouse, spatially explicit habitat model.

Niemuth, N. D. 2011. Spatially explicit habitat models for prairie grouse. Pp. 3–20 *in* B. K. Sandercock, K. Martin, and G. Segelbacher (editors). Ecology, conservation, and management of grouse. Studies in Avian Biology (no. 39), University of California Press, Berkeley, CA.

oss and fragmentation of grassland and shrubland habitat in North America have dramatically reduced the numbers and range of North American prairie grouse (*Tympanuchus* spp.). For example, the Greater Prairie-Chicken (*T. cupido pinnatus*) was once found in portions of approximately 17 U.S. states and 4 Canadian provinces (Ross et al. 2006), but presently is in danger of extirpation in 7 of the 11 states in which it is found (Schroeder and Robb 1993). The Lesser Prairie-Chicken (*T. pallidicinctus*) is still found in all 5 of the states in which it originally occurred (Giesen 1998), but by 1980 its range had been reduced 92% from the 1800s (Taylor and Guthery 1980a). The Sharp-tailed Grouse (*T. phasianellus*) originally was found in 21 states and 8 provinces, but has since been extirpated from 8 states, and populations are small and isolated in much of the southern and eastern portions of its present range (reviewed in Connelly et al. 1998).

The primary cause of the declines for prairie grouse is broad-scale loss of grassland and brushland habitat. Concern about the effects of widespread habitat loss on bird populations has prompted recent bird conservation initiatives to adopt a landscape approach to conservation planning and implementation. The first of these was the North American Waterfowl Management Plan (NAWMP; U.S. Department of Interior and Environment Canada 1986), which, through the action of bird conservation joint ventures guided in part by landscape models, has positively influenced more than 5 million ha of breeding, migration, and wintering waterfowl habitat in North America (Abraham et al. 2007). Following the successes of the NAWMP, other efforts, including the Grassland Conservation Plan for Prairie Grouse (Vodehnal and Haufler 2007), have explicitly adopted a landscape approach to conservation planning.

An appreciation of the importance of landscapes to prairie grouse is not new: lacking radiotelemetry technology to track individuals, early researchers used lek counts, harvest monitoring, field surveys, and incidental observations to note the effects of patch size (Ammann 1957), isolation (Grange 1948), disturbance regimes (Grange 1948, Ammann 1957), and landscape composition and configuration (Grange 1948, Hamerstrom et al. 1957, Westemeier 1971) on the presence, size, and persistence of prairie grouse populations. However, the development of remotely sensed spatial data, geographic information systems (GIS), and statistical modeling techniques provides present-day researchers with unprecedented ability to identify and quantify relationships between landscape characteristics and prairie grouse (Kareiva and Wennergren 1995, Stauffer 2002, Wiens 2002). Increasingly isolated and declining populations of prairie grouse increase the impetus to explore relationships between landscape characteristics and grouse populations and identify the most appropriate locations for conservation.

The effects of habitat loss, fragmentation, and isolation may take place at a scale much broader in extent than the patches or habitat clusters occupied by local populations of prairie grouse, which is the scale at which prairie grouse are often studied and managed. Population dynamics of prairie grouse on managed reserves are often synchronous with adjacent populations off managed areas (Bergerud 1988a, Morrow et al. 1996), indicating that broad-scale as well as local factors influence prairie grouse populations. Because prairie grouse populations may be influenced by landscape factors out of the control or consideration of local efforts, conservation may fail if a landscape context is not considered, particularly if local populations are connected at landscape or regional scales by movements and if prairie grouse exhibit a metapopulation structure or source/sink dynamics. The need to consider landscape ecology and geospatial information in grouse conservation has previously been noted (Braun et al. 1994, Morrow et al. 1996, Samson et al. 2004), but specific relationships, hypotheses, and information needs have rarely been identified as they relate to prairie grouse.

Spatially explicit models provide a means of specifying relationships between landscape characteristics and species in a manner that is intuitive to use in conservation applications. The general class of models that includes species distribution models, spatially explicit population models, conservation design, or spatial planning tools provides a habitat-based context for conservation over broad spatial extents (Beissinger et al. 2006). These models differ from metapopulation models in that the entire landscape is considered in the context of multiple variables describing landscape characteristics rather than the presence or absence of populations in discrete habitat patches (Moilanen and Hanski 2001, Tischendorf and Fahrig 2001). Models are spatially explicit because they use digital landcover data to consider the spatial configuration of habitat and objects and create maps showing modeled characteristics across the area of interest.

Recent improvements in spatial analysis software and availability of spatial data have led to increased interest in using spatially explicit models to direct conservation actions (Wiens 2002). However, although landscape approaches to bird conservation are popular, the development and application of spatially explicit models that result in improved conservation efficiency is a complex process that must consider many aspects of biology, statistics, data quality, scaling, and implementation (Shenk and Franklin 1991, Scott et al. 2002, Millspaugh and Thompson 2008). As is the case with any model, ignoring the complexities of model development can lead to landscape models that are inaccurate and misleading.

Spatially explicit landscape models offer several benefits for conservation. When landscape models are applied to appropriate GIS layers, the resulting maps can be used to guide prairie grouse conservation and management, including translocating prairie grouse or linking prairie grouse populations (McDonald and Reese 1998, Niemuth 2003). When suitable data are available, landscape models can be used to assess the effects of environmental perturbations such as energy development (i.e., wind, oil, and gas), conversion of grassland to cropland, or the benefits of programs such as the Conservation Reserve Program (CRP). Disturbance to prairie grouse can be minimal, as data collection for landscape models based on lek counts does not require trapping or handling of birds. There is considerable precedent for using lek-based landscape analyses to study the spatial ecology of prairie grouse (Westemeier 1971, Pepper 1972, Merrill et al. 1999, Niemuth 2000, Woodward et al. 2001), but there is also potential to apply this approach to conservation.

In this review, I summarize biological characteristics of prairie grouse that make them sensitive to landscape characteristics, review theories important to the landscape ecology of prairie grouse, and present ideas for landscape-scale research, conservation, and management of prairie grouse. My review focuses on analyses using lek location, attendance, and persistence as response variables in spatially explicit habitat models, acknowledging the desirability of incorporating information from more intensive, local studies into models and management. The primary premise of this approach is that conservation efforts should occur over broad areas, so landscape models may better inform conservation actions than detailed studies of local populations.

BIOLOGICAL TRAITS THAT PROMOTE LANDSCAPE SENSITIVITY

Several biological traits of prairie grouse make them sensitive to the amount and configuration of habitat as well as small population size, the effects of which are compounded by loss and fragmentation of habitat. Prairie grouse have fairly narrow habitat requirements and occur at low densities relative to many other gamebirds. In addition, prairie grouse are area sensitive, requiring large blocks or aggregations of habitat to be present (Ammann 1957, Niemuth 2000, Woodward et al. 2001). Area sensitivity is typically associated with increased probability of a species being present in an area, but prairie grouse also may be area sensitive in that reproductive success (Ryan et al. 1998, Manzer and Hannon 2005), density (Pepper 1972, Niemuth 2000), and persistence of leks or populations (Merrill et al. 1999, Woodward et al. 2001) also increase with amount of suitable habitat. Finally, prairie grouse are resident species, which are generally more susceptible to loss and fragmentation of habitat than latitudinal migrants (Bender et al. 1998).

As resident species, prairie grouse generally are not known to migrate or move long distances, even when juveniles disperse in fall. Mean dispersal distance for a brood of six transmitter-equipped juvenile Greater Prairie-Chickens in Kansas was 1.0 km, and maximum recorded dispersal for 24 juveniles was 10.8 km (Bowman and Robel 1977). Maximum recorded dispersal for a transmitter-equipped juvenile Lesser Prairie-Chicken in Texas was 12.8 km (Taylor and Guthery 1980b). Maximum recorded dispersal for a transmitter-equipped juvenile female Sharp-tailed Grouse in Wisconsin was 5.8 km (Gratson 1988), and 59% of banded juvenile Sharp-tailed Grouse reported by hunters in South Dakota were recovered <1 km from the site where they were trapped (Robel et al. 1972). Prairie grouse can make longer total movements (Moe 1999), but intermediate habitat patches are critical for maintaining connectivity between populations and providing "stepping stones" for these movements

(Hamerstrom and Hamerstrom 1973). Some populations of prairie grouse migrated in the past (Grange 1948, Ammann 1957), but the proportion of the population that migrated and distances that birds migrated are unknown. Partial migration, where a portion of the population moves between breeding and wintering areas, is evident in some populations of prairie grouse. In Colorado, female and male Greater Prairie-Chickens showed seasonal movements of 9.2 and 2.7 km, respectively, between breeding and wintering areas, with birds showing fidelity to leks, general nest sites, and wintering areas (Schroeder and Braun 1993). Greater Prairie-Chickens in the Sandhills of Nebraska also showed evidence of migration, apparently to winter in areas with grain for food (Kobriger 1965).

Limited movements by prairie grouse reduce interchange among subpopulations, with subsequent reductions in gene flow, both historically and following recent anthropogenic habitat loss (Johnson et al. 2003, Van den Bussche et al. 2003, Bouzat and Johnson 2004, Ross et al. 2006). Many populations of prairie grouse exhibit limited genetic diversity as a consequence of the lek mating system, low nest success, and historic population bottlenecks; these problems are exacerbated by the small size of many prairie grouse populations and reduced gene flow between populations that are increasingly isolated in the landscape (Bouzat et al. 1998, Westemeier et al. 1998a, Bouzat and Johnson 2004, Johnson et al. 2004). Limited movements among isolated populations reduce the potential for demographic rescue and maintenance of genetic diversity (Westemeier et al. 1998a, Reed 1999, Niemuth 2005). However, as important and problematic as loss of genetic diversity may be, it is largely a symptom of broad-scale habitat loss and isolation.

Many ecological processes that affect prairie grouse are influenced by landscape characteristics. Nesting success is considered the primary driver of grouse population dynamics (Bergerud 1988b, Peterson and Silvy 1996, Wisdom and Mills 1997), and the community composition and behavior of many nest predators are influenced by landscape characteristics (Pedlar et al. 1997, Sovada et al. 2000, Phillips et al. 2004, Manzer and Hannon 2005). Consequently, nesting success of prairie grouse can increase with the proportion of grassland in the surrounding landscape (Ryan et al. 1998, Manzer and Hannon 2005). Landscape

characteristics of sites used by Ring-necked Pheasants (*Phasianus colchicus*) differed from those of sites used by Lesser Prairie-Chickens (Hagen et al. 2007a), suggesting that the potential for aggression and interspecific nest parasitism may vary across the landscape (see Vance and Westemeier 1979, Westemeier et al. 1998b). Large, newly created areas of habitat sometimes support high densities of grouse (reviewed in Bergerud 1988a). The mechanisms for this "big new space" phenomenon are unknown, but may include changes in vegetation structure, increased food availability, low predator densities, the creation of habitat patches that facilitate dispersal, or a lag in the establishment of predator populations (Bergerud 1988a, Niemuth and Boyce 2004). Anthropogenic processes associated with landscape composition can also influence reproductive success, as nests and young are often destroyed by farm equipment when prairie grouse nest in hay fields or stubble (Yeatter 1963, Pepper 1972, Ryan et al. 1998). Similarly, the distribution of fences and power lines, which can influence habitat use and cause substantial mortality of prairie grouse, is also associated with landscape composition and land use (Patten et al. 2005, Wolfe et al. 2007, Hagen et al., this volume, chapter 5). Population dynamics may be particularly sensitive to mortality if breeding females are more vulnerable than other sex–age classes, as is the case with loss of hens on nests (Hagen et al. 2007b).

DEVELOPMENT OF LANDSCAPE MODELS FOR CONSERVATION

Several key concepts underlie the development and application of spatially explicit habitat models for conservation. First, the approach assumes that habitat selection is a hierarchical process where birds first consider regional and landscape characteristics before selecting habitat at a finer scale, such as the home range, nest, or foraging site (Wiens 1973). Conservation planning therefore focuses on the landscape scale, and provides context for local management actions. If habitat is purchased or otherwise selected for management based on landscape characteristics, then local characteristics of the grassland, such as vegetation composition and structure, can be modified relatively easily. Conversely, it is more difficult to modify the landscape around a patch with suitable local characteristics in an unsuitable landscape matrix. Local characteristics such as vegetation height, density, and composition will vary from year to year with precipitation, land use, grazing intensity, fire, and other edaphic factors; a landscape approach focuses on maintaining appropriate landscape conditions so that species can persist through time and flourish when local conditions are good.

Types of spatially explicit models vary, but generally cost of development and usefulness for conservation actions are positively related. Models using lek data to relate prairie grouse presence or lek attendance to landscape characteristics will not be as expensive or useful as models relating landscape characteristics and demographic parameters such as nesting success or adult survival, which require intensive study involving radio telemetry. Interestingly, because of the limited dispersal of prairie grouse, lek fidelity, and response to landscape characteristics, models relating long-term persistence of leks to landscape characteristics may provide an indication of demographic performance, although vital rates and specific mechanisms affecting long-term population persistence will be unknown. Methods for developing landscape models to guide conservation can range from simple conceptual models to complex statistical models that incorporate demographic processes and the spatial structure of prairie grouse populations, with the type of model depending on its intended purpose and available information and resources. The technical aspects of developing statistical models are relatively straightforward once a clear and explicitly stated purpose has been articulated and appropriate data collected. However, the quality and success of landscape models depend on several critical assumptions and details. Here, I will focus on specific factors related to the development of spatially explicit habitat models for prairie grouse. General discussions of modeling approaches, and landscape models in particular, can be found in Shenk and Franklin (2001), Scott et al. (2002), Beissinger et al. (2006), and Millspaugh and Thompson (2008).

The availability of digital data sets has increased greatly in recent years, but not all digital data sets are suitable for spatial planning. Therefore, the spatial and thematic accuracy of spatial data should be verified before use, as even coarse-scale range maps can suffer from large errors of omission and commission (Niemuth et al. 2009). Similarly, digital landcover data should reflect what is actually present on the ground at an acceptable level of spatial and thematic accuracy. Most digital landcover data that cover large spatial extents are based on satellite imagery that has been processed to separate digital signatures that can be associated with various landcover classes. However, classification of satellite imagery is subject to considerable error caused by variation in land use and vegetation, shading, atmospheric conditions, sensor variation, choice of landcover classes, timing, spatial error, and differences in phenology, soil types, and soil moisture (Lillesand and Kiefer 2000, Gallant 2009). Acceptable levels of error in any data set will be determined by the goals and intended use of the conservation assessment, but accuracy of landcover data should be reported for any spatially explicit habitat model.

Many prairie grouse habitat models are developed by characterizing landscapes around leks. Lek locations, however, may be poorly defined or may simply shift within or among years. Locations must be sufficiently accurate to link leks with the landscape the birds are using, but the objective of the model is usually to describe the landscape surrounding leks at broad scales (>1.6 km), rather than the actual lek locations. Positional errors of 100–200 m will have relatively little effect on parameter estimates, but accuracy of estimates will decline as positional error increases. Leks included in model development should be representative of the population of interest. Biases may be introduced if areas in which surveys were conducted were selected in a non-random manner, such as from roadside surveys (see Anderson 2001). For some small, isolated populations a complete census of leks may be possible, but this will be the exception. The timing of surveys can also introduce bias, as leks may be less likely to be detected and apparent lek attendance may be reduced in areas that are sampled late in the day or season.

The spatial scale at which models are created also must be considered, as it will affect the intended application of the model. Coarse-grained analyses that use watersheds, major land resource areas (MLRAs), EMAP hexagons, or county-level summaries are of little value for targeting specific acquisitions or treatments, as the proportion of the landscape occupied by prairie grouse may be small relative to the size of the reporting unit. Even if the proportion of a coarse-grained reporting unit occupied by prairie grouse is high, coarse-grained occurrence records do not provide insights about biological relationships or allow precise targeting of conservation actions. Also, coarse-grained analyses can only provide crude measures of proximity or connectivity among subpopulations.

Spatially explicit models must have spatial resolution that is sufficiently fine to allow targeting of specific sites, but should also be developed at scales that accommodate the large expanses of habitat occupied by prairie grouse and environmental factors that operate at different spatial scales (Fuhlendorf et al. 2002, Mayor et al. 2009). One approach to determining the proper scale for spatial analysis is to consider previous research describing home range sizes, daily movements, brood ranges, and the distance from the lek within which most females nest (Merrill et al. 1999, Niemuth 2005). Another is to characterize the landscape within different buffer distances from leks and assess model fit for the various buffers (Hanowski et al. 2000, Niemuth 2000, Fuhlendorf et al. 2002). Selecting the proper scale for data analysis may be complicated if proximity to other populations, habitat selection, and amount of habitat are confounded. The issue of scaling is further complicated if landscape characteristics influence distance metrics used as measures of proximity or permeability (Moilanen and Nieminen 2002).

Often, the output of a spatially explicit model is used to determine optimal sites for the study species, and sites with sufficient habitat and proximity to other populations may then be targeted for preservation. However, sites can also be identified for other conservation treatments. Sites that are close to existing populations but have insufficient habitat would be suitable for habitat restoration, whereas sites with sufficient habitat that are far from other populations may benefit from efforts to link populations. Restoration and connection of sites with little habitat that are distant from other populations may be cost-prohibitive or receive little use by local prairie grouse populations.

TOPICS NEEDING ADDITIONAL RESEARCH

There are many information gaps and untested principles related to the ecology and conservation of grouse (Braun et al. 1994, Wisdom et al. 2002, Applegate et al. 2004); I review specific information needs and assumptions related to the spatial ecology of prairie grouse. Sensitivity analyses will be useful for determining the relative influence of these factors on prairie grouse and corresponding priorities for research and conservation (Wisdom and Mills 1997, Hagen et al. 2009).

Habitat Relationships

The relationship between habitat area and prairie grouse is almost certainly more complex than a simple, linear relationship between area of grassland or shrubland and the presence, density, reproductive success, or persistence of prairie grouse populations. The relationship may be non-linear, with a threshold below which local prairie grouse populations are not present or are destined for extirpation. Extinction thresholds are most likely to occur in species that are habitat specialists (Andrén 1994), occur in metapopulations (Kareiva and Wennergren 1995), or have limited dispersal capabilities (With and King 1999), all of which likely apply to prairie grouse. Spatial configuration of habitat may be problematic only when the amount of habitat drops below a threshold (Andrén 1994, Fahrig 1998), especially if the costs of dispersal vary among habitat types in the fragmented landscape. Consequently, identification of threshold levels, especially in relation to habitat fragmentation, is a critical information need, especially as landscapes presently harboring prairie grouse are increasingly subjected to conversion of native habitat to cropland and increased development of energy production facilities.

Non-linear relationships between area of grassland habitat and prairie grouse populations may differ among population metrics. Conversion of grassland habitat to crop fields has been the greatest factor contributing to the decline of prairie grouse populations, yet small amounts of cropland can have a positive effect on numbers of prairie chickens by providing additional food resources (Hamerstrom et al. 1957, Crawford and Bolen 1976, Christisen 1985). Consequently, numbers of prairie chickens at leks or long-term persistence of populations might be highest with some small amount of cropland in a grass-dominated landscape (Fig. 1.1A). However, nesting success of prairie grouse and other grassland nesting birds generally increases with the amount of grass in the landscape (Ryan et al. 1998, Herkert et al. 2003, Manzer and Hannon 2005; Fig. 1.1B). Therefore, the benefit afforded by providing additional grassland could vary depending on the

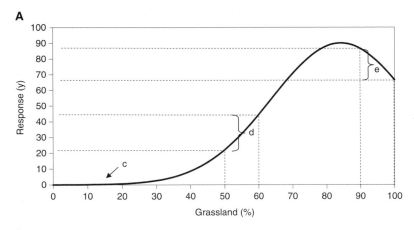

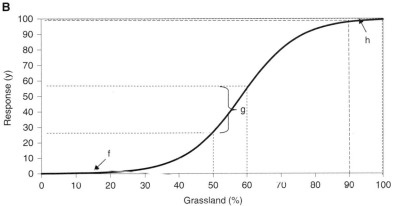

Figure 1.1. The type of relationship between landscape characteristics and prairie grouse can influence the degree and direction of response by prairie grouse to conservation efforts. In these hypothetical examples, an unspecified response by prairie grouse (y) such as density or probability of presence (heavy dark line) is scaled from 0 to 100 and varies quadratically (A) and asymptotically (B) with the amount of grass in the surrounding 800-ha landscape. (A) Prairie grouse response (y) increases by 0.5 when 80 ha of grass are added to a landscape comprised of 10% grass (c); increases by 22.5 when 80 ha of grass are added to a landscape comprised of 50% grass (d); and decreases by 19.8 when 80 ha of grass are added to a landscape comprised of 90% grass (e). (B) Prairie grouse response (y) increases by 0.7 when 80 ha of grass are added to a landscape comprised of 10% grass (f); increases by 28.1 when 80 ha of grass are added to a landscape comprised of 50% grass (g); and increases by 1.5 when 80 ha of grass are added to a landscape comprised of 90% grass (h). If the relationship is linear (not shown), prairie grouse response (y) is the same when 80 ha of grass are added regardless of the amount of habitat in the surrounding landscape. Size of the sampling window was based on landscape analyses (Merrill et al. 1999, Niemuth 2000) and distance of Greater Prairie-Chicken nests from leks (Schroeder 1991).

nature of the relationship (i.e., linear vs. curvilinear) and landscape context (Fig. 1.1). Because the degree and even direction of the response by prairie grouse to changes in the amount of habitat in the landscape varies, management treatments such as grassland restoration should explicitly consider landscape context as well as the population metrics the treatments are intended to address (Fig. 1.1). Responses to landscape characteristics can differ among populations as well as species, depending on the availability of different cover types in the landscape (Niemuth 2005). In all cases, the scale at which the landscape is assessed should be appropriate for the species, population metric, and conservation treatments being considered.

Similarly, the effects of habitat loss relative to habitat fragmentation are poorly understood. Problems include inconsistent definitions of fragmentation (Fahrig 2003), which are exacerbated by the numerous fragmentation metrics available in GIS software packages. Quantitative models of prairie grouse response to landscape composition under a broad range of landscape characteristics should be developed; candidate models should consider biologically appropriate curvilinear relationships. Assessments of prairie grouse habitat should explicitly define metrics used to assess fragmentation, how the metrics differ from simple habitat loss, and how birds respond to habitat fragmentation versus habitat loss.

Many investigations have focused on minimum area requirements of prairie grouse (Samson 1980, Winter and Faaborg 1999). However, a landscape perspective is preferable because landscape characteristics can influence metapopulation dynamics and modify patterns of area sensitivity (Wiens 1997, Ribic et al. 2009). The Great Plains provides an opportunity to assess the role of landscape characteristics on prairie grouse at varying levels of habitat loss and fragmentation. Grasslands of the Great Plains follow a gradient of habitat loss, with tallgrass prairie in the east showing the greatest loss and short-grass prairie in the west showing the least loss, with intermediate loss in the mixed-grass region (Samson et al. 2004). Regional studies that span this gradient and incorporate varying levels of fragmentation will provide more information about the relative effects of habitat loss and fragmentation than localized studies where landscape characteristics show less variation. Of course, landscape characteristics may be confounded with other factors, as precipitation, grass height, grass density, and litter depth will also likely decrease from east to west. Therefore, potential confounding factors should be sampled and assessed in the framework of landscape models when possible.

Previous analyses of the spatial ecology of prairie grouse focused on populations at the periphery of the species' range (McDonald and Reese 1998, Merrill et al. 1999, Niemuth 2000). Focusing on peripheral populations is understandable, given the vulnerability of these populations and the high levels of management associated with them (Bergerud 1988a). Similar efforts are needed throughout the extant range of all prairie grouse. No populations of prairie grouse, even in the core of their range, are immune from the increasing pressures of agricultural conversion and energy development. In general, prairie grouse have been lost from places where other land uses were valued more highly than grazing, so remaining grouse populations may persist in marginal lands relative to sites where birds have been extirpated. Consequently, focusing conservation efforts on large, extensive populations where land values are relatively low will likely be more cost effective than trying to preserve small, isolated populations. Spatially explicit habitat models pertinent to core populations of prairie grouse should be developed, as the spatial ecology of large prairie grouse populations in extensive areas of habitat may differ from that of grouse in small, isolated populations (Braun et al. 1994, Fuhlendorf et al. 2002). However, small, isolated populations may be local management priorities or important to maintain connectivity among populations. Spatially explicit models can help determine cost and provide context when assessing habitat and populations for prioritization and conservation triage (Wisdom et al. 2005).

Prairie grouse habitat selection in the context of predation and nest parasitism also needs further investigation. For example, the distribution, nest site selection, and nesting success of Ring-necked Pheasants are influenced by landscape composition and configuration (Clark et al. 1999, Leif 2005). If the abundance of pheasants and, by extension, potential for interactions between prairie grouse and pheasants can be modeled (Hagen et al. 2007a), managers can better tailor treatments to benefit prairie grouse while minimizing or avoiding possible negative effects associated with Ring-necked Pheasants. Similar approaches could be used to guide conservation treatments with regard to predators that are influenced by landscape characteristics.

Movement

Connectivity among populations is an important component of prairie grouse ecology and conservation that will likely become even more important as grassland habitats continue to be converted to other uses and prairie grouse populations become more isolated. Because the population dynamics of prairie grouse often exhibit spatial structure, landscape models can benefit from inclusion of ideas derived from metapopulation theory and source–sink dynamics. Euclidean distance has been used as an index of connectivity and proximity to other populations, but more complex measures of movement distances may be needed to better determine the influence of landscape composition and configuration on patterns of dispersal, colonization of new habitat, and, eventually, gene flow of prairie grouse (Moilanen and Nieminen 2002, Manel et al. 2003, Wang et al. 2008). Options include GIS-based friction or cost analyses of movements across landscapes with different compositions or configurations (Chetkiewicz et al. 2006, Kindlmann and Burel 2008). The response variable in models can be actual movement, assessed with radio-marked birds; observed colonization of new habitat; or evidence of movement, assessed with genetic analysis of samples from known locations (Manel et al. 2003). Information-theoretic methods can be used to evaluate competing models of movement cost (e.g., Burnham and Anderson 1998). Landscape models should incorporate measures of connectivity; information is needed on how landscape composition and configuration affect prairie grouse dispersal so movements can best be incorporated into models. Assessments of movement should consider changes in population size and amount of available habitat over time, as these can influence dispersal and connectivity. Important as connectivity may be to prairie grouse populations, though, minimum levels of habitat are the foundation of conservation efforts and must be preserved to attract dispersing birds and maintain local populations.

What Do Leks Represent?

Leks are often considered a focal point for prairie grouse ecology and management (Westemcier 1971, Hamerstrom and Hamerstrom 1973, Giesen and Connelly 1993), and the number of males attending leks can be used as an index of habitat quality (Hamerstrom and Hamerstrom 1973). Consequently, lek data are frequently used in the development of landscape models. However, lek-based landscape models—as well as non-spatial population models—make a variety of assumptions about what leks represent to prairie grouse populations. For example, the number of males attending a lek at any one time may represent only a portion of the males associated with that lek (Robel 1970, Rippin and Boag 1974, Clifton and Krementz 2006, but see Schroeder and Braun 1992). Incomplete attendance by all the males associated with a lek is not a problem *per se*, as the number of males present at a lek can be a useful index to the total number of males associated with a lek if the proportion of males attending a lek is constant. However, lek attendance varies with weather, daily and seasonal timing, changes in land use, lek age, and the presence of predators; the number of birds present that are detected can vary with observer ability, topography, vegetation, survey methodology, and time spent observing the lek (Robel 1970, Clifton and Krementz 2006, Haukos and Smith 1999, McNew et al., this volume, chapter 15). Maximum counts of birds observed at leks during multiple visits have the potential to reduce the influence of unusually low counts, for instance, where birds were disturbed by predators, but maximum counts will introduce bias if the number of visits varies among leks. Species-, time-, and region-specific estimates of the proportion of males attending leks and how this proportion varies are necessary to calibrate lek counts relative to the populations the leks represent.

Nesting success is considered the primary driver of grouse populations (Bergerud 1988b, Peterson and Silvy 1996, Wisdom and Mills 1997), but the relationship between counts of males at leks and the number of females associated with the lek, their survival, or their reproductive success is unknown or poorly understood. Several models of lek formation have been posed, but the balance of evidence indicates that males establish leks in areas where they can encounter many females (Schroeder and White 1993), which suggests that counts of males on leks may be correlated with number of females in the vicinity. Habitat quality is a function of density, survival, and reproduction vital rates (Van Horne 1983); landscape models predicting density provide an important component of that definition, but information on survival and reproductive success is necessary to ensure that sites with high densities are not population sinks (but see Bock and Jones 2004). Nevertheless, density models can help ensure that expenditure of limited conservation resources consider many, rather than few, birds and may be especially useful where conservation treatments enhance survival or reproductive success. Persistence of leks with many males over time suggests that survival and reproductive success of females in the vicinity are at or above maintenance levels, but this assumes a closed population. Male prairie grouse do not disperse as far as females and show strong fidelity to leks (Hamerstrom et al. 1957, Robel et al. 1972,

Nooker and Sandercock 2008), so the presence or number of males at a site might also reflect past conditions (Knick and Rotenberry 2000, Fuhlendorf et al. 2002). Similar landscape characteristics have been associated with lek presence, lek attendance, lek persistence, and nesting success (Ryan et al. 1998, Merrill et al. 1999, Fuhlendorf et al. 2002, Niemuth 2005, Aldridge et al. 2008, Gregory et al., this volume, chapter 2), which suggests some potential for lek-based analyses to identify areas that are attractive for nesting or have high nesting success. However, it has been shown with radio-marked Greater Sage-Grouse (*Centrocercus urophasianus*) that attractive nest sites may experience low nesting success (Aldridge and Boyce 2007). Relationships between numbers of males and females associated with leks must be identified to determine if lek counts are useful predictors of reproductive potential. Similarly, lek attendance and persistence should be related to hen survival and nesting success over time and across a broad range of landscape characteristics.

Lek counts do not consider the spatial distribution of leks or the effects of scale that are intertwined with the ecology of prairie grouse. For example, if lek size is considered an index of habitat quality, two leks in a given area, each with eight males, will be considered to indicate lower-quality habitat than one lek in the same area with 16 males. In a comparison using data from four long-term studies of prairie chickens, Cannon and Knopf (1981) found that the number of leks in an area was more strongly correlated with the density of displaying males than average lek size; therefore, number of leks may be a better index to populations than counts of males. However, because Cannon and Knopf's (1981) focus was population trends, their analysis treated all study sites as homogeneous units and did not consider differences in attendance among leks, which is the primary interest in a lek-based density model. Changes in numbers of birds and leks across years may reflect the presence or absence of temporary leks, which can have different timing, age structure, and attendance than permanent leks (Schroeder and Braun 1992, Haukos and Smith 1999). Lek dynamics can influence reproductive potential, as nesting success varies with age of females and nest initiation date (reviewed in Bergerud 1988b). The spatial distribution of leks, inter-lek distance, or lek density can be explicitly

incorporated into landscape models, but the implications of lek attendance, lek type (temporary vs. stable), and lek density to the population ecology of prairie grouse need further research.

Finally, lek-based models assume that the landscapes (at some broad scale) surrounding leks are sufficient to meet the annual needs of prairie grouse. However, lek-based models will not include wintering habitat for those populations that move between breeding and wintering areas (Kobriger 1965, Schroeder and Braun 1993), particularly if prairie grouse move greater distances than has been documented. If wintering habitat is a limiting factor, research, modeling, and conservation efforts will have to be adjusted accordingly.

Broad-scale patterns of habitat use can provide information about underlying ecological relationships (Arthur et al. 1996), which can guide future, local studies of mechanisms responsible for observed patterns. For example, landscape-level analysis of Sharp-tailed Grouse leks in northern Wisconsin indicated that attendance was higher at leks in open landscapes created through clearcut harvest of insect-damaged timber relative to leks on landscapes managed for Sharp-tailed Grouse using prescribed fire (Niemuth and Boyce 2004). Additional research with radio-marked birds showed that Sharp-tailed Grouse nesting success and hen survival were also higher in clearcuts relative to landscapes managed with prescribed fire (Connolly 2001). Similar approaches could be used to assess broad-scale patterns and focus research regarding effects of other perturbations such as energy development on prairie grouse.

The reliability of information about grouse could be improved through monitoring and research that incorporates field experiments in an adaptive management framework (Holling 1978, Walters 1986). It is rarely possible to experimentally manipulate landscape characteristics in an active adaptive management context, but spatial information could be incorporated into sampling frameworks and study designs in a passive adaptive management context for both landscape-level and local research and monitoring (Aldridge et al. 2004, Powell et al., this volume, chapter 25). In addition to providing information about effects of management manipulations, monitoring programs also can provide baseline data useful for Before-After-Control-Impact (BACI; Green 1979, Stewart-Oaten et al. 1986) studies following perturbations or changes to portions of the landscape.

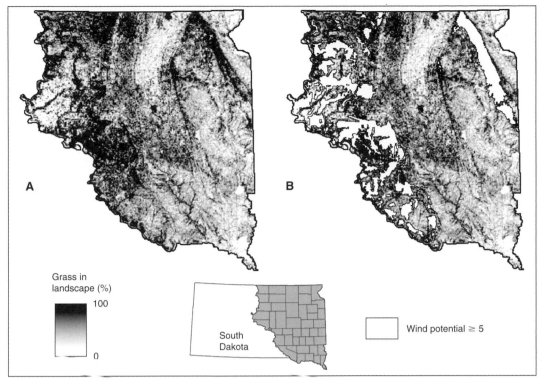

Figure 1.2. (A) Percent of landscape within 800 m comprised of grassland, hay, and Conservation Reserve Program grasslands in South Dakota east of the Missouri River. (B) Same region as (A) but overlain with areas of wind potential ≥ 4 where potential is rated on a scale of 1 (lowest) to 7 (highest). Landcover data described by Niemuth et al. (2008); wind data provided by the National Renewable Energy Laboratory (http://www.nrel.gov/gis/wind.html).

For example, areas in eastern South Dakota with high potential for wind development overlap substantially with remnant landscapes containing large amounts of grassland (Fig. 1.2); sampling Sharp-tailed Grouse and Greater Prairie-Chickens in these areas prior to and after installation of wind-generation towers would increase the strength of inferences about broad-scale effects of wind development on prairie grouse populations. Studies using a BACI approach to assess the effects of wind power development on Greater Prairie-Chickens in Kansas are in progress (B. K. Sandercock, pers. comm.).

MANAGEMENT RECOMMENDATIONS

Managers cannot change the biological traits that make prairie grouse sensitive to landscape characteristics, but managers can and should consider these characteristics when assessing and managing prairie grouse populations. A landscape perspective might require a paradigm shift from a local-management focus to management that incorporates local and landscape scales and that actively pursues reliable quantitative information (Braun et al. 1994, Applegate et al. 2004, Wisdom et al. 2005). Management does not take place in isolation; loss of satellite populations surrounding reserves managed for prairie chickens has likely contributed to declines in prairie chicken populations at core reserves (Morrow et al. 1996, Westemeier et al. 1998a). Concerns regarding the scale of Northern Bobwhite (*Colinus virginianus*) management were expressed by Williams et al. (2004:861), who stated that "traditional management principles currently are incompatible with the spatial scale necessary to address the nationwide decline in bobwhite abundance." Similarly, Wisdom et al. (2005) advocated a landscape approach to management of Greater Sage-Grouse habitat and populations, using landscape models to inventory resources, estimate costs, prioritize, and perform triage. Most prairie grouse conservation problems are land-use problems, and maps resulting from spatially explicit habitat models can provide the context necessary for planners, politicians, and managers to better conserve prairie grouse.

Products from the North American Waterfowl Management Plan are a good example of how this can be accomplished. The "Thunderstorm Map" shows the breeding distribution and density of five species of dabbling ducks across several states in the U.S. Prairie Pothole Region (Reynolds et al. 2006). Copies of the Thunderstorm Map are found in resource agency offices throughout the northern prairies; these maps are used to prioritize and target landscapes for expenditures of approximately $13 million annually. These efforts have permanently protected >1.1 million ha of wetlands and grasslands in the U.S. Prairie Pothole Region, with tremendous benefits for other species in addition to waterfowl (Beyersbergen et al. 2004). The primary sources of funding for these efforts are the federal Migratory Bird Hunting Stamp Act ("Duck Stamp") and private groups such as Ducks Unlimited. Prairie grouse do not have comparable programs directly channeling millions of dollars into habitat, but similar spatial modeling techniques could be used to better identify and prioritize lands for conservation of prairie grouse and possibly garner additional support for conservation. For example, the science behind the Thunderstorm Map was sufficiently strong that the U. S. Department of Agriculture initiated a separate CRP practice that dedicated 40,000 ha of wetlands and grasslands to priority areas identified by the model.

The development of spatially explicit models will have to shift, at least in part, from the realm of research to management. A management focus would have several benefits, particularly in the continuity and long-term commitment provided by agencies charged with legal responsibility for a species as well as the synergy that arises when field biologists and modelers combine their expertise (Beissinger et al. 2006). Engagement of agency staff will help ensure the continuous feedback essential to adaptive management and the improvement of landscape models. Adopting a spatial approach to management and data collection will likely require changes in how prairie grouse are surveyed. Typically, prairie grouse are surveyed to assess population trends over time, often with little or no attention given to spatial balance of sampling to avoid geographic bias,

stratification by land use and land cover, the range of land cover characteristics surrounding survey areas, or changes in land use over time. Spatial data should be used to identify and stratify areas for sampling (e.g., Fig. 1.2), similar to methodology in the recently developed national sampling framework for secretive marshbirds (Johnson et al. 2009). Precise locations and lek attendance will have to be recorded during surveys; the value of survey data will increase if common standards and methodology are adopted by multiple agencies (Connelly and Schroeder 2007). Repeated visits per season will enable estimation of probability of detection. Finally, many of the questions regarding what leks actually represent can be addressed through sentinel-lek surveys, where lek counts are adjusted using the results of additional monitoring and analysis conducted on a subset of the leks that are surveyed (Garton et al. 2007).

Developing landscape models that are useful across large regions will require cooperative, regional efforts, but the models can provide many benefits beyond identification of sites for preservation. For example, development of wind energy infrastructure already may be affecting prairie grouse and will continue to spread (Pruett et al. 2009a, 2009b; Brennan et al. 2009). Considerable uncertainty exists about the immediate and cumulative effects of stressors such as wind energy; management agencies could approach such issues proactively by stratifying the landscape by existing or potential land development, surveying leks, and evaluating the effects of stressors using landscape models and information-theoretic methods. Similar approaches have been used to evaluate the response of Sharp-tailed Grouse to ecological disturbances (Niemuth and Boyce 2004), as well as the response of Greater Sage-Grouse to energy development (Aldridge and Boyce 2007, Walker et al. 2007).

Incorporating spatially explicit habitat models into prairie grouse conservation efforts will not solve all prairie grouse problems. Spatial models are only tools to increase conservation efficiency and do not alter the root problem of human use and conversion of native habitats. Additional information, such as the risk of grassland conversion and the costs of conservation treatments will also have to be considered in conservation decisions. Environmental and ecological factors beyond human control also influence prairie grouse populations, but spatially explicit models provide a scientifically and biologically sound means of assessing landscapes, identifying appropriate conservation actions, and demonstrating the benefits of those actions. However, landscape-level models and plans are of little value unless they are accompanied by a landscape-level commitment to on-the-ground action. Successful conservation of prairie grouse will require an increased commitment from society and the conservation community to broad-scale conservation of prairie grouse habitat.

ACKNOWLEDGMENTS

I thank the many colleagues who have discussed landscape-level conservation of prairie grouse with me over the years, particularly M. E. Estey, J. R. Keir, L. H. Niemuth, R. E. Reynolds, and J. A. Shaffer. This paper was greatly improved by comments from K. E. Church, J. S. Gleason, B. K. Sandercock, and two anonymous reviewers. The findings and conclusions in this article are those of the author and do not necessarily represent the views of the U.S. Fish and Wildlife Service.

LITERATURE CITED

Abraham, K. F., M. G. Anderson, R. Clark, L. Colpitts, E. Reed, R. Bishop, J. Eadie, M. Petric, F. Rohwer, M. Tome, and A. Rojo. 2007. Final report: North American Waterfowl Management Plan continental progress assessment. U.S. Fish and Wildlife Service, Arlington, VA. <http://www.fws.gov/birdhabitat/NAWMP/files/FinalAssessmentReport.pdf> (24 February 2009).

Aldridge, C. L., and M. S. Boyce. 2007. Linking occurrence and fitness to persistence: habitat-based approach for endangered greater sage-grouse. Ecological Applications 17:508–526.

Aldridge, C. L., M. S. Boyce, and R. K. Baydack. 2004. Adaptive management of prairie grouse: how do we get there? Wildlife Society Bulletin 32:92–103.

Aldridge, C. L., S. E. Nielsen, H. L. Beyer, M. S. Boyce, J. W. Connelly, S. T. Knick, and M. A. Schroeder. 2008. Range-wide patterns of greater sage grouse persistence. Diversity and Distributions 14:983–994.

Ammann, G. A. 1957. The prairie grouse of Michigan. Technical Bulletin. Michigan Department of Conservation, Lansing, MI.

Anderson, D. R. 2001. The need to get the basics right in wildlife field studies. Wildlife Society Bulletin 29:1294–1297.

Andrén, H. 1994. Effects of habitat fragmentation on birds and mammals in landscapes with different proportions of suitable habitat: a review. Oikos 71:355–366.

Applegate, R. D., C. K. Williams, and R. R. Manes. 2004. Assuring the future of prairie grouse: dogmas, demagogues, and getting outside the box. Wildlife Society Bulletin 32:104–111.

Arthur, S. M., B. F. J. Manly, L. L. McDonald, and G. W. Garner. 1996. Assessing habitat selection when availability changes. Ecology 77:215–227.

Beissinger, S. R., J. R. Walters, D. G. Catanzaro, K. G. Smith, J. B. Dunning Jr., S. M. Haig, B. R. Noon, and B. M. Smith. 2006. Modeling approaches in avian conservation and the role of field biologists. Ornithological Monographs 59:1–56.

Bender, D. J., T. A. Contreras, and L. Fahrig. 1998. Habitat loss and population decline: a meta-analysis of the patch size effect. Ecology 79:517–533.

Bergerud, A. T. 1988a. Increasing the numbers of grouse. Pp. 686–731 in A. T. Bergerud and M. W. Gratson (editors), Adaptive strategies and population ecology of northern grouse. University of Minnesota Press, Saint Paul, MN.

Bergerud, A. T. 1988b. Population ecology of grouse. Pp. 578–685 in A. T. Bergerud and M. W. Gratson (editors), Adaptive strategies and population ecology of northern grouse. University of Minnesota Press, Saint Paul, MN.

Beyersbergen, G. W., N. D. Niemuth, and M. R. Norton. 2004. Northern Prairie and Parkland Waterbird Conservation Plan. Prairie Pothole Joint Venture, Denver, CO.

Bock, C. E., and Z. F. Jones. 2004. Avian habitat evaluation: should counting birds count? Frontiers in Ecology and the Environment 2:403–410.

Bouzat, J. L., H. H. Cheng, H. A. Lewin, R. L. Westemeier, J. D. Brawn, and K. N. Paige. 1998. Genetic evaluation of a demographic bottleneck in the Greater Prairie Chicken. Conservation Biology 12:836–843.

Bouzat, J. L. and K. Johnson. 2004. Genetic structure among closely spaced leks in a peripheral population of Lesser Prairie-Chickens. Molecular Ecology 13:499–505.

Bowman, T. J., and R. J. Robel. 1977. Brood break-up, dispersal, mobility, and mortality of juvenile prairie chickens. Journal of Wildlife Management 41:27–34.

Braun, C. E., K. Martin, T. E. Remington, and J. R. Young. 1994. North American grouse: issues and strategies for the 21st century. Transactions of the North American Wildlife and Natural Resources Conference 59:428–438.

Brennan, L. A., R. M. Perez, S. J. DeMaso, B. M. Ballard, and W. P. Kuvlesky, Jr. 2009. Potential impacts of wind farm energy development on upland game birds: questions and concerns. Pp. 179–183 *in* T. D. Rich, C. D. Thompson, D. Demarest, and C. Arizmendi, (editors), Tundra to tropics: Connecting birds, habitats, and people. Proceedings of the fourth international Partners in Flight conference, 13–16 February 2008, McAllen, TX.

Burnham, K. P., and D. R. Anderson. 1998. Model selection and inference: a practical information-theoretic approach. Springer-Verlag, New York, NY.

Cannon, R. W., and F. L. Knopf. 1981. Lek numbers as a trend index to prairie grouse populations. Journal of Wildlife Management 45:776–778.

Chetkiewicz, C. B., C. C. St. Clair, and M. S. Boyce. 2006. Corridors for conservation: integrating pattern and process. Annual Review of Ecology, Evolution, and Systematics 37:317–342.

Christisen, D. M. 1985. The Greater Prairie Chicken and Missouri's land-use patterns. Missouri Department of Conservation Terrestrial Series No. 15.

Clark, W. R., R. A. Schmitz, and T. R. Bogenschutz. 1999. Site selection and nest success of Ring-necked Pheasants as a function of location in Iowa landscapes. Journal of Wildlife Management 63:976–989.

Clifton, A. M., and D. G. Krementz. 2006. Estimating numbers of Greater Prairie-Chickens using mark-resight techniques. Journal of Wildlife Management 70:479–484.

Connelly, J. W., M. W. Gratson, and K. P. Reese, 1998. Sharp-tailed Grouse (*Tympanuchus phasianellus*). A. Poole and F. Gill (editors), The birds of North America No. 354. The Birds of North America, Inc., Philadelphia, PA.

Connelly, J. W., and M. A. Schroeder. 2007. Historical and current approaches to monitoring Greater Sage-Grouse. Pp. 3–9 *in* K. P. Reese and R. T. Bowyer (editors), Monitoring populations of sage-grouse. University of Idaho College of Natural Resources Experiment Station Bulletin 88, Moscow, ID.

Connolly, T. T. 2001. Reproductive ecology of Sharp-tailed Grouse in the pine barrens of northwestern Wisconsin. M.S. thesis, University of Wisconsin–Stevens Point, Wisconsin.

Crawford, J. A., and E. G. Bolen. 1976. Effects of land use on Lesser Prairie Chickens in Texas. Journal of Wildlife Management 40:96–104.

Fahrig, L. 1998. When does fragmentation of breeding habitat affect population survival? Ecological Modelling 105:272–292.

Fahrig, L. 2003. Effects of habitat fragmentation on biodiversity. Annual Review of Ecology and Systematics 34:487–515.

Fuhlendorf, S. D., A. J. W. Woodward, D. M. Leslie, Jr., and J. S. Shackford. 2002. Multi-scale effects of habitat loss and fragmentation on Lesser Prairie Chicken populations of the US southern Great Plains. Landscape Ecology 17:617–628.

Gallant, A. L. 2009. What you should know about land-cover data. Journal of Wildlife Management 73:796–805.

Garton, E. O., D. D. Musil, K. P. Reese, J. W. Connelly, and C. L. Anderson. 2007. Sentinel lek-routes: an integrated sampling approach to estimate Greater Sage-Grouse population characteristics. Pp. 31–41 *in* K. P. Reese and R. T. Bowyer (editors), Monitoring populations of sage-grouse. University of Idaho College of Natural Resources Experiment Station Bulletin 88, Moscow, ID.

Giesen, K. M. 1998. Lesser Prairie-Chicken (*Tympanuchus pallidicinctus*). A. Poole and F. Gill (editors), The birds of North America No. 364. The Academy of Natural Sciences, Philadelphia, PA, and the American Ornithologists' Union, Washington, DC.

Giesen, K. M., and J. W. Connelly. 1993. Guidelines for management of Columbian Sharp-tailed Grouse habitats. Wildlife Society Bulletin 21:325–333.

Grange, W. B. 1948. Wisconsin grouse problems. Wisconsin Conservation Department, Madison, WI.

Gratson, M. W. 1988. Spatial patterns, movements, and cover selection by Sharp-tailed Grouse. Pp. 158–192 *in* A. T. Bergerud and M. W. Gratson (editors), Adaptive strategies and population ecology of northern grouse. University of Minnesota Press, Saint Paul, MN.

Green, R. H. 1979. Sampling design and statistical methods for environmental biologists. Wiley Inter-science, Chichester, England.

Hagen, C. A., J. C. Pitman, R. J. Robel, T. M. Loughin, and R. D. Applegate. 2007a. Niche partitioning by Lesser Prairie-Chicken *Tympanuchus pallidicinctus* and Ring-necked Pheasant *Phasianus colchicus* in southwestern Kansas. Wildlife Biology 13(Suppl. 1): 34–41.

Hagen, C. A., J. C. Pitman, B. K. Sandercock, R. J. Robel, and R. D. Applegate. 2007b. Age-specific survival and probable causes of mortality in female Lesser Prairie-Chickens. Journal of Wildlife Management 71:518–525.

Hagen, C. A., B. K. Sandercock, J. C. Pitman, R. J. Robel, and R. D. Applegate. 2009. Spatial variation in Lesser Prairie-Chicken demography: a sensitivity analysis of population dynamics and management alternatives. Journal of Wildlife Management 73:1325–1332.

Hamerstrom, F. N., Jr., O. E. Mattson, and F. Hamerstrom. 1957. A guide to prairie chicken management. Wisconsin Conservation Department Technical Wildlife Bulletin 15, Madison, WI.

Hamerstrom, F. N., Jr., and F. Hamerstrom. 1973. The prairie chicken in Wisconsin: highlights of a 22-year study of counts, behavior, movements, turnover, and habitat. Wisconsin Department of Natural Resources Technical Bulletin 64, Madison, WI.

Hanowski, J. M., D. P. Christian, and G. J. Niemi. 2000. Landscape requirements of prairie Sharp-tailed Grouse *Tympanuchus phasianellus campestris* in Minnesota, USA. Wildlife Biology 6:257–263.

Haukos, D. A., and L. M. Smith. 1999. Effects of lek age on age structure and attendance of Lesser Prairie Chickens (*Tympanuchus pallidicinctus*). American Midland Naturalist 142:415–420.

Herkert, J. R., D. L. Reinking, D. A Wiedenfield, M. Winter, J. L. Zimmerman, W. E. Jensen, E. J. Finck, R. R. Koford, D. H. Wolfe, S. K. Sherrod, M. A. Jenkins, J. Faaborg, and S. K. Robinson. 2003. Effects of prairie fragmentation on the nest success of breeding birds in the midcontinental United States. Conservation Biology 17:587–594.

Holling, C. S. 1978. Adaptive environmental assessment and management. John Wiley and Sons, New York, NY.

Johnson, D. H. Johnson, D. H., J. P. Gibbs, M. Herzog, S. Lor, N. D. Niemuth, C. A. Ribic, M. Seamans, T. L. Shaffer, G. Shriver, S. Stehman, and W. L. Thompson. 2009. A sampling design framework for monitoring secretive marshbirds. Waterbirds 32:203–215.

Johnson, J. A., M. R. Bellinger, J. E. Toepfer, and P. Dunn. 2004. Temporal changes in allele frequencies and low effective population size in Greater Prairie-Chickens. Molecular Ecology 13:2617–2630.

Johnson, J. A., J. E. Toepfer, and P. O. Dunn. 2003. Contrasting patterns of mitochondrial and microsatellite population structure in fragmented populations of Greater Prairie-Chickens. Molecular Ecology 12:3335–3347.

Kareiva, P., and U. Wennergren. 1995. Connecting landscape patterns to ecosystem and population processes. Nature 373:299–302.

Kindlmann, P., and F. Burel. 2008. Connectivity measures: a review. Landscape Ecology 23:879–890.

Knick, S. T., and J. T. Rotenberry. 2000. Ghosts of habitats past: contribution of landscape change to current habitats used by shrubland birds. Ecology 81:220–227.

Kobriger, G. D. 1965. Status, movements, habitats, and foods of prairie grouse on a Sandhills refuge. Journal of Wildlife Management 29:788–800.

Leif, A. P. 2005. Spatial ecology and habitat selection of breeding male pheasants. Wildlife Society Bulletin 33:130–141.

Lillesand, T. M., and R. W. Kiefer. 2000. Remote sensing and image interpretation. John Wiley & Sons, New York, NY.

Manel, S., M. K. Schwartz, G. Luikart, and P. Taberlet. 2003. Landscape genetics: combining landscape ecology and population genetics. Trends in Ecology and Evolution 18:189–197.

Manzer, D. L., and S. J. Hannon. 2005. Relating grouse nest success and corvid density to habitat: a multi-scale approach. Journal of Wildlife Management 69:110–123.

Mayor, S. J., D. C. Schneider, J. A. Schaefer, and S. P. Mahoney. 2009. Habitat selection at multiple scales. EcoScience 16:238–247.

McDonald, M. W., and K. P. Reese. 1998. Landscape changes within the historical distribution of Columbian Sharp-tailed Grouse in eastern Washington: is there hope? Northwest Science 72:34–41.

Merrill, M. D., K. A. Chapman, K. A Poiani, and B. Winter. 1999. Land-use patterns surrounding Greater Prairie-Chicken leks in northwestern Minnesota. Journal of Wildlife Management 63:189–198.

Millspaugh, J. J., and F. R. Thompson, III (editors). 2008. Models for planning wildlife conservation in large landscapes. Elsevier Science. Burlington, MA.

Moe, M. 1999. Status and management of the Greater Prairie Chicken in Iowa. Pp.123–127 *in* W. D. Svedarsky, R. H. Hier, and N. J. Silvy (editors). The Greater Prairie Chicken: a national look. University of Minnesota Agricultural Experiment Station Miscellaneous Publication 99-1999, Saint Paul, MN.

Moilanen, A., and I. Hanski. 2001. On the use of connectivity measures in spatial ecology. Oikos 95:147–151.

Moilanen, A., and M. Nieminen. 2002. Simple connectivity measures in spatial ecology. Ecology 83:1131–1145.

Morrow, M. E., R. S. Adamcik, J. D. Friday, and L. B. McKinney. 1996. Factors affecting Attwater's Prairie-Chicken decline on the Attwater Prairie Chicken National Wildlife Refuge. Wildlife Society Bulletin 24:593–601.

Niemuth, N. D. 2000. Land use and vegetation associated with Greater Prairie Chickens in an agricultural landscape. Journal of Wildlife Management 64:278–286.

Niemuth, N. D. 2003. Identifying landscapes for Greater Prairie Chicken translocation using habitat models and GIS: a case study. Wildlife Society Bulletin 31:145–155.

Niemuth, N. D. 2005. Landscape composition and Greater Prairie-Chicken lek attendance: implications for management. Prairie Naturalist 37:127–142.

Niemuth, N. D., and M. S. Boyce. 2004. Influence of landscape composition on Sharp-tailed Grouse lek location and attendance in Wisconsin pine barrens. Ecoscience 11:209–217.

Niemuth, N. D., M. E. Estey, and R. E. Reynolds. 2009. Data for developing spatial models: Criteria for effective conservation. Pp. 396–411 *in* T. D. Rich, C. D. Thompson, D. Demarest, and C. Arizmendi, (editors), Tundra to tropics: connecting birds, habitats, and people. Proceedings of the fourth international Partners in Flight conference, 13–16 February 2008, McAllen, TX.

Niemuth, N. D., R. E. Reynolds, D. A. Granfors, R. R. Johnson, B. Wangler, and M. E. Estey. 2008. Landscape-level planning for conservation of wetland birds in the U.S. Prairie Pothole Region. Pp. 533–560 *in* J. J. Millspaugh and F. R. Thompson, III (editors), Models for planning wildlife conservation in large landscapes. Elsevier Science, Burlington, MA.

Nooker, J. K., and B. K. Sandercock. 2008. Phenotypic correlates and survival consequences of male mating success in lek-mating Greater Prairie-Chickens (*Tympanuchus cupido*). Behavioral Ecology and Sociobiology 62:1377–1388.

Patten, M. A., D. H. Wolfe, E. Shochat, and S. K. Sherrod. 2005. Habitat fragmentation, rapid evolution, and population persistence. 2005. Evolutionary Ecology Research 7:235–249.

Pedlar, J. H., L. Fahrig, and H. G. Merriam. 1997. Raccoon habitat use at two spatial scales. Journal of Wildlife Management 61:102–112.

Pepper, G. W. 1972. The ecology of Sharp-tailed Grouse during spring and summer in the aspen parklands of Saskatchewan. Saskatchewan Department of Natural Resources Wildlife Report Number 1, Regina, SK.

Peterson, M. J., and N. J. Silvy. 1996. Reproductive stages limiting productivity of the endangered Attwater's Prairie Chicken. Conservation Biology 10:1264–1276.

Phillips, M. L, W. R. Clark, S. M. Nusser, M. A. Sovada, and R. J. Greenwood. 2004. Analysis of predator movement in prairie landscapes with contrasting grassland composition. Journal of Mammalogy 85:187–195.

Pruett, C. L., M. A. Patten, and D. H. Wolfe. 2009a. It's not easy being green: wind energy and a declining grassland bird. BioScience 59:257–262.

Pruett, C. L., M. A. Patten, and D. H. Wolfe. 2009b. Avoidance behavior by prairie grouse: Implications for development of wind energy. Conservation Biology 23:1253–1259.

Reed, J. M. 1999. The role of behavior in recent avian extinctions and endangerments. Conservation Biology 13:232–241.

Reynolds, R. E., T. L. Shaffer, C. R. Loesch, and R. R. Cox, Jr. 2006. The farm bill and duck production in the Prairie Pothole Region: Increasing the benefits. Wildlife Society Bulletin 34:963–974.

Ribic, C. A., R. R. Koford, J. R. Herkert, D. H. Johnson, R. B. Renfrew, N. D. Niemuth, D. E. Naugle, and K. K. Bakker. 2009. Area sensitivity in North American grassland birds: patterns, processes, and research needs. Auk 126:233–244.

Rippin, A. B., and D. A. Boag. 1974. Recruitment to populations of male Sharp-tailed Grouse. Journal of Wildlife Management 38:616–621.

Robel, R. J. 1970. Possible role of behavior in regulating Greater Prairie Chicken populations. Journal of Wildlife Management 34:306–314.

Robel, R. J., F. R. Henderson, and W. Jackson. 1972. Some Sharp-tailed Grouse population statistics from South Dakota. Journal of Wildlife Management 36:87–98.

Ross, J. D., A. D. Arndt, R. F. C. Smith, J. A. Johnson, and J. L. Bouzat. 2006. Re-examination of the historical range of the Greater Prairie Chicken using provenance data and DNA analysis of museum collections. Conservation Genetics 7:735–750.

Ryan, M. R., L. W. Burger, Jr., D. P. Jones, and A. P. Wywialowski. 1998. Breeding ecology of Greater Prairie Chicken (*Tympanuchus cupido*) in relation to prairie landscape configuration. American Midland Naturalist 140:111–121.

Samson, F. B. 1980. Island biogeography and the conservation of nongame birds. Transactions of the North American Wildlife and Natural Resources Conference 45:245–251.

Samson, F. B., F. L. Knopf, and W. R. Ostlie. 2004. Great Plains ecosystems: past, present, and future. Wildlife Society Bulletin 32:6–15.

Schroeder, M. A. 1991. Movement and lek visitation by female Greater Prairie-Chickens in relation to predictions of Bradbury's female preference hypothesis of lek evolution. Auk 108: 896–903.

Schroeder, M. A., and C. E. Braun. 1992. Greater Prairie-Chicken attendance at leks and stability of leks in Colorado. Wilson Bulletin 104:273–284.

Schroeder, M. A., and C. E. Braun. 1993. Partial migration in a population of Greater Prairie-Chickens in northeastern Colorado. Auk 110:21–28.

Schroeder, M. A., and L. A. Robb. 1993. Greater Prairie-Chicken (*Tympanuchus cupido*). A. Poole and F. Gill (editors), The birds of North America No. 36. The Birds of North America, Inc., Philadelphia, PA.

Schroeder, M. A., and G. C. White. 1993. Dispersion of Greater Prairie Chicken nests in relation to lek location: evaluation of the hot-spot hypothesis of lek evolution. Behavioral Ecology 4:266–270.

Scott, M. J., P. J. Heglund, M. L. Morrison, J. B. Haufler, M. G. Raphael, W. A. Wall, and F. B. Samson. 2002. Predicting species occurrences: Issues of accuracy and scale. Island Press, Washington, DC.

Shenk, T. M., and A. B. Franklin. 2001. Modeling in natural resource management: development, interpretation, and application. Island Press, Washington, DC.

Sovada, M. A., M. C. Zicus, R. J. Greenwood, D. P. Rave, W. E. Newton, R. O. Woodward, and J. A. Beiser. 2000. Relationships of habitat patch size to predator community and survival of duck nests. Journal of Wildlife Management 64:820–831.

Stauffer, D. F. 2002. Linking populations and habitats: Where have we been? Where are we going? Pp. 53–61 in J. M. Scott, P. J. Heglund, M. L. Morrison, J. B. Haufler, M. G. Raphael, W. A. Wall, and F. B. Samson (editors), Predicting species occurrences: Issues of accuracy and scale. Island Press, Washington, DC.

Stewart-Oaten, A. W., M. Murdoch, and K. R. Parker. 1986. Environmental impact assessment: "Pseudoreplication" in time? Ecology 67:929–940.

Taylor, M. A., and F. S. Guthery. 1980a. Status, ecology, and management of the Lesser Prairie-Chicken. USDA Forest Service General Technical Report RM-77. USDA Forest Service, Rocky Mountain Research Station, Ft. Collins, CO.

Taylor, M. A., and F. S. Guthery. 1980b. Fall-winter movements, ranges, and habitat use of Lesser Prairie Chickens. Journal of Wildlife Management 44:521–524.

Tischendorf, L., and L. Fahrig. 2001. On the use of connectivity measures in spatial ecology: a reply. Oikos 95:152–155.

U.S. Department of the Interior and Environment Canada. 1986. North American Waterfowl Management Plan. Washington, DC.

Vance, D. R., and R. L. Westemeier. 1979. Interactions of pheasants and prairie chickens in Illinois. Wildlife Society Bulletin 7:221–225.

Van Den Bussche, R. A., S. R. Hoofer, D. A. Wiedenfield, D. H. Wolfe, and S. K. Sherrod. 2003. Genetic variation within and among fragmented populations of Lesser Prairie-Chickens (*Tympanuchus pallidicinctus*). Molecular Ecology 12:675–683.

Van Horne, B. 1983. Density as a misleading indicator of habitat quality. Journal of Wildlife Management 47:893–901.

Vodehnal, W. L., and J. B. Haufler. 2007. A grassland conservation plan for prairie grouse. North American Grouse Partnership, Fruita, CO.

Walker, B. L., D. E. Naugle, and K. E. Doherty. 2007. Greater Sage-Grouse population response to energy development and habitat loss. Journal of Wildlife Management 71:2644–2654.

Walters, C. J. 1986. Adaptive management of renewable resources. McGraw Hill, New York, NY.

Wang, Y., K. Yang, C. L. Bridgman, and L. Lin. 2008. Habitat suitability modeling to correlate gene flow with landscape connectivity. Landscape Ecology 23:989–1000.

Westemeier, R. L. 1971. The history and ecology of prairie chickens in central Wisconsin. University of Wisconsin Research Bulletin 281, Madison, WI.

Westemeier, R. L, J. D. Brawn, S. A. Simpson, T. L. Esker, R. W. Jansen, J. W. Walk, E. L. Kershner, J. L. Bouzat, and K. N. Paige. 1998a. Tracking the long-term decline and recovery of an isolated population. Science 282:1695–1698.

Westemeier, R. L., J. E. Buhnerkempe, W. R. Edwards, J. D. Brawn, and S. A. Simpson. 1998b. Parasitism of Greater Prairie-Chicken nests by Ring-necked Pheasants. Journal of Wildlife Management 62:854–863.

Wiens, J. A. 1973. Pattern and process in grassland bird communities. Ecological Monographs 43:237–270.

Wiens, J. A. 1997. Metapopulation dynamics and landscape ecology. Pp. 43–62 in I. Hanski and M. Gilpin (editors), Metapopulation biology: ecology, genetics, and evolution. Academic Press, London, UK.

Wiens, J. A. 2002. Predicting species occurrences: Progress, problems, and prospects, Pp. 739–749 in J. M. Scott, P. J. Heglund, M. L. Morrison, J. B. Haufler, M. G. Raphael, W. A. Wall, and F. B. Samson (editors), Predicting species occurrences: Issues of accuracy and scale. Island Press, Washington, DC.

Williams, C. K., F. S. Guthery, R. D. Applegate, and M. J. Peterson. 2004. The Northern Bobwhite decline: scaling our management for the twenty-first century. Wildlife Society Bulletin 32:861–869.

Winter, M., and J. Faaborg. 1999. Patterns of area sensitivity in grassland-nesting birds. Conservation Biology 13:1424–1436.

Wisdom, M. J., and L. S. Mills. 1997. Sensitivity analysis to guide population recovery: prairie-chickens as an example. Journal of Wildlife Management 61:302–312.

Wisdom, M. J., M. M. Rowland, and R. J. Tausch. 2005. Effective management strategies for sage-grouse and sagebrush: a question of triage? Transactions of the North American Wildlife and Natural Resources Conference 70:206–227.

Wisdom, M. J., B. C. Wales, M. M. Rowland, M. G. Raphael, R. S. Holthausen, T. D. Rich, and V. A. Saab. 2002. Performance of Greater Sage-Grouse models for conservation assessment in the interior Columbia Basin, U.S.A. Conservation Biology 16:1232–1242.

With, K. A., and A. W. King. 1999. Dispersal success on fractal landscapes: a consequence of lacunarity thresholds. Landscape Ecology 14:73–82.

Wolfe, D. H., M. A. Patten, E. Shochat, D. L. Pruett, and S. K. Sherrod. 2007. Causes and patterns of mortality in Lesser Prairie-Chickens *Tympanuchus pallidicinctus* and implications for management. Wildlife Biology 13(Suppl. 1):95–104.

Woodward, A. J. W., S. D. Fuhlendorf, D. M. Leslie, Jr., and J. Shackford. 2001. Influences of landscape composition and change on Lesser Prairie-Chicken populations. American Midland Naturalist 145:261–274.

Yeatter, R. E. 1963. Population responses of prairie chickens to land-use changes in Illinois. Journal of Wildlife Management 27:739–757.

CHAPTER TWO

Hierarchical Modeling of Lek Habitats of Greater Prairie-Chickens

Andrew J. Gregory, Lance B. McNew, Thomas J. Prebyl,
Brett K. Sandercock, and Samantha M. Wisely

Abstract. Greater Prairie-Chickens (*Tympanuchus cupido*) are a lek-mating prairie grouse of the central Great Plains. Males gather each spring at communal display grounds or leks to compete for mating opportunities with females, and lek sites are essential for the reproductive biology of prairie-chickens. We obtained geographic coordinates for 166 active leks located in eastern Kansas. Using GIS analysis, we developed a spatially explicit model to identify landcover and geomorphological variables associated with lek locations. We used a hierarchical approach to model selection to identify the best predictor variables at three spatial scales (0 m, 200 m, and 5 km), and then combined factors from the best models into a global multiscale model. We found that a synthetic variable, weighted elevation or the point elevation standardized by the elevation of the surrounding landscape, best explained lek occurrence at a lek point scale of 0 m. At broader spatial scales of 200 m and 5 km, avoidance of agricultural, urban, and

forest habitats, avoidance of high densities of roads, and a preference for grassland cover were the best predictors of lek site locations. Next, we created an entropy model based on factors from our minimum Bayesian Information Criterion global model to create an index of suitable lek habitat across the Flint Hills, Smoky Hills, and Osage Plains ecoregions of eastern Kansas. The entropy model showed that >85% of lek sites were in habitat strata that comprised <20% of the regional landscape, suggesting that prairie-chickens may be utilizing areas that are of marginal quality. Our research results have important implications for conservation because Kansas prairies are the core of extant distribution of Greater Prairie-Chickens and include the largest remaining intact grasslands in the United States.

Key Words: entropy modeling, landscape ecology, lek habitat, niche modeling, tallgrass prairie, *Tympanuchus cupido.*

Gregory, A. J., L. B. McNew, T. J. Prebyl, B. K. Sandercock, and S. M. Wisely. 2011. Hierarchical modeling of lek habitats of Greater Prairie-Chickens. Pp. 21–32 *in* B. K. Sandercock, K. Martin, and G. Segelbacher (editors). Ecology, conservation, and management of grouse. Studies in Avian Biology (no. 39), University of California Press, Berkeley, CA.

Conversion of native grasslands to agriculture has caused dramatic declines in prairie habitats since European settlement, and tallgrass prairie is one of the most highly endangered ecosystems in North America, with <5% of the original area remaining (Samson and Knopf 1994). Eastern Kansas includes >90% of the tallgrass prairie ecosystem left in North America, and the Smoky Hills, Flint Hills, and Osage Plains ecoregions have been recognized as ecologically important because they are core areas for grassland birds, an avian community of conservation concern (Fitzgerald et al. 2000, Pashley et al. 2000, Brennan and Kuvlevsky 2005). Unfortunately, long-term changes in land use and rangeland management practices may be negatively impacting the regional population viability of grassland birds in Kansas (Powell 2006, With et al. 2008, Rahmig et al. 2009).

Greater Prairie-Chickens (*Tympanuchus cupido*, hereafter "prairie-chickens") are a prairie grouse that are native to the grasslands of North America (Schroeder and Robb 1993). Prairie-chickens have been extirpated from much of their historical range, and historic losses were likely due to anthropogenic conversion of grasslands to row crop agriculture. The core of the remaining range of the species is in eastern Kansas and adjacent states, and populations in Kansas have been declining for over 30 years (Svedarsky et al. 2000, Rodgers 2008). The underlying causes for ongoing population declines are poorly understood but may be related to changes in land use practices or predator communities. Regardless of the cause, ongoing population declines are a serious conservation concern. Kansas is the core of the remaining range, and translocations of birds from source populations in Kansas have been used to bolster population numbers and increase genetic diversity within relict populations of prairie-chickens in Illinois and Missouri (Bouzat et al. 1998, B. E. Jamison, pers. comm.). A better understanding of the distribution and habitat requirements of prairie-chickens will aid conservation for this species and the associated community of grassland birds.

Wildlife habitat use is hierarchical, and animals make decisions about which areas to use at multiple spatial scales (Johnson 1980). At broad scales of ~10 km, prairie-chickens may avoid unsuitable habitats within their large home ranges (Hamerstrom and Hamerstrom 1960, Prose 1985).

At finer scales of ~1 km, prairie-chickens may use different vegetative cover types for different purposes such as nesting, feeding, and roosting (Svedarsky 1988, McCarthy et al. 1994, Ryan et al. 1998). When engaged in these activities, prairie-chickens may select patches to reduce predation risk, to optimize their thermal environment, or to forage on important food plants (Buhnerkempe et al. 1984, Ryan et al. 1998). Heterogeneity among patches within landcover types provides different resources, and consequently some patches may be more desirable than others. At each spatial scale, prairie-chickens must make decisions about where to allocate time and energy, and habitat preferences at broader scales likely impact the choices available at finer spatial scales.

We evaluated the suitability of the Flint Hills, Smoky Hills, and Osage Plains ecoregions of Kansas based on multiscale geospatial modeling of lek site locations for Greater Prairie-Chickens. Leks or booming grounds are communal display sites where male prairie-chickens congregate to display and mate with females. Male prairie-chickens show high site fidelity to leks from one breeding season to the next, and lek locations can be relatively stable over time (Robel 1970, Nooker and Sandercock 2008). Most mating is thought to occur at lek sites, and consequently suitable lekking sites are a necessary component of prairie-chicken habitat (Hamerstorm and Hamerstrom 1960, Schroeder and Robb 1993). Female prairie-chickens usually nest in the vicinity of leks (≤ km; Hamerstrom 1939, Schroeder 1991), and lek site location ought to serve as a proxy for the occurrence of suitable nesting habitat at a landscape scale. Indeed, one proposed mechanism for lek evolution (the hot spot hypothesis) hypothesizes that leks evolved as males settled and clustered on pathways used preferentially by females to travel between needed resources (Beehler and Foster 1988, Schroeder and White 1993).

The primary goal of our landscape model was to identify suitable versus unsuitable habitat based on the location attributes of lek sites. We used a hierarchical modeling approach with three spatial scales of 0 m, 200 m, and 5 km, based on the movements and space use of prairie-chickens in Kansas (Robel et al. 1970). When modeling habitat suitability, even the most refined spatial scales are a coarse-grained approach to conservation, because we must assume that microhabitat features within identified habitat patches have the potential to be improved with management practices. Prairie-chickens could be considered an umbrella species for grassland communities because the species requires large tracts of grasslands (Svedarsky 1988, Poiani et al. 2001). Our main goal was to identify areas in need of conservation or enhancement for prairie-chickens, but our modeling approach and research results also have conservation implications for other sensitive species of grassland birds (Herkert 1994, Brennan and Kuvlesky 2005).

METHODS

To create an index of suitable prairie-chicken habitat for our study region, we performed a geospatial analysis of 166 lek locations distributed across the Flint Hills, Smoky Hills, and Osage Plains ecoregions of Kansas (Fig. 2.1; Griffith et al. 2008). Geographical coordinates of leks were collected as part of a 3-year population study of prairie-chickens in eastern Kansas (2006–2008; L. B. McNew et al., this volume, chapter 15) and from lek surveys conducted by the Kansas Department of Wildlife and Parks (KDWP, 2005–2007). KDWP survey routes were originally established in the late 1950s at a sampling density of 1 route

surveying 57.8 km² per county, but sampling efforts are being continually expanded. KDWP survey routes were not established in targeted areas with known prairie-chicken populations, but rather were selected based on the presence of large tracts of grassland habitat and relatively good access via county roads. For the purposes of KDWP surveys, leks were defined as ≥3 males displaying in an area, and were located by listening for prairie-chicken booming at 1-mile intervals along the survey routes and by performing flush counts on located leks (R. D. Rodgers, pers. comm.). Our intensive population study was primarily conducted in Cloud, Geary, and Elk counties. For the intensive surveys, leks were also defined as ≥3 males displaying in an area and were located via listening along all county roads within the identified counties. We also sought landowner permission to survey large road-free tracts of land either on foot or with all-terrain vehicles.

We used Arc Info 9.2 (Environmental Systems Research Institute, Redlands, CA) for all geospatial analysis and data extraction. We acquired all data sets from the Kansas Geospatial Community Commons (www.kansasgis.org). For land cover analyses, we used the 30-m resolution, 2005 land cover map of the State of Kansas (Kansas Applied

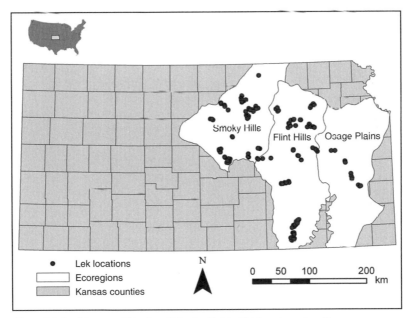

Figure 2.1. Study area and lek sites of Greater Prairie-Chickens in eastern Kansas, 2005–2008. Ecoregions represent areas of similar ecosystems and geomorphological characteristics. Black dots are locations of leks used for model development and validation. Inset map indicates the location of Kansas within the United States.

Remote Sensing Program 2005; Whistler et al. 2006), which we reclassified to an Anderson Level I classification scheme, depicting five biologically relevant landcover classes: grassland, row crop agriculture, urban, forested, and water (Anderson et al. 1971). Grasslands included all CRP (Conservation Reserve Program) lands, or grazed and ungrazed pastures of native prairie. Row crop agriculture included croplands plus all tillable acres. Urban areas were defined as all cities, towns, roads, and human dwellings. Forested lands included gallery forests and riparian corridors, whereas water included rivers, streams, stock ponds, and reservoirs. For geomorphological analyses, we used the 30-m resolution 1999 National Elevation Dataset (U.S. Geological Survey, EROS Data Center). We also included a 1991 Riparian Inventory data set for the state of Kansas (U.S. Department of Agriculture, Natural Resources Conservation Service) and a roadway data set that combined the 2006 Kansas State and Non-State Road System data sets (Kansas Department of Transportation, Bureau of Transportation Planning). Each land cover data set was aggregated to 100-m grain size prior to landscape analysis.

To assess differences in landscape and habitat features of lek sites versus potentially available landscape features and habitats, 132 random points were generated within the same spatial extent as lek locations using Arc Info 9.2 and were later used in logistic regression model fitting. Prior to model fitting procedures, we randomly selected 34 of 166 (20%) lek locations and an additional 34 randomly generated locations and withheld them from model development to be used for model validation.

To evaluate characteristics of the area surrounding lek sites at a landscape spatial scale, we buffered each lek site with a 5-km neighborhood radius which evaluated landscape patterns at large spatial scale. Females typically choose nesting sites within 2 km of lek locations (Hamerstrom 1939, Schroeder 1991), and the average home range size of a prairie-chicken in Kansas is 500 ha (Robel 1970). Thus a 5-km buffer was selected to encompass possible nesting habitats around lek sites. For the intermediate spatial scale, we analyzed lek habitat characteristics within overlapping neighborhoods of 200 m radii, a distance that would likely characterize the habitat used for lekking itself. Analysis at these two spatial scales tested whether characteristics of the landscape surrounding lek sites influenced the presence or absence of leks. We used neighborhood statistics to calculate the percent area for each of the six landcover types and Fragstats 3.3 to calculate the total core area of grassland patches using the eight neighbor patch rule and 100 m edge depth (McGarigal and Marks 1995). Within neighborhoods, we calculated the density of all roads (km per km^2) as an index of disturbance, and the density of 10-m elevational contour lines as an index of habitat complexity or topographic relief.

At the finest spatial scale at 0 m or the point of the lek, we measured attributes of the geographic center of the lek. We recorded four variables: distance to riparian areas, distance to urban areas, distance to roads, and weighted elevation. Weighted elevation was a synthetic variable that compared the absolute elevation of the lek site relative to the surrounding landscape, calculated as the elevation of the lek location divided by the average elevation of all grid centroid points within 1 km of the location. Use of weighted elevations standardized the topographic positions of leks within our study region to values ranging from 0.7 to 2.0. All measured variables were extracted from landscape data for both known lek sites and an equal number of random points.

Prior to model construction, all variables were standardized by z-transformations to normal distributions with a mean of zero and a standard deviation of 1, so that β coefficients from the resulting models were in the same units and would be directly comparable. We employed a hierarchical approach to model selection. Factors from each spatial scale were first entered into separate logistic regression models, and then significant factors were combined into a global model that pooled important variables across multiple scales. Our hierarchical model selection process consisted of Bayesian model selection at each of three spatial scales, followed by a second round of model selection for models with factors at multiple spatial scales (Schwarz 1978, Hosmer et al. 1997). Hierarchical procedures were used to avoid spatial autocorrelation between data sets. Spatial autocorrelation between scales can occur because scales are nested within each other hierarchically. Hierarchical procedures adjust for spatial autocorrelation by allowing models to be developed for each scale independently and then concatenated across scales. During the concatenation process, if variables are correlated across scales they are unlikely to be included because of the penalty associated with adding extra parameters. Bayesian model selection (BIC) procedures were used for model selection (Anderson et al. 2000, Johnson and Omland 2004), because these statistics tend to be more conservative and less likely to over-fit data than Akaike's Information Criterion (Burnham and Anderson 2004). A conservative approach to model selection was desirable to compensate for highly spatially correlated data sets. Principal components analysis or factor analysis could have been used to address this issue, but we did not use multivariate techniques because we were primarily interested in the effects of our original landscape variables. Use of the untransformed landscape data was important because we wanted to apply model predictions directly to spatially explicit ecological niche modeling. Improved GIS analysis techniques allow many landscape metrics to be calculated, but our goal was to ensure that only biologically relevant and statistically meaningful metrics were included in our analysis (McGarigal and Marks 1995).

Landscape variables from the minimum BIC multiscale model were used as data inputs for ecological niche modeling using program MaxEnt. Program MaxEnt uses entropy theory to model landscape suitability based only on presence data and to integrate analyses across spatial scales (Phillips et al. 2004); it has several advantages compared to other software for ecological niche modeling, including program GARP (Phillips et al. 2006, Austin 2007). However, program MaxEnt and other niche modeling software packages tend to overestimate landscape suitability when many environmental variables are used, but a conservative model selection procedure based on BIC should have ameliorated this possibility (Phillips et al. 2006). Ecological niche modeling yielded a preliminary index of suitable prairie-chicken lek habitat across the Flint Hills, eastern Smoky Hills, and Osage Plains regions of Kansas. We validated our model by using a random 20% subset of our lek points that were withheld from model development, and assessed the proportion of leks that mapped onto each of the suitability categories of our index. We also compared the suitability of the landscape as predicted from our hierarchical model to the suitability of the landscape as predicted by models based on each of the single spatial scales.

RESULTS

Environmental Covariates

At each spatial scale, our analysis indicated different features of the landscape were influencing lek presence. At the broadest scale, which described the area adjacent to leks in a 5-km neighborhood, five of six competing models each received some support ($w_i > 0.11$; Table 2.1). In general, the broad-scale models indicated that lek occurrence was negatively associated with percent forest area, road density, and urban area within the region, but was weakly and positively associated with percent grassland cover and the total core grassland area in a 5-km neighborhood. All possible candidate models for these variables were considered. Using a logistic model describing relative probability of lek occurrence conditional on habitat variables, the minimum BIC model included three variables: percent urban cover, road density, and percent forest cover: leks = $1.08 - 5.92 \times$ percent urban area $- 1.39 \times$ road density $- 0.76 \times$ percent forest.

At the lek habitat scale of a 200-m neighborhood, the global model included percent land coverage for forest, agriculture, urban, grassland, crop, and an estimate of topographic relief based on density of contour lines. All possible candidate models for the variables included in the global model were evaluated. Our selection procedure indicated that a single candidate model received 99% of the model support (Table 2.1), which indicated that lek habitat at a neighborhood of 200 m was most strongly influenced by a negative association with cover of row crop agriculture: leks = $0.01 - 0.46 \times$ percent agriculture.

At a lek point scale of 0 m, the global model included four variables: distance to roads, urban areas, forest, and weighted elevation. Of all possible candidate models for these four variables, two models received similar levels of support ($w_i > 0.45$; Table 2.1). The minimum BIC model, which received 49% of the model support, modeled lek sites as a function of both the weighted elevation and the distance from urban centers: leks = $0.003 + 1.34 \times$ weighted elevation $+ 0.36 \times$ distance to urban areas. A second model, which received 46% of the model support, modeled lek site location as a function of weighted elevation.

To understand the importance of different spatial scales in habitat selection, we reran the model selection procedure combining different scales.

TABLE 2.1

Bayesian model selection to identify landscape attributes associated with lek sites of Greater Prairie-Chickens in eastern Kansas, 2005–2008.

Model	K	$-2 \ln (K)$	BIC	ΔBIC	w_i	Hosmer-Lemeshow C	P
Habitat Models 5 km							
%Urban, % Frst, Rd. Den	4	286.9	309.3	0.0	0.33	0.80	0.03
%Grass, %Urban, %Frst	4	287.7	310.0	0.7	0.24	0.78	0.07
%Frst, Rd. Den	3	293.8	310.6	1.3	0.18	0.79	0.04
%Urban, %Frst, Rd. Den, ALGP	5	283.1	310.9	1.6	0.14	0.80	0.04
%Grass, %Urban, %Frst, Rd. Den	5	283.5	311.4	2.1	0.12	0.80	0.04
C-Den, %Ag, %CRP, &Grass, %Urban, %H$_2$O, %Frst, Rd. Den, ALGP	10	270.5	326.2	16.9	<0.01	0.82	0.86
Habitat Models 200 m							
%Ag	2	227.3	271.9	0.0	0.99	0.66	0.01
C-Den, %Ag, %CRP, %Grass, %Urban, %H$_2$O	8	353.0	364.1	92.0	0.01	0.86	0.02
%Frst							
Points Models 0 m							
D-Urb, Wt-Elev	3	281.3	298.1	0.0	0.49	0.82	0.01
Wt-Elev	2	287.0	298.1	0.1	0.46	0.82	0.01
Multi-Scale Model							
%Grass 5 km, %Urban 5 km, Wt-Elev	4	230.2	262.7	0.0	0.42	0.85	0.05
%Urban 5 km, %Frst 5 km, Rd. Den 5 km, Wt-Elev	5	259.3	263.2	0.6	0.32	0.88	0.06
%Urban 5 km, %Frst 5 km, Rd. Den 5 km, ALGP, Wt-Elev	6	259.2	263.6	0.9	0.26	0.88	0.09
%Grass 5 km, %Urban 5 km, %Frst 5 km, Rd. Den 5 km, D-Urb, ALGP, Wt-Elev	8	223.4	273.6	10.9	<0.01	0.89	0.30

NOTE: Variables are defined as follows: ALGP = area in m^2 of the largest contiguous grassland patch in the 5-km neighborhood, C-Den = contour line density, D-RIP = distance to riparian area, D-URB = distance to urban town or city, D-Road = distance to nearest road, %Ag = % of the neighborhood in row crop agriculture, %CRP = % of the neighborhood Conservation Reserve Program, %Frst = % of the neighborhood in forest, %Grass = % of the neighborhood in grassland, %H$_2$O = % of the neighborhood in water, %Urban = % of the neighborhood in urban cover, Rd.Den = road density in km per km^2 of the neighborhood, and Wt-Elev = weighted elevation. Column heading labels are as follows: K = number of parameters, $-2Ln(K)$ = maximum likelihood estimate from logistic model, BIC is the Schultz Criterion, and Hosmer-Lemeshow statistics are a goodness-of-fit test for the logistic model.

From the first set of analyses at different spatial scales (0 m, 200 m, and 5 km), we identified a set of eight landscape attributes from the subset of models that had high BIC weights and were equally parsimonious (ΔBIC $\leq$ 2). The global model combining factors from multiple spatial scales included eight factors: percent grassland, urban or forest at 5 km, road density at 5 km, total core grassland area at 5 km, percent agriculture at 200 m, distance to urban areas, and weighted elevation. We included all possible combinations of variables in the candidate models in the selection procedure, and three of these models received strong model support ($w_i > 0.25$; Table 2.1). The minimum BIC model, which received 42% of the total model support, indicated strong avoidance of urban areas and preference for relatively high sites with grassland cover: leks = 1.02 − 6.62 × percent urban area at 5 km + 1.21 × weighted elevation + 0.43 × percent grassland at 5 km. Other parsimonious models had similar coefficients for these three factors, but also included weak effects for avoidance of areas with high road density or forests and a preference for lek sites close to large grassland patches.

Niche Modeling

We used the minimum BIC multiscale model and program MaxEnt to create an index of suitable lek habitat for the three ecoregions in our study area (Fig. 2.2). Our niche model predicted that highly suitable habitat for leks was found at or near the highest point on the surrounding landscape. The average weighted elevation was 1.07 ± 0.07 SE, and highly suitable areas were comprised of 90% grassland and 8% agriculture, with <2% of the landscape surrounding leks being forest, water, or urban habitats. Moderately suitable areas were similar to highly suitable areas in having an average weighted elevation of 1.02 ± 0.07 SE, but had less grassland (77%) and more agriculture

(20%) than highly suitable areas; other habitats accounted for ~3% of the landscape. In contrast, low suitability areas were usually distributed in low-lying areas of the landscape and had an average weighted elevation of 0.87 ± 0.22 SE. Low-elevation habitats in our study area frequently included gallery forests, river beds and flood plains, row crop agricultural areas, and urbanized developments. Overall, low-suitability areas were comprised of 51% grassland, 33% agriculture, 9% forests, 4% urban, and 3% water.

Overall, our model predicted that lek sites would occur at or near the highest point on the landscape away from forests, large bodies of water, or urban centers, in areas comprised primarily of grassland with slight to moderate amounts of row crop agriculture. We created similar indices of lek habitat suitability for the minimum BIC models at each single spatial scale and compared the area of suitable habitat predicted by these indices to the area predicted by the multiscale grand model. Predictions from models based on a spatial scale of 0 m and 5 km were most similar to the predictions of the multiscale model (Fig. 2.3), presumably because environmental covariates at these spatial scales had the strongest effects on lek occurrence, as measured by the slope coefficients for z-transformed landscape covariates.

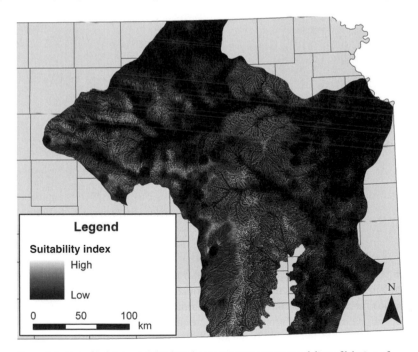

Figure 2.2. Map of habitat suitability based on maximum entropy modeling of lek sites of Greater Prairie-Chickens in eastern Kansas, 2005–2008.

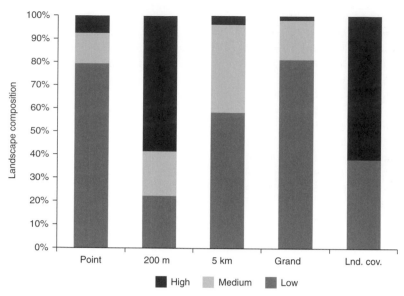

Figure 2.3. Comparisons of habitat suitability of our study area as predicted by models for three spatial scales (0 m, 200 m, and 5 km), a multi-scale model (Grand), and a model based solely on land cover (Lnd. cov.).

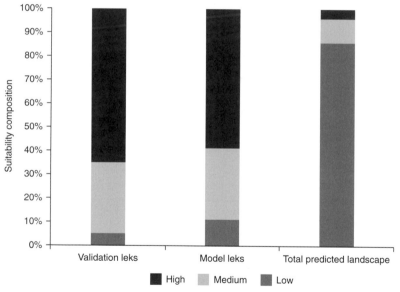

Figure 2.4. Comparisons of habitat suitability among a random set of leks used for model validation (*n* = 34), lek locations used for model development (*n* = 132), and the entire study area (Landscape).

During validation procedures, the final multiscale model was able to distinguish between randomly selected lek sites and randomly plotted points ($t = -3.9$, df = 35, $P \leq 0.001$). Moreover, 88.2% of the randomly selected validation leks (*n* = 34) were correctly classified into suitable habitat. We next plotted geographic coordinates of random leks onto the multiscale lek habitat suitability index to determine what proportion of these leks occurred in each of the three suitability categories, and compared the proportions to the frequency of occurrence of each habitat type in the landscape as a whole (Fig. 2.4). The top three models gave similar predictions, and in all cases the leks had substantially higher levels of suitability than the entire landscape. A total of 85–90% of the lek sites occurred in habitat areas of moderate to high suitability, yet only 5–20% of the total landscape included habitats that met these criteria for Greater Prairie-Chickens (Fig. 2.4).

DISCUSSION

Our hierarchical approach to ecological niche modeling showed that >85% of lek sites of Greater Prairie-Chickens were in habitat strata that comprised <20% of the regional landscape in eastern Kansas. Our results are somewhat discouraging because Kansas prairies are considered to be the last remaining strongholds for conservation of grassland birds in the U.S. (Fitzgerald et al. 2000, Pashley et al. 2000). We expected that the environmental correlates of lek site selection might vary across the geographic distribution of prairie-chickens if differences in the degree of habitat fragmentation affected lek placement. An estimated 36–45% of the landscape of our Kansas study area is comprised of intact grasslands (Applegate et al. 2003, A. J. Gregory and D. G. Goodin, unpubl. data). Previous analyses of lek habitat suitability have been conducted in more fragmented landscapes in Minnesota and Wisconsin, where grasslands are part of a patchy mosaic in a matrix of forest, wetlands, and row crop agriculture (Merrill et al. 1999, Niemuth 2003).

Despite potential differences in landscape configuration, our major findings were consistent with previous analyses of lek site selection for prairie-chickens. Grassland cover at 5 km was a relevant factor in our multiscale model for Kansas, which is consistent with previous studies that have identified cover or size of grassland patches as important factors in determining the presence of prairie-chicken leks elsewhere in their range (Hamerstrom et al. 1957, Kirsch 1974, Merrill et al. 1999, Niemuth 2003). Although grassland cover was relevant in Kansas, we found that relative elevation at the lek location, a lack of agriculture within 200 m, and a lack of urban areas, forest, and roads within 5 km were better predictors of lek presence. Grassland cover alone has previously been found to be a poor predictor of lek location in fragmented landscapes, and avoidance of areas with residential development and forest cover appears to be a general finding for lek site selection by prairie-chickens

(Merrill et al. 1999, Niemuth 2003). In the prairie-dominated landscapes of eastern Kansas, the amount of grassland available may be of less concern for prairie-chicken habitat conservation than the degree of fragmentation and configuration of remaining grassland habitats.

Our multiscale approach to modeling prairie-chicken habitat had two advantages over single-scale niche modeling approaches. First, by using data at different spatial scales and in a combined model, we obtained different suitability estimates of the landscape for the study species. Second, animal habitat use is predicted to be a hierarchical set of decisions (Johnson 1980), and our modeling indicated that different habitat attributes were preferred at different spatial scales. Our point model indicated that prairie-chicken lek sites were located at the highest portion of the surrounding landscape and distant from urban areas. Thus, prairie-chickens may choose display grounds with high visibility, good auditory projection, or areas free from ambient noise (Hamerstrom and Hamerstrom 1960, Aspbury and Gibson 2004, Slabbekoorn and Ripmeester 2008). This combination of features may assist females in locating lek sites or males in detecting and avoiding approaching predators. At the spatial scales of 200 m and 5 km, we observed lek sites being placed in areas that avoided urban areas, row crop agriculture, and roads, indicating avoidance of anthropogenic disturbance at spatial scales relevant to grassland conservation (Schroeder and Robb 1993).

Our model offers insights into the current suitability of the Kansas tallgrass prairies for prairie-chickens. Eastern Kansas represents >90% of the remaining tallgrass prairie in the U.S. (Samson and Knopf 1994). However, >80% of this area was predicted by our model to be of relatively low suitability (Figs. 2.2 and 2.3). Moreover, ~15–20% of the active lek sites in our analysis were in areas of low suitability, which may indicate use of marginal habitats. Prairie-chickens show high site fidelity to lek sites, but landscapes and habitat suitability can change rapidly. Site fidelity may result in a lag period between the time of landscape degradation and habitat abandonment or local extirpation. Thus, males may continue to display at lek sites that are effectively demographic sinks before the population is eventually extirpated (Schroeder and Robb 1993, Nooker and Sandercock 2008). Consequently, lek count surveys that are routinely used for population monitoring of prairie-chickens may be slow to reveal the impacts of environmental change. Our lek habitat suitability index does not account for the effects of land management on the demographic performance of prairie-chickens attending leks in marginal habitats, and caution should be used when interpreting our map. However, if the habitat requirements for lek and nest sites are closely associated, then our model suggests that much of the landscape in eastern Kansas is unsuitable for prairie-chickens, and habitat may be a limiting factor contributing to ongoing population declines (Rodgers 2008).

Spatial models for prairie-chickens in eastern Kansas, Minnesota, and Wisconsin (Merrill et al. 1999, Niemuth 2003, this study) have shown that lek sites are usually associated with grassland cover and negatively associated with anthropogenic disturbance and forest habitats. One goal for conservation of prairie-chickens should be to preserve large remaining tracts of natural grasslands with little development. A second goal should be to expand connectivity among unfragmented grassland habitats by removal of hedgerows and encroaching woody plants, and by enrollment of agriculture fields into the Conservation Reserve Program. The microhabitats required by prairie-chickens were not identified by our landscape approach, but the suitable habitats identified by our model could represent sites where improved land management would be beneficial. Rangeland management in eastern Kansas frequently includes use of early season burning to enhance forage quality for cattle production (With et al. 2008, Rahmig et al. 2009). Spring burning removes the vegetation that provides nesting cover for female prairie-chickens during the breeding season (L. B. McNew et al., this volume, chapter 15). Changes in land management from annual spring burns to a patch-burn rotational system could benefit Greater Prairie-Chickens and associated species of grassland birds by providing additional cover for ground-nesting species in a more heterogeneous landscape.

ACKNOWLEDGMENTS

The authors thank R. D. Rodgers and J. C. Pitman of the Kansas Department of Wildlife and Parks for generously providing lek survey data for prairie-chickens in eastern Kansas. We also thank T. Cikanek, V. Hunter, K. Rutz, W. White, and other seasonal research technicians who helped to locate and monitor lek sites in our study region. The Conservation Genetics and Molecular Ecology Lab at Kansas State University provided access to SAS statistical software and a dedicated GIS work station for analysis of spatial data. Financial support was provided by the Division of Biology at Kansas State University and by wind industry partners, state and federal agencies, and conservation groups under the National Wind Coordinating Collaborative and a grant from the National Fish and Wildlife Foundation (2007-0117-000).

LITERATURE CITED

Anderson, D. R., K. P. Burnham, and W. L. Thompson. 2000. Null hypothesis testing: problems, prevalence, and an alternative. Journal of Wildlife Management 64:912–923.

Anderson, J. R., E. E. Hardy, J. T. Roach, and R. E. Witmier. 1971. A land use and land cover classification system for use with remote sensor data. U.S. Geological Survey Professional Paper 964.

Applegate, R. D., B. E. Flock, and E. J. Finck. 2003. Changes in land use in eastern Kansas, 1984–2000. Transactions of the Kansas Academy of Science 106:192–197.

Aspbury, A. S., and R. M. Gibson. 2004. Long-range visibility of Greater Sage-Grouse leks: a GIS-based analysis. Animal Behavior 67:1127–1132.

Austin, M. 2007. Species distribution models and ecological theory: a critical assessment and some possible new approaches. Ecological Modeling 200:1–19.

Beehler, B. M., and M. S. Foster. 1988. Hotshots, hotspots, and female preferences in the organization of lek mating systems. American Naturalist 131:203–219.

Bouzat, J. L., H. H. Cheng, H. A. Lewin, R. L. Westemeier, J. D. Brawn, and K. N. Paige. 1998. Genetic evaluation of a demographic bottleneck in the Greater Prairie-Chicken. Conservation Biology 12:836–843.

Brennan, L. A., and W. P. Kuvlesky. 2005. North American grassland birds: an unfolding conservation crisis? Journal of Wildlife Management 69:1–13.

Buhnerkempe, J. E., W. R. Edwards, D. R. Vance, and R. L. Westemeier. 1984. Effects of residual vegetation on prairie-chicken nest placement and success. Wildlife Society Bulletin 12:382–386.

Burnham, K. P., and D. R. Anderson. 2004. Understanding AIC and BIC model selection. Sociological Methods and Research 33:261–304.

Fitzgerald, J., B. Busby, M. Howery, R. Klatske, D. Reinking, and D. Pashley. 2000. Partners in

Flight Bird Conservation Plan for the Osage Plains (Physiogeographic Area 33), Version 1.0. American Bird Conservancy, The Plains, VA.

Griffith, G. E., J. M. Omernik, and M. McGinley. 2008. Ecoregions of Kansas and Nebraska (EPA). *In* C. J. Cleveland (editor), Encyclopedia of earth. Environmental Information Coalition, National Council for Science and the Environment, Washington, DC.

Hamerstrom, F. N. 1939. A study of Wisconsin prairie-chicken and Sharp-tailed Grouse in Wisconsin. Wilson Bulletin 51:105–120.

Hamerstrom, F. N, O. E. Matson, and F. Hamerstrom. 1957. A guide to prairie-chicken management. Wisconsin Department of Natural Resources Technical Bulletin 15. Wisconsin Conservation Department, Game Management Division, Madison, WI.

Hamerstrom, F. N., Jr., and F. Hamerstrom. 1960. Comparability of some social displays of grouse. Transactions of the International Ornithological Congress 12:274–293.

Herkert, J. R. 1994. The effects of habitat fragmentation on Midwestern grassland bird communities. Ecological Applications 4:461–471.

Hosmer, D. W., T. Hosmer, S. Le Cessie, and S. Lemeshow. 1997. A comparison of goodness-of-fit tests for the logistic regression model. Statistics in Medicine 16:965–980.

Johnson, D. 1980. The comparison of usage and availability measurements for evaluating resource preference. Ecology 6:65–71.

Johnson, J. B., and K. S. Omland. 2004. Model selection in ecology and evolution. Trends in Ecology and Evolution 19:101–108.

Kirsch, L. M. 1974. Habitat considerations for prairie-chickens. Wildlife Society Bulletin 2:123–129.

McCarthy, C., T. Pella, G. Link, and M. A. Rumble. 1994. Greater Prairie-Chicken nesting habitat, Sheyenne National Grassland, North Dakota. Proceedings of the North Dakota Academy of Science 48:13–18.

McGarigal, K., and B. J. Marks. 1995. FRAGSTATS: spatial pattern analysis program for quantifying landscape structure. General Technical Report PNW-GTR-351. USDA Forest Service, Pacific Northwest Research Station, Portland, OR.

Merrill, M. D., K. A. Chapman, K. A. Poiani, and B. Winter. 1999. Land-use patterns surrounding Greater Prairie-Chicken leks in northwestern Minnesota. Journal of Wildlife Management 63:189–198.

Niemuth, N. D. 2003. Identifying landscapes for Greater Prairie-Chicken translocation using habitat models and GIS: a case study. Wildlife Society Bulletin 31:145–155.

Nooker, J. K., and B. K. Sandercock. 2008. Correlates and consequences of male mating success in lek-mating Greater Prairie-Chickens (*Tympanuchus cupido*). Behavioral Ecology and Sociobiology 62:1377–1388.

Pashley, D. N., C. J. Beardmore, J. A. Fitzgerald, R. P. Ford, W. C. Hunter, M. S. Morrison, and K. V. Rosenberg. 2000. Partners in flight: conservation of the landbirds of the United States. American Bird Conservancy, The Plains, VA.

Phillips, S. J., R. P. Anderson, and R. E. Schapire. 2006. Maximum entropy modeling of species geographic distributions. Ecological Modeling 190:231–259.

Phillips, S. J., M. Dudik, and R. E. Scharpire. 2004. A maximum entropy approach to species distribution modeling. Pp. 83 *in* Proceedings of the Twenty-first International Conference on Machine Learning. Association for Computing Machinery, Banff, AB, Canada.

Poiani, K. A., M. D. Merrill, and K. A. Chapman. 2001. Identifying conservation-priority areas in a fragmented Minnesota landscape based on the umbrella species concept and selection of large patches of natural vegetation. Conservation Biology 2:513–522.

Powell, A. F. L. A. 2006. Effects of prescribed burning and bison (*Bos bison*) grazing on breeding bird abundances in tallgrass prairie. Auk 123:183–197.

Prose, B. L. 1985. Habitat suitability index models: Greater Prairie-Chicken (multiple levels of resolution). U.S. Fish and Wildlife Service, Biological Report 82:(10.102).

Rahmig, C. J., W. E. Jensen, and K. A. With. 2009. Grassland bird responses to land management in the largest remaining tallgrass prairie. Conservation Biology 23:420–432.

Robel, R. J. 1970. Possible role of behavior in regulating Greater Prairie-Chicken populations. Journal of Wildlife Management 34:306–312.

Robel, R. J., J. N. Briggs, J. J. Cebula, N. J. Silvy, C. E. Viers, and P. G. Watt. 1970. Greater Prairie-Chicken ranges, movements, and habitat usage in Kansas. Journal of Wildlife Management 34:286–306.

Rodgers, R. D. 2008. Prairie-chicken lek survey—2008. Performance report statewide wildlife research and surveys, May 2008. Kansas Department of Wildlife and Parks, Pratt, KS.

Ryan, M. R., L. W. Burger, and D. P. Jones. 1998. Breeding ecology of Greater Prairie-Chickens (*Tympanuchus cupido*) in relation to prairie landscape configuration. American Midland Naturalist 140:111–121.

Samson, F., and F. Knopf. 1994. Prairie conservation in North America. Bioscience 44:418–421.

Schroeder, M. A. 1991. Movement and lek visitation by female Greater Prairie-Chickens in relation to predictions of Bradbury's female preference hypothesis of lek evolution. Auk 108:896–903.

Schroeder, M. A., and L. A. Robb. 1993. Greater Prairie-Chicken (*Tympanuchus cupido*). Birds of North America No. 36.

Schroeder, M. A., and G. C. White. 1993. Dispersion of Greater Prairie-Chicken nests in relation to lek location: evaluation of the hot-spot hypothesis of lek evolution. Behavioral Ecology 4:266–270.

Schwarz, G. 1978. Estimating the dimension of a model. Annals of Statistics 6:461–464.

Slabbekoorn, H., and E. A. P. Ripmeester. 2008. Birdsong and anthropogenic noise: Implications and applications for conservation. Molecular Ecology 17:72–83.

Svedarsky, W. D. 1988. Reproductive ecology of female Greater Prairie-Chickens in Minnesota. Pp. 193–239 *in* A. T. Bergerud and M. W. Gratson (editors), Adaptive strategies and population ecology of northern grouse. University of Minnesota Press, Minneapolis, MN.

Svedarsky, W. D., R. L. Westemeier, R. J. Robel, S. Gough, and J. E. Toepfer. 2000. Status and management of the Greater Prairie-Chicken *Tympanuchus cupido pinnatus* in North America. Wildlife Biology 6:277–284.

Whistler, J. L., B. N. Mosiman, D.L. Peterson, and J. Campbell. 2006. The Kansas satellite image database 2004–2005. Landsat Thematic Map Imagery Final Report , No. 127.

With, K. A., A. W. King, and W. E. Jensen. 2008. Remaining large grasslands may not be sufficient to prevent grassland bird declines. Biological Conservation 141:3152–3167.

CHAPTER THREE

Estimating Lek Occurrence and Density for Sharp-tailed Grouse

Stephen Hamilton and Doug Manzer

Abstract. Agricultural expansion in native prairie may damage or destroy vegetation normally used for nesting, brood-rearing, and winter habitat for Sharp-tailed Grouse (*Tympanuchus phasianellus*). Understanding the interaction between Sharp-tailed Grouse and their human-modified prairie habitat is vital for predicting potential population declines and working toward preventing losses similar to what has occurred with sage grouse, *Centrocercus* spp. We designed a rigorous method to survey for Sharp-tailed Grouse leks over broad spatial extents in east central Alberta. We used historic lek location data from 1958–2005 to build two complementary models that first predict lek occurrence and then estimate the density of leks among stratified areas of the region. We used a resource selection function (RSF) to predict lek occurrence and a distance sampling approach to estimate lek density. The RSF was based on the availability of habitat features (i.e., grassland, crop, shrubs, trees) and

enabled us to stratify our area into high, medium, and low classes by the likelihood of a lek being present. We surveyed 630 random sites near sunrise in the springs of 2006–2008, roughly 18% of the 26,000 km² region, and located 146 new leks. Our predictions of where leks would occur were roughly validated by the proportion of leks newly discovered among the three classes. Our distance function was derived using a modified point-count method allowing us to estimate the density of leks in each stratum, ranging from 0.017 leks/km² to 0.048 leks/km² in the low and high classes, respectively. Our approach provides an efficient means for predicting lek occurrence based on habitat features and for using this information to estimate lek density across vast spatial extents.

Key Words: density estimate, distance sampling, lek occurrence, RSF, Sharp-tailed Grouse, survey techniques.

Hamilton, S., and D. Manzer. 2011. Estimating lek occurrence and density for Sharp-tailed Grouse. Pp. 33–49 *in* B. K. Sandercock, K. Martin, and G. Segelbacher (editors). Ecology, conservation, and management of grouse. Studies in Avian Biology (no. 39), University of California Press, Berkeley, CA.

Many grouse populations are in decline worldwide. While a number of factors contribute to such declines, anthropogenic influences on habitat over large areas are often cited as the ultimate factor (Storch 2000). Sharp-tailed Grouse (*Tympanuchus phasianellus*) occur in grasslands throughout much of North America (Connelly et al. 1998). In Alberta, grassland habitat has experienced substantial disturbance due to agricultural expansion and oil and gas development (Timoney and Lee 2001, Fritcher et al. 2004, Manzer 2004), which has likely led to declines in numbers of Sharp-tailed Grouse (Connelly et al. 1998). In order to conserve Sharp-tailed Grouse populations, it is important to understand the link between animal numbers and the dynamic prairie landscape measured over broad regional extents. Sharp-tailed Grouse have a broad range through Alberta, and this combined with their cryptic behavior makes this species difficult to inventory. Sharp-tailed Grouse use leks in their sexual selection strategy, and the number of leks in an area can be used to positively correlate with Sharp-tailed Grouse populations (Berger and Baydack 1992, Hanowski et al. 2000).

Given the expanse of the various grassland ecoregions in Alberta (nearly 100,000 km²; Government of Alberta 2005), there is a real limitation in terms of personnel time and the costs of doing exhaustive surveys to monitor Sharp-tailed Grouse leks over a large area of interest. There have been many efforts to model landscapes relative to animal use or density in an effort to understand the effects of related limiting factors without sacrificing scientific rigor (Stillman and Brown 1995, Carlson and Schmiegelow 2002). Manly et al. (2002) propose that resource selection functions (RSFs) are a viable model for statistically describing the relationships between available habitat resources and animal selection. Manley et al. (2002) define a RSF as any function that is proportional to the probability of use of a given resource by an organism. Commonly, the RSF uses logistic regression models with known locations of resource use plotted against available locations where use is not known, referred to as a use-available design (Boyce et al. 2002).

Sharp-tailed Grouse range in Alberta extends from the southern prairies to the western foothills, and north through open habitat within the boreal forest. Nevertheless, typical habitat of Sharp-tailed Grouse in the prairies is characterized by open grasslands, and in some areas patches of trees or tall shrubs occur (Baydack and Hein 1987, Hanowski et al. 2000, Manzer and Hannon 2005). Grasslands are commonly used for nesting and loafing, while mesic areas provide dense escape and brood-rearing habitat, as well as winter cover. Human influenced landscapes, including agricultural crops, hay fields, and heavy shrub or treed habitats on a prairie landscape may each provide sought-after resources, although the positive effects of agricultural disturbance may create ecological traps if predation rates are elevated (Andrén 1992, Manzer and Hannon 2005). It is reasonable to assume that all of these factors may contribute to the selection of lek locations, especially if leks tend to be located within a certain distance of where females prefer to nest and raise broods. Nests are commonly located within 1.6 km of leks (Manzer and Hannon 2005). Assuming leks are primarily located due to female preferences, we would expect even in the highest quality habitat to see leks rarely occurring >1/1.6 km radius, or at densities >0.13 leks/km². Since this threshold value assumes good habitat conditions and an abundance of Sharp-tailed Grouse, we predict this to be the upper end of lek density over broad areas. In habitat of lesser value, we predict that lek densities would range lower in proportion to the resources needed for nesting, brood-rearing, forage, and escape cover.

In order to assess populations, it is not necessary to rely on individual bird counts, nor is it necessary to count all the leks at a regional scale. While leks may be relatively easy to detect when nearby to surveyors, the vast extent and variability of the landscape where this species lives makes the survey process challenging. Buckland et al. (2001) demonstrated that it is possible to survey only a small subset of a total population but still arrive at reasonable estimates of population size while minimizing survey effort (Cassey et al. 2007, Reinkensmeyer et al. 2008).

We elected to use an area of east central Alberta as our study area since it retains large tracts of relatively intact native prairie landscape, along with a gradient of lands disturbed by cropland, fire suppression, and oil and gas extraction. There was also an abundance of historical Sharp-tailed Grouse lek location data in this region from the last 50 years to enable preliminary model development. Moreover, the continued pressure from human activities suggests that the long-term trend for Sharp-tailed Grouse in this region is uncertain and that the population may be vulnerable to continued human change.

METHODS

Study Area

The study area is approximately 26,000 km², bounded on the south by the Red Deer River, on the north by the Battle River, and on the east by the Alberta–Saskatchewan border (Fig. 3.1). Most (51%) of the region is in the dry mixed-grass prairie, with 31% in northern fescue, 16% in central parkland, and 2% in mixed-grass prairie (Government of Alberta 2005). Land use varies from rangeland used for cattle grazing to vast areas in agricultural crop production. We used historical data of lek locations from the study area collected between 1958 and 2005, and surveyed leks in April of 2006–2008.

Field Surveys for Detecting Leks

In 2006–2008, we conducted field surveys for the detection or nondetection of leks, as well as the

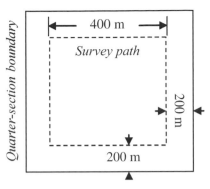

Figure 3.2. A sketch of the survey path. The surveyor walks the inside perimeter of a quarter-section, stopping to look and listen for leks at each corner.

location of leks, to validate the lek occurrence RSF and to provide data for estimating lek density. We conducted surveys using a modified point count design by walking a 400 m × 400 m square centered on the middle of a quarter-section of land, approximately 200 m from the boundaries of the quarter-section (Fig. 3.2). Surveyors began at any corner between 30 minutes before sunrise and 1 hour after sunrise, stopping at each corner to listen and look for lek activity for between 2 and 3 minutes. The time constraint for beginning the point count was to account for a lack of visibility at sunrise and declines in lek activity by late morning. If we detected a lek at any point along the survey, we noted the location of detection and then sought out the lek to record the actual location at a later time.

The time when the surveyor left the line to search for the lek was modified from year to year. In 2006, we prioritized completing the maximum number of point counts possible in a given morning. To do so, we recorded lek detection locations with a compass bearing toward the detected lek, but then continued surveying the remainder of that point and perhaps an additional random point as time allowed until 1 hr after sunrise. We then returned to the detection location on the survey line and began the search for the actual lek by walking in the direction of the compass bearing. Returning to locate leks following the completion of the survey resulted in a high proportion of leks initially detected but not located and therefore less "use" data for developing the lek occurrence and density estimates. Accordingly, we changed the design in 2007, when we searched for leks immediately upon detection, improving our leks found-to-detected ratio, but resulting in fewer random point counts

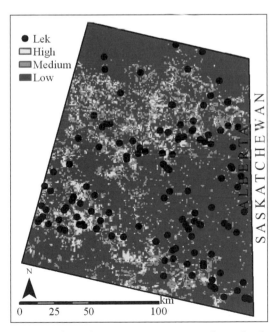

Figure 3.1. The study area encompasses a sparsely populated region of east-central Alberta primarily used for agricultural crops and pasture. Sharp-tailed Grouse leks were found more often in areas predicted by our resource selection model as "high"-quality habitat than in the lower-quality regions.

being completed per day. More point counts were incomplete because it took more time to find the lek and then return to the point count to complete the survey. If we did not return in time to complete the survey within the survey time slot, the point count was deemed incomplete as it was surveyed inconsistently from the remainder of the samples. In 2008, we combined the two methods by first completing the current point count (approximately 25 min) and then returning to a detection location to search for the lek while walking in the direction of the compass bearing.

The location of each survey point (sample unit) was chosen by random stratification according to the corresponding year's RSF. The RSF scores were stratified into 33% bins (high, medium, and low) to create survey strata for the field surveys. We randomly selected 400 points at a 1:2:1 (high/medium/low) ratio within the strata, with no constraints on minimum distance in 2006 and a minimum distance between survey points of 3 km in 2007 and 2008. A minimum distance of 3 km was chosen to limit the amount of potential overlap in lek detection during a nearby survey.

As an indicator of model performance, we used a chi-square goodness-of-fit test comparing lek detection events to the habitat class predicted for the survey by the lek occurrence model. The test compared the difference between observed and expected surveys where at least one lek was detected and no leks were detected for each RSF stratum. We performed the test both by stratifying by the point count location itself (i.e., the stratum of the quarter-section surveyed) as well as by the most prevalent RSF category <1.6 km of the survey center.

Predictive Covariates for Lek Occurrence

Sharp-tailed Grouse may select lek sites based on a number of criteria. It is not fully understood whether lek sites are chosen by males responding to the presence of females (i.e., females selecting ideal nesting habitat and thereby attracting males to the vicinity) or whether males simply establish leks based on their habitat preferences for display sites that minimize predation risk. In either case, a starting assumption is that healthy, densely covered expanses of grasslands are a strong attractant to Sharp-tailed Grouse (Connelly et al. 1998, Niemuth and Boyce 2004). Agricultural cropland may be a source of forage, which could represent a positive influence on lek site selection, but it may have an overall negative effect given that during much of the year there is little available cover, especially for selecting nest sites or winter cover. Tree and shrub cover could also influence selection in different ways: on the one hand offering escape and loafing cover, but on the other providing a resource for corvids and birds of prey seeking perch and nest sites in a landscape historically devoid of these features. There is some evidence that female Sharp-tailed Grouse avoid potential predator perch sites at small scales (Manzer and Hannon 2005), but the relationship between trees and lek locations is less clear for a prairie landscape.

Vegetation Classification

A 2004 Landsat 5 TM image of the study area was preclassified in PCI Geomatica 8 using an isodata spectral classifier on bands 1–5 and 7 to break the image into 50 distinct spectral clusters. Clusters that clearly belonged to cloud, haze, shadow, or water were grouped accordingly and the remaining vegetation/nonvegetation classes were isolated. Ground-truthing took place in October 2005–2007 to verify the landcover and improve classification accuracy. Ground-truth locations were randomly stratified to comprise at least two of each of the remaining spectral classes (minimum contiguous area: 1 ha) and categorize classes according to the following four landcover types: water, urban/anthropogenic, tree/shrub, and non-woody vegetation. Non-woody vegetation was further divided into six land use categories, including agricultural crops (including hay) and five categories of perennial grassland differentiated by decreasing 20% gradients of bare soil visible to the satellite sensor (sparse to dense grass cover). The resulting pixels were randomly selected within each cluster category, with half used for classification and the other half for validation. For cases where an odd number of ground-truth points existed within a cluster, the majority of points were assigned to classification, resulting in a slightly greater number of classification points relative to the validation points.

Spectral clusters were assigned ground cover values based on the majority of coincidental ground-truth points of a given class. When no majority of values was available, the decision was made based on a combination of examining surrounding spectral classes for consistency and photo interpretation. The resulting clusters were amalgamated into a classified image with the above-defined classes and validated with the remaining ground-truth data. Results of the validation process were reported in overall classification error with a Kappa statistic.

Model Development

Lek Occurrence Models

There were four main objectives for our efforts to develop a lek occurrence model for Sharp-tailed Grouse: (1) successfully predict habitat features associated with lek occurrence across the study area; (2) optimize the effort and cost associated with detecting leks across a broad spatial area; (3) provide a means to validate if the lek occurrence model could reliably predict where leks occur within years; and (4) provide a means to stratify the area for conducting lek detection field surveys based on distance sampling.

The lek occurrence model was initially developed using historical lek locations from roadside surveys (Allen 1987) in 1958–2005 as training sites (the Training RSF) but was iteratively updated each year with new data from annual lek detection surveys. Surveys performed in 2006 were based on the Training RSF, and the leks found in 2006 were used to develop the 2006 RSF. In turn, the surveys performed in 2007 were based on the 2006 RSF, and leks found in 2006 and 2007 were the basis of the 2007 RSF. Surveys were performed again in 2008 to evaluate the 2007 RSF, but no further lek occurrence models were developed.

Under the conservative assumption that locations in the study area without detected leks do not necessarily represent areas where no leks are present, we adopted the use-available design for lek occurrence RSFs. The use-available approach limits our model to an estimation function, with model results being relative scores that are not true probabilities (Boyce et al. 2002). Our "use" points were known locations of leks with >5 birds attending, taken either from historical data (Training RSF) or from our own field survey results (2006 RSF, 2007 RSF). "Available" points were randomly selected locations (omitting water and anthropogenic features) from within the study area based on the current model for a given year. The use-available design required an initial number of lek locations along with their associated habitat characteristics. A higher number of

use points for a constant availability of habitat would tend toward an asymptote where the model more accurately represents the relationship between the leks and their surrounding habitat. Available data were selected randomly using ArcGIS and represented resource units where use is unknown. We used a ratio of three available points to each use point in order to provide a sufficient sample size for analysis, while accepting some potential contamination. Contamination is defined by Johnson et al. (2006) as the mixing of used and unused locations when sampling available resource units of unknown use. Overuse of random points could bias the results of the logistic regression in a case-control study, but is not a concern in a use-availability design where the purpose of the model is to estimate the true resource selection function. Ratios greater than 3:1 can be used when evaluating small-scale patterns in animal movement (Fortin et al. 2005), but lower ratios are used when a large number of used locations already exist, such as data based on satellite telemetry (Hebblewhite et al. 2005).

In all cases, our RSFs to predict lek occupancy were tested using five candidate models, four based on what we consider female-based preferences, and one on male-based preferences (Table 3.1). Candidate models related landscape-based data to known lek distribution data, representing competing hypotheses to best describe the relationship between habitat covariates and lek occurrence. The first model represented leks occurring within short distances of areas preferred by females for nesting (Manzer and Hannon 2005), while the second model expanded on this concept to include the brood-rearing potential of habitats near the lek as well (Manzer 2004). The third model represented avoidance of potential perch and nest areas for raptors and corvids by incorporating the presence of trees and shrubs. The fourth model focused on a preference for leks being near a good source of food for females, while the final model assumed that males chose the location of the lek based primarily on forage. Each year, the candidate models were evaluated and a best model was chosen using Akaike Information Criteria (AIC). AIC is described in detail by Burnham and Anderson (1998) and is a tool based on the principle of parsimony, penalizing candidate models for using too many covariates to describe the variation in the data. All models were created and evaluated in R 2.5.1 (R Development Core Team 2007).

TABLE 3.1

Candidate models for the selection of lekking habitat by Sharp-tailed Grouse.

Covariate	Nesting	Nesting + brood rearing	Nesting + brood rearing + forage	Security	Male-based
Grass	X	X	X	X	X
Crop[a]	X		X	X	X
Tree				X	X
Water[b]		X	X	X	

[a] The "crop" covariate represented the total available area of crop in the Training and 2006 lek occurrence models, but in 2007 it represented the distance to the nearest agricultural field of 32 ha or greater.

[b] The water covariate was measured as the total available area comprised of water in the first model, but in the last two models it represented the perimeter of water bodies.

TABLE 3.2

Covariates used in the development of the lek occurrence resource selection function.

Covariate	Description
Grass (training)[a]	Area of high-density grassland (40–100% grass cover) within 1.6 km from center of quarter-section
Grass (2006, 2007)	Area of high-density grassland clusters[b] (up to 8 km²) from center of quarter-section
Crop (training)	Area of crop (including hay) 1,600 m from center of quarter-section
Crop (2006)	Area of crop clusters (up to 8 km²) from center of quarter-section
Distance to crop (2007)	Distance from center of quarter-section to nearest 32 ha cluster of crop
Tree/shrub (training)	Area of trees and shrubs within 1,600 m from center of quarter-section
Tree/shrub (2006, 2007)	Area of tree and shrub clusters (up to 8 km²) from center of quarter-section
Water (training)	Area of water within 1,600 m from center of quarter-section
Water (2006, 2007)	Sum of total water edge within 1,600 m of center of quarter-section

[a] Perennial grassland was classified into 5 categories ranging from sparse (0–20% and 20–40%) grass cover to dense (40–60%, 60–80%, and 80–100%) grass cover. Because sparse and dense grasslands were inversely correlated, we used only dense cover in our models.
[b] A cluster is defined as a contiguous region of one landcover class greater than 1 ha.

The three models varied slightly in design in the way the covariates were defined (Table 3.2). In our Training RSF, we initially calculated land-cover characteristics in terms of hectares of each cover class within a 1,600-m radius of use and available points. Because the habitat classes, when summed, were potentially 100% of the space for each use or random point, the results lacked independence. We recognized this shortcoming, and in successive RSFs (2006 and 2007) landcover areas were calculated differently: We weighted all landcover polygons within 1.6 km of the point by the total area of each cover polygon intersecting the buffer. In the case of large polygons (>8 km²), area weighting was limited to 8 km² based on the assumption that further increases in area cease to be appealing to Sharp-tailed Grouse, given nesting preferences of females (Manzer and Hannon 2005). Our approach differed from that proposed by Aitchison (1986) by directly removing the constraint of a constant area and therefore did not require us to transform our data.

Covariates for the Training RSF and 2006 RSF differed only in that we replaced water area with the perimeter of water bodies smaller than 2 ha. We judged the perimeter to be a more useful covariate because area of water was, by default, unavailable as lek habitat, whereas a shoreline may offer a variety of opportunities to Sharp-tailed Grouse lek establishment. Finally, the crop covariate in the Training RSF and 2006 RSF was replaced in the 2007 RSF by distance to nearest agricultural field (32 ha of crop cover, or roughly half of a quarter-section) as a metric. We believed the new metric would better represent Sharp-tailed Grouse avoidance of agricultural fields than our original area metric.

There are multiple ways to select the best-fitting model in a use-available design; however, Boyce et al. (2002) recommend withholding a random portion of the data from the model for comparison (test data) while using the other, larger portion of the data to build the model (training data). This idea is the basic premise behind the k-fold cross-validation procedure (Boyce et al. 2002), which is robust in a use-available design. We followed Boyce et al. (2002) and withheld approximately 20% of the use points and 20% of the available points, using the remainder of the data to build the lek occupancy models. The procedure was repeated k times (five times in our study) and the congruence of the results was tested using Spearman's correlation. One drawback to this approach is that if there are a limited number of use points, removing 20% of the data may result in low correlations among iterations, and thus a poor fit. We performed the k-fold cross-validation procedure in R 2.5.1 for each of the RSFs we developed to gauge the relative strength of the models relative to one another among years.

For all years, the lek occupancy RSF was reported at a quarter-section (0.25 mi² or ~64 ha) scale (one RSF score per quarter-section). Reporting RSF scores in this way tied our results to land use in the area because, in the majority of cases, land use coincided with land ownership and a quarter-section is managed homogeneously within its bounds. Furthermore, it complemented our lek field survey scheme (described below), which was also performed on a quarter-section scale. The RSF score attributed to a given quarter-section was the mean of all RSF scores within its boundaries.

Lek Density Estimates

When a study area is large or animals are difficult to detect with certainty, it can be challenging to adequately estimate population density if the goal is to achieve a complete census. Accurate estimates of many bird populations can be challenging and alternative methods introduce different sources of error (Alldredge et al. 2007, Cassey et al. 2007). Several approaches estimate density in ways seeking efficient use of resources, many of which are based on the distance sampling methods proposed by Buckland et al. (2001). Our primary objective was to balance the effort and cost of surveys with the data necessary to estimate lek densities over a broad area. We used a modified point count design as the sample unit, and equations from Buckland et al. (2001) to derive detection functions and distance sampling estimates for lek densities across the study area and for the three RSF classes.

Buckland et al. (2001) emphasized several assumptions when performing point count distance sampling analysis. First, objects on the point should be detected with certainty. As applied to our technique, we made the reasonable assumption that any leks within the boundaries of the point were detected with certainty. Second, we assumed that all leks are detected at the initial location. This is not a problem because scat and feathers provide physical evidence to identify lek locations. Last, we assume that our lek location measurements are exact. Using handheld GPS units allowed us to be within a small (<15 m) amount of spatial error both at the point of detection and the location of the lek itself. Given that leks are typically hundreds of meters from the point of detection, we considered the accuracy of our GPS units to be acceptable.

In addition to the three assumptions listed above, we added two more to interpret our surveys as a point transect. First, the corners of our transects were ~300 m from the center and our "point" was assumed to be 300 m in radius or 28 ha in area and not a single coordinate in space. A second assumption was that all leks were detected from the center of the point rather than the actual point of detection. We did not include data from any leks detected away from survey points such as incidental detections recorded while observers were en route to a point.

Our modified point count design (Fig. 3.2, described below) may resemble a line transect, with each edge of the survey point being considered a line transect in the analysis. However, this

TABLE 3.3
Comparison of candidate models (using ΔAIC) for each annual iteration of the resource selection function.

Model	Training	2006	2007
Nesting	22.9	0	0
Nesting + brood rearing	24.5	3.2	5.2
Nesting + brood rearing + forage	9.2	3.4	1.8
Security	1.5	5.0	3.8
Male-based	0	2.1	2.0

NOTE: We selected the model with ΔAIC = 0 in each case as the best model.

would generate a case where the center of the square, and especially the insides of the corners, would be oversampled relative to any region outside of the square. As a result, we chose to use a point transect sampling design with the point being the center of the square of each quarter-section.

We stratified our detected leks by the RSF categories of low, medium, and high relative probabilities of use for their respective quarter-sections, and also by year. We then plotted the frequency of these data at varying distance intervals and chose intervals of 300 m as an effective way to bin the data. Based on a visual analysis of the distribution of our lek detection distances, we right-truncated approximately 10% of the data to eliminate leks found at long distances from the point center (Buckland et al. 2001). We also examined the shape of the distribution and compared goodness of fits using half-normal and hazard-rate curves with and without cosine series expansions (Buckland 1985), and selected the best model based on GOF and AIC using program Distance 5.0 (release 2). We developed a regional density model that included data from all years and all RSF strata, as well as a model for each survey year (2006, 2007, and 2008) and a model for each stratum (low, medium, and high).

RESULTS

Vegetation Classification

The vegetation classification was improved each year with additional ground-truthing. The final overall classification accuracy was 88% ($K = 0.78$) when distinguishing between crop and perennial grassland, and 73% ($K = 0.62$) when attempting to discern between sparsely and densely covered grassland. Lower accuracy was primarily due to the introduction of classification error when sparse grassland (0–40% grass cover) and dense grassland (40–100% grass cover) categories were misclassified, with most errors coming from sparse grassland: 57% accuracy on the ground relative to the classification, versus 83% accuracy for dense grassland. Furthermore, errors inherent in distinguishing between cropland and perennial grassland were entirely due to the sparse grassland category.

Lek Occurrence Models

Our best candidate model (male-based) selected for the Training RSF, based on 233 lek locations in 1958 to 2005, predicted that Sharp-tailed Grouse preferentially choose lek locations in areas with high abundance of dense grassland while avoiding large areas of cropland and forest (Table 3.3). Validation using the k-fold analysis was strong, with $r = 0.91$ ($k = 5$). The 2006 RSF had only 43 lek locations to derive the model, and this reduced set of use points suggested a female-based nesting model as the best fit, predicting greater lek occurrence in areas with dense grassland commonly associated with nesting cover (Table 3.3). The 2006 RSF had much poorer predictive power than the Training RSF ($r = 0.50$, $k = 5$). The 2007 RSF was based on 93 unique lek locations accumulated from field surveys completed in 2006 and 2007. The best model again predicted leks occurring in areas with greater proportions of dense grassland <1.6 km and avoidance of agricultural fields. Our k-fold

TABLE 3.4

Lek surveys completed, amount of area surveyed, and proportion of area in each stratum surveyed by year.

Survey Year	Surveys				Surveyed area (km²)			% Surveyed		
	n	L	M	H	L	M	H	L	M	H
2006	209	57	100	52	878	444	253	4	11	17
2007	204	45	101	58	986	407	133	5	9	15
2008	217	49	109	59	832	490	314	4	16	22

TABLE 3.5

Field survey results from 2006 to 2008 detailing the number of surveys performed and the resulting number of leks detected and percent found.

Year	n	Survey		Total	
		Detected	% Found	Detected	% Found
2006	209	51	80	60	75
2007	204	54	87	58	83
2008	212	53	89	77	69

NOTE: The "Survey" column represents the number of surveys performed with at least one detection or location result, while the "Total" column includes instances of multiple leks per survey.

validation ($r = 0.67$, $k = 5$) performed better than the 2006 RSF, but not as well as the original Training RSF.

Field Surveys for Detecting Leks

We surveyed 630 stratified random points in April of 2006–2008, with a similar number of surveys completed each year (Table 3.4). Our survey effort represented 1.5% of the total number of quarter-sections available for survey in the region and a total of 18% of the study area (approximately 4,680 km²), assuming a 1.8-km radius from the center of the survey was within detection range of the surveyor. A 1.8-km radius was consistent with the detection function estimated for each of the three survey years. We therefore surveyed approximately 6% of the region each year, while among years each stratum was surveyed with consistent effort (Table 3.4). The highest proportion of habitat surveyed each year was in the high stratum and the lowest in the low stratum, which followed the availability of these classes within the region (Fig. 3.1).

We detected approximately 50 new leks each year, with an increasing proportion of leks found per survey from year to year as our survey methods were improved (Table 3.5). We detected proportionately more leks in habitat predicted as higher value, with surveys in low-quality habitat resulting in the lowest detection per survey rate (Table 3.6). The exception was in 2007, which was likely related to the poor predictive power of the underlying lek occurrence model (2006 RSF). It had no discernible impact on the results of our analysis if we used the habitat identified at the center of the survey as the basis for the analysis or the most prevalent class of the 1.6-km radius surroundings.

Lek Density Estimate

Analysis by Stratum

As predicted, the density estimates decreased from high to low among our three lek density classes, though there was some overlap in the 95% confidence intervals, especially with the medium

TABLE 3.6
Goodness-of-fit test results with P-values comparing lek detection events to predicted habitat class.

Stratum	2006		2007		2008	
	Center	Majority	Center	Majority	Center	Majority
Low	9 (57)	22 (127)	12 (45)	42 (152)	8 (49)	28 (111)
Medium	23 (100)	17 (48)	33 (101)	9 (39)	32 (109)	16 (59)
High	16 (52)	11 (34)	10 (58)	4 (13)	23 (59)	19 (47)
χ^2	3.45	7.85	4.46	0.43	6.68	2.13
P	0.18	0.02	0.11	0.81	0.04	0.15

NOTE: Numbers represent individual surveys with at least one detection (total surveys in parentheses) given by the transect center's habitat category, as well as the majority habitat class within a 1.6-km radius.

TABLE 3.7
Inputs and results for the lek density model for the entire region as well as by RSF stratum.

	Regional	High	Medium	Low
n	129	27	62	40
$n < 300$ m	25 (19%)	8 (30%)	9 (15%)	8 (20%)
$n > 1800$ m	11 (9%)	2 (7%)	9 (15%)	0 (0%)
10% truncation distance (km)	1.8	1.2	2.0	1.2
$n >$ truncation distance	11 (9%)	3 (11%)	7 (11%)	3 (8%)
Maximum distance (km)	3.7	2.8	3.7	1.8
Density (leks/km^2)	0.026	0.048	0.030	0.017
Lower 95% CI	0.016	0.023	0.018	0.0090
Upper 95% CI	0.043	0.10	0.049	0.032

NOTE: The longest 10% of the detections were censored from the data (see Buckland et al. 2001). A comparison of the number of leks truncated using 1.8 km (the regional truncation distance) vs. a truncation distance specific to each stratum is shown in rows 3–5.

category (Table 3.7). The total number of leks estimated to occur in the low stratum was greater than in the high stratum (Fig. 3.3), and followed the greater availability of low than high habitat (Table 3.4). The number of leks in the study area was estimated at 676 leks based on the regional model, with the sum of the individual stratum models being 537 leks, within the 95% confidence intervals of our regional model.

Of the four candidate models we tested for the distance function, we found that the hazard-rate curves had identical fit regardless of series expansion and that the hazard-rate curves were similar to the half-normal with cosine series expansion curve (ΔAIC = 1.39). However, the goodness-of-fit test was better for the half-normal/cosine curve than for the hazard-rate curve ($P = 0.06$ and $P = 0.10$, respectively). We found similar results when testing the different functions by RSF stratum. Following recommendations of Buckland et al. (2001), we examined the effects of using a common right-truncation distance of 1.8 km against a truncation distance specific to each stratum. In all cases, truncating the data based on the individual stratum's data was a superior model.

Analysis by Year

The density estimates for 2006 and 2007 were similar, with estimates of 0.034 and 0.038 leks/km^2,

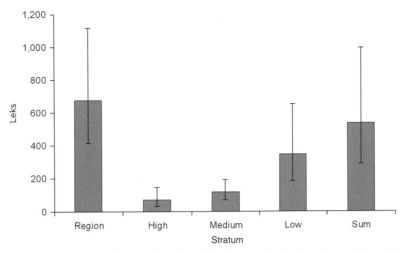

Figure 3.3. The number of leks in the study area based on a regional estimation as well as each stratum (95% CI). The Sum column represents the total number of leks for all strata, while the error bars for Sum are the sum of the upper and lower confidence limits of each stratum.

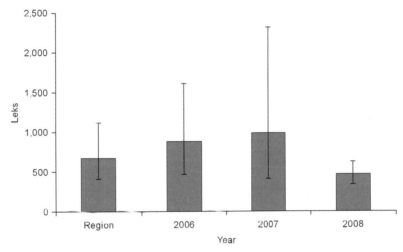

Figure 3.4. The number of leks in the study area based on the regional estimation as well as by year (95% CI). Each year's estimates are based on the RSF from the previous year.

respectively, putting them in line with the overall regional estimate for all years. However, the 2008 estimate was lower, at 0.018 leks/km², but within the lower confidence limits of the other years (Fig. 3.4). Stratification of the data by year resulted in better model fit using the hazard rate key function with the cosine series expansion, despite there being little difference in AIC scores (ΔAIC = 0.04 and 1.36 for 2006 and 2007, respectively), with the exception of the 2008 surveys. In 2008, the half-normal curve with cosine series expansion fit the data better, though again AIC

was not helpful in distinguishing among models (ΔAIC = 0.69).

DISCUSSION

In our study, we combined lek surveys and spatial modeling for Sharp-tailed Grouse and were successful at predicting lek occurrence and density in a large area of Alberta. Lek occurrence was positively related to the abundance of high-quality grasslands and negatively correlated to the presence or proximity of cropland and forest, and our predictions were validated by the locations of new leks discovered in three strata of different habitat quality. Our distance function model successfully illustrated a relationship between habitat quality and density of leks, and estimated density within expected boundaries. Our surveys enabled us to collect reliable data for multiple models, while gaining efficiencies for the effort and cost of logistics over a vast spatial area during a limited time period.

We found the most significant covariate in each candidate model tested was the abundance of dense grassland. Indeed, in the 2006 RSF it was the only variable in the successful model. Our results are consistent with previous work demonstrating the importance of dense grassland to Sharp-tailed Grouse (Hanowski et al. 2000, Niemuth 2003, Niemuth and Boyce 2004). Nevertheless, our strongest model among years indicated lek occurrence was also negatively associated with crop abundance, trees, and shrubs. While Niemuth and Boyce (2004) identified shrubby cover as a positive factor in attracting Sharp-tailed Grouse, they found that forests are not preferred, although forest in their study area was coniferous dominated and thus different from the aspen-parkland forest in the northern portion of our study area. Because of the limitations of our classification, we were unable to distinguish reliably between forest and shrub categories. Where low shrubs may have been positively correlated with Sharp-tailed Grouse, trees and tall shrubs can serve as potential predator perches, and thus may have been negatively correlated with Sharp-tailed Grouse. The abundance of crop in the Training RSF was negatively associated with lek occurrence, which may have been related to the lack of cover provided for brood-rearing and nesting females (Manzer and Hannon 2005). Similarly, in the 2007 RSF we found that the distance to ≥ 32 ha of cropland was positively correlated with lek occurrence, meaning that leks were more likely to occur when agricultural fields were at greater distances.

The variation in the lek occurrence models selected among years may be explained, at least partly, by the number of lek locations used to drive model development after the Training RSF. Although the best models in all years had highly significant covariates in all cases ($P < 0.01$), the k-fold cross-validation did not strongly support any but the Training RSF. If the primary leks in a region were relatively stable and they persisted from year to year with only slight changes in population, it is feasible that our survey efforts each spring located a subset of the same primary leks that existed over the past 50 years. A reduction in "use" points between the training RSF and later RSFs was likely responsible for the weaker validation.

Our strongest model based on k-fold validation suggested lek occurrence was positively associated with dense grassland and negatively associated with crop, trees, and shrubs. The analogous model in 2006 and 2007, while not the minimum AIC model, was equally parsimonious, with $\Delta AIC = 2.1$ and 2.0, respectively (Table 3.3). Although it is not clear whether male or female behavior drives the location of Sharp-tailed Grouse leks, we predicted that female preferences would drive the model. It is unclear whether the combination of covariates we chose to distinguish a male- from a female-based selection was adequate to determine gender roles in lek location selection. It is possible the same set of covariates may be equally attractive to the opposite sex in the candidate models used. Our covariates were based on broad landcover categories and do not consider disturbance features, which are present in the study area and associated with diminishing grouse populations (Connelly et al. 1998, Storch 2000, Manzer 2004). Nevertheless, leks were detected based on the presence of displaying males but not all leks successfully attract females (Baydack and Hein 1987). Given the concerns over disturbance-related effects in the area (Manzer and Hannon 2005), it is plausible that a subset of leks detected were in disturbed areas where males continued to attend the site but with few or no females attending, although we tried to limit this factor by building our models using leks where no fewer than 6 birds attended.

Historical lek surveys were conducted by driving along roads and listening for lek activity every mile (Allen 1987). Such an approach was effective for systematically surveying in a relatively confined area, especially where road access was consistently good, and may add more area to the overall survey effort because more survey points could be obtained per surveyor-day than in our method. Nevertheless, the time and effort at a site is reduced significantly and could substantially limit the distance at which lek detection occurred.

Biased distances would weaken the overall fit of the distance function, and, because survey routes were not prestratified by habitat quality, limit the number of leks that could potentially be detected. We believe our method provided a more rigorous means of detecting leks without significantly sacrificing the time spent covering a sufficient proportion of the study area.

The amount of effort in terms of surveys completed was similar from year to year, with slightly more transects being completed in each successive year as survey crews and methods became more efficient. In 2008, the results of the distance analysis exhibited the smallest coefficient of variance (16%, compared to 31% and 44% in 2006 and 2007, respectively), likely connected to the changes in survey methods, which focused on detecting leks without sacrificing the remainder of the survey. Further reduction in confidence intervals could have been achieved using a different approach to survey effort, focusing more time in higher-quality habitat classes. Our survey ratio was chosen to emphasize work in the transition areas between good and poor habitat, which could be useful for monitoring change over time, and understanding where marginal levels of habitat with potential for habitat improvement and establishment of leks. To optimize density estimates, survey ratios could be stratified for achieving low confidence intervals among habitat classes. Furthermore, greater effort in strata with higher expected lek density would help to improve both the lek occurrence model and the lek density model.

Estimates of lek density for the study area fell within expectations, given the size of the area and the variation in habitat throughout. Our maximum predicted estimate of 0.13 leks/km^2 represented a density that could exist under ideal circumstances. Given the relatively low percentage of high-quality habitat in the study area (<6%), it is sensible that the true density would be much lower. When examining the results by stratum, however, we noted a number of apparent differences in density: First, the mean density estimate in the high class was above that of the low stratum, but with some overlap in the respectively lower and upper confidence limits (Table 3.7). Given the small sample size used to produce our estimate, we believe there is good reason to assume the density of leks differs among RSF strata, especially when the densities

are expressed as numbers of leks per stratum, where the overlap was much less pronounced (low and high estimates do not overlap at all). It may be wise to use density differences as a basis for stratifying the RSF categories in the future, rather than the arbitrary 33% bins used in our study. Second, while there was no trend in the percentage of leks detected within 300 m of the survey point, there was a loose association in terms of leks detected at great distances. In the low stratum, the maximum distance to a lek from the transect center was ~1.8 km, >1 km less than the maximum distances in the medium and high strata (>2.8 km), which both had multiple detections at greater distances than the maximum in the low stratum. The difference in maximum detection distances corroborated the lowest density estimate being in the low stratum, as leks may be so much farther apart in locations with a low likelihood of use that they are more often farther than surveyors can reasonably detect compared with superior habitat. Detection distances may also be related to the geography typical of low-ranked regions in the study area; expanses of cropland typically have less prominent topography, which may make detection easier up to a certain distance, but some of our surveyors observed that the funneling and echoing of sounds of lek activity across shallow valleys often led them to discover the farthest leks from the survey point.

Summing all the estimates for the three strata resulted in a total count of leks for the region well within the 95% confidence limits of the regional estimate, with a peak almost identical to the upper limit of the regional density estimate (Fig. 3.3). The pooled results suggest that either approach to stratifying the data within the model (either as a single region with strata or as three strata handled independently) can lead to reasonable results, although the coefficient of variation for the regional model was the lowest of the four, likely due to the larger sample size. While minimizing the confidence intervals of the density estimate is an important goal, it is perhaps more important to understand the variation in densities among strata. Given the link between habitat strata and density that can be derived using the latter approach, we recommend using both analyses but reporting density estimates based on the regional model to best explain variation. When considering the lek density functions on a year-by-year basis, the results were all within range of each other's confidence limits, although the lower density for the 2008 surveys is difficult to explain.

Potential Improvements

Our study provides insight into the iterative process we followed to estimate Sharp-tailed Grouse lek occurrence and density across a large segment of Alberta's prairie landscape. While our methods were successful, we offer suggestions for further improvements on them. As the fundamental basis for the lek occurrence models, it is imperative that the habitat data used from the landscape be well defined. Using a Landsat TM image for our habitat data limited us to broad spectral categories and made it difficult to distinguish among grass cover types. Furthermore, there may be a difference in how Sharp-tailed Grouse perceive the advantages of low shrubs over tall shrubs, although our habitat classification did not discriminate among deciduous cover types, regardless of species. While discrimination of strata is a common issue with using regional-scale remotely sensed data, satellite and aerial images are often an efficient way to categorize the landscape. However, additional data, such as reliable elevation maps, may have helped us improve the classification accuracy further. For a smaller study area, analysts may prefer to use more large-scale data than we did to more accurately map different classes of shrubs, for example (see Aldridge and Boyce 2007).

Our covariates were broad representations of factors that may influence lek occurrence, based either on the abundance of a vegetation class or on the distance to a different vegetation class. We did not have a reliable source to identify changing anthropogenic features that may affect lek occupancy, such as the negative association displayed between lek use by sage grouse and oil and gas activity (Lyon and Anderson 2003, Holloran 2005). Our models assumed lek density would decline with decreasing habitat quality, but lek density could remain constant with a drop in lek attendance. Low lek attendance may have reduced our ability to detect leks, leading us to underestimate density in these areas. Lek attendance data (males and females per lek) was inconsistently recorded among surveyors,

eliminating the possibility of reliably examining the connection between attendance and detection. Moreover, because we restricted our mapping to clusters of habitat smaller than 1 ha, we were limited in interpreting the potential importance of small patches of shrubs within perennial grassland classes (Hanowski et al. 2000, Manzer and Hannon 2005). Further to this point, we did not investigate any candidate models with non-linear relationships. It is possible that some of the uncertainty in our landscape classes could have been addressed with non-linear relationships or interactions.

Our decision to develop the lek occurrence model at the quarter-section scale may have had implications for the utility of our RSFs. Landscape metrics were summarized for a region 1.6 km from the center of each quarter-section, and due to the patterns of land management in our study area, many quarter-sections were influenced by the landcover at the center of our analysis window. Had we used a moving window approach to model development, and centered the window on each pixel rather than each quarter-section center, our RSFs may have been more sensitive to subtle changes in the landscape or smaller areas of habitat. A pixel-based approach could have resulted in a higher proportion of the landscape being classified in the medium and high strata. Furthermore, these subtleties in the landscape that were missed in our evaluation may have accounted for some of the variance in our models, potentially improving our RSFs, distance functions, and inferences for conservation of Sharp-tailed Grouse.

The lek occurrence models developed over the course of the study underwent several key changes, all designed to improve the reliability of the RSF. However, the key factor with the model predictability seemed to be the amount of training data available to develop it. Our findings emphasized locating sufficient leks during the survey season to make robust models. A power analysis prior to selecting survey sites would indicate an optimal number of surveys in each stratum. Minor changes to the survey method among years improved our ability to detect leks; however, without additional manpower it is improbable that more surveys could be completed in a spring season. Two possible changes could be made to increase sampling effort: (1) a wholesale change to the survey method which would require less time per survey; and (2) a change from a use-available design to a use/non-use design (i.e., Resource Selection Probability Function).

From our analysis of distance functions in our study, we are comfortable with the assumption that any leks within 600 m of the survey center were detected with certainty. It may therefore be possible to reduce the length of the transect. It may also be worthwhile exploring a change to a line-transect sampling design, which has been shown to provide more reliable density estimates with passerines (Cassey et al. 2007). However, a line transect is less efficient than our square transect method if surveyors must walk back to the beginning of the transect after a survey while collecting no additional data. Switching to a true point-sampling approach where the surveyor would simply stand in one location may also introduce inefficiency if random stations reduce the proportion of leks detected. Instead, any changes to the sampling method should focus on increasing the number of detected leks attainable by surveyors, as a large sample size of leks is vital to building robust lek occurrence models.

If we were to change our approach to a use/non-use design, however, our focus would not need to be on maximizing the number of detected leks, as locations without leks would be just as important to the model as those where leks were found. In a use/non-use case, adapting a line-transect design may indeed increase survey efficiency over the existing method. Because we are confident that leks within 600 m of the survey center are detected with certainty, we could focus our surveys on passing through the centers of multiple quarter-sections, yielding use and non-use points for each quarter-section visited. A line-transect approach would present us with the ability to report true probabilities of use rather than relative likelihoods (Boyce et al. 2002), meaning we would have a real connection between habitat characteristics and a probability of lek use rather than a relative estimate of occurrence. Coupled with the inclusion of disturbance-related covariates, such a model would more clearly indicate the impacts of development on Sharp-tailed Grouse lek establishment and persistence.

Our methods quantified the density of leks among strata, rather than estimating the population density for individuals. Estimation of abundance is the most common approach for wildlife conservation, although particularly challenging for cryptic, ground-dwelling galliformes. A few key pieces of information are needed to convert lek densities to population estimates: the proportion of males in the population that attend lek sites during the survey period; the male to female sex ratio; and, of course, the number of males attending a lek. With a reasonable understanding of the variation associated with each of these parameters, lek density estimates could be translated to population estimates needed for allocation and legislative purposes.

The tools we have developed in our study have proven effective at estimating Sharp-tailed Grouse lek occurrence as well as lek density over a broad spatial scale. Our model is an important step toward categorizing large regions in terms of the abundance of quality habitat and, perhaps more importantly, toward providing a means of estimating Sharp-tailed Grouse population status in a large prairie ecosystem without attempting exhaustive, time-consuming, and expensive census-oriented methods. Furthermore, we now have the capacity to link lek occurrence likelihoods with lek density estimates by comparing survey results from a given year to the expected habitat quality for that year. Given enough observations in time, it should be possible to predict probable changes in lek density based on hypothetical land use scenarios, enabling an evaluation of the tradeoffs associated with anthropogenic disturbance. As such, our study provides a useful framework of tools for understanding interactions between Sharp-tailed Grouse and their habitat and for planning conservation efforts for an important species of prairie grouse in Alberta.

ACKNOWLEDGMENTS

We gratefully acknowledge the anonymous reviewers who commented on earlier versions of our manuscript. We also recognize the support and funding of the Alberta Conservation Association and the many ACA staff members who contributed their time and effort to the project design and the field surveys. We further acknowledge staff members of Alberta Sustainable Resource Development for their logistic support and manpower during the field seasons, and from the Special Areas Board for their logistic support.

We also thank the Alberta Sport, Recreation, Parks and Wildlife Foundation for their generous financial contributions.

LITERATURE CITED

Aitchison, J. 1986. The statistical analysis of compositional data. Monographs on statistics and applied probability. Chapman and Hall Ltd. London, UK.

Aldridge, C. L., and M. S. Boyce. 2007. Linking occurrence and fitness to persistence: habitat based approach for endangered Greater Sage Grouse. Ecological Applications 17:508–526.

Alldredge, M. W., T. R. Simons, and K. H. Pollock. 2007. A field evaluation of distance measurement error in auditory avian point count surveys. Journal of Wildlife Management 71:2759–2766.

Allen, J. R. 1987. Esther and Rowley study areas: Sharp-tailed Grouse dancing ground census. Alberta Forestry, Lands and Wildlife. Government of Alberta. Edmonton, Alberta, Canada.

Andrén, H. 1992. Corvid density and nest predation in relation to forest fragmentation: a landscape perspective. Ecology 73:794–804.

Baydack, R. K., and D. A. Hein. 1987. Tolerance of Sharp-tailed Grouse to lek disturbance. Wildlife Society Bulletin 15:535–539.

Berger, R. P., and R. K. Baydack, 1992. Effects of aspen succession on Sharp-tailed Grouse, *Tympanuchus phasianellus*, in the interlake region of Manitoba. Canadian Field-Naturalist 106:185–191.

Boyce, M. S., P. R. Vernier, S. E. Nielsen, and F. K. A. Schmiegelow. 2002. Evaluating resource selection functions. Ecological Modelling 157:281–300.

Buckland, S. T. 1985. Perpendicular distance models for line transect sampling. Biometrics 41:177–195.

Buckland, S. T., D. R. Anderson, K. P. Burnham, J. L. Laake, D. L. Borchers, and L. Thomas. 2001. Introduction to distance sampling: estimating abundance of biological populations. Oxford University Press, New York, NY.

Burnham, K. P., and D. R. Anderson. 1998. Model selection and multimodel inference: a practical information-theoretic approach. 2nd ed. Springer Science. New York, NY.

Carlson, M., and F. Schmiegelow. 2002. Cost-effective sampling design applied to large-scale monitoring of boreal birds. Conservation Ecology 6:11. <http://www.consecol.org/vol6/iss2/art11>.

Cassey, P., J. L. Craig, B. H. McArdle, and R. K. Barraclough. 2007. Distance sampling techniques compared for a New Zealand endemic passerine (*Philesturnus carunculatus rufusater*). New Zealand Journal of Ecology 31:223–231.

Connelly, J. W., M. W. Gratson, and K. P. Reese. 1998. Sharp-tailed Grouse. A. Poole and F. Gill (editors), The birds of North America No. 354. Birds of North America, Inc., Philadelphia, PA.

Fortin, D., H. L. Beyer, M. S. Boyce, D. W. Smith, T. Duchesne, and J. S. Mao. 2005. Wolves influence elk movements: behavior shapes a trophic cascade in Yellowstone National Park. Ecology 86:1320–1330.

Fritcher, S.C., M. A. Rumble, and L. D. Flake. 2004. Grassland bird densities in seral stages of mixed-grass prairie. Journal of Range Management 57:351–357.

Government of Alberta. 2005. 2005 natural regions and subregions of Alberta. Alberta Sustainable Resource Development, Alberta Environment, Alberta Community Development, and Agriculture and Agri-Food Canada, June 2005.

Hanowski, J. M., D. P. Christian, and G. J. Niemi. 2000. Landscape requirements of prairie Sharp-tailed Grouse *Tympanuchus phasianellus campestris* in Minnesota, USA. Wildlife Biology 6:257–263.

Hebblewhite, M., E. H. Merrill, and T. McDoncald. 2005. Spatial decomposition of predation risk using resource selection functions. Oikos 111:101–111.

Holloran, M. J. 2005. Greater Sage-Grouse (*Centrocercus urophasianus*) population response to natural gas field development in western Wyomping. Ph.D. dissertation, University of Wyoming, Laramie, WY.

Johnson, C. J., S. E. Nielsen, E. H. Merrill, T. L. McDonald, and M. S. Boyce. 2006. Resource selection functions based on use-availability data: theoretical motivation and evaluation methods. Journal of Wildlife Management 70:347–357.

Lyon, A. G., and S. H. Anderson. 2003. Potential gas development impacts on sage-grouse nest initiation and movement. Wildlife Society Bulletin 31:486–491

Manly, B. F. J., L. L. McDonald, D. L. Thomas, T. L. McDonald, and W.P. Erickson. 2002. Resource selection by animals: statistical design and analysis for field studies. Kluwer Academic Publishers, Dordrecht, The Netherlands.

Manzer, D. L. 2004. Sharp-tailed Grouse breeding success, survival, and site selection in relation to habitat measured at multiple scales. Ph.D. dissertation, University of Alberta, Edmonton, Alberta, Canada.

Manzer, D. L., and S. J. Hannon. 2005. Relating grouse nest success and corvid density to habitat: a multi-scale approach. Journal of Wildlife Management 69:110–123.

Niemuth, N. D. 2003. Identifying landscapes for Greater Prairie Chicken translocation using habitat models and GIS: a case study. Wildlife Society Bulletin 31:145–155.

Niemuth N. D., and M. S. Boyce. 2004. Influence of landscape composition on Sharp-tailed Grouse lek location and attendance in Wisconsin pine barrens. Ecoscience 11:209–217.

R Development Core Team. 2007. R: a language and environment for statistical computing. R Foundation for Statistical Computing, Vienna, Austria. ISBN 3-900051-07-0. <http://www.R-project.org>.

Reinkensmeyer, D. P., R. F. Miller, R. G. Anthony, V. E. Marr, and C. M. Duncan. 2008. Winter and early spring bird communities in grasslands, shrubsteppe, and juniper woodlands in central Oregon. Western North American Naturalist 68:25–35.

Stillman, R. A., and A. F. Brown. 1995. Minimizing effort in large-scale surveys of terrestrial birds: an example form the English uplands. Journal of Avian Biology 26:124–134.

Storch, I. 2000. Conservation status and threats to grouse worldwide: an overview. Wildlife Biology 6:195–204.

Timoney, K. and P. Lee. 2001. Environmental management in resource-rich Alberta, Canada: first world jurisdiction, third world analogue? Journal of Environmental Management 63:387–405.

CHAPTER FOUR

Home Range Size and Movements of Greater Prairie-Chickens

Michael A. Patten, Christin L. Pruett, and Donald H. Wolfe

Abstract. Size of a home range is key to a species' conservation and management. Estimates of home range size vary with movement patterns, which in turn vary with sex, age class, season, time of day, and habitat configuration, particularly extent of fragmentation. We describe variation in home range and movements in a grouse endemic to North American prairie, the Greater Prairie-Chicken (*Tympanuchus cupido pinnatus*). Our study area included a large, contiguous block of tallgrass prairie. We found that daylight movements varied with time of day: typically, birds were least active in the heat of midday and most active in the relative cool of morning and evening, a pattern consistent with sunrise and sunset, particularly in autumn, winter, and spring. The species' lek and nesting biology predicted observed lulls in male movement in spring and female movement in summer; sexes are equally mobile at other seasons. Females had larger home ranges than

males, moved more frequently between activity centers, and moved greater maximum distances; therefore, females may be more susceptible to the negative effects of habitat fragmentation. Yearlings of both sexes tended to move more than adults. A synthesis of home range estimates from our work and past studies suggests there may be an inverse relationship between habitat continuity and home range sizes. Our results underscore the need to consider various environmental and other factors when estimating home range size. We also present preliminary evidence that habitat fragmentation may force prairie grouse to expand their home range, potentially decreasing survivorship through increased mortality from predation risk or energy expenditure.

Key Words: circadian rhythm, fragmentation, home range, movement, seasonality, *Tympanuchus cupido*.

Patten, M. A., C. L. Pruett, and D. H. Wolfe. 2011. Home range size and movements of Greater Prairie-Chickens. Pp. 51–62 *in* B. K. Sandercock, K. Martin, and G. Segelbacher (editors). Ecology, conservation, and management of grouse. Studies in Avian Biology (no. 39), University of California Press, Berkeley, CA.

The home range—the amount of physical space individuals need, on average, to survive, grow, and reproduce—is a fundamental aspect of a species' ecology and is crucial to an understanding of a species' place in the ecosystem (i.e., its ecological niche). Moreover, it is difficult to develop meaningful management and conservation strategies for rare species if we lack a basic knowledge of their spatial needs (Belovsky 1987). Both movements and home ranges depend on a variety of endogenous and exogenous factors, including demographic status and local habitat and conditions (Southwood 1977). From the organism's view, habitat can be continuous, patchy, or isolated, and large-bodied organisms generally require larger home ranges (Kelt and Van Vuren 1999, Peery 2000).

Few habitats in North America are more fragmented and depleted than tallgrass prairie: Only ~4% of this biome remains (Samson and Knopf 1996), with most remnants being small, widely scattered, and altered by human activity. As a result of this extensive alteration, the Greater Prairie-Chicken (*Tympanuchus cupido pinnatus*)—a species emblematic of tallgrass prairie—now survives on native grassland embedded in a matrix of pastures, cultivated fields, roads, fences, homesteads, and woodlands. Most prairie chicken populations are of conservation concern, underscoring the need for a clear understanding of the species' home range requirements and movement patterns (Niemuth, this volume, chapter 1). In this species, movement frequency and distance varies temporally with season or time of day (Robel et al. 1970, Hamerstrom and Hamerstrom 1973, Drobney and Sparrowe 1977); endogenously with demographic factors such as sex, age, or breeding status (see Toepfer 1988); and exogenously with habitat extent and fragmentation.

Early studies of movements and home ranges of birds relied on band recoveries, fortuitous sighting of marked individuals, or following individual flocks through a day (Hamerstrom and Hamerstrom 1949). These methods produced important data but only allowed study of short-term, short-distance movement. The advent of lightweight radio transmitters in the 1960s greatly benefited the study of both the frequency and the distance of animal movement. Still, there have been relatively few telemetry studies on prairie grouse, most of them involving prairie chickens in fragmented habitat or small blocks of prairie (Hamerstrom and Hamerstrom 1949, Burger 1988, Toepfer 1988, Schroeder and Braun 1992a).

Our objective was to estimate home range size for Greater Prairie-Chickens on a large block of unfragmented tallgrass prairie. We further sought to examine movement patterns at several time scales, ranging from within a day to among seasons to over the life span of an individual. This last effort allowed us to identify whether a prairie chicken's center of activity—by which we mean the extent of the principal area used—was stable or changed over a bird's life. We compared our findings to results from previous studies elsewhere in the species' range, allowing us to postulate how home range size might be affected by habitat fragmentation. Given that our study area was a contiguous block of tallgrass prairie, we predicted our home range estimates would be smaller than in previous studies because in fragmented areas energetic needs for maintenance, growth, and reproduction can be met equally in either a smaller contiguous block of suitable habitat or in a larger mosaic of suitable and unsuitable habitats (Reiss 1988). We also examined how movements vary with sex, age, season, and time of day, thus generating a better understanding of the spatial ecology of the Greater Prairie-Chicken.

METHODS

Study Area

Our study area encompassed ~450 km^2 of tallgrass prairie in the Flint Hills of north-central Osage County, Oklahoma, its north edge abutting Kansas (36°46'–37°00' N, 96°22'–96°40' W). The Flint Hills ecoregion consists largely of unplowed tallgrass prairie, although much of this region is grazed heavily and burned annually (Zimmerman 1997, With et al. 2008). Habitat in our study area was relatively homogenous prairie, with no cultivation (<1% of the area has ever been cultivated), no significant development, and few fences. The few roads were primarily graded dirt or gravel without bordering ditches or embankments. Deciduous woodland (<5% of the area) occupied a small portion of the southeast corner of the area, chiefly occurring in narrow corridors along two creeks.

Prescribed fires burned 60–80% of the area annually (Patten et al. 2007), generally in early spring (March–April). Cattle grazing usually followed burning, the predominant system being early intensive stocking: Steers are brought to the ranches for ~100 days from April to July, allowing the range vegetation to recover in late summer and autumn (Smith and Owensby 1978). Cow–calf operations occupied ~10% of the study area; such operations avoid annual burns and graze at a lower stocking rate throughout the year. A low density of American bison (*Bison bison*) grazed ~5% of the study area year-round, all on the Nature Conservancy's Tallgrass Prairie Preserve (in the southeastern quadrant), and blocks (<100 ha) of this preserve burned sporadically. A small fraction (1–5%) of the study area was hayed each year, chiefly in August. Rainfall during the study (1997–2000) exceeded ($z = 0.44$–1.42) the long-term (1949–2003) average, but annual temperature centered on the average ($z = -0.45$–1.40).

Tracking

We radio-tagged and tracked Greater Prairie-Chickens year-round for a 3-year period, from April 1997 to July 2000. Birds were trapped at leks using walk-in funnel traps (Schroeder and Braun 1991) connected by 8-m zigzags of plastic drift fence. At first capture, birds were fitted with a bib-mounted radio transmitter and a loop antenna (AVM and Telemetry Solutions Inc.®) weighing 18 g, which was ~2% of body mass (800–1000 g). We used feather wear and replacement to identify age classes (Wright and Hiatt 1943, Ammann 1944). Yearlings were birds <1 year old (we treated 1 July as the "birthday" for all birds), adults were birds >1 year old, some of which may have been 3–4 years old. Tracking equipment consisted of five-element, handheld Yagi antennas and ATS® model R-2000 or R-4000 receivers. We began tracking a bird the day it was captured; it was tracked thereafter as often as possible, averaging once every 3 days, at varying times of day. Over 99% of bird locations were from direct homing (<1% triangulation) and from <50 m distance. Roughly 17% of birds flushed, the majority during night trapping or being females at nests or with broods; no birds were pushed into new areas, and we have no reason to suspect that a flushed bird left its home

range. Two person-days per week were devoted to finding "lost" birds, defined as individuals not detected for two weeks, and we conducted study area–wide aerial transects for lost birds 5–6 times per year, extending 3–8 km beyond any tracked bird and often >25 km from previous locations.

There are several sources of bias when estimating home range and maximum distances moved. Unless a marked bird is found dead, a bird's life span is unknown, particularly if it is lost from the study area; range size and distance moved for lost birds may be larger than estimated (Sharp 2009). We determined if estimates of home ranges for birds eventually found dead were systematically smaller than estimates for lost or dispersed birds, in each case using movement data up to our last confirmed location. Birds were also tracked for different durations. We evaluated sources of potential bias by regressing range size against two tracking measures: (1) the number of tracking records and (2) days elapsed between when a bird was first and last tracked. We log_{10}-transformed home range estimates before performing each regression. We evaluated tracking records versus home range by fitting a loess smooth ($f = 0.5$) to the data (see Cleveland and Devlin 1988).

Movements

We analyzed daily movement data from a subset of individuals tracked at 30-min intervals from within 1 hour of sunrise to within 1 hour of sunset. Because sunrise and sunset vary through the year, we obtained a variable number of locations—generally between 20 and 30—per bird-day of tracking. Of 185 tagged birds, we used only those 32 individuals with extensive data: 22 males (17 adult, 5 yearling) and 10 females (8 adult, 2 yearling). Linear distances were calculated between consecutive tracking locations. Maximum distance refers to the greatest distance between any two consecutive locations that a given individual moved.

Because movement distances are not comparable when tracking intervals vary widely, we used only locations from nonoverlapping 3-day intervals. Our approach eliminates bias by restricting data to equal intervals over which distances can be calculated between consecutive locations.

For these analyses, we selected only individuals with >25 records, used only individuals for which there were at least four 3-day tracking intervals in a season ($n = 617$ tracking events), and used only the first location of a day if a bird was tracked more than once that day. We avoided bias in first locations by tracking a given set of birds across different periods in a 5-day rotation: day 1, 1400–2300 H; day 2, 1200–2100 H; day 3, 1000–1900 H; day 4, 0800–1700 H; and day 5, 0600–1500 H.

In some cases, we divided analyses into four 3-month seasons, with prairie-chicken biology defining the seasons: Spring corresponded with lekking activity (15 February–14 May), summer with nesting and brood rearing (15 May–14 August), autumn with late-season lekking (15 August–14 November), and winter accounting for the rest (15 November–14 February). We averaged movements for individual birds within a season and smoothed hourly and seasonal data using loess regression with $f = 0.5$ (Cleveland and Devlin 1988). Differences in movements were assessed with the Mann–Whitney U-test.

Annual Home Range

We estimated annual home range using both a 100% minimum convex polygon (MCP) and a kernel density estimator (see Powell 2000). We used ArcGIS 9.2 (ESRI, Redlands, CA) and program Abode (P. N. Laver 2005, http://filebox.vt.edu/users/ plaver/abode/contact.html) to estimate MCP of all points, centered on their median, to compare our estimates of home range size to MCP values reported in prior studies of Greater Prairie-Chickens (Robel et al. 1970, Newell 1987, Toepfer 1988, Burger 1988, Schroeder and Braun 1992a). For other comparisons, home ranges for each bird were estimated using kernel methods at 95% and 50% isopleths, smoothed with least-squares cross-validation. We considered the 50% isopleth to be a center of primary activity that may be spread over more than one area. We minimized temporal autocorrelation by including only one tracking location per day, selected at random.

We restricted initial analyses to individuals tracked at 20 unique locations (all had >50 tracking records across >70 days), yielding a set of 100 individuals, 29 females (14 adult, 6 yearling, 9 unknown) and 71 males (27 adults, 36 yearlings, 8 unknown). The number of locations at which an individual was recorded ranged from 20 to 159, for a sum of 4,925 locations across all birds. The mean ($\pm$SE) duration between which bird was first and last tracked was 354.6 $\pm$ 21.6 days. Differences in home range size were assessed with the Wilcoxon two-sample test.

Habitat Continuity

In order to place our estimate of annual home range size in context, we compiled other MCP-based estimates for the Greater Prairie-Chicken

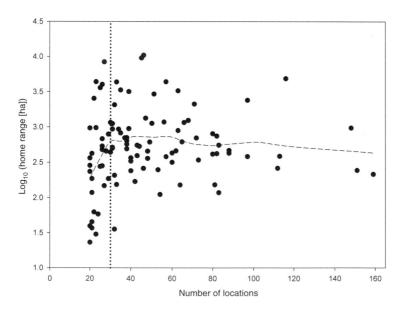

Figure 4.1. The relationship of range size and number of tracking locations for the Greater Prairie-Chicken. A loess smooth (dashed line; $f = 0.5$) of these data shows no relationship after ~30 locations have accrued.

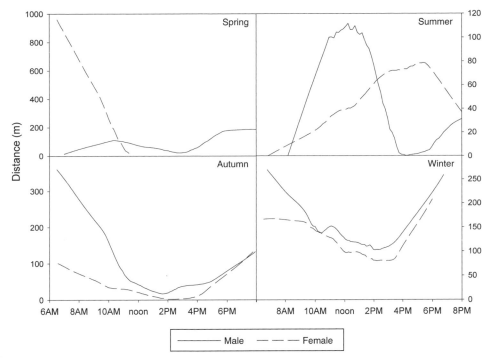

Figure 4.2. Mean hourly movement of the Greater Prairie-Chicken across seasons (see text for definitions). Curves were smoothed using a loess procedure ($f = 0.5$). Note that the pattern of summer movements differs fundamentally from patterns in other seasons, and that male and female behavior differs markedly in spring and summer.

(*T. c. pinnatus* subspecies only) from Newell (1987), Toepfer (1988), and Burger (1988); the last two studies present estimates from two sites each. Neither Robel et al. (1970) nor Schroeder and Braun (1992a) reported annual home range sizes, but we used their data on seasonal home range size. For each study site, we used the largest block of unbroken prairie-chicken habitat as an index for habitat continuity. We assumed that the relationship between this continuity index and home range size would be an inverse polynomial curve of the form $f(x) = y_0 + a/x$, where y_0 marks the inflection point and a the slope parameter.

RESULTS

Potential Sources of Bias

Estimated home range size, regardless of method, was not related to the number of tracking locations (95% kernel: $r^2 = 0.01$, $F_{1,98} = 0.85$, $P > 0.35$; MCP: $r^2 = 0.001$, $F_{1,98} = 0.10$, $P > 0.70$), although the slope flattened only after ~30 locations had accumulated (Fig. 4.1). Accordingly, our

home range estimates below were based on individuals for which >30 tracking locations were available ($n = 71$). The estimate of home range size increased with the number of days a bird was tracked (95% kernel: $r^2 = 0.08$, $F_{1,98} = 8.02$, $P < 0.01$; MCP: $r^2 = 0.04$, $F_{1,98} = 4.36$, $P < 0.05$), up to one year, after which the relationship flattened to a slope of zero. Range size estimates thus increase over the short term, but our long-term data should mitigate this problem. Estimates of home range size did not differ for birds lost from the study (1,302 ± 172 ha) versus those eventually found dead (1,496 ± 395 ha; Wilcoxon $C = 1506$, $P > 0.40$).

Hourly Movements

At the latitude of our study site there are ~5 hours more daylight at summer solstice than at winter solstice, so movements per hour of individuals in a single day should be computed only within season. In autumn, winter, and spring, prairie chickens moved most often in early morning and late evening, and least often at midday and early afternoon (Fig. 4.2). The same pattern

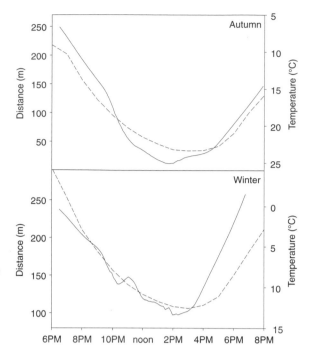

Figure 4.3. Mean movement (solid line) of the Greater Prairie-Chicken in relation to ambient temperature (dashed line) during autumn and winter (sexes combined; see text for definitions of seasons). Bird activity decreases as temperature increases (note the inverse scale for temperature).

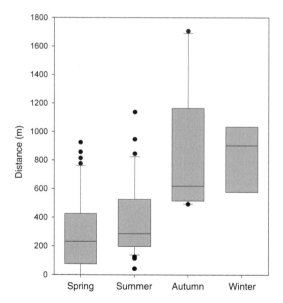

Figure 4.4. Movement (average distance among locations, collected 3 d apart) of the Greater Prairie-Chicken in spring ($n = 43$ individuals, sexes and ages combined), summer ($n = 38$), autumn ($n = 10$), and winter ($n = 8$). Note the sharp increase in movements in autumn and winter. Apart from autumn vs. winter, all pairwise comparisons differ significantly (Mann–Whitney U tests). Box plots represent boundaries of the 25th and 75th percentiles, whiskers the 10th and 90th percentiles; dots are outliers and the internal line is the median.

held when we used sunrise and sunset as common reference points, suggesting movements may track ambient temperature, at least during non-breeding seasons (Fig. 4.3). Male movements in spring were low in the morning, an expected consequence of lek attendance. Although the pattern within days was similar, birds moved farther in winter than in autumn and spring (see below); activity in summer was different, with a male peak near midday and a female peak in early afternoon (Fig. 4.2).

Movements by Season

Individual movements across the year averaged 429 m (range 0–1,706 m) per 3-day period. Distances varied substantially among four seasons (Fig. 4.4). Average autumn ($\pm SE = 839 \pm 142$ m) and winter (845 ± 121 m) movements were 2–3 times greater than those from spring (289 ± 40 m) and summer (391 ± 43 m). Males tended to move less (266 ± 41 m) than females (461 ± 132 m) in spring (Fig. 4.5; $U_{38,5} = 138$, $0.05 < P < 0.10$) but more (469 ± 57 m) than females (272 ± 55 m) in summer (Fig. 4.5; $U_{23,15} = 265$, $P < 0.01$). Because our sample size of females was small, we restricted our analysis of movements

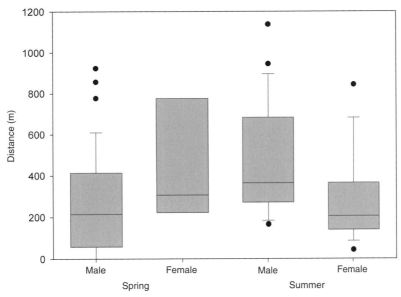

Figure 4.5. Movements (average distance among locations, collected 3 d apart) of male and female Greater Prairie-Chickens in spring (males: $n = 36$, females: $n = 5$) and in summer (males: $n = 23$, females: $n = 15$). Box plots represent boundaries of the 25th and 75th percentiles, whiskers the 10th and 90th percentiles; dots are outliers and the internal line is the median.

by age classes to males. Average movements of adults (375 m) and yearlings (426 m) did not differ, but adult movement tended to vary less (0–1,138 m, SE = 43 m) than yearling movement (0–1,405 m, SE = 102 m; variance test: $F_{17,45} = 2.05$, $P < 0.06$). None of the pairwise comparisons between seasons was significant.

Annual Home Range Size

Greater Prairie Chickens moved, on average, 5.0 km from one end of their MCP home range to the other (Fig. 4.6); the minimum distance was only 1.1 km, whereas the maximum was >15 km. Females moved twice as far as males on average (Fig. 4.6; 8 vs. 4 km; $U_{33,12} = 313$, $P < 0.01$); minimum and maximum distances moved by females were also twice those of males, half of the females having a maximum distance greater than that for all but a few males.

Annual 95% kernel home range size averaged 1,203 ± 219 ha ($n = 71$; median = 554 ha; range 36–10,433 ha). Many of the largest home ranges were split into multiple activity centers, although the largest with a single activity center was 4898 ha. The average annual home range for females (2,593 ha) was >3 times larger than that for males (731 ha; Fig. 4.6; $C = 722$,

$P < 0.001$). By contrast, the proportion of females and males with split ranges—in which a bird moved between two or more clearly defined activity centers—did not differ for 95% kernel home ranges (males: 0.793; females : 0.704; $\chi_3^2 = 4.40$, $P > 0.20$); however, on the basis of 50% kernels, significantly more males had a single core area (males: 0.103, females: 0.366; $\chi_4^2 = 9.81$, $P < 0.05$), presumably their lek. We detected neither a seasonal nor a daily pattern to movements among activity centers, and home range size was not associated with the number of activity centers ($r^2 = 0.02$, $F_{3,96} = 0.49$, $P > 0.60$). In both sexes, annual home range did not vary with body mass (males: $r^2 < 0.01$, $F_{1,17} = 0.11$, $P > 0.70$; females: $r^2 < 0.01$, $F_{1,51} = 0.05$, $P > 0.80$) and did not differ between adults and yearlings (males: $C = 39$, $P > 0.10$; females: $C = 423$, $P > 0.30$), although yearlings tended to have larger home ranges (males: 4792 ha vs. 1614 ha; females: 801 ha vs. 637 ha).

Habitat Continuity

The annual 100% MCP home range for the Greater Prairie-Chickens in our contiguous block of prairie averaged 1,371 ha, a value almost exactly at the mean for six other studies (=1,370 ha)

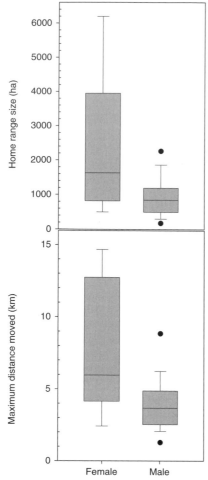

Figure 4.6. Size of 95% kernel home range (males: $n = 53$, females: $n = 18$) and maximum distance moved (males: $n = 33$, females: $n = 11$) by Greater Prairie-Chickens with >30 tracking locations. Box plots represent the 25th and 75th percentiles, whiskers the 10th and 90th percentiles, and dots 5th and 95th percentiles; the internal line is the median.

across a range of fragmented prairie (Fig. 4.7). An inverse polynomial ($y_0 = 1,025.4$, $a = 986,465.0$) indicated that extent of fragmentation explained much of the variance in home range size ($r^2 = 0.83$). The curve's transition point implies that home range size is stable at >4,000 ha of contiguous habitat but increases sharply below ~3,000 ha.

DISCUSSION

On average, the Greater Prairie-Chicken's pattern of movements across a day is like that of most diurnal endotherms: extensive morning and evening activity bracketing a midday lull (Rensing

and Ruoff 2002). Hamerstrom and Hamerstrom (1949) noted a similar pattern, with movement in morning from the feeding area to a midday loafing area, then back to the feeding area in evening shortly before going to roost. We found diurnal patterns to be fairly consistent even when day length varied among seasons. Accordingly, it seems reasonable to conclude that, at least outside the breeding season, the movements are driven partly by thermoregulation and, if so, should track ambient temperature (Fig. 4.3), which climbs from dawn until early afternoon, then drops to a low in the middle of the night. Yet it is possible that the birds respond directly to photoperiod, meaning the association between temperature and movement is spurious—both measures simply vary with time of day. Still, we noted more movement in winter, perhaps because food is dispersed more widely and cold temperatures reduce the risk of heat stress.

Seasonal Movements

Variation in seasonal movements cannot be attributed solely to climate—it depends also on the Greater Prairie-Chicken's biology and their environment. For example, movement by both sexes is reduced during the late summer molt of flight feathers (Schroeder and Braun 1992a). Burger (1988) noted that hourly movement peaked in autumn and winter, a pattern we also documented (Fig. 4.4). Between spring and summer, males and females reverse their propensity to move (Fig. 4.5). A male prairie grouse spends most of his time at a lek in spring (Hamerstrom and Hamerstrom 1949, Giesen 1997), limiting his movement. Lekking male sage grouse lose a great deal of fat during the energetically demanding courtship period (Vehrencamp et al. 1989), so in summer they primarily forage.

Females visit leks only briefly in spring, but by summer a female is tied to her nest or brood, limiting her movement at that season. Greater Prairie-Chickens have precocial young, but chicks cannot fly until two weeks of age, so even after completing incubation a female's movements are restricted by the distance her chicks can walk safely. Burger (1988) reported a similar reduction in female movement in summer, about half as much as in autumn and winter. He further reported a significant difference in summer movement between females with and without

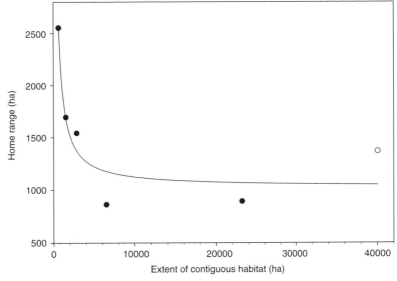

Figure 4.7. Relationship between approximate extent of habitat fragmentation, estimated as the largest block of contiguous prairie in the study and home range size (100% MCP) for Greater Prairie-Chickens using our data and results reported by Newell (1987), Toepfer (1988), and Burger (1988). The open circle is for the present study.

broods; Schroeder and Braun (1992a) also found that females with broods tended to have smaller home ranges than females without broods.

Robel et al. (1970) reported a somewhat different pattern of seasonal movement, with maximum movements in February, corresponding to the early part of our spring. Some studies of the Lesser Prairie-Chicken (*T. pallidicinctus*), which occurs in arid shortgrass prairie, have also reported the most movement in early spring (March; Jamison 2000). Our data suggest that prairie chickens moved the least in spring, although this result may be partly the result of different definitions of the seasons, as our results correspond well with reports of reduced movement in summer and increased movement in autumn and winter for both the Greater (Hamerstrom and Hamerstrom 1949, Robel et al. 1970) and Lesser Prairie-Chickens (Taylor and Guthery 1980, Jamison 2000). Likewise, our data were consistent with previous findings that female prairie chickens move less than males in summer and early autumn (Robel et al. 1970, Riley et al. 1994).

Finally, we did not detect a difference between movements of adult and yearling males of the Greater Prairie-Chicken, a finding consistent with other studies of prairie chickens (Robel et al. 1970, Taylor and Guthery 1980). As with our study, Robel et al. (1970) noted that movement of yearling males was more variable than movement of adult males, perhaps the result of yearling males visiting multiple leks before settling to establish a territory (Rippen and Boag 1974, Schroeder and Braun 1992b).

Migration

Schroeder and Braun (1993) reported that Greater Prairie-Chickens in northeastern Colorado were partly migratory. We found no evidence that the population in northern Oklahoma migrated in the strict sense (*sensu* Gauthreaux 1985), that is, moved between separate geographic locations or habitats. Birds in our study were sedentary, remaining in their home range until they died or were "lost," either because they moved out of the study area or their transmitter ceased to work (as evidenced by recaptures). A few females moved among different activity centers, but we feel that such movement would be better termed *reciprocal dispersal*, a tendency to move among local sites regularly but without distinct seasonality. We had no birds disappear completely only to reappear later. Milder winters in northern Oklahoma than in northern Colorado may account for lack of migratory behavior. Still, even in northern populations that ought to be prone to migration, most Greater Prairie-Chickens appear to spend their entire lives in a small area (Hamerstrom and Hamerstrom 1949).

Annual Home Range

In interspecific comparisons of lekking grouse, male territory size is inversely related to female home range size (Bradbury et al. 1986). If home range size is proportional to territory size, then this relationship may imply that a male's home range is smaller than a female's. In our study, female Greater Prairie-Chickens had larger home ranges than males. In Wisconsin and northeastern Colorado, female Greater Prairie-Chickens likewise tended to have larger home ranges than males (Toepfer 1988, Schroeder and Braun 1992a); however, Giesen (1997) found no difference in home range size between sexes of the Sharp-tailed Grouse (*T. phasianellus*).

Relative to males, female Greater Prairie-Chickens moved significantly farther and more often between activity centers, a finding in agreement with past research (Hamerstrom and Hamerstrom 1949, 1973; Toepfer 1988; Schroeder and Braun 1993). Increased movements may have both costs and benefits. Annual survival and reproductive success of female prairie chickens may be related negatively to their amount of movement (Burger 1988), although a female may place her nest farther from the lek where she was inseminated than from the nearest lek (Wakkinen et al. 1992, Aldridge and Brigham 2001). Neither we nor Schroeder and Braun (1992a, 1993) found significant differences in home range size among age classes, although in each case yearlings tended to have larger home ranges.

Habitat Continuity

Fragmentation depends on scale and on the organism under study. We follow Franklin et al. (2002) in defining habitat fragmentation as "the discontinuity, resulting from a given set of mechanisms, in the spatial distribution of resources and conditions present in an area at a given scale that affects occupancy, reproduction, or survival in a particular species." Under this view, anthropogenic habitat alteration that results in discontinuity can have a profound effect on an animal's home range requirements (Haskell et al. 2002). Habitat alteration and fragmentation may unduly influence an organism's movements, either by restricting them or by forcing the organism to move farther to find food, shelter, or a mate (McNab 1963, Herfindal

et al. 2005). This restriction includes features that affect an organism's perception of contiguity, such as erection of high-tension power lines across prairie (Pruett et al. 2009a, 2009b).

We could only approximate extent of fragmentation in other studies, but our survey suggests a negative relationship between extent of contiguous habitat and size of the home range (Fig. 4.7). Excepting when blocks become too small to sustain a population (Winter and Faaborg 1999), we posit that a prairie chicken requires a larger home range in fragmented habitat because it must move farther (see Ryan et al. 1998) and more often to locate food, suitable cover, and safe nesting and roosting sites. This pattern has been implied elsewhere for the Greater Prairie-Chicken: Toepfer (1988) noted that home range was smaller with a "closer year-round proximity of food and cover"; Schroeder and Braun (1992a) reported a tendency for increased home range with greater distance between cover and nests or leks; Svedarsky and Van Amburg (1996) attributed larger egg-laying ranges to wider spacing of cover, and Ryan et al. (1998) observed that "greater dispersion of native prairie in the mosaic area was associated with . . . greater movement of broods." Beyond energetic costs, extensive movement may increase mortality risk, whether from increased exposure to predators or increased probability of collision with fence lines or other anthropogenic structures (e.g., Patten et al. 2005a, Wolfe et al. 2007, McNew et al., this volume, chapter 19).

CONCLUSIONS

An apparent negative relationship between habitat continuity and home range size suggests a possible mechanism by which populations of the Greater Prairie-Chicken decline as habitat fragmentation increases: It is an indirect result of increased mortality imposed on the birds because they must expand their home ranges. Females, the more peripatetic sex (Fig. 4.6), may be particularly susceptible. On the basis of seasonal and circadian movements (Figs. 4.2–4.5; see also Robel et al. 1970), we predict that the dangers of increased movement in fragmented landscapes may be magnified in winter, in the morning and evening, and for females. A substantial increase in female mortality could doom a population to extinction (Patten et al. 2005a). Indeed, a preliminary analysis of our data implies that survival probability decreases as home range increases (Cox regression: Wald

$\chi^2 = 5.98$, $P < 0.02$). Increasingly fragmented habitat could also impair a yearling male's ability to locate leks with available territories, jeopardizing long-term lek persistence, and may limit the extent to which a brooding female can locate enough food and shelter for growing chicks.

Management strategies should focus on minimizing further fragmentation of the tallgrass prairie, including fragmentation from the birds' perspective (e.g., Pruett et al. 2009b). Limiting direct loss of habitat is an obvious step, but continuity may be reduced by other means, such as extensive spring burning or an increase in fences (Patten et al. 2005a, Reinking 2005). Our data also suggest a potential impact of reducing cover: Thermoregulation may play a key role in habitat selection by prairie grouse (figure 3 of Patten et al. 2005b). Habitat restoration may address these potential problems by ensuring that there is adequate vegetative cover, particularly at key seasons, for long-term survival and reproduction.

ACKNOWLEDGMENTS

Research was funded in part by Wildlife Restoration Program Grant W146R from the Oklahoma Department of Wildlife Conservation (ODWC) and a grant from the U.S. Fish and Wildlife Service. We received additional financial support from World Publishing, Conoco-Phillips, the Kenneth S. Adams Foundation, Arrow Trucking, and various private donors. We appreciate the efforts of the field technicians who collected data under difficult field conditions. We thank the land owners—especially The Nature Conservancy, ODWC, National Farms, Sooner Land Company, Paul Jones, and Lincoln Robinson Ranch—who allowed us to conduct research on their property. We thank David A. Wiedenfeld for preliminary analyses of some of the data and Robert J. Robel, Brett K. Sandercock, Michael A. Schroeder, and several referees for critical reviews of earlier drafts of this paper.

LITERATURE CITED

Aldridge, C. L., and R. M. Brigham. 2001. Nesting and reproductive activities of Greater Sage-Grouse in a declining northern fringe population. Condor 103:537–543.

Ammann, G. A. 1944. Determining age of Pinnated and Sharp-tailed Grouses. Journal of Wildlife Management 8:170–172.

Belovsky, G. E. 1987. Extinction models and mammalian persistence. Pp. 35–57 in M. E. Soulé (editor), Viable populations for conservation. Cambridge University Press, Cambridge, UK.

Bradbury, J., R. Gibson, and I. M. Tsai. 1986. Hotspots and the evolution of leks. Animal Behaviour 34:1694–1709.

Burger, L. W., Jr. 1988. Movements, home range, and survival of female Greater Prairie-Chickens in relation to habitat pattern. M.S. thesis, University of Missouri, Columbia, MO.

Cleveland, W. S., and S. J. Devlin. 1988. Locally weighted regression: an approach to regression analysis by local fitting. Journal of the American Statistical Association 83:596–610.

Drobney, R. D., and R. D. Sparrowe. 1977. Land use relationships and movements of Greater Prairie Chickens in Missouri. Transactions of the Missouri Academy of Science 10–11:146–160.

Franklin, A. B., B. R. Noon, and T. L. George. 2002. What is habitat fragmentation? Studies in Avian Biology 25:20–29.

Gauthreaux, S. A., Jr. 1985. Migration. Pp. 232–258 in O. S. Pettingill, Jr. (editor), Ornithology in laboratory and field. Academic Press, San Diego, CA.

Giesen, K. M. 1997. Seasonal movements, home ranges, and habitat use by Columbian Sharp-tailed Grouse in Colorado. Colorado Division of Wildlife Special Report 72.

Hamerstrom, F. N., Jr., and F. Hamerstrom. 1949. Daily and seasonal movements of Wisconsin Prairie Chickens. Auk 66:313–337.

Hamerstrom, F. N., Jr., and F. Hamerstrom. 1973. The prairie chicken in Wisconsin: highlights of a 22-year study of counts, behavior, movements, turnover and habitat. Wisconsin Department of Natural Resources Technical Bulletin 64.

Haskell, J. P., M. E. Ritchie, and H. Olff. 2002. Fractal geometry predicts varying body size scaling relationships for mammal and bird home ranges. Nature 418:527–530.

Herfindal, I., J. D. C. Linnell, J. Odden, E. B. Nilsen, and R. Andersen. 2005. Prey density, environmental productivity and home-range size in the Eurasian lynx (Lynx lynx). Journal of Zoology 265:63–71.

Jamison, B. E. 2000. Lesser Prairie-Chicken chick survival, adult survival, and habitat selection and movements of males in fragmented rangelands of southwestern Kansas. M.S. thesis, Kansas State University, Manhattan, KS.

Kelt, D. A., and D. H. Van Vuren. 1999. Energetic constraints and the relationship between body size and home range area in mammals. Ecology 80:337–340.

McNab, B. K. 1963. Bioenergetics and the determination of home range size. American Naturalist 97:133–140.

Newell, J. A. 1987. Nesting and brood rearing ecology of the Greater Prairie Chicken in the Sheyenne National Grasslands, North Dakota. M.S. thesis, Montana State University, Bozeman, MT.

Patten, M. A., D. H. Wolfe, E. Shochat, and S. K. Sherrod. 2005a. Habitat fragmentation, rapid evolution and population persistence. Evolutionary Ecology Research 7:235–249.

Patten, M. A., D. H. Wolfe, E. Shochat, and S. K. Sherrod. 2005b. Effects of microhabitat and micro-climate selection on adult survivorship of the Lesser Prairie-Chicken. Journal of Wildlife Management 69:1270–1278.

Patten, M. A., E. Shochat, D. H. Wolfe, and S. K. Sherrod. 2007. Lekking and nesting response of the Greater Prairie-Chicken to burning of tallgrass prairie. Proceedings of the Tall Timbers Fire Ecology Conference 23:149–155.

Peery, M. Z. 2000. Factors affecting interspecies variation in home-range size of raptors. Auk 117:511–517.

Powell, R. A. 2000. Animal home ranges and territories and home range estimators. Pp. 65–110 in L. Boitani and T. K. Fuller (editors), Research techniques in animal ecology: controversies and consequences. Columbia University Press, New York, NY.

Pruett, C. L., M. A. Patten, and D. H. Wolfe. 2009a. It's not easy being green: wind energy and a declining grassland bird. BioScience 59:257–262.

Pruett, C. L., M. A. Patten, and D. H. Wolfe. 2009b. Avoidance behavior by prairie grouse: implications for wind energy development. Conservation Biology 23:1253–1259.

Reinking, D. L. 2005. Fire regimes and avian responses in the central tallgrass prairie. Studies in Avian Biology 30:116–126.

Reiss, M. 1988. Scaling of home range size: body size, metabolic needs and ecology. Trends in Ecology and Evolution 3:85–88.

Rensing, L., and P. Ruoff. 2002. Temperature effect on entrainment, phase shifting, and amplitude of circadian clocks and its molecular bases. Chronobiology International 19:807–864.

Riley, T. Z., C. A. Davis, M. A. Candelaria, and H. R. Suminski. 1994. Lesser Prairie-Chicken movements and home ranges in New Mexico. Prairie Naturalist 26:183–186.

Rippen, A. B., and D. A. Boag. 1974. Recruitment to populations of male Sharp-tailed Grouse. Journal of Wildlife Management 38:616–621.

Robel, R. J., J. N. Briggs, J. J. Cebula, N. J. Silvy, C. E. Viers, and P. G. Watt. 1970. Greater Prairie Chicken ranges, movements, and habitat usage in Kansas. Journal of Wildlife Management 34:286–306.

Ryan, M. R., L. W. Burger, Jr., D. P. Jones, and A. P. Wywialowski. 1998. Breeding ecology of Greater Prairie-Chickens (Tympanuchus cupido) in relation to prairie landscape configuration. American Midland Naturalist 140:111–121.

Samson, F. B., and F. L. Knopf (editors). 1996. Prairie conservation: preserving North America's most endangered ecosystem. Island Press, Washington, DC.

Schroeder, M. A., and C. E. Braun. 1991. Walk-in traps for capturing Greater Prairie-Chickens on leks. Journal of Field Ornithology 62:378–385.

Schroeder, M. A., and C. E. Braun. 1992a. Seasonal movement and habitat use by Greater Prairie-Chickens in northeastern Colorado. Colorado Division of Wildlife Special Report 68.

Schroeder, M. A., and C. E. Braun. 1992b. Greater Prairie-Chicken attendance at leks and stability of leks in Colorado. Wilson Bulletin 104:273–284.

Schroeder, M. A., and C. E. Braun. 1993. Partial migration in a population of Greater Prairie-Chickens in northeastern Colorado. Auk 110:21–28.

Sharp, S. P. 2009. Bird ringing as a tool for behavioural studies. Ringing & Migration 24:213–219.

Smith, E. F., and C. E. Owensby. 1978. Intensive early stocking and season-long stocking of Kansas Flint Hills range. Journal of Range Management 31:14–17.

Southwood, T. R. E. 1977. Habitat, the template for ecological strategies? Journal of Animal Ecology 46:337–365.

Svedarsky, W. D., and G. L. Van Amburg. 1996. Integrated management of the Greater Prairie Chicken and livestock on the Sheyenne National Grassland. Game and Fish Department, Bismark, ND.

Taylor, M. A., and F. S. Guthery. 1980. Fall-winter movements, ranges, and habitat use of Lesser Prairie-Chickens. Journal of Wildlife Management 44:521–524.

Toepfer, J. E. 1988. Ecology of the Greater Prairie-Chicken as related to reintroduction. Ph.D. dissertation, Montana State University, Bozeman, MT.

Vehrencamp, S. L., J. W. Bradbury, and R. M. Gibson. 1989. The energetic cost of display in male sage grouse. Animal Behaviour 38:885–896.

Wakkinen, W. L., K. P. Reese, and J. W. Connelly. 1992. Sage grouse nest locations in relation to leks. Journal of Wildlife Management 56:381–383.

Winter, M., and J. Faaborg. 1999. Patterns of area sensitivity in grassland-nesting birds. Conservation Biology 13:1424–1436.

With, K. A., A. W. King, and W. E. Jensen. 2008. Remaining large grasslands may not be sufficient to prevent grassland bird declines. Biological Conservation 141:3152–3167.

Wolfe, D. H., M. A. Patten, E. Shochat, C. L. Pruett, and S. K. Sherrod. 2007. Causes and patterns of mortality in Lesser Prairie-Chickens and implications for management. Wildlife Biology 13(Supplement 1):95–104.

Wright, P. L., and R. W. Hiatt. 1943. Outer primaries as age determiners in gallinaceous birds. Auk 60:265–266.

Zimmerman, J. L. 1997. Avian community responses to fire, grazing, and drought in the tallgrass prairie. Pp. 167–180 in F. L. Knopf and F. B. Samson (editors), Ecology and conservation of Great Plains vertebrates. Springer, New York, NY.

CHAPTER FIVE

Impacts of Anthropogenic Features on Habitat Use by Lesser Prairie-Chickens

Christian A. Hagen, James C. Pitman, Thomas M. Loughin,
Brett K. Sandercock, Robert J. Robel, and Roger D. Applegate

Abstract. Suitable habitat for the Lesser Prairie-Chicken (*Tympanuchus pallidicinctus*) has been reduced markedly over the past 100 years. The remaining habitat is widely used for petroleum exploration and extraction, cattle grazing, power line easements, and the generation of electricity. Given the tenuous status of the species and recent demands on land use in remaining habitat, it is imperative that Lesser Prairie-Chicken avoidance behavior of anthropogenic features be quantified for impact assessment and conservation planning. We examined the relationship of several anthropogenic features as they pertained to habitat use of radiomarked female Lesser Prairie-Chickens ($n = 226$) in southwestern Kansas from 1997 to 2002. We used Poisson rate regression and contingency tables to examine spatial use patterns of monthly home ranges (95% fixed kernels, $n = 539$) and estimated the likelihood that anthropogenic features (i.e., power lines, wells, roads, and buildings) occurred within aggregates of all monthly home ranges (monthly use ranges). We calculated the distance from the centroids of home ranges to anthropogenic features and, using Monte Carlo simulations, evaluated whether or not they were farther than would be expected at random. There was temporal variation in the average odds of each feature occurring within monthly use ranges, but generally there was a pattern of avoidance. Monte Carlo simulations of expected distances indicated that the nearest 90% of Lesser Prairie-Chicken centers of use were farther from anthropogenic features than would be expected at random. We also had the opportunity to evaluate changes in habitat use (or avoidance behavior) 1 year post-construction of a power line using a Before–After-Control-Impact design. Post–power line construction analysis indicated that Lesser Prairie-Chicken monthly use areas were less likely to include power lines than non-use areas. However, centers of use were closer to power lines than would be expected at random in the impact area. The discrepancies between short- and long-term results suggest a lag period between power line construction and avoidance by Lesser Prairie-Chickens, possibly due to site fidelity of the species. Our study provides some minimum behavioral avoidance distances for mitigating energy developments in Lesser Prairie-Chicken habitats, and we recommend clustering these features to maximize available habitats.

Key Words: avoidance behavior, energy development, habitat use, Monte Carlo randomizations, *Tympanuchus pallidicinctus.*

Hagen, C. A., J. C. Pitman, T. M. Loughin, B. K. Sandercock, R. J. Robel, and R. D. Applegate. 2011. Impacts of anthropogenic features on habitat use by Lesser Prairie-Chickens. Pp. 63–75 *in* B. K. Sandercock, K. Martin, and G. Segelbacher (editors). Ecology, conservation, and management of grouse. Studies in Avian Biology (no. 39), University of California Press, Berkeley, CA.

Suitable habitat for the Lesser Prairie-Chicken (*Tympanuchus pallidicinctus*) has been markedly reduced over the past 100 years. What remains is highly fragmented throughout the species' range, and the remaining habitat is widely used for fossil fuel exploration and extraction, cattle grazing, power line easements, and generation of electricity (Hagen and Giesen 2005). The cumulative loss of habitat, declining population trends, and imminent threats led to a recent increase in priority ranking of the 1995 "warranted but precluded" listing under the Endangered Species Act (USFWS 2008). Recent demand on domestic energy production has provided additional human disturbances to the southern Great Plains (Pruett et al. 2009b), and the intermountain West (Naugle et al. 2011). Specifically, increased development of oil and natural gas reserves and electrical generation from wind turbines are creating new challenges to wildlife conservation (Ingelfinger and Anderson 2004, Sawyer et al. 2006, Arnett et al. 2008). In some cases, the rate of development has outpaced conservationists' ability to evaluate the impacts of these disturbances to wildlife. Recent work on Greater Sage-Grouse (*Centrocercus urophasianus*) and prairie chickens (*Tympanuchus* spp.) has demonstrated avoidance behavior, and in some cases negative demographic consequences of development (Pruett et al. 2009a, Naugle et al. 2011). Negative consequences suggest there is a larger ecological footprint associated with energy development than the immediate area disturbed by infrastructure. A clearer understanding of these impacts is necessary for more effective conservation measures and mitigation efforts for grouse populations.

The Lesser Prairie-Chicken's dependency on native rangeland has been well studied, at micro- (Jones 1963, Crawford and Bolen 1974, Riley et al. 1992, Jamison et al. 2002, Pitman et al. 2005, Patten et al. 2005a), and macroscales (Jamison 2000, Woodward et al. 2001, Fuhlendorf et al. 2002, Hagen et al. 2007b). However, limited information exists on the effects of anthropogenic features on Lesser Prairie-Chicken habitat use, most of which has focused on nesting and lek locations (Robel et al. 2004, Pitman et al. 2005, Pruett et al. 2009a). Habitat use studies from southwestern Kansas indicated that generally, Lesser Prairie-Chickens utilized sand sagebrush (*Artemisia filifolia*) throughout the year, but exhibited higher selection ratios for this cover type during the summer months (Jamison 2000, Hagen et al. 2007b). Despite the apparent importance of sagebrush, there were several areas within the prairie fragment (~5,000 ha) which appeared suitable for prairie chickens, but where radiomarked birds were not located, and unmarked individuals were not flushed or observed in these areas (Jamison 2000). The proximity of anthropogenic features, the apparent adequacy of vegetative cover, and the concomitant lack of habitat use in these areas afforded us an opportunity to test hypotheses of behavioral avoidance of landscape features by Lesser Prairie-Chickens in southwestern Kansas.

Our objectives were to: (1) examine if Lesser Prairie-Chickens avoided anthropogenic features (i.e., power lines, paved roads, oil wells, and buildings) in their monthly use of habitat; (2) estimate minimum distances from features to the center of monthly activity; and (3) in cases of apparent avoidance, estimate minimum distances and provide them as conservation guidelines setback distances in siting of future anthropogenic infrastructure.

METHODS

Study Area

The study region was comprised of two 5,000-ha fragments of native sandsage prairie near Garden City, Finney County, Kansas (37°52′ N, 100°59′ W). We began work on site I (southwest of Garden City) in 1997, and we expanded trapping and monitoring efforts to include site II (southeast of Garden City) in 2000. Prior to 1970, these two areas were part of a large contiguous tract of native sandsage prairie (Robel et al. 2004). The development of center pivot irrigation for row-crop agriculture left these areas as two fragments with about 19 km of non-habitat between patch centroids (Waddell and Hanzlick 1978). Shrub and grass vegetation in the prairie fragments was comprised of sand sagebrush, yucca (*Yucca* spp.), sandreed grasses (*Calamovilfa* spp.), bluestem grasses (*Andropogon* spp.), sand dropseed (*Sporobolus cryptandrus*), and sand lovegrass (*Eragrostis trichodes*). Primary forb species in the region included ragweed (*Ambrosia* spp.), sunflower (*Helianthus* spp.), and Russian thistle (*Salsola tragus*) (Hullett et al. 1988, Pitman et al. 2005, Hagen et al. 2007b). Land use was primarily livestock grazing, oil and natural gas extraction, and electricity transmission. Livestock grazing was generally season-long. There were periodic sagebrush thinning treatments on various pastures.

Field Methods

We captured Lesser Prairie-Chickens at leks using walk-in funnel traps during March and April 1997–2002 (Haukos et al. 1990). We fitted each captured bird with a lithium battery–powered transmitter with a mass <12-g (battery life 12–14 months). During phase I (1997–2000), we tracked females daily from April to September. We tracked females daily from April to April during phase II (2000–2003). A truck-mounted null-peak twin-Yagi telemetry system was used to triangulate locations of individuals remotely from tracking stations georeferenced using global positioning systems.

Home Range Estimation and GIS

Study area polygons were delineated from native rangeland areas as classified by GAP data layers using ARCview 3.1 (ESRI 1998). We quantified the proportion of landscape features in monthly use ranges of radiomarked females for location data from 1997 to 1999 at site I and at both study sites from 2000 to 2002. Azimuths from fixed stations were entered into Locate II triangulation software, and locations of prairie chickens were estimated using Lenth maximum likelihood estimators (Nams 2002). Location data were then imported into a GIS database of the study area and home ranges were estimated using a 95% fixed kernel estimator in ARCview 3.1 (Worton 1989). We limited our sample to those birds that had ≥15 locations per month. While this criterion reduced the number of individuals, it ensured that observed range sizes were likely representative of the area covered in a month. Our data selection limited the seasons for which such an analysis could be conducted, because the number of individuals tracked decreased from premature battery failure and mortality as the year progressed. Thus, we focused our analyses on a 6-month period (April to September) when samples were relatively large and coincided with a peak usage of sagebrush habitats (Hagen et al. 2007b). All female locations at nest sites were excluded from home range estimates because the

relationship between nest site selection and distance to anthropogenic feature has already been examined (Pitman et al. 2005).

Locations of oil and gas wells (hereafter wells), paved roads (hereafter roads), power lines, and buildings were digitized into GIS from aerial photos and GPS locations recorded in the field. We included wells with pumping units powered by electric, natural gas, or diesel motors, but did not distinguish among these in our analyses. We included human dwellings, gas compressor stations, and a 380-MW coal-fired electric generating station in our GIS layer for buildings. For the power line layer, we primarily included 125-, 138-, and 345-kV double circuit conductors that distributed electricity from the generating station, but we also included all smaller power lines to homes and wellheads. Roads and power lines were digitized as line features in the GIS, and we buffered these features by 100 m and 20 m, respectively, to more accurately depict the disturbed area these features occupied. We calculated distances from home range centroids to features in the GIS using Nearest Features (version 3.8) in ARCview.

Monthly home ranges (hereafter home range) for each bird were computed separately by year and month (e.g., April 2000) to control for yearly and biological variation in monthly habitat use. We combined home ranges for all birds in a given year and month into a single overall "monthly use range" which was overlaid onto the study area. Each monthly use range was used as a sampling frame to calculate the number of wells, roads, power lines, and buildings occurring within (i.e., "success") and outside of Lesser Prairie-Chicken monthly use ranges. Areas not overlapped by a monthly use range were referred to as non-use ranges, and served as the "failure" response (or expected frequencies) in our binomial sampling design. The number of wells occurring in use versus non-use ranges could be counted, but roads and power lines could not. Thus, following the procedure of Marcum and Loftsgaarden (1980), we generated 200,000 random points (site I = 100,000; site II = 100,000) across the study area, and tallied each point that occurred within or outside of a monthly use range and within or outside of a road or power line (see Pitman et al. 2005). We tabulated the numbers of points occurring in these four subsets. This approach resulted in count data that could be input into contingency table or Poisson rate regression analyses.

Analysis

Monthly Use Ranges and Odds of Use

Phase I of our study only included one of the two study areas; thus we analyzed each phase as a separate data set. We chose month as our sampling unit as it provided reasonable sample sizes (i.e., spatially and temporally), enabled us to examine temporal changes in habitat use during spring and summer, and reduced the potential for serial correlation of daily telemetry locations. We expected that the multiple locations recorded for an individual bird would be serially correlated, and that the correlation would decrease as the time between locations increased. It was therefore anticipated that use of monthly ranges as the observational unit for each bird (as described above) would have the effect of reducing correlations between consecutive measurements, so that treating them approximately as independent was reasonable. Because the goal of the analysis was to examine general patterns of use with regard to anthropogenic landscape features, and not to differentiate the variation in usage across months, we controlled for the monthly variability by stratifying on month. There were six months in 1998–2001, five months in 1997, and four months in 2002.

We used contingency table analyses to examine the association between use of habitat and presence of anthropogenic features, stratified by month. We used the Breslow–Day test for homogeneity of odds ratios to evaluate whether the association was similar across months within each year (Agresti 1996). In 3 of 6 years no monthly use range overlapped with buildings, and this occurrence was rare in the other three years; thus we did not include a separate analysis of buildings. We used Poisson rate regression to model the number of wells present per hectare of monthly use and non-use ranges [offset = log (monthly use range ha)] as a function of use, month, and year. Starting from a saturated model in these three factors, backward elimination (using 0.05 significance level) was used to find the most parsimonious model, and the fit of the chosen model was assessed by examining the scale parameter (deviance/df) and residual plots.

Permutations of Distance to Anthropogenic Features

Because anthropogenic features were not randomly or uniformly located on the landscape in relation to home ranges (here we refer to ranges individually rather than as a collective distribution), we used a modified Monte Carlo simulation (Manly 1997) to test (one-tailed) whether the centroids of home ranges were farther from anthropogenic features than would be expected by chance (Pitman et al. 2005). If anthropogenic features had no impact on the birds' monthly home ranges (the null hypothesis), then one would expect the distances between these centroids and anthropogenic features to follow the same distribution as distances between randomly placed points and these features. If, on the other hand, the birds demonstrated some avoidance of these anthropogenic features, as hypothesized, then the centroids of home ranges should be farther away from anthropogenic features than random points.

We compared the lower 10th percentile of the distribution of observed distances to the random distances as follows. First, the 10th percentile was chosen because it represents that portion of the study population that tended to come closest to the anthropogenic features, but is not so extreme as to be subject to excessive random variation in its estimation. Under the null hypothesis, the observed 10th percentile should behave like the 10th percentile from an equal number of randomly placed points. We approximated the null sampling distribution for this 10th percentile from: (1) a random sample ($n = 539$) was drawn from the distribution of 200,000 random points and distances; (2) the lower 10th percentile of distances was recorded for the sample; (3) this was repeated 1,000 times; (4) the observed distribution of these 1,000 random 10th percentiles estimated the required sampling distribution. If the observed 10th percentile distance was too extreme in the upper tail of this sampling distribution, then the null hypothesis (features do not affect home range placements) was rejected. We computed a P-value as the proportion of estimated sampling distribution that lay above the observed 10th percentile of all centroids. The observed percentiles and the expected (mean) value of that percentile under randomness were compared. We chose to analyze distances to features separately for each

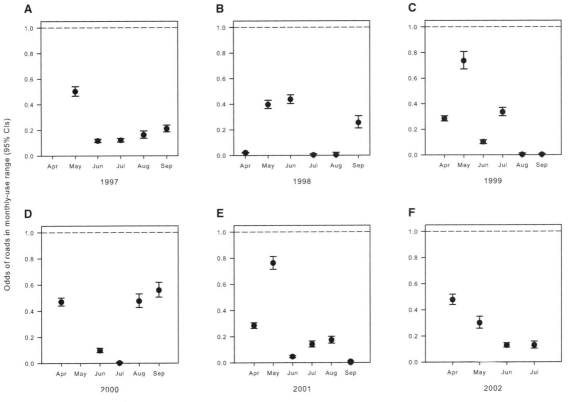

Figure 5.1. Odds ratios (95% CI) for roads occurring in Lesser Prairie-Chicken monthly use ranges in 1997 (A), 1998 (B), 1999 (C), 2000 (D), 2001 (E), and 2002 (F). The dashed line indicates odds of 1, and confidence limits intersecting this line indicate odds not different than expected by chance.

study site for ease of comparison, especially for power lines pre- and post-construction.

Impacts of Power Line Construction

A 138-kV power line was constructed on site I early in 2002; thus we employed an informal Before–After-Control-Impact (BACI) design to examine how the construction of the line may have affected monthly use. In this design, we used site I as the impact area and site II as the control and compared the differences in distances from centroids to power lines pre- and post-construction. Additionally, we estimated mean odds ratios by year and site to examine the potential avoidance of the new power line on habitat use.

RESULTS

We captured and radiomarked 226 female Lesser Prairie-Chickens; because of right censoring, we only included 190 in our analyses (phase I = 64,

phase II = 126). We recorded 21,047 (phase I = 7,524; phase II = 13,523) daily telemetry locations from 1997 to 2002. However, our effective sample size was reduced to 15,903 after removing nest locations, bird-months with <15 locations, and winter months (October–March). Thus, we estimated 539 home ranges (95% fixed kernel) and centroids to calculate avoidance distances from anthropogenic features, and used these home ranges to estimate the odds of these features occurring in 33 monthly use ranges.

Proportional Use

Phase I

The likelihood of roads occurring in monthly use ranges were significantly different across months in all 3 years of phase I (Breslow–Day test; 1997: $\chi^2 = 794.41$, df = 4, $P < 0.0001$; 1998: $\chi^2 = 4,881.69$, df = 5, $P < 0.0001$; 1999: $\chi^2 = 1,064.61$, df = 5, $P < 0.0001$; Fig. 5.1). The annual average odds

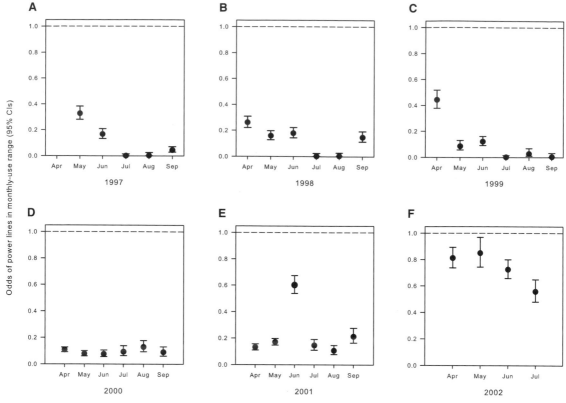

Figure 5.2. Odds ratios (95% CI) for power lines occurring in Lesser Prairie-Chicken monthly use ranges in 1997 (A), 1998 (B), 1999 (C) 2000 (D), 2001 (E), and 2002 (F). The dashed line indicates odds of 1, and confidence limits intersecting this line indicate odds not different than expected by chance.

of roads occurring in a monthly use range were between 4.1 and 5.3 times less likely than that of non-use ranges. The likelihood of power lines occurring in monthly use ranges was significantly different across months in all 3 years of phase I (1997: $\chi^2 = 307.25$, df = 4, $P < 0.0001$; 1998: $\chi^2 = 148.04$, df = 5, $P < 0.0001$; 1999: $\chi^2 = 396.19$, df = 5, $P < 0.0001$). The annual average odds of a power line occurring in a monthly use range were 11.1 to 15.9 times less likely than in a non-use range (Fig. 5.2).

Phase II

The Breslow–Day test for the odds of roads occurring in monthly use ranges indicated that the odds ratios were significantly different across months in all years of phase II (2000: $\chi^2 = 1,830.3$, df = 5, $P < 0.0001$; 2001: $\chi^2 = 2,005.3$, df = 5, $P < 0.0001$; 2002: $\chi^2 = 373.46$, df = 3, $P < 0.0001$; Fig. 5.1). The annual average odds of roads occurring in a monthly use range were 2.1 to 6.5 times less

likely than that of non-use range. The Breslow–Day test for the odds of power lines occurring in monthly use ranges indicated that the odds ratios varied throughout the sampling period in 2 of 3 years (2000: $\chi^2 = 9.61$, df = 5, $P = 0.087$; 2001: $\chi^2 = 403.73$, df = 5, $P < 0.0001$; 2002: $\chi^2 = 21.52$, df = 3, $P < 0.0001$). The annual average odds of a power line occurring in a monthly use range were between 6.0 and 10.8 times less likely than in a non-use ranges in 2000 and 2001, respectively (Fig. 5.2). Additionally, the odds of power lines occurring in monthly use ranges increased in 2002 relative to previous years, but on average was 1.4 times less likely to occur than in non-use ranges.

Wells

Backward selection of Poisson rate regression models of counts of wells per hectare of monthly use range as a function of prairie chicken use indicated that a model containing a month × use

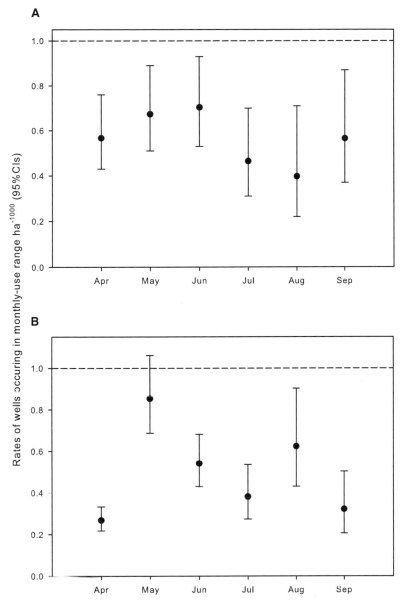

Figure 5.3. Odds ratios (95% CI) of wells ha$^{-1,000}$ between monthly and non-use ranges as determined from Poisson rate regression for phase I (A) and phase II (B) in Finney County, Kansas, 1997–2002. The dashed line indicates odds of 1, and confidence limits intersecting this line indicate observed values did not differ from those expected by chance.

interaction (deviance/df − 0.623) best described the data for phase I, and a model including year × month and month × use interaction and all main effects was the best model for phase II (deviance/df = 0.448). Estimated rates of number of wells per 1,000 ha of monthly use varied among months and ranged between 26 to 86% times less than non-use ranges (Fig. 5.3). Generally, the number of wells ha$^{-1,000}$ decreased in late summer with fewer wells

in monthly use ranges than non-use ranges in both phases of our study.

Distance to Features

Monte Carlo simulations of distances to features indicated that 90% of monthly centroids were farther than expected by chance for wells (242–320 m), buildings (1,132–1,666 m), and roads (715–990 m)

TABLE 5.1
Monte Carlo simulation tests.

Area/feature	Summary statistics				Randomization test		
	Observed		Random		Observed	Expected	
	$\bar{x}$ (m)	SE	$\bar{x}$ (m)	SE	10th P(m)	10th P(m)	$P \leq$ value[a]
Site I ($n = 369$)							
Power lines	1,494	41	594	3	709	172	0.001
Post—new line[b]	751	90	582	3	123	273	0.434
Wells	559	14	490	1	242	240	0.011
Buildings	1,929	38	1,987	4	1,132	1,092	0.005
Roads	1,712	48	1,547	6	715	564	0.025
Site II ($n = 119$)							
Power lines	1,388	50	552	3	662	176	0.001
Post—new line[b]	1,015	137	567	6	272	269	0.026
Wells	559	17	551	1	320	230	0.007
Buildings	2,374	46	2,179	4	1,666	1,458	0.001
Roads	2,695	141	2,019	6	990	946	0.016

NOTE: Tests of distances (M) of the nearest 10th percentile (P) of all Lesser Prairie-Chicken monthly (Apr–Sep) home range centroids to anthropogenic features on two sand-sagebrush prairie areas in southwestern Kansas, USA, 1997–2002. Overall mean distances (SE) from observed centroids and random points (N = 1,000) to features are also reported.

[a] We calculated P-values as the proportion of distances in the random distribution that were ≥ observed 10th P for a given feature.

[b] Sample sizes were 28 and 23 monthly home range centroids for sites I and II, respectively, post construction of a new power line at site I.

at both sites (Table 5.1). Similarly, distances from home range centroids to power lines were farther than expected at both sites (662–702 m) prior to the new power line construction in 2001.

Post–Power Line Construction

Using a BACI design, the observed distance to power lines decreased at both sites in 2002. However, the observed distance (173 m) was closer than expected (272 m) only at site I, where the power line was constructed (Table 5.1), despite the fact that the average observed distance (751 m) was greater than would be expected by chance (582 m). We summarized odds ratios by year and site to further examine the potential avoidance of power line construction on habitat use (Fig. 5.4). Power lines were less likely to occur in monthly use ranges on site II than site I for years that birds were monitored on both sites (Fig 5.4). On

average, odds ratios indicated that power lines were less likely to occur in monthly use ranges than non-use areas for all years and sites (range of odds ratios = 1.1–17.1 times less likely), but the likelihood of occurrence did increase the last 2 years of our study (1.1 and 1.6 times less likely).

DISCUSSION

Our study adds to growing evidence that anthropogenic features may act as barriers to use of otherwise suitable habitat for Lesser Prairie-Chickens and other grassland or shrubsteppe grouse (Robel et al. 2004, Pitman et al. 2005, Pruett et al. 2009b, Naugle et al. 2011). Notably, we found support for avoidance for power lines, paved roads, buildings, and wells and identified minimum footprints that are likely to be avoided in habitat selection. Specifically, of the vertical features at our study sites, power lines and buildings appeared to be the least likely

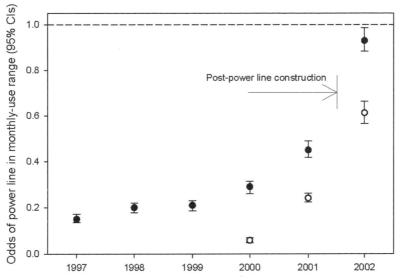

Figure 5.4. Odds ratios (95% CI) for power lines occurring in Lesser Prairie-Chicken average monthly use ranges pre- and post-construction (2002) of a power line at site I (filled circles) and site II (open circles), in Finney County, Kansas, 1997–2002. Power line construction did not occur at site II. The dashed line indicates odds of 1, and confidence limits intersecting this line indicate odds not different than expected by chance.

to occur within monthly use ranges and were the furthest from monthly centroids. We summarize our findings first in the context of seasonal variation and then in regard to each anthropogenic feature.

Annual variation in avoidance distances is described by individual heterogeneity of marked prairie chickens and their different monthly patterns of habitat use and movement. Generally, female movements are more wide-ranging during the breeding season (April and May) as they seek mates and a nest site (Hagen and Giesen 2005). Movements from breeding areas to summer habitat often occur in June and may result in larger monthly ranges and increase the likelihood of a range including a feature. However, from July to September, movements are minimized as Lesser Prairie-Chickens shift into summer use areas that are abundant with invertebrates (Jamison et al. 2002). Regardless of season of use, most habitats used by Lesser Prairie-Chickens tended to avoid anthropogenic features.

Roads

The presence of paved highways does not appear to serve as a barrier to movement by prairie chickens (Pruett et al. 2009a), but the noise and disturbance associated with roads can alter nest site selection, habitat use, and lek persistence (Lyon and Anderson 2003, Holloran 2005, Pitman et al. 2005, this study). Direct mortality associated with vehicle collisions can occur year-round, but is a relatively small percentage of overall mortality (Aldridge and Boyce 2007, Wolfe et al. 2007). Perhaps of greater concern is the potential for displaced habitat use by the breeding cohort in a population. Nesting female Lesser Prairie-Chickens placed nest sites farther from paved roads than would be expected at random, even though vegetation characteristics were similar near these edges (Pitman 2003, Pitman et al. 2005). Lesser Prairie-Chickens exhibited greater tolerance (avoidance distances of >100 m) to one of two paved highways in Oklahoma than we documented in our study (Pruett et al. 2009a). Similarly, yearling female Greater Sage-Grouse appear to select nest sites farther from main haul roads in natural gas fields, as they do not exhibit the same site fidelity as older females (Holloran 2005). If net productivity is affected by displacement, lek persistence and patch occupancy may diminish. Greater Sage-Grouse lek persistence declined as a function of traffic volumes on haul roads in natural gas developments and resulted in lek abandonment in some cases (Holloran 2005). Gunnison Sage-Grouse (*Centrocercus minimus*) patch occupancy was best described by models of landscape scale features (relative to models with microhabitat variables), and occupancy was positively correlated with distance from paved roads (Oyler-McCance 1999). Thus, roads may be one of a cumulative set of factors that impact population persistence of Lesser Prairie-Chickens (Patten et al. 2005b, Hagen et al. 2009).

Wells

Well density can affect habitat use by wildlife species (Sawyer et al. 2006, Walker et al. 2007, Doherty et al. 2008). The presence of oil and gas wells may also affect productivity and population viability through habitat displacement or changes in predator communities (Coates 2007). If suitable habitat is adjacent to such developments and the fragmentation effects increase mortality rates outside the range of natural variation, the result is an ecological trap (Aldridge and Boyce 2007). Alternatively, if critical habitats (e.g., winter range, nesting) are no longer used because of well density, it may prove detrimental to demographic rates (Holloran 2005, Doherty et al. 2008). The likelihood that Greater Sage-Grouse would use otherwise suitable winter range was diminished by 10% with the development of 1 well per 4 km^2 (Doherty et al. 2008). Our study indicated that number of wells affected habitat selection and nest site selection at one of two sites (Pitman et al. 2005), but this and other work from southwestern Kansas did not document any direct effects to demographic rates as a function of well density or distance.

Power Lines

Power lines serve as a barrier to prairie chicken movement, nest site selection, and general habitat use (Pitman et al. 2005, Pruett et al. 2009a). Lesser and Greater Prairie-Chickens (*Tympanuchus cupido*) avoided crossing power line right-of-ways even though these features frequently occupied their home ranges (Pruett et al. 2009a). However, Lesser Prairie-Chickens did cross power lines during winter to feed in agricultural fields in our study. Collisions with power lines resulted in approximately 5% of known mortalities in our study areas (Hagen et al. 2007b), similar to collision rates reported for Lesser (3%) and Greater Prairie-Chickens (5%) in Oklahoma (Wolfe et al. 2007, Pruett et al. 2009a). General avoidance of

these tall structures may result from prairie chickens experiencing predation attempts or predator presence with these features. Perching on power line structures increases a raptor's range of vision, allowing for greater speed and effectiveness in searching for and acquiring prey in habitats with low vegetation and relatively flat terrain (Ellis 1984, Steenhof et al. 1993). Thus, raptors may preferentially seek out transmission structures in areas where natural perches and nesting sites are limited. Increased abundance of raptors within occupied grouse habitats may result in elevated predation rates above the range of natural variation (Ellis 1984, Coates 2007). However, only 20% of Lesser Prairie-Chicken mortality was associated with raptor predation in our study areas (Hagen 2003), compared to >30% in Oklahoma and New Mexico (Wolfe et al. 2007). Differences in mortality from raptors between these studies suggest that either the observed avoidance distances in our study are adequate to alleviate excessive raptor predation from these structures, or raptor communities differ considerably. We do not have data on raptor populations to substantiate either of these scenarios, except that raptors were most abundant in winter months (Hagen 2003).

The presence of a power line may fragment grouse habitats even if raptors are not present. Our minimum avoidance distances (662 and 726 m) for Lesser Prairie-Chickens are similar to those observed for Greater Prairie-Chicken and Greater Sage-Grouse (Braun 1998, Pruett et al. 2009a). Braun (1998) documented use of suitable habitat within 600 m of a power lines, and habitat use increased as a function of distance from the power line.

A novel finding in our study was a reduction in avoidance distance after construction of a power line on site I, which suggested minimal behavioral avoidance. Our data were limited to one year post-construction and likely were not collected over a long enough period to fully assess long-term effects. Adult Lesser Prairie-Chickens exhibit strong site fidelity to breeding areas and nest sites (Hagen et al. 2005, Pitman et al. 2006), and we should expect a post-disturbance lag effect (Holloran 2005, Walker et al. 2007). The lag effect may be less than has been documented in Greater Sage-Grouse (3–5 years), because of shorter generation times in Lesser Prairie-Chickens (Hagen et al. 2009). However, avoidance of existing power lines (pre-1997) on our study areas may provide future insight to likelihood of use and associated avoidance distances once the population has redistributed its habitat use relative to the new power line.

SUMMARY

Previous work indicated that changes in habitat composition at large scales (7,200 ha) explain most of the variability in declining Lesser Prairie-Chicken populations (Woodward et al. 2001, Fuhlendorf et al. 2002). Increases in habitat edge density can be a significant factor in declining populations at relatively small scales, and such changes may have negative impacts on breeding activity (Fuhlendorf et al. 2002). If increased edge density was correlated to power lines, wells, or buildings (in our study), it may help explain the declining population trends observed on our study sites (Hagen et al. 2009). While direct habitat loss has not occurred on our study areas in 15–20 years, the avoidance of anthropogenic features by prairie chickens may result in the functional elimination of habitat (Robel et al. 2004), potentially further explaining population declines.

Future impact assessments should consider the construction of new anthropogenic features as a potential detriment to habitat suitability for Lesser Prairie-Chickens. We offer the following siting guidelines to protect 90% of breeding and summer habitats of Lesser Prairie-Chickens: power lines ≥700 m, wells ≥300 m, buildings ≥1,400 m, and paved roads ≥850 m. We hypothesize that future development of wind generation facilities likely will have effects on avoidance behavior similar to that of power lines and buildings, and we would recommend >1.4 km setback for wind generation until empirical data are available. Although our setback distances do not account for densities of anthropogenic features, clustering developments such that undisturbed native habitats are maximized would likely benefit populations. Additionally, the distances reported from our study areas are relative to populations that occupy habitats already fragmented by agriculture. Thus, additional measurements should be gathered from across the range to identify tolerance thresholds of feature densities within different patch sizes. Future work should evaluate population viability and the probability of patch occupancy as they relate to the density and configurations of anthropogenic features.

ACKNOWLEDGMENTS

We thank J. O. Cattle Co., Sunflower Electric Corp., Brookover Cattle Co., and P. E. Beach for property access. C. G. Griffin, B. E. Jamison, G. C. Salter, T. G. Shane, T. L. Walker, Jr., and T. J. Whyte assisted with trapping and tracking of prairie chickens. Financial and logistical support was provided by Kansas Department of Wildlife and Parks (Federal Aid in Wildlife restoration projects W-47-R and W-53-R), with special thanks to R. D. Applegate (formerly of KDWP) for securing funding of phase II, Kansas Agricultural Experiment Station (Contribution No. 04–223), and Division of Biology at Kansas State University.

LITERATURE CITED

Agresti, A. 1996. An introduction to categorical data analysis. Wiley & Sons Inc. New York, NY.

Aldridge, C. L. and M. S. Boyce. 2007. Linking occurrence and fitness to persistence: habitat based approach for endangered Greater Sage-Grouse. Ecological Applications 17:508–526.

Arnett, E. B., W. K. Brown, W. P. Erickson, J. K. Fielder, B. L. Hamilton, T. H. Henry, A. Jain, G. D. Johnson, J. Kerns, R. R. Koford, C. P. Nicholson, T. J. O'Connell, M. D. Piorkowski, R. D. Tankersley. 2008. Patterns of bat fatalities at wind energy facilities in North America. Journal of Wildlife Management 72:61–78.

Braun, C. E. 1998. Sage Grouse declines in western North America: what are the problems? Proceedings of the Western Association of State Fish and Wildlife Agencies 78:139–156.

Coates, P. S. 2007. Greater Sage-Grouse (*Centrocercus urophasianus*) nest predation and incubation behavior. Ph.D. dissertation, Idaho State University, Pocatello, ID.

Crawford, J. A. and E. G. Bolen. 1976. Effects of land use on Lesser Prairie-Chickens in Texas. Journal of Wildlife Management 40:96–104.

Doherty, K. E., D. E. Naugle, B. L. Walker, and J. M. Graham. 2008. Greater Sage-Grouse winter habitat selection and energy development. Journal of Wildlife Management 72:187–195.

Ellis, K. L. 1984. Behavior of lekking Sage-Grouse in response to a perched Golden Eagle. Western Birds 15:37–38.

ESRI. 1998. Arc View GIS 3.1, online users manual. Environmental Systems Research Institute, Redlands, CA.

Fuhlendorf, S. D., A. J. Woodward, D. M. Leslie, and J. Shackford. 2002. Multi-scale effects of habitat loss and fragmentation on Lesser Prairie-Chicken populations in US southern Great Plains. Landscape Ecology 17:617–628.

Hagen, C. A. 2003. A demographic analysis of Lesser Prairie-Chicken populations in southwestern Kansas: survival, population viability, and habitat use. Ph.D. dissertation, Kansas State University, Manhattan, KS.

Hagen, C. A., and K. M. Giesen. 2005. Lesser Prairie-Chicken (*Tymapanuchus pallidinctus*). F. Gill and A. Poole (editors), The birds of North America No. 364. The Birds of North America Inc., Philadelphia, PA.

Hagen, C. A., B. E. Jamison, K. M. Giesen, and T. Z. Riley. 2004. Guidelines for managing Lesser Prairie-Chicken populations and their habitats. Wildlife Society Bulletin 32:69–82.

Hagen, C. A., J. C. Pitman, B. K. Sandercock, R. J. Robel, and R. D. Applegate. 2007a. Age-specific survival and probable causes of mortality in female Lesser Prairie-Chickens. Journal of Wildlife Management 71:518–525.

Hagen, C. A., J. C. Pitman, R. J. Robel, T. M. Loughin, and R. D. Applegate. 2007b. Niche partitioning by Lesser Prairie-Chickens *Tympanuchus pallidicinctus* and Ring-necked Pheasants *Phasianus colchicus* in southwestern Kansas. Wildlife Biology 13(Suppl. 1):51–58.

Hagen, C. A., J. C. Pitman, B. K. Sandercock, R. J. Robel, and R. D. Applegate. 2005. Age specific variation in apparent survival rates of male Lesser Prairie-Chickens. Condor 107:78–86.

Hagen, C. A., B. K. Sandercock, J. C. Pitman, R. J. Robel, and R. D. Applegate. 2009. Spatial variation in Lesser Prairie-Chicken demography: a sensitivity analysis of population dynamics and management alternatives. Journal of Wildlife Management 73:1325–1332.

Haukos, D. A., L. M. Smith, and G. S. Broda. 1990. Spring trapping of Lesser Prairie-Chickens. Journal of Field Ornithology 61:20–25.

Holloran, M. J. 2005. Greater Sage-Grouse (*Centrocercus urophasianus*) population response to natural gas field development in western Wyoming. Ph.D. dissertation, University of Wyoming, Laramie, WY.

Holloran, M. J., B. J. Heath, A. G. Lyon, S. J. Slater, J. L. Kuipers, and S. H. Anderson. 2005. Greater Sage-Grouse nesting habitat selection and success in Wyoming. Journal of Wildlife Management. 69:638–649.

Hulett, G. K., J. R. Tomelleri, and C. O. Hampton. 1988. Vegetation and flora of a sandsage prairie site in Finney County, southwestern Kansas. Transactions of the Kansas Academy of Science 91:83–95.

Ingelfinger, F., and S. Anderson. 2004. Passerine response to roads associated with natural gas extraction in a sagebrush steppe habitat. Western North American Naturalist 64:385–395.

Jamison, B. E. 2000. Lesser Prairie-Chicken chick survival, adult survival, and habitat selection and movement of males in fragmented landscapes of southwestern Kansas. M.S. Thesis, Kansas State University, Manhattan, KS.

Jamison, B. E., R. J. Robel, J. S. Pontius, and R. D. Applegate. 2002. Invertebrate biomass: associations with Lesser-Prairie Chicken habitat use and sand sagebrush density in southwestern Kansas. Wildlife Society Bulletin 30:517–526.

Jones, R. E. 1963. Identification and analysis of Lesser and Greater Prairie-Chicken habitat. Journal of Wildlife Management 27:757–778.

Lyon, L. A. and S. H. Anderson. 2003. Potential gas development impacts on Sage Grouse nest initiation and movement. Wildlife Society Bulletin 31:486–491.

Manly, B. F. 1997. Randomization, bootstrap and Monte Carlo methods in biology. Chapman & Hall, London, UK.

Marcum, C. L., and D. O. Loftsgaarden. 1980. A non-mapping technique for studying habitat preferences. Journal of Wildlife Management 44:963–968.

Nams, V. 2000. Locate II: user's guide. Pacer Computer Software, Truro, Nova Scotia, Canada.

Naugle, D. E., K. E. Doherty, B. L. Walker, M. J. Holloran, and H. E. Copeland. 2011. Energy development and Greater Sage-Grouse. Studies in Avian Biology. 38:489–504.

Oyler-McCance, S. J. 1999. Genetic and habitat factors underlying conservation strategies for Gunnison Sage-Grouse. Ph.D. dissertation, Colorado State University, Fort Collins, CO.

Patten, M. A., D. H. Wolfe, E. Shochat, and S. K. Sherrod. 2005a. Effects of microhabitat and microclimate selection on adult survivorship of the Lesser Prairie-Chicken. Journal of Wildlife Management 69:1270–1278.

Patten, M. A., D. H. Wolfe, E. Shochat, and S. K. Sherrod. 2005b. Habitat fragmentation, rapid evolution, and population persistence. Evolutionary Ecology 7:235–249.

Pitman, J. C. 2003. Lesser Prairie-Chicken nest site selection and nest success, juvenile gender determination and growth, and juvenile survival and dispersal in southwestern Kansas. M.S. Thesis, Kansas State University, Manhattan, KS.

Pitman, J. C., C. A. Hagen, R. J. Robel, T. M. Loughin and R. D. Applegate. 2006. Nesting ecology of the Lesser Prairie-Chicken in Kansas. Wilson Journal of Ornithology 118:23–35.

Pitman, J. C., C. A. Hagen, R. J. Robel, T. M. Loughin, and R. D. Applegate. 2005. Location and success of Lesser Prairie-Chicken nests in relation to human disturbance. Journal of Wildlife Management 69:1259–1269.

Pruett, C. L., M. A. Patten, and D. H. Wolfe. 2009a. Avoidance behavior by prairie grouse: implications for wind energy development. Conservation Biology 23:1253–1259.

Pruett, C. L., M. A. Patten, and D. H. Wolfe. 2009b. It's not easy being green: wind energy and a declining grassland bird. Bioscience 59:257–262.

Riley, T. Z., C. A. Davis, M. Oritz, and M. J. Wisdom. 1992. Vegetative characteristics of successful and unsuccessful nests of Lesser Prairie-Chicken. Journal of Wildlife Management 56:383–387.

Robel, R. J., J. A. Harrington, C. A. Hagen, J. C. Pitman, and R. R. Recker. 2004. Effect of energy development and human activity on the use of sand sagebrush habitat by Lesser Prairie-Chickens in southwestern Kansas. Transactions of the North American Natural Resources Conference 69:251–266.

Sawyer, H., R. M. Nielson, F. Lindzey, and L. L. Mcdonald. 2006. Winter habitat selection of mule deer before and during development of a natural gas field. Journal of Wildlife Management 70:396–403.

Steenhof, K., M. N. Kochert, and J. A. Roppe. 1993. Nesting by raptors and Common Ravens on electrical transmission line towers. Journal of Wildlife Management 57:271–281.

U.S. Department of the Interior, Fish and Wildlife Service. 2008. Endangered and threatened wildlife and plants; review of species that are candidates that are proposed for listing as endangered or threatened: the Lesser Prairie-Chicken. Federal Register 73:75719–75180.

Waddell, B. and B. Hanzlick. 1978. The vanishing sandsage prairie. Kansas Fish and Game Magazine 35:1–4.

Walker, B. L., D. E. Naugle, and K. E. Doherty. 2007. Greater Sage-Grouse population response to energy development and habitat loss. Journal of Wildlife Management 71:2644–2654.

Wolfe, D. H., M. A. Patten, E. Shochat, C. L. Pruett, and S. K. Sherrod. 2007. Causes and patterns of mortality in Lesser Prairie-Chickens *Tympanuchus pallidicinctus* and implications for management. Wildlife Biology 13(Suppl. 1):95–104.

Woodward, A. J., S. D. Fuhlendorf, D. M. Leslie, and J. Shackford. 2001. Influence of landscape composition and change on Lesser Prairie-Chicken (*Tympanuchus pallidicinctus*) populations. American Midland Naturalist 145:261–274.

Worton, B. J. 1989. Kernel methods for estimating the utilization distribution in home-range studies. Ecology 70:164–168.

CHAPTER SIX

Landscape Fragmentation and Non-breeding Greater Sage-Grouse

Jay F. Shepherd, Kerry P. Reese, and John W. Connelly

Abstract. This study assessed patterns of habitat use during summer by non-breeding Greater Sage-Grouse (*Centrocercus urophasianus*). From 1999 to 2002 we examined two study areas in southwestern Idaho with different levels of recent wildfire activity, and hence different levels of sagebrush shrubsteppe fragmentation. The Jarbidge study area was 69.5% sagebrush cover (*Artemisia* spp.) at the beginning of the study in 1999 and declined to 63.5% in 2002. The Grasmere study area declined from 84.9% to 83.3% sagebrush cover from 1999 to 2002. The combination of existing grasslands and recent wildfires (open grass–forb dominated areas with varying amounts of bare ground) was 31.9% and 7.6% for the Jarbidge and Grasmere study areas, respectively, at the end of the study in 2002. In both study areas, we analyzed landscape metrics within circular buffers of 150- and 450-m radii centered on each Greater Sage-Grouse location as well as the same number of random locations; each extent was analyzed separately. The amount of sagebrush land cover was not significantly different at Greater Sage-Grouse locations (use points) and available habitat (random points) at either analysis extent, regardless of study area. At the 150-m analysis extent, Greater Sage-Grouse in the Jarbidge study area used areas with higher edge density of grass–forb dominated land cover than was randomly available and than was used or available in the Grasmere study area. In the Jarbidge study area, sagebrush patches adjacent to large, abrupt patches of grass–forb dominated habitat were effectively smaller (extent actually used is smaller than areal extent) than patches with more interspersed habitats on their perimeter, since the patches with abrupt edges receive less use on their periphery. At the 450-m extent, Greater Sage-Grouse used areas similar to those randomly available with respect to all landscape metrics in both study areas.

Key Words: Centrocercus urophasianus, fragmentation, Greater Sage-Grouse, landscape metrics, landscape-scale, sagebrush cover, wildfire.

Shepherd, J. F., K. P. Reese, and J. W. Connelly. 2011. Landscape fragmentation and non-breeding Greater Sage-Grouse. Pp. 77–88 *in* B. K. Sandercock, K. Martin, and G. Segelbacher (editors). Ecology, conservation, and management of grouse. Studies in Avian Biology (no. 39), University of California Press, Berkeley, CA.

reater Sage-Grouse (*Centrocercus uropha-sianus*), a shrubsteppe obligate species, evolved within an environment subject to disturbances that occasionally revert sagebrush communities to early successional stages. Wildfire is the primary natural disturbance event determining the spatial and temporal distribution of successional stages, as well as the plant community composition, in the sagebrush shrubsteppe ecosystem (Harniss and Murray 1973, Young and Evans 1978, Humphrey 1984, Akinsoji 1988, Whisenant 1990). Prior to European settlement, Greater Sage-Grouse habitat was characterized by large areas of sagebrush cover with relatively isolated areas in earlier seral stages due to wildfire. Currently, sagebrush shrubsteppe communities can be subject to a more frequent disturbance regime than existed before European settlement, consisting of a mosaic of patches of various cover types, sizes, and successional stages. Plant community composition and the relative frequency, and more recently, the relative permanency, of disturbance has been altered by influences such as exotic invasive plants, livestock production, agricultural conversion, increased human-caused wildfire ignitions, and various types of human settlement (Mack 1981, Whisenant 1990, Saab et al. 1995, West 1996, Knick and Rotenberry 1997, Leonard et al. 2000, Wisdom et al. 2002, Connelly et al. 2004, Baker 2006).

Given that Greater Sage-Grouse are mature sagebrush shrubsteppe obligate species, the clear implication is that anthropogenic causes of disturbance promoting loss and fragmentation of sagebrush plant communities will negatively affect Greater Sage-Grouse. However, the general effects of sagebrush shrubsteppe fragmentation on Greater Sage-Grouse have been implicated as a cause in range-wide declines (Connelly et al. 2004), yet these have not been studied and are not well understood. Given the scope and urgency of the problem, response of sage grouse to the increasing magnitude, frequency, and permanency of habitat fragmentation should be investigated at intermediate spatial scales, such as an extent reflecting daily habitat use or that includes adjacent landscape patterns that may promote predation. Our purpose is to provide a more complete understanding of the effects of sagebrush shrubsteppe fragmentation on Greater Sage-Grouse. We hypothesized that (1) Greater Sage-Grouse would use habitat in similar proportion as available in relatively unfragmented sagebrush shrubsteppe; (2) Greater Sage-Grouse would modify habitat use, and therefore use habitat differently than available in fragmented sagebrush shrubsteppe; and (3) habitat use represented by amount of different landcover types, such as sagebrush or grasslands, and other landscape metrics may therefore be similar in areas that have different levels of sagebrush shrubsteppe fragmentation.

METHODS

Study Areas

The Jarbidge study area extends from the Idaho–Nevada border north to the Snake River plain, and east of the Bruneau River canyon to agricultural lands near State Route 93, south of Twin Falls (Fig. 6.1). The Jarbidge study area is approximately 1,500 km^2 and ranges from approximately 1,200 to 1,800 m in elevation.

The Grasmere study area is approximately 56 km east of the Idaho–Oregon border east to the Bruneau River canyon, and from the Idaho–Nevada border north 64 km to the breaks of the Snake River plain (Fig. 6.1). The study area is approximately 3,000 km^2 and ranges from approximately 1,300 to 2,200 m in elevation. State Route 51 bisects the Grasmere study area. The Bruneau and Jarbidge River canyons which separate the study areas are approximately 1–2 km wide and several hundred meters deep, representing potential barriers to Greater Sage-Grouse movement.

Both study areas are dominated by varieties of the big sagebrush complex (*A. tridentata*). At lower elevations, basin big sagebrush (*A. t. tridentata*) or Wyoming big sagebrush (*A. t. wyomingensis*) are the dominant shrubs, with mountain big sagebrush (*A. t. vaseyana*) more prevalent at higher elevations. On foothill slopes in the southern portion of both study areas, low sagebrush (*A. arbuscula*) dominates the vegetation community where the soil is more shallow and rocky (Hironaka et al. 1983). The understory is primarily native perennial grasses and forbs, with some encroachment of exotic annual grasses, primarily cheatgrass (*Bromus tectorum*). Some areas disturbed by wildfire are entirely without sagebrush and are dominated by non-native grasses such as cheatgrass or crested wheatgrass (*Agropyron cristatum*). Other non-sagebrush areas are agricultural and generally

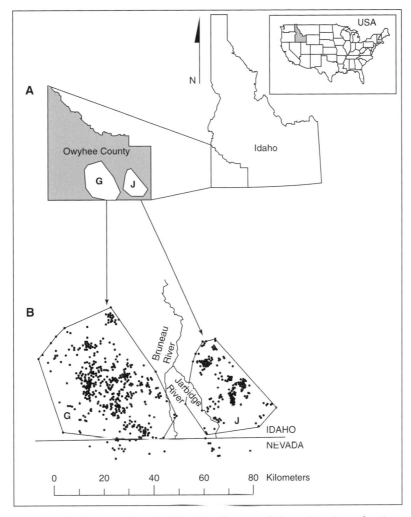

Figure 6.1. The Grasmere (G) and Jarbidge (J) study areas and Greater Sage-Grouse locations in relation to Owyhee County (A) and the Bruneau and Jarbidge River canyons (B) in southwestern Idaho, USA, 1 March–1 October 1999–2002 (scale bar in reference to study areas with Greater Sage-Grouse locations).

used to grow alfalfa (*Medicago sativa*). Both study areas are approximately 3–5% privately owned, 5–10% state-owned, with the remainder of the land managed by the Bureau of Land Management (BLM), and therefore primarily managed for multiple uses including sustainable wildlife populations.

Precipitation and temperature in both study areas are similar and the areas can be viewed as adjacent in terms of broad-scale processes such as weather. Average yearly precipitation is approximately 24 cm, with high temperatures in January ranging between 0 and 5°C and high temperatures in August between 35 and 40°C (State Climate Services, Biological and Agricultural

Engineering Department, University of Idaho). The average minimum temperature for January in the Jarbidge study area is −10.9°C from 1971 to 2000 (Three Creek Site, Western Regional Climate Center, National Oceanographic and Atmospheric Administration).

Sage Grouse Capture and Monitoring

Field crews captured Greater Sage-Grouse in the early spring during the mating display period when birds were on or near leks. We used a spotlighting technique, modified by the use of binoculars, to capture Greater Sage-Grouse nocturnally roosting near leks (Giesen et al. 1982, Wakkinen et al. 1992). We marked Greater Sage-Grouse with an aluminum leg band and then equipped each with a necklace-mounted 15-g battery-powered transmitter with a 4-hour mortality sensor. Locations were obtained using ATS receivers (Advanced Telemetry Systems, Isanti, MN), hand-held Yagi and vehicle-mounted antennas, and recorded with GPS units. We classified all grouse by age and gender and released each bird at the point of capture (Crunden 1963, Dalke et al. 1963). We monitored grouse using receivers from Advanced Telemetry Systems, Isanti, MN, and handheld Yagi and vehicle-mounted antennas, determined locations with Global Positioning System (GPS; Garmin) units, and recorded locations in Universal Transverse Mercator (UTM) coordinates (zone 11).

Sample Sizes and Locations

Field crews located each Greater Sage-Grouse ≥3 times per week, with each successive location recorded in an alternating period of the day to obtain observations during potentially different activity patterns. For individual Greater Sage-Grouse, locations were separated by >24 hours. Sage grouse locations obtained in both study areas were precise ground (non-flight) radiotelemetry locations of males and females.

Sage grouse locations were obtained from 1 June to 30 September 1999–2002, the summer habitat period. Females on nests or with broods were not used in this analysis because habitat used for reproduction often differs from nonbreeding habitat and therefore should be analyzed separately (Klebenow 1969, Dunn and Braun 1986, Klott and Lindzey 1990, Gregg et al. 1994, Sveum et al. 1998). Locations before 1 June were not used because the Grasmere population was a migratory population and used different seasonal habitats: Males and females moved an average of 23.9 and 17.0 km from spring to summer habitat, respectively (Wik 2002). Differential use of seasonal habitats in the Jarbidge study area was unknown.

Remotely Sensed Data

Information used to obtain patch cover type, configuration, and size was obtained from Landsat Thematic Mapper Imagery data classified by the 1998 Idaho GAP Analysis Project (Scott et al. 2002). The Landsat imagery was from spring/ summer satellite passes (June–September), when vegetation is in an active growth stage. Due to cloud cover, data from several years was used to produce a statewide data layer. Idaho GAP Analysis metadata indicated satellite passes occurred from 1991 to 1996, and the Landsat image was then updated with the appropriate annual BLM fire information to create a landcover layer for each year of the Greater Sage-Grouse study from 1999 to 2002 (Scott et al. 2002). For example, to analyze Greater Sage-Grouse location data collected in 2001, we combined the original 1996 landcover layer and fire data for 1996–2000 to create a landcover map that had the most recent wildfire extent information, reflecting the dynamic nature of the landscape. There were 81 landcover classification categories (Shepherd 2006) in the original landcover map and seven new classification categories for wildfire by year (for fires during 1996–2002). The pixel size was 30 m, or 0.09 ha, and the accuracy of the landcover classification for the 1998 Idaho Gap Analysis was estimated to be 69.3% with a range of 63.6% to 79.3% for nine different subsections of southern Idaho.

We reclassified the cover type map and aggregated the 81 land cover classifications into 12 categories. For example, the reclassification process combined seven sagebrush cover types from the original Idaho GAP Analysis cover type map into one composite sagebrush cover category, as well as all grassland types into one composite grassland category. The composite grassland category was then combined with wildfire-affected areas representing open areas with few if any shrubs (referred to as grass–forb dominated cover from this point forward). Reclassification should increase accuracy of the cover type map and applicability of the results to other areas, as well as be better suited for addressing the basic sagebrush shrubsteppe fragmentation issue and the dynamic nature of the landscape. Reclassification should also increase the statistical validity of the sample by increasing the number of non-zero data points: By aggregating cover types, more buffers of Greater Sage-Grouse locations have values for grassland–forb cover, sagebrush, or other simplified categories as opposed to data extracted from highly categorized cover type maps, which may produce more entries of zero.

In both study areas, we analyzed landscape metrics within circular buffers of 150- and 450-m radii centered on each Greater Sage-Grouse location and the same number of random locations;

each extent was analyzed separately. The 150-m buffer extent was the maximum radiotelemetry error found (mapped using triangulation followed by prompt flushing of radio-collared Greater Sage-Grouse), and therefore represents the highest degree of accuracy and the smallest usable buffer extent. The 450-m buffer extent encompassed the average daily distance moved for male and female Greater Sage-Grouse in both study areas, and therefore represents habitat available per day (unpublished data). We used an ArcInfo Macro Language (AML) routine for ArcInfo GIS (ESRI 1999) to extract landcover data for each buffer extent from the appropriate yearly landcover map, depending on the date a Greater Sage-Grouse location was obtained. Therefore, two raster data files were produced for each location, one for each analysis extent. If two vegetation buffers of the same spatial extent spatially overlapped, we randomly removed one of the buffers from the analysis. This procedure eliminated the overlap of vegetation sampling, and hence reduced spatial correlation of the data.

To define study areas, we derived a 100% minimum convex polygon (MCP) of the Idaho locations for each study area (Fig. 6.1B) using the Animal Movement extension for ArcView (ver. 2.04; Hooge and Eichenlaub 2000). Radio-marked Greater Sage-Grouse also used areas south of the Idaho–Nevada border, with 18 Nevada locations of Greater Sage-Grouse trapped in the Jarbidge study area and 17 in the Grasmere study area; these locations were excluded from analyses. Within each MCP, we derived the same number of random points as Greater Sage-Grouse locations for both study areas, and for each random point we created nonoverlapping 150- and 450-m buffers completely within the boundaries of the study areas.

The vegetation cover type buffers for random points were extracted from the yearly cover type maps in the same proportion as the yearly distribution of Greater Sage-Grouse locations.

Data Analysis

We used FRAGSTATS 3.3 to derive a suite of landscape metrics for each vegetation buffer (McGarigal et al. 2001). For Greater Sage-Grouse locations, we obtained a mean for each landscape metric by individual grouse. We paired each random point with a Greater Sage-Grouse location and aggregated random data in a similar manner to obtain an equal sample size. We calculated mean daily distance moved for individual Greater Sage-Grouse. The distance between locations was divided by the numbers of intervening days and averaged for each Greater Sage-Grouse.

We used multivariate analysis of variance (MANOVA) to assess differences in landscape variables between study areas, plot types (random and location), and sex. If MANOVA results were significant, ANOVA was used to examine univariate effects of various interaction levels. The distribution and normal probability plots of the data were not significantly improved by transformation, and the data were not transformed (Zar 1999). We used the statistical software SYSTAT (SPSS, Inc. 2000) and SPSS (SPSS, Inc. 1993) for data configuration and analysis.

RESULTS

Six hundred thirty-eight locations of non-breeding Greater Sage-Grouse were obtained in the Grasmere and Jarbidge study areas from 1 June to 30 September 1999–2002 (Table 6.1). There were

TABLE 6.1

Summary of Greater Sage-Grouse location data (male, female) by study area, Owyhee County: southwestern Idaho, 1999–2002.

Data type	Jarbidge	Grasmere
Non-breeding points	228 (181, 47)	410 (147, 263)
Points after reduced spatial correlation, 150-m scale	127 (85, 42)	267 (107, 160)
Individual Greater Sage-Grouse, 150-m scale	37 (27, 10)	111 (47, 64)
Points after reduced spatial correlation, 450-m scale	69 (43, 26)	195 (87, 108)
Individual Greater Sage-Grouse, 450-m scale	32 (22, 10)	94 (43, 51)

TABLE 6.2

Cumulative percent landcover by year and study area using reclassified Idaho GAP vegetation cover map:
Owyhee County, southwestern Idaho, 1999 and 2002.

Cover Type	Grasmere		Jarbidge	
	1999	2002	1999	2002
Sagebrush	84.9	83.3	69.5	63.5
Grassland	6.0	5.9	16.1	15.9
Other shrubs	5.8	5.8	1.2	1.0
Agriculture	1.4	1.4	1.5	1.5
Recently burned area (1996–2002)	0	1.7	9.2	16.0
Rabbitbrush	0.1	0.1	2.0	1.7
Water/riparian areas	1.3	1.3	0.3	0.3
Deciduous/coniferous Trees	0.4	0.4	0	0
Barren ground	0	0	0.1	0.1

394 locations available for analysis after reduction of possible spatial correlation at the 150-m extent and 264 at the 450-m extent (Table 6.1). Therefore, use of nonoverlapping buffers for each buffer extent caused a reduction in data used by 38% and 59% for 150-m and 450-m buffers, respectively.

Comparison of Study Areas

As defined by 100% minimum convex polygons of all Greater Sage-Grouse locations within each study area, the Grasmere study area was larger (287,070 ha) than the Jarbidge study area (125,526 ha), and the areas also differed in landcover proportions throughout the study period. The Grasmere study area had 15.4% and 19.8% more sagebrush landcover than the Jarbidge study area at the beginning (1999) and end (2002) of the study period, respectively (Table 6.2). During the relatively short period of this study, sagebrush cover declined by 6% within the Jarbidge study area MCP compared to 1.6% for the Grasmere study area MCP. Approximately 60% of the Jarbidge study area MCP burned from 1970 to 1995, with approximately 20% burning in each decade, in comparison to only 4.4% for the Grasmere MCP during the same time period.

Percent landcover of shrubs other than sagebrush was almost 5% higher in the Grasmere study area than in the Jarbidge study area, while percent landcover of agriculture was similar for the two areas throughout the study period (Table 6.2). Over 90% and 95% of the Grasmere and Jarbidge study areas, respectively, were composed of sagebrush or grass–forb dominated land cover.

Sage Grouse Locations and Random Points: 150- and 450-m Buffer Extents

Landscape metrics were similar for males and females at the 150- and 450-m buffer extents ($F = 0.46$, $P = 0.90$ and $F = 0.07$, $P = 1.00$, respectively). The interaction of gender by area and plot type also did not differ at the 150-m and 450-m buffer extents ($F = 0.57$, $P = 0.82$ and $F = 1.01$, $P = 0.43$, respectively).

For both analysis extents, study area was a significant MANOVA effect ($F = 11.87$, $P < 0.01$ and $F = 10.90$, $P < 0.01$, respectively). At 150- and 450-m buffer extents, percent land cover of sagebrush (PLSB) and grass–forb dominated cover (PLGF), edge density (m/ha) of sagebrush cover (EDSB) and edge density of grass-forb dominated cover (EDGF), and number of grass–forb dominated patches (NPGF) differed between study areas for both plot types (buffers of random and use points) combined ($P < 0.01$). At the 450-m extent, number of sagebrush patches (NPSB) did not differ

TABLE 6.3

*Means (SE) for landscape metrics of 150-m random and Greater Sage-Grouse location buffers for the Grasmere
and Jarbidge study areas: Owyhee County, southwestern Idaho, 1 June–30 September 1999–2002.*

Landscape metric	Grasmere random points	Grasmere use points	Jarbidge random points	Jarbidge use points
PLSB[a]	82.0 (2.6)	82.5 (2.2)	56.6 (5.0)	69.8 (4.3)
PLGF[a,b,c]	6.2 (1.5)	4.4 (0.8)	39.3 (5.0)	24.5 (4.0)
PLAG	2.1 (1.0)	6.4 (2.0)	1.3 (1.0)	0.8 (0.8)
NPSB[b,c]	1.1 (0.0)	1.1 (0.0)	0.9 (0.1)	1.2 (0.1)
NPGF[a]	0.2 (0.1)	0.3 (0.1)	0.7 (0.1)	0.7 (0.1)
NPAG	0.0 (0.0)	0.1 (0.0)	0.0 (0.0)	0.0 (0.0)
EDSB[a,c]	21.0 (3.1)	27.4 (2.8)	30.5 (4.5)	45.7 (5.5)
EDGF[a,b,d]	10.7 (2.2)	12.6 (2.1)	24.6 (4.0)	38.6 (5.8)
EDAG	0.1 (0.1)	1.9 (0.7)	0.1 (0.1)	0.4 (0.4)

NOTE: Landscape metrics include Percent Landcover (PL), Number of Patches (NP), and Edge Density (ED; m/ha) for the cover types Sagebrush (SB), Grass–Forb Dominated Areas (GF), and Agricultural Areas (AG).

[a] Significantly different by area, $P \le 0.01$.

[b] Significantly different by plot type, $P \le 0.01$.

[c] Significantly different by area–plot type interaction, $P \le 0.01$.

[d] Significantly different by area–plot type interaction, $P \le 0.05$.

between the study areas despite plot type ($F = 0.01$, $P = 0.90$). For both analysis extents, percent land cover of agricultural areas (PLAG), edge density of agricultural areas (EDAG), and number of agricultural patches (NPAG) did not differ between study areas despite plot type. At the 150- and 450-m buffer extents, PLSB was higher in the Grasmere study area regardless of plot type, while PLGF, NPGF, EDSB, and EDGF were all higher in the Jarbidge study area despite plot type. At the 450-m extent, the NPSB was higher in the Jarbidge study area regardless of plot type.

For both analysis extents, the effect of plot type (use locations versus random regardless of study area) was significant ($F = 4.38$, $P < 0.01$ and $F = 1.93$, $P = 0.05$, respectively). At the 150-m extent, EDGF, EDSB, and NPSB were higher at Greater Sage-Grouse locations ($P < 0.05$) and PLGF was higher at random locations ($P < 0.01$). At the 450-m extent, NPSB and NPGF were higher at Greater Sage-Grouse locations than at random locations ($P = 0.05$ and $P = 0.02$, respectively). At both buffer extents, the landscape metrics PLSB, PLAG, EDAG, and NPAG did not differ between plot types despite gender and study area.

The interaction between plot type and study area was significant at the 150-m extent ($F = 4.17$, $P < 0.01$) and moderately significant at the 450-m extent ($F = 1.86$, $P = 0.06$). At the 150-m extent, PLGF, EDGF, and NPSB differed for the plot type by study area interaction ($F = 6.67$, $P = 0.01$; $F = 5.58$, $P = 0.02$; and $F = 8.48$, $P < 0.01$, respectively). At the 450-m extent, NPGF differed for plot type by study area interaction ($F = 4.08$, $P = 0.04$). At the 150-m extent in the Jarbidge study area, EDGF and NPSB were higher, and PLGF was lower, at Greater Sage-Grouse locations than at random points (Table 6.3). In the Grasmere study area, these landscape metrics were similar between Greater Sage-Grouse locations and random points (Table 6.3). At the 450-m extent, NPGF was higher at Greater Sage-Grouse locations than at random points in the Jarbidge study area and similar between plot types in the Grasmere study area (Table 6.4).

In general, at the 150-m extent, Greater Sage-Grouse used less open habitat with more patches and edge density than was randomly available in the Jarbidge study area. However, Greater Sage-Grouse in the Jarbidge study area used habitat with proportions of the sagebrush land cover

TABLE 6.4

Means (SE) for landscape metrics of 450-m random and Greater Sage-Grouse location buffers for the Grasmere and Jarbidge
study areas: Owyhee County, southwestern Idaho, 1 June–30 September 1999–2002.

Landscape metric	Grasmere random points	Grasmere use points	Jarbidge random points	Jarbidge use points
PLSB[a]	85.5 (1.9)	81.7 (2.0)	65.1 (5.0)	69.1 (4.5)
PLGF[a]	5.7 (0.9)	5.6 (0.8)	30.7 (5.0)	25.3 (4.4)
PLAG	2.3 (1.3)	5.1 (1.8)	0.9 (0.9)	1.0 (1.0)
NPSB[a,b]	1.4 (0.1)	1.6 (0.1)	1.9 (0.2)	2.1 (0.2)
NPGF[a,b,c]	1.0 (0.1)	1.1 (0.1)	1.6 (0.2)	2.3 (0.3)
NPAG	0.1 (0.0)	0.1 (0.0)	0.0 (0.0)	0.0 (0.0)
EDSB[a]	25.7 (2.6)	32.2 (2.6)	45.6 (5.4)	45.8 (4.9)
EDGF[a]	15.2 (2.1)	16.3 (2.1)	34.5 (5.6)	37.2 (4.8)
EDAG	0.6 (0.3)	1.6 (0.5)	0.7 (0.6)	0.7 (0.7)

NOTE: Landscape metrics include Percent Landcover (PL), Number of Patches (NP), and Edge Density (ED; m/ha) for the cover types
Sagebrush (SB), Grass–Forb Dominated Areas (GF), and Agricultural Areas (AG).

[a] Significantly different by area, $P \leq 0.01$.
[b] Significantly different by plot type, $P \leq 0.05$.
[c] Significantly different by area–plot type interaction, $P \leq 0.05$.

similar to that randomly available. In the Grasmere study area, Greater Sage-Grouse used areas
more similar to those randomly available with
respect to all landscape metrics.

DISCUSSION

The types of plant communities occurring across
the landscape generally characterize the amount
of habitat available to particular wildlife species.
The relative proportions and interspersion levels
of landcover types within our study areas indicate a distinct difference in available sagebrush
habitat between study areas. The implications
for Greater Sage-Grouse in each study area are
distinct as well, as different cover type and interspersion levels are available, which may affect
productivity and survival and have eventual consequences for population size. The difference in
available mature sagebrush habitat, or at least
areas dominated by sagebrush cover, occurring
within the study areas was reflected by both cover
type levels within MCPs of the study areas and
buffered random points at both analysis extents.
Sagebrush obligate species such as sage grouse
will continue to find proportionally less suitable

sagebrush habitat in the Jarbidge study area over
time in comparison to the more consistent and
homogeneous habitat within the Grasmere study
area. The ongoing decline of sagebrush cover in
the Jarbidge study area is comparable with other
findings concerning the decline of sagebrush
shrubsteppe communities (Leonard et al. 2000,
Connelly et al. 2004).

We hypothesized that sagebrush cover at
Greater Sage-Grouse locations in the Jarbidge
study area would be similar to that found at
Greater Sage-Grouse locations in the Grasmere
study area (80–83%). Greater Sage-Grouse in the
Jarbidge study area either could not move to areas
where the percent cover of sagebrush in their
immediate vicinity was similar to levels used by
grouse within the Grasmere study area due to the
degraded habitat condition, or their minimum life
requisites were being met without doing so, or
perhaps both of these situations were occurring.
This was not a sampling artifact, as trapped leks
had 71.6% sagebrush cover within a 450-m radius
in comparison to 59.6% for unsampled leks and
63.9% for all leks in the Jarbidge study area. Also
by 1 June, Greater Sage-Grouse can move from
the vicinity of leks to better habitat. This implies

that within areas of lower sagebrush landcover at the landscape scale, such as the 60–70% range in the Jarbidge study area, Greater Sage-Grouse do not need or have a limited ability to modify habitat use that increases sagebrush cover in their immediate surroundings relative to the general landscape.

In both study areas, Greater Sage-Grouse used the sagebrush cover type in similar proportion to the amount randomly available at both 150- and 450-m buffer extents. Although proportion of sagebrush was not statistically different between used and random locations at either analysis extent in the Jarbidge study area, at the 150-m extent sagebrush cover used appeared to be greater than randomly available (69.8 vs. 56.5%, respectively). However, random locations also had lower sagebrush landcover than the total amount of sagebrush cover found in the 100% MCPs for each year of the study, including the lowest proportion observed (63.5%, at the end of the study period in 2002). Greater Sage-Grouse locations were more similar, with respect to sagebrush cover, to the 100% MCP representing the highest proportion of sagebrush cover at the beginning of the study in 1999, 69.5%. Therefore, similar amounts of sagebrush cover at Greater Sage-Grouse locations and proportions within the MCPs tend to support the statistical evidence of similarity between used and randomly available sagebrush habitat, despite apparent numerical differences. Our general findings suggest non-breeding Greater Sage-Grouse used the same proportion of sagebrush landcover as was available within the delineated study areas. Studies indicate other avian species that use sagebrush shrubsteppe communities, such as Golden Eagles (*Aquila chrysaetos*), have also adapted so that the proportion of sagebrush cover used reflects the available sagebrush cover within areas that have become increasingly fragmented (Marzluff et al. 1997). In our study, Greater Sage-Grouse adapted to use sagebrush landcover levels that represented the areas they inhabited, whether similar to pre–European settlement Greater Sage-Grouse habitat conditions such as the Grasmere study area, or to the fragmented sagebrush shrubsteppe areas recently affected by anthropogenic influences such as the Jarbidge study area.

The Jarbidge study area had less sagebrush and more grass–forb dominated cover across the land-scape, and Greater Sage-Grouse used the more fragmented landscape within the Jarbidge study area differently than was randomly available. In contrast, overall habitat use by the Grasmere population was similar to randomly available habitat. The primary differences in habitat use between study areas were higher edge density and less landcover of grass–forb dominated areas at use versus randomly available points in the Jarbidge study at the 150-m analysis extent; this did not occur in the Grasmere study area. This suggests that Greater Sage-Grouse in the Jarbidge study area were modifying habitat use by avoiding open grass–forb areas and selecting more interspersed habitat than was randomly available, and this habitat shift was not necessary, or available, in the Grasmere study area. A major difference in habitat use relative to availability was the level of cover type interspersion, and not the proportion of sagebrush land cover. A large proportion of the landscape with 60–70% sagebrush land cover may be used if it is well interspersed with the grass–forb dominated cover type. Areas with abrupt, straight edges adjacent to large, distinct grass–forb dominated patches appear to have been used less; for example, sagebrush patches adjacent to recent wildfires will thus have less use on their periphery. If Greater Sage-Grouse do not use less interspersed areas equally as well-interspersed areas, given similar overall cover type proportions, the effective patch size of sagebrush cover in disturbed and degraded sagebrush shrubsteppe with low interspersion may be smaller than the actual patch size in terms of Greater Sage-Grouse habitat. There may be less Greater Sage-Grouse habitat in an area depending on interspersion and juxtaposition of habitat than may be apparent when only measuring the areal extent of sagebrush landcover. Therefore, interspersion levels are also indicative of the relative amount of potential use within an area, and proportion of sagebrush cover in an area may not be the singular or primary landscape-scale factor determining use.

There are numerous management implications from our results. Use of sagebrush and other cover types was variable and differed by study area. Relative proportions of different shrubsteppe seral stages may continue a process of long-term change, particularly in more disturbed areas. These trends may be effectively monitored annually, or at longer intervals given management needs, with remotely sensed vegetation data updated with wildfire information.

Areas of mature sagebrush shrubsteppe landcover may be less effective habitat if the interspersion level of grass–forb dominated cover is low. Measuring the direct areal extent of sagebrush landcover may overestimate available habitat in areas with low interspersion. The extent of analysis may have an effect on determining suitable levels of interspersion that will still allow Greater Sage-Grouse use: Interspersion levels at a smaller extent may be more or less suitable than the same habitat interspersion at larger extents.

Our data were collected from early summer through early fall, a period in which Greater Sage-Grouse use relatively open areas more than during winter or spring, when sagebrush cover provides a higher percentage of the diet and is used for cover and thermoregulation (Eng and Schladweiler 1972, Wallestad et al. 1975, Beck 1977, Remington and Braun 1985, Connelly et al. 2000). Our findings should not be applied to habitats used by grouse during the breeding season or winter. If Greater Sage-Grouse use areas of fragmented sagebrush shrubsteppe differently than unfragmented shrubsteppe during summer, a period of reduced habitat specificity, there may be stronger effects of habitat fragmentation during periods of greater dependence on sagebrush cover, such as winter.

Knowledge of demographic parameters within areas of differing levels of fragmentation is necessary. If a population is stable, it may be irrelevant that Greater Sage-Grouse do not modify habitat use, even if capable of doing so. However, a stable Jarbidge population does not necessarily imply healthy population levels: The current population level may be stable yet suppressed due to lower amounts of sagebrush habitat, or acting as a sink population sustained by adjacent populations in more suitable habitat (Pulliam 1988, Pulliam and Danielson 1991). Without knowing population densities as well as fecundity, mortality, and movement rates within the Jarbidge and Grasmere study areas, the effect of habitat fragmentation and associated behavioral responses on demographic parameters remains speculative. Population trends are unknown in both study areas.

Our primary conclusion is that during summer and early fall, Greater Sage-Grouse response to fragmentation and habitat reduction can be measured by interspersion, possibly by an interspersion and juxtaposition index, and not necessarily by amount of sagebrush cover. Therefore, when quantifying Greater Sage-Grouse habitat, measuring the amount of sagebrush cover without considering interspersion levels can lead to erroneous conclusions concerning the amount of habitat available. When mapping Greater Sage-Grouse habitat at the landscape level, understanding how Greater Sage-Grouse respond to a variety of landscape metrics is crucial for an accurate and realistic assessment of Greater Sage-Grouse habitat.

ACKNOWLEDGMENTS

We thank Michelle Commons-Kemner, Dave Musil, Randy Smith, and Ron Klimes of Idaho Department of Fish and Game for their field, logistical, and data assistance. We also thank Oz Garton, Steve Bunting, and Eva Strand of the College of Natural Resources, University of Idaho for their technical and data assistance. This is contribution number 1041 of the University of Idaho Forest, Wildlife and Range Experiment Station and Idaho Federal Aid in Wildlife Restoration Project W-160-R.

LITERATURE CITED

Akinsoji, A. 1988. Postfire vegetation dynamics in a sagebrush steppe in southeastern Idaho, USA. Vegetatio 78:151–155.

Baker, W. L. 2006. Fire and restoration of sagebrush ecosystems. Wildlife Society Bulletin 34:177–185.

Beck, T. D. I. 1977. Sage grouse flock characteristics and habitat selection during winter. Journal of Wildlife Management 41:18–26.

Connelly, J. W., S. T. Knick, M. A. Schroeder, and S. J. Stiver. 2004. Conservation assessment of Greater Sage-Grouse and sagebrush habitats. Unpublished Report. Western Association of Fish and Wildlife Agencies, Cheyenne, WY.

Connelly, J. W., K. P. Reese, R. A. Fischer, and W. L. Wakkinen. 2000. Response of a sage grouse breeding population to fire in southeastern Idaho. Wildlife Society Bulletin 28:90–96.

Crunden, C. W. 1963. Age and sex of sage grouse from wings. Journal of Wildlife Management 27:846–850.

Dalke, P. D., D. B. Pyrah, D. C. Stanton, J. E. Crawford, and E. F. Schlatterer. 1963. Ecology, productivity and management of sage grouse in Idaho. Journal of Wildlife Management 27:810–841.

Dunn, P. O., and C. E. Braun. 1986. Late summer-spring movements of juvenile sage grouse. Wilson Bulletin 98:83–92.

Eng, R. L., and P. Schladweiler. 1972. Sage grouse winter movements and habitat use in central Montana. Journal of Wildlife Management 36:141–146.

ESRI (Environmental Systems Research Institute, Inc.). 1999. ARC/INFO Version 8.01. Redlands, CA.

Giesen, K. M., T. J. Schoenberg, and C. E. Braun. 1982. Methods for trapping sage grouse in Colorado. Wildlife Society Bulletin 10:224–231.

Gregg, M. A., J. A. Crawford, M. S. Drut, and A. K. DeLong. 1994. Vegetational cover and predation of sage grouse nests in Oregon. Journal of Wildlife Management 58:162–166.

Harniss, R. O., and R. B. Murray. 1973. 30 years of vegetal change following burning of sagebrush-grass range. Journal of Range Management 26:322–325.

Hironaka, M., Fosberg, M. A., and A. H. Winward. 1983. Sagebrush-grass habitat types of southern Idaho. Bulletin 35. College of Forestry, Wildlife, and Range Sciences, Moscow, ID.

Hooge, P. N., and B. Eichenlaub. 2000. Animal movement extension to Arcview. Ver. 2.0. Alaska Science Center, Biological Science Office, U.S. Geological Survey, Anchorage, AK.

Humphrey, L. D. 1984. Patterns and mechanisms of plant succession after fire on *Artemisia*-grass sites in southeastern Idaho. Vegetatio 57:91–101.

Klebenow, D. A. 1969. Sage grouse nesting and brood habitat in Idaho. Journal of Wildlife Management 33:649–661.

Klott, J. H., and F. G. Lindzey. 1990. Brood habitats of sympatric sage and Sharp-tailed Grouse in Wyoming. Journal of Wildlife Management 54:84–88.

Knick, S. T., and J. T. Rotenberry. 1997. Landscape characteristics of disturbed shrubsteppe habitats in southwestern Idaho (U.S.A.). Landscape Ecology 12:287–297.

Leonard, K. M., K. P. Reese, and J. W. Connelly. 2000. Distribution, movements and habitats of sage grouse *Centrocercus urophasianus* on the upper Snake River plain of Idaho: changes from 1950s to the 1990s. Wildlife Biology 6:265–270.

Mack, R. N. 1981. Invasion of *Bromus tectorum* L. into western North America: an ecological chronicle. Agro-ecosystems 7:145–165.

Marzluff, J. M., S. T. Knick, M. S. Vekasy, L. S. Schueck, and T. J. Zarriello. 1997. Spatial use and habitat selection of Golden Eagles in southwestern Idaho. Auk 114:673–687.

McGarigal, K., B. J. Marks, C. Holmes, and E. Ene. 2001. FRAGSTATS: spatial pattern analysis program for quantifying landscape structure, version 3.3.

Pulliam, H. R. 1988. Sources, sinks, and population regulation. American Naturalist 132:652–661.

Pulliam, H. R., and B. J. Danielson. 1991. Sources, sinks, and habitat selection: a landscape perspective on population dynamics. American Naturalist 137:S50–S66.

Remington, T. E., and C. E. Braun. 1985. Sage grouse food selection in winter, North Park Colorado. Journal of Wildlife Management 49:1055–1061.

Saab, V. A., C. E. Bock, T. D. Rich, and D. S. Dobkin. 1995. Livestock grazing effects in western North America. Pp. 311–353 *in* T. E. Martin and D. M. Finch (editors), Ecology and management of neotropical migratory birds: a synthesis and review of critical issues. Oxford University Press, New York, NY.

Scott, J. M., C. R. Peterson, J. W. Karl, E. Strand, L. K. Svancara, and N. M. Wright. 2002. A gap analysis of Idaho: final report. Idaho Cooperative Fish and Wildlife Research Unit, Moscow, ID.

Shepherd, J. F. 2006. Landscape-scale habitat use by greater sage-grouse (*Centrocercus urophasianus*) in southern Idaho. Ph.D. dissertation, University of Idaho, Moscow, ID.

SPSS, Inc. 1993. SPSS for Windows base system user's guide, release 6.0. SPSS, Inc., Chicago, IL.

SPSS, Inc. 2000. SYSTAT user's guide, release 10.0. SPSS, Inc., Chicago, IL.

Sveum, C. M., J. A. Crawford, and W. D. Edge. 1998. Use and selection of brood-rearing habitat by sage grouse in south central Washington. Great Basin Naturalist 58:344–351.

Wakkinen, W. L., K. P. Reese, J. W. Connelly, and R. A. Fischer. 1992. An improved spotlighting technique for capturing sage grouse. Wildlife Society Bulletin 20:425–426.

Wallestad, R. O., J. G. Peterson, and R. L. Eng. 1975. Food habits of adult sage grouse in central Montana. Journal of Wildlife Management 39:628–630.

West, N. E. 1996. Strategies for maintenance and repair of biotic community diversity on rangelands. Pp. 326–346 *in* R. C. Szaro and D. W. Johnston (editors), Biodiversity in managed landscapes. Oxford University Press, New York, NY.

Whisenant, S. G. 1990. Changing fire frequencies on Idaho's Snake River plains: ecological and management implications. Pp. 4–10 *in* E. D. McArthur, E. M. Romney, S. D. Smith, and P. T. Tueller (editors), Proceedings of a symposium on cheatgrass invasion, shrub die-off, and other aspects of shrub biology and management. U.S.

Forest Service General Technical Report INT-276. Intermountain Forest and Range Experiment Station, Ogden, UT.

Wik, P. A. 2002. Ecology of Greater Sage-Grouse in south-central Owyhee County, Idaho. M.S. thesis, University of Idaho, Moscow, ID.

Wisdom, M. J., B. C. Wales, R. S. Holthausen, W. J. Hann, M. A. Hemstrom, and M. M. Rowland. 2002. A habitat network for terrestrial wildlife in the interior Columbia Basin. Northwest Science 76:1–13.

Young, J. A., and R. A. Evans. 1978. Population dynamics after wildfires in sagebrush grasslands. Journal of Range Management 31:283–289.

Zar, J. H. 1999. Biostatistical analysis. 4th ed. Prentice Hall, Upper Saddle River, NJ.

CHAPTER SEVEN

Natal Dispersal Affects Population Dynamics of Hazel Grouse in Heterogeneous Landscapes

Marc Montadert and Patrick Léonard

Abstract. Hazel Grouse (*Bonasa bonasia*) are considered to be highly vulnerable to habitat fragmentation and are poor colonizers, owing to the poor dispersal ability of juveniles. However, these considerations are mainly circumstantial as few data are available on dispersal behavior of this species. The limited telemetry data available on natal dispersal has revealed that some juvenile males are able to cross gaps between forest patches and to disperse long distances. To understand this apparent contradiction, we radio-equipped 38 juvenile Hazel Grouse in autumn in southeastern France. Mean natal dispersal distances were greater for males (6.3 km, $n = 11$) than for females (1.9 km, $n = 14$). Moreover, the shape of dispersal distance distribution tended to be different between sexes, with all juvenile females ($n = 16$) dispersing <6 km, whereas 21% (4 of 19) of juvenile males dispersed >10 km (range 12–25 km). Our observation of philopatric females and longer dispersing males is unusual among grouse. Sex-biased natal dispersal could explain why the species seems particularly sensitive to forest fragmentation and a poor colonizer. Demographic rescue of small habitat patches and colonization rates may be restricted by low dispersal abilities of females. Our data also suggest that use of unfamiliar sites during dispersal period could increase vulnerability to avian predators and potentially reduce juvenile fitness.

Key Words: Bonasa bonasia, breeding dispersal, habitat fragmentation, juvenile survival, natal dispersal, radio-tracking, range expansion.

With increasing fragmentation of natural habitat by human activities (Saunders et al. 1991), there is an increasing need to assess the dispersal pattern of animal populations in different landscapes. In sedentary species, where most adults stay in the same restricted home range for their whole lives, it is only juvenile movements that allow demographic and genetic flow between adjacent populations. For these species, a knowledge of natal dispersal, including distances moved between natal and adult breeding site, is crucial for understanding the spatial functioning of populations (Greenwood and Harvey 1982, Fahrig and Merriam 1994, Paradis et al. 1999,

---

Montadert, M., and P. Léonard. 2011. Natal dispersal affects population dynamics of Hazel Grouse in heterogeneous landscapes. Pp. 89–103 *in* B. K. Sandercock, K. Martin, and G. Segelbacher (editors). Ecology, conservation, and management of grouse. Studies in Avian Biology (no. 39), University of California Press, Berkeley, CA.

Walters 2000, McDonald and Johnson 2001, Lidicker 2002).

Individuals face two challenges during the dispersal process: to stay alive during movement and to reach a suitable patch to settle in and finally reproduce. High mortality rates of dispersers are often considered as the main cost of dispersal (Lidicker 2002), but empirical data are limited owing to difficulties in the study of causes of mortality in relation to individual behavior. In Ruffed Grouse (*Bonasa umbellus*), Yoder et al. (2004) showed that dispersing juveniles suffered a higher predation rate than sedentary individuals, and that higher risk was mainly due to use of unfamiliar habitats rather than increased mobility. Success in achieving dispersal is also related to landscape characteristics (patch pattern, matrix quality) and species-specific dispersal skills (Wiens 2001). Habitat specialists are thought to be more vulnerable to habitat fragmentation if these species are reluctant to cross unsuitable matrix habitat (Brooker et al. 1999). Thus, landscape structure can have varied demographic consequences depending on a species' ecological requirements (Bellamy et al. 2003).

Among grouse species, Hazel Grouse is a strict forest specialist which is virtually never found in open lands (Bergmann et al. 1996). Studies of Hazel Grouse patch occupancy in fragmented forested landscape have concluded that this species was highly vulnerable to patch isolation, as open gaps of more than 150–250 m seemed to act as barriers for dispersers (Aberg et al. 1995, Klaus and Sewit 2000, Sahlsten 2007). These observations, and previous radio-tracking data on juvenile movements, have led to the view that Hazel Grouse have the shortest dispersal distances among grouse species (Swenson 1991). Recent observations of a range expansion in the southern French Alps also suggested that Hazel Grouse have low colonization rates even in continuous habitat (Montadert and Léonard 2006a). Yet telemetry data revealed that some juvenile males can disperse long distances (>10 km) and may cross open gaps during natal dispersal. Montadert and Leonard (2006a) proposed alternative explanations to account for this apparent contradiction, such as the confounding effect of habitat quality and fragment size on habitat isolation (Saari et al. 1998, Sun et al. 2003) or the possibility that natal dispersal could be male-biased in Hazel Grouse. The sample size of radio-tracked females was insufficient to test the latter hypothesis. Moreover, this dispersal pattern would be unusual, as natal dispersal in birds is usually female-biased, with few exceptions (Greenwood 1980, Clarke et al. 1997).

In this work, we mainly wanted to assess the hypothesis that natal dispersal could be male-biased in the Hazel Grouse. We compare the natal dispersal distances of our study population with estimates of mean dispersal distances and sexual patterns of dispersal in other grouse species. Effects of landscape structure on movement patterns in relation to individual fitness were also investigated to determine their consequences for spatial functioning of Hazel Grouse populations in heterogeneous landscapes.

MATERIAL AND METHODS

Study Site

The study site was located in the southeastern French Alps in the Alpes de Haute Provence department (44°17.6′ N, 6°18.3′ E), in a mountainous region recently colonized by Hazel Grouse at the southeastern limit of the species range in Europe.

The study area, encompassing 98% of all individual telemetry fixes (juveniles and adults pooled), covered 6,400 ha, but most juveniles were captured in two forests, Fissac (300 ha) and Sansenu (200 ha). Natural forests cover 76% of this study area. Forest is dominated by spruce (*Picea abies*), fir (*Abies alba*), and Scots pine (*Pinus sylvestris*), mixed with a variable amount of deciduous trees (including beech, *Fagus sylvatica*; birch, *Betula pubescens*; willow *Salix* spp.; rowan, *Sorbus apricaria* and *S. aria*; see Montadert and Léonard 2003 for more details on habitat). Forest habitats are rather young (<100 years old) and have developed following the decline of agriculture since the middle of the 19th century. Currently, these forests constitute optimal habitat for Hazel Grouse, which attain their highest densities in western Europe at this location—up to 7–8 pairs/100 ha (Bergmann et al. 1996, Montadert et al. 2006).

Landscape structure and composition was described at larger scale (circular area of 5 and 25 km radius around the center of study area) using the typology of the National Forest Inventory of France (NFI) (Table 7.1). This inventory scheme categorizes forest habitat using stand

TABLE 7.1
Landscape cover and forest habitat characteristics of circular areas centered on the study site in the southeastern French Alps.

Diameter	Forest cover in total landscape	No. of forest patches	% biggest forest patch	% of coniferous/ mixed stands	% of deciduous stands	% of open forest
5 km	62%	14	59%	79%	3%	18%
25 km	58%	192	85%	41%	9%	50%

composition and openness but is not suitable to accurately predict Hazel Grouse occurrence or abundance due to variation in the carrying capacity of different IFN habitat types (M. Montadert, unpubl. data). Data on important habitat characteristics for Hazel Grouse (understory structure and composition) were only available for the two studied forests, Fissac and Sansenu, which were too small to encompass localizations of radio-tagged juveniles during dispersal. Yet IFN types classified as mixed or coniferous stands were regularly inhabited by Hazel Grouse at high densities, whereas pure deciduous stands were less occupied but comprised a small part of forest area. High-quality forest habitats represented the major part of forest for the first level spatial domain (5 km diameter). At a larger extent (25 km diameter), open forest covered the main part of the landscape (Table 7.1).

To illustrate connectedness of forest, we pooled IFN habitat types in two classes, forest and open lands. Forest included closed forests and open forests which were not usually permanently occupied but were easily crossed and used temporarily during dispersal. Forest connectivity was high at the 5-km and 25-km diameter domains, with 59% and 85% of the forest cover including only one continuous fragment (Figs. 7.1A–B; Table 7.1). Open lands included valleys, agriculture fields with scattered trees, hedgerows, and scarce human settlements, and at higher elevation, alpine meadows and rocky ridges (max. 3000 m a.s.l.). Owing to the rather broad map scale of the IFN inventory (1:25,000), open lands could contain small woods and hedgerows that were not mapped. Forest cover differed according to geographic directions and spatial domain (Table 7.2). In the 5-km diameter area, the east and south quarters were more forested and less fragmented than the north and west quarters. In the 25-km diameter

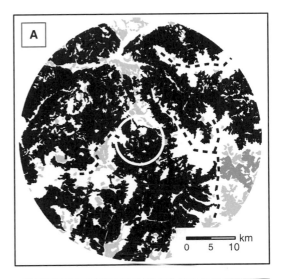

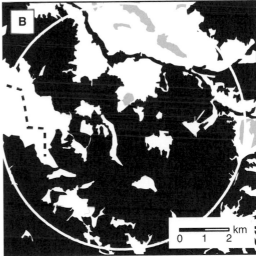

Figure 7.1. Forest cover of the studied landscape in two spatial domains: (A) A circular area of 25 km diameter centered on the study site, and (B) A circular area of 5 km diameter centered on the study site. Key: white = open lands, black and gray = forest, first three dark gray levels = three continuous forest tracks, and broken lines = high mountain ridges.

TABLE 7.2

Landscape cover characteristics of four quarters of two circular areas centered
on the study site in the southeastern French Alps.

Diameter	Quarter	Forest cover (%)	% of biggest forest patch	No. of forest patch
5-km diameter area	North	32%	36%	8
	East	64%	89%	7
	South	75%	100%	2
	West	34%	59%	7
25-km diameter area	North	58%	73%	85
	East	50%	80%	36
	South	66%	91%	38
	West	61%	94%	40

area, differences in forest cover were less important, though the east quarter had less forest cover and the north quarter was more fragmented. At this spatial extent, two main potential barriers for Hazel Grouse movements were high rocky ridges in the east and south, and the Durance River valley in the north (Fig. 7.1A).

Capture and Telemetry

The dispersal behavior of Hazel Grouse was studied using telemetry of radio-equipped birds. Juveniles were captured in autumn between 1998 and 2006. Two techniques were used: capture in walk-in traps in August before the break-up of broods and capture by luring the birds with a decoy whistle into nylon fishing nets in September–October, following the break-up of broods. Apart from juveniles, several adults (>1 year old) and subadults (aged 9–10 months) were caught. Behavioral information gained from these birds was used for comparison purposes. Hazel Grouse were fitted with collar radio transmitters with mortality censors (Holohil Ltd., Carp, ON). The transmitters weighed between 7 and 11 g, corresponding to 1.8–2.8% of bird weight, and had a expected life of 12–16 months.

Throat and flank coloration was used to sex juveniles >3 months old. For younger juveniles, sex identification was impossible with plumage features at capture time. Surviving birds were re-sighted and sexed at >3 months of age after a careful approach using telemetry receivers.

Age determination of juveniles was based on inspection of the first primary feather (Bonczar and Swenson 1992, Stenman and Helminen 1974), with some adaptations to local pattern characteristics (Montadert and Léonard 2009). Radio locations were made 2–3 times per week in spring, summer, and autumn and once per week in winter, with a fix precision of ±25 m. Aircraft were used twice to locate birds that were lost during the dispersal phase.

Data Analysis

Dispersal Distance

Natal dispersal was calculated as the straight-line distance between the brood break-up site (for birds captured in August), or the capture site (for birds captured in September to October), to the center of the spring home range. The spring home range was defined as the minimum convex polygon (100% MCP) of the total locations between 16 March and 15 June and was estimated with movement software (Hooge and Eichenlaub 1997). As spring home range appeared rather elliptic without unused area, MCP was a correct approximation of occupied range.

We used a one-sided Welch test, after square root transformation of distances, for comparison of the mean dispersal distances of males and females. The fit of the transformed data was assessed by a Shapiro test. As some of the radio-tagged juveniles were caught several weeks (up to mid-October) after brood break-up, or were lost several weeks before next spring, the calculated dispersal distance could have been underestimated. Therefore, we excluded these birds from the calculation of the mean dispersal distances and the Welch test.

We used Fisher's Exact test to assess sex differences in the proportion of short (<6 km) and long distance dispersers (>10 km). Because long-distance dispersers could be detected even for an incomplete period of monitoring, we retained all birds alive up to January for this comparison. Confidence intervals for the proportion of long distance dispersers were calculated using a binomial distribution.

Landscape Structure and Movement Pattern

To assess potential links between dispersal behavior and landscape structure or composition, we classified the radio-fixes of juveniles from autumn to spring in two groups, depending on whether the birds were located in unfamiliar (not yet visited) or familiar sites, with the median of familiar site percentage of males as a threshold value. Using IFN habitat descriptions, we classified fixes according to three habitat types: closed forest (pooling coniferous, mixed and deciduous stands), open forest and open lands. We used χ^2 statistics to compare the proportion of familiar/unfamiliar locations between sex and dispersal pattern (long vs. short dispersering males) and to compare IFN habitat selection between sex and dispersal pattern.

Survival

To assess possible consequences of dispersal pattern on individual fitness, we calculated survival rates of different classes of individuals: male versus female and short dispersing males versus long dispersing males. We also classified juveniles in two classes according to the proportions of location in familiar versus unfamiliar sites. Low sample sizes precluded use of complex survival models, such as the Cox model used in Yoder et al. (2004). We simply calculated survival rates following Heisey and Fuller (1985). Calculation of survival and cause-specific mortality rates and standard errors were done using Micromort software. We used χ^2 statistics calculated with Contrast software (Hines

and Sauer 1989), for post hoc comparison of survival rates.

RESULTS

Comparison of Dispersal Distance Between Sexes

A total of 38 juveniles were radio-equipped between 1998 and 2006, including 22 males and 16 females. Three juvenile males were killed by predators in autumn and were excluded from the dispersal distance analysis. Furthermore, 10 juveniles (8 males and 2 females) monitored until winter were excluded from the mean calculation of dispersal distances but retained for calculation of short versus long disperser ratio. Furthermore, 76 subadults and adults were also caught and radio-equipped.

Median natal dispersal distances were 2.8 km (mean: 6.3 km, range: 0.3–25.0 km, $n = 11$) for males and 1.9 km for females (mean: 1.9 km, range: 0.2–5.5 km, $n = 14$). The maximal dispersal distance recorded was 25 km by a male. Natal dispersal distances of males tended to be greater than that of females (one-sided Welch test, $t = -1.7$, $P = 0.057$).

The distribution of female dispersal distances appeared balanced around the mean and none of the juvenile females dispersed farther than 5.5 km (Fig. 7.2). Male dispersal was left-skewed and 21% of juvenile males (4 of 19; 95% CI: 7%–46%) dispersed more than 10 km from their natal site; the others settled <4 km from their natal area. If four males dispersing long distances were excluded from the data set, no difference was apparent between the dispersal distances of males and females (median dispersal distance of short disperser males: 1.6 km). The proportion of long dispersers (21% males, 0% females) was not significantly different between the sexes (Fisher's Exact test, $P = 0.1$).

Behavior Pattern of Juvenile Dispersers

Long dispersal movements by males were characterized by two phases of dispersal. In autumn, males covered the longest distances (median: 11.6 km, range: 10.7–18.8 km) in a rather direct line to their winter home range (median arrival date: 15 November). The departure date was late (median departure date: 17 October) and there were no erratic movements prior to

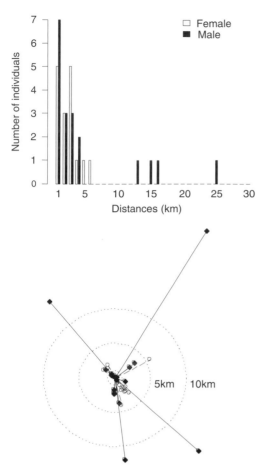

Figure 7.2. Natal dispersal of Hazel Grouse in the southeastern French Alps. Top: distribution of natal dispersal distances. Bottom: dispersion diagram.

range. The occurrence of spring dispersal was more frequent in long than in short dispersing males (long dispersers: 100%, $n = 3$ of 3; short dispersers: 25%, $n = 3$ of 12; Fisher's Exact test, $P = 0.04$). Spring dispersal of females was unusual (7% of females, $n = 1$ of 14). Distances moved in spring by short dispersing males and females were short (median distances, males: 1.1 km, $n = 3$; females: 0.7 km, $n = 1$). The departure date was rather variable, as a proportion of short dispersers undertook erratic movements in autumn and early winter in random directions, up to few kilometers (Figs. 7.3A–B), before finally settling close to their natal site. Arrival date in the spring home range, in the middle of the winter, was earlier than for long dispersers.

Landscape Structure and Dispersal Pattern

From autumn to spring we recorded a total of 1,191 radio-fixes of juvenile females ($n = 16$) and 1,100 fixes of juvenile males (861 fixes for short dispersers, $n = 15$; 239 fixes for long dispersers, $n = 4$) (Figs. 7.4, 7.5). Nearly all locations of juvenile females and short dispersing males were encompassed by the 5-km-diameter circle (96% and 93%, respectively). Thus, we used the proportion of the three habitat types (forest, open forest, open lands) in this domain as available habitat for females and short dispersing males. For the long dispersing males, we calculated habitat availability as the proportion of the three habitat types in the 25-km-diameter circle.

Overall, juvenile females, short dispersing males, and long dispersing males selected closed forest and avoided open forest and open lands (females: $\chi^2 = 806.9$, df = 2, $P < 0.001$; short dispersing males: $\chi^2 = 498.3$, df = 2, $P < 0.001$; long dispersing males: $\chi^2 = 222.8$, df = 2, $P < 0.001$).

The proportion of radio-fixes ($n = 2291$) in familiar or unfamiliar sites differed among the three juveniles groups ($\chi^2 = 213.9$, df = 2, $P < 0.001$). Unfamiliar site proportion for long dispersing males was greater than for short dispersing males (unfamiliar site proportion, long dispersers: 59%; short dispersers: 25%, $\chi^2 = 102.2$, df = 1, $P < 0.001$), which, in turn, were greater than for juvenile females (16%, $\chi^2 = 25.9$, df = 1, $P < 0.001$). Utilization of unfamiliar

departure. After a winter period when birds held a more or less restricted home range, a second phase of dispersal took place in March–April to reach the spring home range (median arrival date: 25 April). Median spring dispersal distances of long-distance males was 4.7 km (range: 2.5–12.4 km). The observed straight-line distances covered in one day was sometimes >2 km (7 records among the 4 males), with maximum daily movements of 5.6 km and 7.5 km in autumn and spring, respectively.

Short dispersers moved mainly in autumn, with median dispersal distances of 1.1 km for males (range: 0.3–4.7 km) and 1.5 km for females (range: 0.2–5.3 km). Spring dispersal was negligible (median spring dispersal: 0.15 km for males and 0.2 km for females), as most short dispersers had their spring home ranges more or less contained within their first winter home

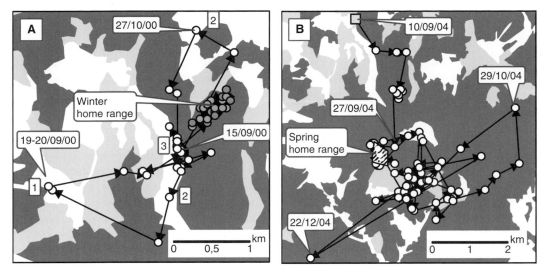

Figure 7.3. Examples of erratic movements of radio-tagged juvenile male Hazel Grouse during dispersal in the southeastern French Alps. A: Male ("Elvis"); locations 1 and 2 were in large hedgerows and small groves not mapped, and 3 on the steep slope of bare ground. Gray dots are fixes during well-delimited first winter home range. B: Male ("Jupon"); intensive erratic movements without definite winter home range. End of erratic movement and arrival in first well-delimited spring home range occurred on 13 March.

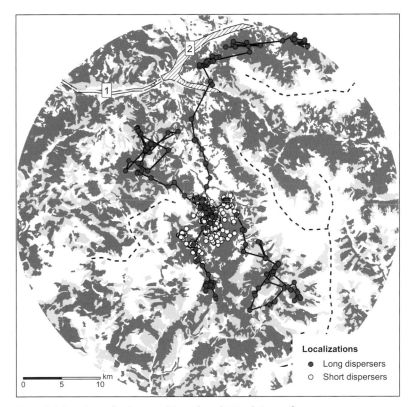

Figure 7.4. Locations of radio-tagged juvenile male Hazel Grouse from autumn (15 September) to spring (15 June) in the southeastern French Alps. Key: 1 = Durance River, 2 = Serre-Poncon dam, white = open lands, light gray = open forest, dark gray = closed forest, broken line = high mountain ridge.

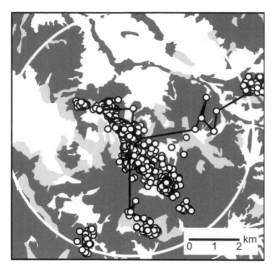

Figure 7.5. Locations of radio-tagged juvenile female Hazel Grouse from autumn (15 September) to spring (15 June) in the southeastern French Alps (see Fig. 7.4 for key). Dark lines represent trajectories of the two longest female dispersals.

sites was associated with increased use of open forest and open lands by short dispersing males ($\chi^2 = 45.8$, df = 2, $P < 0.01$) but not for long dispersing males ($\chi^2 = 0.8$, df = 1, $P = 0.36$), and only use of open lands among juvenile females ($\chi^2 = 16.7$, df = 2, $P < 0.001$).

In the 5-km-diameter area, short dispersing males used more open forest and open lands than did juvenile females ($\chi^2 = 9.8$, df = 2, $P = 0.007$). In this spatial domain, the large proportion and continuity of forest cover permitted most juvenile movements to be under forest cover. The few radio-fixes localized in open lands were locations in small groves or even large hedgerows (Fig. 7.3A). Yet open gaps of 100–200 m and small roads in valley bottoms were easily crossed, probably in flight, even if no direct sightings were made. Existence of erratic movements in some juveniles was not clearly associated with habitat characteristics, as other nonerratic individuals were tracked in same area. However, isolated forest tracks in the north quarter and forest area in the southeastern part (5-km-diameter area) were never reached by short dispersers captured in the center part of this area. For short dispersers, most movements and final settlement were undertaken in the same continuous forest tracks.

For the four long dispersing males, it was not clear to what extent discontinuity in forest cover

impeded movements. For the longest dispersing individual, an agricultural valley with open gaps of 900–1,000 m was crossed successfully, whereas this male could have reached its first winter range without leaving forest cover by a more easterly route (Fig. 7.4). It is even possible that this male flew over water for a distance of 300–600 m at the southeastern part of Serre-Ponçon Dam. We localized the male close to the shore, but there was a gap of 4 days between successive locations when it could have circled the lake. Other long dispersing males did not need to cross open gaps of more than 350 m to achieve their dispersal. If open agriculture land in the valley was not a barrier to movement, open alpine meadows at higher elevation were never crossed by radio-tagged birds.

Survival

From 15 September to 15 June (9 months), survival rate of juvenile males was 0.58 ± 0.11 ($n = 22$) and that of juvenile females 0.77 ± 0.10 ($n = 16$), but the difference was not significant ($\chi^2 = 1.6$, df = 1, $P = 0.21$; Table 7.3). Survival rates of long versus short dispersing males were similar ($\chi^2 = 0.13$, df = 1, $P = 0.71$; Table 7.3). One juvenile was killed early in October and was classified as a short disperser. If this male was considered a long disperser, survival would be 0.37 ± 0.22 for long dispersers and 0.63 ± 0.12 for short dispersers, but still not significantly different ($\chi^2 = 1.02$, df = 1, $P = 0.31$).

Survival rates of juveniles with a high proportion of familiar locations was greater than survival rates of juveniles with a low proportion of familiar locations ($\chi^2 = 5.7$, df = 1, $P = 0.02$; Table 7.3). Among juvenile males, this difference was not quite significant ($\chi^2 = 2.9$, df = 1, $P = 0.08$; Table 7.3). All mortality was caused by predation, with raptors (probably Goshawk, *Accipiter gentilis*) responsible for 70% of mortality events. With both sexes pooled, raptor-induced mortality was 0.26 ± 0.7, and carnivore-induced mortality was 0.08 ± 0.04 ($\chi^2 = 4.2$, df = 1, $P = 0.04$).

Breeding Dispersal of Adults

Of six subadult males monitored from their first spring to the next, two moved in their second autumn to another spring home range (1.1 and 1.4 km between the centers of successive spring

TABLE 7.3

Nine-month survival rates of radio-tagged juvenile Hazel Grouse from autumn to spring in the southeastern French Alps.

	n	Survival	SE	P
Male	22	0.58	0.11	
				0.21
Female	16	0.77	0.10	
Male long dispersers	4	0.48	0.25	
				0.71
Male short dispersers	18	0.58	0.12	
Juvenile low site familiarity	12	0.40	0.14	
				0.02
Juvenile high site familiarity	26	0.78	0.08	
Male low site familiarity	11	0.43	0.15	
				0.09
Male high site familiarity	11	0.73	0.13	

NOTE: n = number of individuals. Survival and SE = survival rates and standard error following Heisey and Fuller (1985). P = Probability from χ^2 test.

ranges). Among 12 adult males (more than 1 year old when captured) monitored during two successive springs, only one moved (0.8 km), just after capture in October. None of seven subadult females monitored from their first spring to the next changed their spring home range. One of nine adult females monitored for two successive springs left the first breeding site after nest predation to settle 3.4 km away in a new home range, where she tried to breed the following spring.

DISCUSSION

Estimated dispersal distances could have been underestimated due to variability in capture dates. Even if birds captured in late autumn were excluded, the true estimate of natal dispersal distance should have been calculated from the nest site, which was not known in most cases. However, our data on brood movements showed that nest to brood break-up site distances were usually less than 1 km (median distance of eight broods, 0.3 km; Montadert and Léonard 2006a). Rates of dispersal should be unbiased because dispersal occurs after capture. The generality of our result could not be assessed due to the scarcity of relevant telemetry data on Hazel Grouse dispersal

compiled by Montadert and Léonard (2006a). Yet a recent telemetry study of Hazel Grouse in South Korea did not find significant differences in juvenile dispersal between males and females (median dispersal distances of males: 1.7 km, n = 19; females: 1.1 km, n = 24) (Rhim and Son 2009). However, the monitoring of radio-tagged juveniles stopped in November, probably too early to encompass the whole dispersing period.

Is the Hazel Grouse a Poor Disperser?

To assess this question, we compiled information on the natal dispersal behavior of eight grouse species from 11 published studies (Table 7.4). Female dispersal distance appeared to be greater in most grouse species, apart from Red Grouse (*Lagopus l. scoticus*) and Dusky Grouse (*Dendragapus obscurus*), which show female dispersal distances similar to our estimate. For male dispersal, our present estimate for Hazel Grouse appears to be greater than for all other grouse species for which information is available. However, if we set aside the few long dispersers in our Hazel Grouse data set, then mean male dispersal was in the range of most other grouse species. However, a global overview of grouse natal dispersal is not yet possible as

TABLE 7.4
Mean and maximum natal dispersal distances (km) of eight species of grouse.

Species	Female		Male		References
	Mean (km)	Max (km)	Mean (km)	Max (km)	
Tetrao urogallus	12.3	(30)			Moss et al. 2006
Tetrao tetrix	8.0	(29)	1.5	(8.2)	Caizergues and Ellison 2002
	9.3	(20)			Warren and Baines 2002
Dendragapus obscurus	1.4*	(11)	0.9*	(2.6)	Hines 1986
Falcipennis canadensis	5.0		0.7		Boag and Schroeder 1992
Bonasa umbellus	5.6		2.8		Yoder in Rusch et al. 2000
	4.9		2.4		Rusch et al. 2000
Lagopus lagopus scoticus	2.0	(10)	0.5	(1)	Hudson 1992
Lagopus lagopus	10.2		3.4		Hörnell-Willebrand 2005
	11.4		2.6		Smith 1997
Lagopus leucura	4.0	(29)	1.2	(7.5)	Giesen and Braun 1993

NOTE:*=median

relevant information is lacking for several poorly known species, such as Chinese Grouse (*Bonasa sewersowi*), Caucasian Grouse (*Tetrao mlokosiewiczi*), Siberian Grouse (*Falcipennis falcipennis*), and even for more intensively studied species such as Rock Ptarmigan (*Lagopus muta*).

Furthermore, natal dispersal may not only be species-specific, but could also be sensitive to environmental factors such as habitat fragmentation (see Whitcomb et al. 1996 for an example in Spruce Grouse, *Falcipennis canadensis*). In our study area, forest habitat was not highly fragmented and most long dispersers could use forested corridors when dispersing, so long dispersal behavior in males was not necessarily related to forest fragmentation. Intraspecific competition could also influence dispersal of individuals belonging to high-density populations (Lambin et al. 2001, Matthysen 2005). The population in our study area was at high density and the sex ratio

is male-biased (Montadert and Léonard 2006b), which could have influenced dispersal distances. Alternatively, contrasting males dispersal pattern, and behavioral differences between long and short dispersing males, could be mainly genetically determined, as suggested in Spruce Grouse (Schroeder and Boag 1988, Keppie and Towers 1992), and Roe Deer (*Capreolus capreolus*) (Wahlström and Liberg 1995).

Sparse telemetry data on Hazel Grouse dispersal in other environmental and geographical conditions limit comprehensive comparisons (Montadert and Léonard 2006a). Recent improvements in genetic techniques provide new tools to study dispersal (Goldstein et al. 1999, Prugnolle and De Meeus 2002). Using genetic techniques, dispersal of a Swedish Hazel Grouse population was estimated to be 0.9–1.5 km in a continuous forested landscape (Sahlsten 2007), similar to estimates of female dispersal based on telemetry (our study, Rhim and Son 2009). Thus, short female natal dispersal may be a common feature of Hazel Grouse populations.

Dispersal, Landscape Pattern, and Fitness

Survival to the first spring to possibly reproduce determines juvenile fitness. With our limited data, differences in survival between different classes of individuals with different dispersal distances (male vs. female, long dispersing male vs. short dispersing male) were not significant. Nevertheless, there were some indications that dispersal phenomena could impact survival. The apparent difference in survival (survival of long dispersers < survival of short dispersers < survival of females) followed the proportion of familiar sites occupied, and lack of familiarity was significantly linked with low survival.

Even if survival was not associated with dispersal distance, survival could be impacted by the spatial behavior of individuals, in particular by the time passed in unfamiliar area (Yoder et al. 2004). A high proportion of unfamiliar sites occupied was often associated with increased use of marginal habitats (open forest, hedgerows) not used by adults. Long-distance dispersers had the longest dispersal period and the highest proportion of unfamiliar sites used; it might be expected that they suffered highest mortality. Likewise, a proportion of short dispersing males undertook erratic movements before they settled, which were clearly associated with increased use of unfamiliar sites and marginal habitats. Thus it would also be foreseeable that they survived less than females, which rarely carry out such movements.

Most juvenile mortality cases were raptor kills, whereas the proportion of raptor/carnivore kills was balanced in adults (Montadert and Léonard 2003). Unsafe behavior in an unfamiliar site may increase the probability of being detected by a raptor, which could be a rather common phenomenon among juvenile grouse (Hannon and Martin 2006).

The possible impact of the landscape structure on movements was not easy to assess, as the landscape studied was mainly covered by good-quality forested habitat. Fine-scale fragmentation of forest did not impede movement of males, either for long dispersers or for erratic short dispersers. For females, the possible impact of small open gaps on movements (<200–300 m) could not be estimated as only one female left the closed forest to cross a road and a small open gap during dispersal (Fig. 7.5). At a larger extent, movement of some long dispersing males seemed to be possible through a rather fragmented landscape which, owing to extensive forest cover in this alpine region, makes connection among distant populations likely. We were unable to evaluate whether high rocky ridges were barriers to dispersal with our sample of radio-tagged birds. Historic reconstruction of colonization routes followed by expanding Hazel Grouse populations in the southern Alps (Montadert and Léonard 2006a) suggests that rocky ridges probably constrained dispersal and explained why remote watersheds have never or only recently been colonized by Hazel Grouse.

Natal Dispersal Versus Breeding Dispersal

Adult Hazel Grouse are considered to be mainly sedentary (Bergmann et al. 1996; this study). Although site fidelity is a common behavioral pattern in birds (Greenwood and Harvey 1982), adults in some grouse populations change breeding sites between years (e.g., White-tailed Ptarmigan, *Lagopus leucura*; Martin et al. 2000). In our study, some Hazel Grouse males moved in their second autumn after having failed to settle in their first spring home range. Anecdotal observations suggested that, in these cases, the first spring home range was already occupied by a pair, which suggests that intrasexual competition could have been responsible for this behavior. We have no evidence that males older than two years changed their territory, but a case was reported in the Jura Mountains where one adult male moved more than 17 km in spring after a complete disappearance of females from the local population (Montadert 1995). Breeding dispersal was not observed in subadult females, which could be connected to a lack of mate competition, resulting from an excess of males in the population. Finally, the only case of adult female breeding dispersal was linked to a reproductive failure, whereas several nest losses among other radio-tagged hens did not lead to change in breeding site. Breeding site fidelity is often connected to reproductive success in birds (Newton 2001) and in grouse (Storch 1993). All reports of adult movements suggest that site fidelity is a typical pattern of adult behavior in Hazel Grouse, but some adult behavioral plasticity exists, which could permit a local population to deal with a changing environment due to inter-annual variation in mate availability, habitat quality, or predation risk.

Dispersal Biases and Consequences on Spatial Dynamics of Populations

The apparent male-biased natal dispersal of our population was unusual, as most banding or telemetry studies on grouse show that females disperse further than males (Table 7.4), or at least that males are more philopatric (Keppie 1979, Martin and Hannon 1987, Beaudette and Keppie 1992). Brøseth et al. (2005) found possible female-biased dispersal in Willow Ptarmigan (*Lagopus lagopus*), but low sample size of females and indeterminacy of gender status of several radio-tagged individuals prevented a solid conclusion. Finally, we know of only one recent telemetry study revealing clear male-biased natal dispersal in a Greater Sage-Grouse (*Centrocercus urophasianus*) population (Thompson et al. 2008). This study revealed that the dispersal pattern is not necessarily species-specific but could vary between different populations (Lambin et al. 2001), as classical female-biased natal dispersal was previously reported in Greater Sage-Grouse (Dunn and Braun 1985). However, we hypothesize that the suggested dispersal pattern, with female philopatry and a proportion of long dispersing males, could be a common phenomenon in Hazel Grouse and may explain two aspects of their spatial dynamics: vulnerability to forest fragmentation and a low colonization ability. Vulnerability to habitat fragmentation could be directly linked to low dispersal performance of females, particularly if females are reluctant to cross open land during dispersal. Even if a noticeable proportion of males cover long distances, and reach isolated forest tracks, long dispersers may not settle in habitats without females where they have no chance of mating before they die.

Likewise, the related low expansion rate in the southern French Alps (1.5 km/year during 30 years; Montadert and Léonard 2006a) could also be bounded by female dispersal. Female philopatry will delay establishment of new populations beyond the colonization front, which facilitates rapid range expansion into new habitat, as seen in the wolf (*Canis lupus*) in the Alps (Fabbri et al. 2007) and roe deer in Scandinavia (Andersen et al. 2004).

The demographic consequences of our observed dispersal pattern could be an increasing asynchrony in the population trajectory of adjacent populations separated by open gaps. Apart from a Moran effect, which could synchronize populations over large areas (Hudson and Cattadori 1999, Koenig 2002), dispersal is a potential mechanism that favors spatial synchrony (Paradis et al. 1999, Bellamy et al. 2003, Kerlin et al. 2007). With low rates of exchange of females among adjacent populations, rescue events will be infrequent, leading to local declines and divergence in local population trends. A similar mechanism was proposed to explain spatial asynchrony in adjacent populations of Pyrenean Grey Partridge (*Perdix perdix hispaniensis*) (Novoa 1998), and our data suggests the potential for similar dynamics in Hazel Grouse.

ACKNOWLEDGMENTS

We are indebted to Kenny Kortland, who kindly checked the manuscript, and to our reviewers for constructive comments which markedly improved the final issue. We also thank Roger Izoard for his continuous and effective support and Jon Swenson for showing us how to capture Hazel Grouse by luring them into nets. For indispensable field assistance, our thanks to F. Kopko, A. Bogtchalian, D. Thiolière, A. Bombaud, J. C. Cauvin, R. Gayraud, D. Michallet, F. Miguel, R. Papet, J. Richelme, M. Teissier, R. Villecrose, E. Du Verdier, P. Michel, G. Chagniau, G. Dubrez, R. Landeau, F. Crozals, and D. Igier of the Office National de la Chasse et de la Faune Sauvage; R. Yonnet and D. Reboul of the Office National des Forêts; F. Normand, R. Clement, and P. Romain of the Fédération Départementale des Chasseurs des Alpes de Haute Provence; and last but not least, J. Lenoir and B. and J. Guillet. We received financial support from the Federation Departementale des Chasseurs des Alpes de Haute Provence, l'Office National de la Chasse et de la Faune Sauvage, and the Leader II-Gal and Objectif II programs of the European Union, locally managed by the Syndicat Intercommunal à Vocation multiple de Seyne-les-Alpes.

LITERATURE CITED

Aberg, J., G. Jansson, J. E. Swenson, and P. Angelstam. 1995. The effect of matrix on the occurrence of Hazel Grouse (*Bonasa bonasia*) in isolated habitat fragments. Oecologia 103:265–269.

Andersen, R., I. Herfindal, and B.-E. Saether. 2004. When range expansion rate is faster in marginal habitats. Oikos 107:210–214.

Beaudette, P. D., and D. M. Keppie. 1992. Survival of dispersing Spruce Grouse. Canadian Journal of Zoology 70:693–697.

Bellamy, P. E., P. Rothery, and S. A. Hinsley. 2003. Synchrony of woodland bird populations: the effect of landscape structure. Ecography 26:338–348.

Bergmann, H. H., S. Klaus, F. Muller, W. Scherzinger, J. E. Swenson, and J. Wiesner. 1996. Die Hazelhühner: Bonasa bonasia und B. swerzowi. Die Neue Brehm-Bücherei, Magdeburg.

Boag, D. A., and M. A. Schroeder. 1992. Spruce Grouse, A. Poole, P. Stettenheim, and F. Gill (editors), The Birds of North America No. 5. Academy of Natural Sciences, Philadelphia, PA.

Bonczar, Z., and J. E. Swenson. 1992. Geographical variation in spotting patterns on Hazel Grouse Bonasa bonasia primary feathers: consequences for age determination. Ornis Fennica 69:193–197.

Brooker, L., M. Brooker, and P. Cale. 1999. Animal dispersal in fragmented habitat: measuring habitat connectivity, corridor use, and dispersal mortality. Conservation Ecology 5:9. <http://www.consecol.org/vol3/iss1/art4>.

Brøseth, H., J. Tufto, H. C. Pedersen, H. Steen, and L. Kastdalen. 2005. Dispersal patterns in a harvested Willow Ptarmigan population. Journal of Applied Ecology 42:453–459.

Caizergues, A., and L. N. Ellison. 2002. Natal dispersal and its consequences in Black Grouse Tetrao tetrix. Ibis 144:478–487.

Clarke, A. L., B. E. Saether, and E. Røskaft. 1997. Sex biases in avian dispersal: a reappraisal. Oikos 79:429–438.

Dunn, P. O., and C. E. Braun. 1985. Natal dispersal and lek fidelity of sage grouse. Auk 102:621–627.

Fabbri, E., C. Miquel, V. Lucchini, A. Santini, R. Caniglia, C. Duchamp, J.-M. Weber, B. Lequette, F. Marucco, L. Boitani, L. Fumagalli, P. Taberlet, and E. Randi. 2007. From the Apennines to the Alps: colonization genetics of the naturally expanding Italian wolf (Canis lupus) population. Molecular Ecology 16:1661–1671.

Fahrig, L., and G. Merriam. 1994. Conservation of fragmented populations. Conservation Biology 8:50–59.

Giesen, K. M., and C. E. Braun. 1993. Natal dispersal and recruitment of juvenile White-tailed Ptarmigan in Colorado. Journal of Wildlife Management 57:72–77.

Goldstein, D. B., G. W. Roemer, D. A. Smith, D. E. Reich, A. Bergmen, and R. Wayne. 1999. The use of microsatellite variation to infer population structure and demographic history in a natural model system. Genetics 151:797–801.

Greenwood, P. J. 1980. Mating systems, philopatry and dispersal in birds and mammals. Animal Behavior 28:1140–1162.

Greenwood, P. J., and P. H. Harvey. 1982. The natal and breeding dispersal of birds. Annual Review of Ecology and Systematics 13:1–21.

Hannon, S. J., and K. Martin. 2006. Ecology of juvenile grouse during the transition to adulthood. Journal of Zoology 269:422–433.

Heisey, D. M., and T. K. Fuller. 1985. Evaluation of survival and cause-specific mortality rates using telemetry data. Journal of Wildlife Management 49:668–674.

Hines, J. E. 1986. Survival and reproduction of dispersing Blue Grouse. Condor 88:43–49.

Hines, J. E., and J. R. Sauer. 1989. Program CONTRAST: a general program for the analysis of several survival or recovery rate estimates. Fish and Wildlife Technical Report 24:1–7.

Hooge, P. N., and B. Eichenlaub. 1997. Animal movement extension to Arcview. ver. 1.1. Alaska Biological Science Center, U.S. Geological Survey, Anchorage, AK.

Hörnell-Willebrand, M. 2005. Temporal and spatial dynamics of willow grouse Lagopus lagopus. Ph.D. dissertation, Swedish University of Agricultural Sciences, Umea, Sweden.

Hudson, P. J. 1992. Grouse in space and time: the population biology of a managed gamebird. Game Conservancy Trust, Fordingbridge, UK.

Hudson, P. J., and I. M. Cattadori. 1999. The Moran effect: a cause of population synchrony. Trends in Ecology and Evolution 14:622–637.

Keppie, D. M. 1979. Dispersal, overwinter mortality, and recruitment of Spruce Grouse. Journal of Wildlife Management 43:717–727.

Keppie, D. M., and J. Towers. 1992. A test on social behavior as a cause of dispersal of spruce grouse. Behavioral Ecology and Sociobiology 30:343–346.

Kerlin, D. H., D. T. Haydon, D. Miller, N. J. Aebischer, A. A. Smith, and S. J. Thirgood. 2007. Spatial synchrony in Red Grouse population dynamics. Oikos 116:2007–2016.

Klaus, S., and A. Sewit. 2000. Ecology and conservation of Hazel Grouse Bonasa bonasia in the Bohemian forest (Sumava, Czech Republic). Pp. 138–146 in P. Malkova (editor), Proceedings of the International Conference in Ceské Budejovice, Czech Republic.

Koenig, W. D. 2002. Global patterns of environmental synchrony and the Moran effect. Ecography 25:183–288.

Lambin, X., J. Aars, and S. B. Piertney. 2001. Dispersal, intraspecific competition, kin competition and kin facilitation: a review of the empirical evidence. Pp. 110–122 in J. Clobert, E. Danchin, A. A. Dhondt, and J. D. Nichols (editors), Dispersal. Oxford University Press, New York, NY.

Lidicker, W. Z. 2002. From dispersal to landscapes: progress in the understanding of population dynamics. Acta Theriologica 47(Suppl.):23–37.

Martin, K., and S. J. Hannon. 1987. Natal philopatry and recruitment of Willow Ptarmigan in north-western Canada. Oecologia 71:518–524.

Martin, K., P. B. Stacey, and C. E. Braun. 2000. Recruitment, dispersal and demographic rescue in spatially-structured White-tailed Ptarmigan. Condor 102:503–516.

Matthysen, E. 2005. Density-dependent dispersal in birds and mammals. Ecography 28:403–416.

McDonald, D. W., and D. D. P. Johnson. 2001. Dispersal in theory and practice: consequences for conservation biology. Pp. 358–372 in J. Clobert, E. Danchin, A. A. Dhondt, and J. D. Nichols (editors). Dispersal. Oxford University Press, New York, NY.

Montadert, M. 1995. Occupation de l'espace par des mâles de Gélinotte des bois (Bonasa bonasia) dans le Doubs (France). Gibier Faune Sauvage [Game & Wildlife] 12:197–211.

Montadert, M., and P. Léonard. 2003. Survival in an expanding Hazel Grouse Bonasa bonasia population in the southeastern French Alps. Wildlife Biology 9:357–364.

Montadert, M., and P. Léonard. 2006a. Post-juvenile dispersal of Hazel Grouse Bonasa bonasia in an expanding population of the southeasten French Alps. Ibis 148:1–13.

Montadert, M., and P. Léonard. 2006b. Skewed sex ratio and differential adult survival in the Hazel Grouse Bonasia bonasa. Acta Zoologica Sinica 52:655–662.

Montadert, M., and P. Léonard. 2009. Age determination of Hazel Grouse in the south-western limit of its European range. Grouse News 37:7–14.

Montadert, M., P. Léonard, and P. Longchamp. 2006. Les méthodes de suivi de la Gélinotte des bois: Analyse comparative et proposition alternative. Faune Sauvage 271:28–35.

Moss, R., N. Picozzi, and D. C. Catt. 2006. Natal dispersal of Capercaillie Tetrao urogallus in northeast Scotland. Wildlife Biology 12:227–232.

Newton, I. 2001. Causes and consequences of breeding dispersal in the Sparrowhawk Accipiter nisus. Ardea 89(special issue):143–154.

Novoa, C. 1998. La perdrix grise dans les Pyrénées-Orientales. Utilisation de l'habitat, éléments de démographie, incidence des brûlages dirigés. Ph.D. dissertation, Université Paris VI, Paris, France.

Paradis, E., S. R. Baillie, W. J. Sutherland, and R. D. Gregory. 1999. Dispersal and spatial scale affect synchrony in spatial population dynamic. Ecology Letters 2:114–120.

Prugnolle, F., and T. De Meeus. 2002. Inferring sex-biased dispersal from population genetic tools: a review. Heredity 88:161–165.

Rhim, S. J., and S. H. Son. 2009. Natal dispersal of Hazel Grouse Bonasa bonasia in relation to habitat in a temperate forest of South Korea. Forest Ecology and Management 258:1055–1058.

Rusch, D. H., S. Destefano, M. C. Reynolds, and D. Lauten. 2000. Ruffed Grouse (Bonasa umbellus). A. Poole and F. Gill (editors), The birds of North America No. 515. The Birds of North America, Inc., Philadelphia, PA.

Saari, L., J. Aberg, and J. E. Swenson. 1998. Factors influencing the dynamics of occurrence of the Hazel Grouse in a fine-grained managed landscape. Conservation Biology 12:586–592.

Sahlsten, J. 2007. Impact of geographical and environmental structures on habitat choice, metapopulation dynamics and genetic structure for Hazel Grouse (Bonasa bonasia). Acta Universitatis Upsaliensis 314, Uppsala, Sweden.

Saunders, D. A., R. J. Hobbs, and C. R. Margules. 1991. Biological consequences of ecosystem fragmentation: a review. Conservation Biology 5:18–32.

Schroeder, M. A., and D. A. Boag. 1988. Dispersal in Spruce Grouse: is inheritance involved? Animal Behavior 36:305–306.

Smith, A. A. 1997. Dispersal and movements in a Swedish Willow Grouse Lagopus lagopus population. Wildlife Biology 3:279.

Stenman, O., and M. Helminen. 1974. Pyyn ikäluokan määritys siiven perusteella [In Finnish with English summary: Aging method for Hazel Grouse (Tetrastes bonasia) based on wings]. Suomen Riista 25:90–96.

Storch, I. 1993. Habitat use and spacing of capercaillie in relation to forest fragmentation patterns. Ph.D. dissertation, University of Munich, Munich, Germany.

Sun, Y.-H., J. E. Swenson, Y. Fang, S. Klaus, and W. Scherzinger. 2003. Population ecology of the Chinese Grouse Bonasa sewerzowi, in a fragmented landscape. Biological Conservation 110:177–184.

Swenson, J. E. 1991. Is the Hazel Grouse a poor disperser? Pp. 347–352 in S. Csanyi and J. Ernhaft (editors), Proceedings of the 20th Congress of the International Union of Game Biologists, Gödöllö, Hungary.

Thompson, T., K. P. Reese, and A. D. Apa. 2008. From the nest to the lek: survival, natal dispersal, and recruitment of juvenile Greater Sage-Grouse in northwestern Colorado. International Grouse Symposium 11, Whitehorse, Canada.

Wahlström, K., and O. Liberg. 1995. Contrasting dispersal patterns in two Scandinavian roe deer *Capreolus capreolus* populations. Wildlife Biology 1:159–164.

Walters, J. R. 2000. Dispersal behavior: an ornithological frontier. Condor 102:479–481.

Warren, P., and D. Baines. 2002. Dispersal, survival and causes of mortality in Black Grouse *Tetrao tetrix* in northern England. Wildlife Biology 8:91–97.

Whitcomb, S. D., F. A. Servello, and A. F. J. O'Connell. 1996. Patch occupancy and dispersal of Spruce Grouse on the edge of its range in Maine. Canadian Journal of Zoology 74:1951–1955.

Wiens, J. A. 2001. The landscape context of dispersal. Pp. 96–109 *in* J. Clobert, E. Danchin, A. A. Dhondt, and J. D. Nichols (editors), Dispersal. Oxford University Press, New York, NY.

Yoder, J. M., E. A. Marschall, and D. A. Swanson. 2004. The cost of dispersal: predation as a function of movement and site familiarity in Ruffed Grouse. Behavioural Ecology 15:469–476.

PART TWO

Habitat Relationships

CHAPTER EIGHT

Nesting Success and Resource Selection of Greater Sage-Grouse

Nicholas W. Kaczor, Kent C. Jensen, Robert W. Klaver,
Mark A. Rumble, Katie M. Herman-Brunson,
and Christopher C. Swanson

Abstract. Declines of Greater Sage-Grouse (*Centrocercus urophasianus*) in South Dakota are a concern because further population declines may lead to isolation from populations in Wyoming and Montana. Furthermore, little information exists about reproductive ecology and resource selection of sage grouse on the eastern edge of their distribution. We investigated Greater Sage-Grouse nesting success and resource selection in South Dakota during 2006–2007. Radio-marked females were tracked to estimate nesting rates, nest success, and habitat resources selected for nesting. Nest initiation was 98.0%, with a maximum likelihood estimate of nest success of 45.6 ± 5.3%. Females selected nest sites that had greater sagebrush canopy cover and visual obstruction of the nest bowl compared to random sites. Nest survival models indicated that taller grass surrounding nests increased nest survival.

Tall grass may supplement the low sagebrush cover in this area in providing suitable nest sites for Greater Sage-Grouse. Land managers on the eastern edge of Greater Sage-Grouse range could focus on increasing sagebrush density while maintaining tall grass by developing range management practices that accomplish this goal. To achieve nest survival rates similar to other populations, predictions from our models suggest 26 cm grass height would result in approximately 50% nest survival. Optimal conditions could be accomplished by adjusting livestock grazing systems and stocking rates.

Key Words: Centrocercus urophasianus, Greater Sage-Grouse, nest initiation, nest success, renesting, resource selection, sagebrush, South Dakota.

Kaczor, N. W., K. C. Jensen, R. W. Klaver, M. A. Rumble, K. M. Herman-Brunson, and C. C. Swanson. 2011. Nesting success and resource selection of Greater Sage-Grouse. Pp. 107–118 *in* B. K. Sandercock, K. Martin, and G. Segelbacher (editors). Ecology, conservation, and management of grouse. Studies in Avian Biology (no. 39), University of California Press, Berkeley, CA.

reater Sage-Grouse (*Centrocercus urophasianus*; hereafter sage grouse) are a sensitive species for state and federal resource management agencies due to declining populations and degradation and loss of nesting habitat (Aldridge and Brigham 2001, Connelly et al. 2004, Schroeder et al. 2004). Estimated trends of male sage grouse lek counts in South Dakota declined steadily from 1973 to 1997. From 1997 to 2004, sage grouse populations may have increased slightly (Connelly et al. 2004). Isolation from populations in neighboring states raises additional concerns for sage grouse persistence in South Dakota (Aldridge et al. 2008).

Declines in sage grouse populations have resulted in several petitions to list sage grouse under the Endangered Species Act (ESA) of 1973 (Connelly et al. 2004). Currently, federal land management agencies are responsible for approximately 66% of the sagebrush landscape in the United States. Federal agencies such as the U.S. Bureau of Land Management (BLM) and U.S. Forest Service (USFS) are directed by administrative policy to manage public lands for sustained multiple use under the Federal Land Policy and Management Act (1976) and the Public Rangelands Improvement Act (1978). Currently, sage grouse are managed as a sensitive species by BLM and USFS, and their management should not result in further population declines of sage grouse, which could lead to listing under ESA. The South Dakota Department of Game, Fish, and Parks has identified sage grouse as a species of special concern (South Dakota Department of Game, Fish, and Parks 2006). Listing of sage grouse under the ESA could have major ramifications on the use and management of public lands in the western United States (Knick et al. 2003).

Nest success is one factor that can determine whether sage grouse populations increase or decrease (Braun 1998, Schroeder et al. 1999, Dinsmore and Johnson 2005). Yet information is lacking on the ecological requirements of nesting sage grouse in western South Dakota. The objectives of this study were to develop an understanding on the nesting ecology, success, and resource selection of sage grouse at the eastern edge of their range.

STUDY AREA

The study was conducted within a 3,500 km² area in Butte and Harding counties, South Dakota; Crook County, Wyoming; and Carter County,

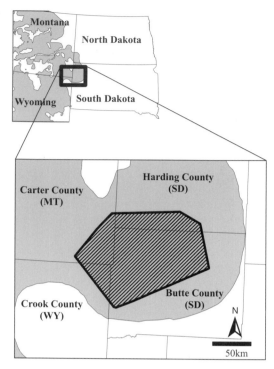

Figure 8.1. Location of study area for Greater Sage-Grouse in Butte, Carter, Crook, and Harding counties, 2006–2007. The hatched area encompasses all locations; the gray area is the current range of Greater Sage-Grouse (Schroeder et al. 2004).

Montana (44°44′ N to 45°20′ N, 103°15′ W to 104°21′ W; Fig. 8.1). Approximately 75% of the area was privately owned. The remaining 25% of the study area was managed by the BLM and State of South Dakota School and Public Lands Division. The area was predominately used for grazing, although small grain production also occurred. Open-pit mining for bentonite occurred at the south end of the study site on Pierre soils (C. Berdan, pers. comm.).

Vegetation consisted of short shrubs, mostly Wyoming big sagebrush (*Artemisia tridentata* spp.) and plains silver sagebrush (*A. cana* spp.). Other shrubs included broom snakeweed (*Gutierrezia sarothrae*), greasewood (*Sarcobatus vermiculatus*), and saltbushes (*Atriplex* spp.) (Johnson and Larson 1999). Common grasses included western wheatgrass (*Pascopyrum smithii*), Junegrass (*Koeleria macrantha*), bluegrass species (*Poa* spp.), green needle-grass (*Nassella viridula*), and Japanese brome (*Bromus japonicus*). Common forbs included western yarrow (*Achillea millefolium*), common dandelion (*Taraxacum officinale*), pepperweed (*Lepidium densiflorum*), and field pennycress (*Thlaspi arvense*) (Johnson and Larson 1999).

Temperatures in summer (May–August) averaged 20.1°C but can reach highs of 43.3°C (South Dakota State Climate Office 2007). During the months of March through June 2006 and 2007, the study area received approximately 14 cm and 22 cm of precipitation, 33% less and 5% more than the 58-year average of 21 cm (1956–2007; South Dakota State Climate Office 2007). Elevation ranges from 840 to 1,225 m above sea level with nearly level to moderately steep clayey soils over clay shale (Johnson 1976).

METHODS

Data Collection

We captured female sage grouse at or near six leks using large nets and spotlighting them from all-terrain vehicles each year between March and mid-April 2006 and 2007 (Giesen et al. 1982, Wakkinen et al. 1992). Females were weighed and equipped with a 22-g necklace style transmitter; transmitters were approximately 1.4% of mean female sage grouse body mass and had a life expectancy of 434 days. Transmitters could be detected from a distance of approximately 2–5 km from the ground and were equipped with an 8-hour mortality switch. Females were classified as yearlings (<1 yr old) or adults (>1 yr old) based on primary wing feather characteristics (Eng 1955, Crunden 1963). The South Dakota State University Institutional Animal Care and Use Committee approved trapping and handling techniques, as well as study design (Protocol #07-A032).

We located radio-marked female sage grouse twice each week during the breeding, laying, and incubation periods. In the event we could not locate an individual from the ground, we searched the study area from a fixed-wing aircraft to obtain an approximate location. Once a female was believed to be incubating, we recorded four coordinates approximately 15 m away from the nest in the four cardinal directions with a Global Positioning System (GPS) receiver. We confirmed nest presence/absence during the subsequent visit. If a female was present on the second visit, we flushed her to determine clutch size. Our use of this method did not decrease nest survival for the immediate interval after the female was flushed from the nest. Nests were considered successful

if ≥1 egg hatched. We calculated distances from nearest active display ground to nests, renests, and previous nests by the same bird using Hawth's Analysis Tool (Beyer 2004).

We characterized vegetation at nest sites after their fate was determined. Four 50-m transects were established radiating in the four cardinal directions from the nest bowl and four additional 5 m transects were established at the 45° intervals. A modified Robel pole was used to estimate visual obstruction (VOR) and maximum grass height at 1-m intervals from 0 m to 5 m ($n = 21$), and at 10-m intervals out to 50 m ($n = 20$) along each 50 m transect (Robel et al. 1970, Benkobi et al. 2000). We estimated sagebrush (*A. tridentata* spp. and *A. cana* spp.) density and height at 10-m intervals ($n = 80$) using the point-centered quarter method (Cottam and Curtis 1956). Vegetation canopy cover was estimated using a 0.10 m^2 quadrat at 1-m intervals to 5 m ($n = 44$) and at 2-m intervals along the long transects to 30 m ($n = 52$). We estimated percent canopy cover for total vegetation, grass, forb, shrub, litter, bare ground, and individual shrub and grass species (Daubenmire 1959). This method is amenable to collecting data on windy days and yields data that are similar (<3% difference for sagebrush) to the line-intercept method, but may provide more accurate estimates of cover (Floyd and Anderson 1987, Booth et al. 2006).

We measured an equal number of random sites within a 3-km buffer of capture leks to estimate resource selection. We navigated to the coordinates of random sites with a GPS and located the center of the transects over the nearest sagebrush because sage grouse usually nest beneath a shrub.

Data Analyses

Nesting Parameters

We used the multi-response permutation procedure (MRPP; Mielke and Berry 2001) to test the null hypothesis that there were no differences between mass of female age-classes, clutch size of female age-classes, clutch size between first nests and renests, nest initiation date between years, distance among nests within a year, distance between nests between years (nest site fidelity), and distance to display grounds between years and age-classes of females. To avoid biasing estimates of nesting and renesting rates, we randomly selected one observation for females that nested both years. Chi-square goodness-of-fit tests were used to test for differences in nest initiation rates between years and between age-classes of females. Statistical significance was set at $\alpha \leq 0.05$. Egg hatchability was the proportion of eggs hatching from successful clutches.

Average grass height and VOR were calculated for each 1-m interval away from the nest to 5 m, at 10-m intervals from 10 to 50 m, and for the site at 0 to 50 m. We used a maximum likelihood estimator to estimate sagebrush density (Pollard 1971). We calculated average sagebrush height for each site from the sagebrush plants that were measured to estimate density. Canopy coverage values were recoded to midpoint values of categories, and these were summarized to an average for 0 to 5 m, 6 to 30 m, and for the site at 0 to 30 m (Daubenmire 1959). To reduce the number of variables in the vegetative dataset to a manageable level and identify biologically important variables to carry forward in the analyses, we used MRPP to identify variables that exhibited differences ($\alpha \leq 0.15$) between nest and random sites, and again between successful and failed nests (Boyce et al. 2002, Stephens et al. 2005). Two separate screen processes were conducted as some variables could be important for nest selection but may not have a measurable effect on nest success.

Resource Selection

We identified ten habitat variables from the nest site selection MRPP analyses (Table 8.1). We used these and a year effect to investigate sage grouse nesting resource selection. Variables included: percent total vegetation cover, grass cover, sagebrush cover, and litter; site averages for sagebrush height, grass height, and visual obstruction; grass height 0–5 m from the nest; visual obstruction at the nest; and visual obstruction 1 m from nest.

Year was included as a design variable in all resource selection candidate models. To reduce potential variable interaction in our models, variables that were correlated to one another ($r > 0.70$) were not included in the same model (e.g., total vegetation cover plus grass cover). We used an information theoretic approach with logistic regression to estimate the support for models evaluating resource selection at nest sites (Burnham and Anderson 2002, SAS Institute Inc. 2007). Due to a small sample size with respect to

the number of parameters estimated ($n/K < 40$); we used the small-sample adjustment for Akaike's Information Criterion (AIC_c) to evaluate models (Burnham and Anderson 2002). We ranked our models based on differences between AIC_c for each model and the minimum AIC_c model (ΔAIC_c), and Akaike weights (w_i) to assess the weight of evidence in favor of each model and the sum AIC_c weight for each variable (Beck et al. 2006). In addition, we investigated the slope of the coefficient estimates (β) to determine variable effect. We evaluated the predictive strength of our models using a receiver operation characteristic curve (ROC); values between 0.7 and 0.8 were considered acceptable predictive discrimination and values higher than 0.8 were considered excellent predictive discrimination. Model goodness-of-fit was determined using a Hosmer–Lemeshow test (Hosmer and Lemeshow 2000).

Nest Success

We used the nest survival procedure in program MARK to evaluate environmental and biological factors that might influence nest survival (White and Burnham 1999, Dinsmore et al. 2002). We standardized nesting dates among years by using the earliest date we discovered a nest as the first day of the nesting season. We monitored nests over a 59-day period beginning 23 April and ending 20 June, which comprised 58 daily intervals of observations to be used in estimating daily survival rate (DSR) for the 27-day incubation period. We identified four variables from the MRPP analyses of nest success as having potential to impact nest success. These variables included: grass height at the site level, visual obstruction at the site level, litter cover at the site level, and forb cover at the nest bowl. The variables were then combined with daily precipitation, daily minimum temperature, bird age, stage of incubation, and year. We did not model nest survival associated with nesting attempt because of a small number of renests ($n = 10$), although they were included in the analysis to test for seasonal variation. Daily weather variables were obtained from the nearest daily weather station located at Nisland, South Dakota, ~50 km from the center of the study area (South Dakota State Climate Office 2007). To reduce the effect of variable interaction in our models, variables that were correlated ($r > 0.70$) were not included in the same model.

TABLE 8.1

Mean vegetation characteristics of nest sites and random sites between years for Greater Sage-Grouse in northwestern South Dakota, 2006–2007.

Variable	Nest			Random			Pooled		
	2006 (n = 34)	2007 (n = 39)	P ≤	2006 (n = 35)	2007 (n = 39)	P ≤	Nest (n = 73)	Random (n = 74)	P ≤
Total cover (%)[1]	61.1 (2.3)	75.1 (2.0)	0.01	55.8 (2.4)	66.1 (2.4)	0.01	68.6 (1.7)	61.2 (1.8)	0.01
Litter (%)	7.6 (0.8)	7.1 (0.6)	0.79	6.5 (0.7)	6.1 (0.4)	0.88	7.4 (0.5)	6.3 (0.4)	0.01
Grass cover (%)[1]	24.2 (1.9)	31.4 (1.8)	0.01	21.1 (1.9)	25.8 (2.0)	0.21	28.1 (1.4)	23.6 (1.4)	0.01
Max grass hgt. (cm)[2]	23.4 (0.9)	29.5 (1.6)	0.01	20.4 (0.8)	25.0 (1.1)	0.01	26.7 (1.0)	22.8 (0.7)	0.01
Max grass hgt. 0–5 m (cm)[2]	25.7 (0.9)	30.9 (2.0)	0.02	20.3 (0.8)	24.3 (1.1)	0.01	28.5 (1.2)	22.4 (0.8)	0.01
Visual obstruction (cm)	5.5 (0.6)	11.1 (1.0)	0.01	3.7 (0.4)	5.1 (0.6)	0.14	8.5 (0.7)	4.4 (0.4)	0.01
Visual obstruction 0 m (cm)[3]	20.8 (1.7)	29.4 (1.8)	0.01	10.5 (1.1)	8.9 (1.0)	0.13	25.4 (1.3)	9.6 (0.7)	0.01
Visual obstruction 1 m (cm)[3]	7.3 (0.9)	13.7 (1.7)	0.01	3.7 (0.5)	4.1 (0.6)	0.45	10.7 (1.0)	3.9 (0.4)	0.01
Sagebrush cover (%)	10.3 (0.8)	10.1 (0.8)	0.75	5.3 (0.8)	6.3 (0.7)	0.98	10.2 (0.6)	6.2 (0.5)	0.01
Sagebrush hgt. (cm)	25.8 (1.2)	29.7 (1.6)	0.04	23.8 (1.0)	24.0 (1.0)	0.97	27.9 (1.7)	23.9 (1.3)	0.01

NOTE: All values are reported as $\bar{x} \pm$ (SE). Variables with the same superscript number were correlated (r > 0.70) and not modeled together.

We used an information theoretic approach to evaluate support for models that influenced DSR (Burnham and Anderson 2002). We began by developing base models that included female age-classes, year, and constant survival. From these base models, we further explored the degree to which habitat and weather variables improved model fit. We used back-transformed estimates of DSR to estimate effects of variables on nest survival for the best supported models (Dinsmore et al. 2002). We then plotted DSR versus simulated values of variables to determine the effect of variables independently from one another. Estimated standard error for nest survival over the 27-day nesting cycle was calculated using the delta method (Seber 1982).

RESULTS

Nesting Parameters

We captured and attached transmitters to 53 female sage grouse (28 yearlings and 25 adults); 29 individuals were included both years for the resource selection analyses. Adults weighed (1,664 ± 14 g, $\bar{x}$ ± SE; $n = 43$) more than yearlings (1,524 ± 16, $n = 24$; $P < 0.01$). There were no differences in female mass between years ($P = 0.20$; $n = 67$). Nest initiation rate for all females was 98.0% and did not differ significantly between years ($P = 0.96$; $n = 67$) or with female age-class ($P = 0.92$; $n = 67$). Renest initiation rate was 25.8% (8/31) and did not differ significantly between years ($P = 0.19$; $n = 31$) or female age-class ($P = 0.62$; $n = 31$). Females were more likely to renest if their first nest was lost early in the incubation period ($P = 0.02$; $n = 31$). The number of nest observation days for first nests was 7.9 ± 1.3 SE days ($n = 8$) for females that renested and 14.6 ± 1.8 SE ($n = 23$) days for females that did not renest.

Average date of nest initiation for successful first nests was 24 April ± 1.6 SE ($n = 30$) days, with adults initiating egg laying approximately 6.7 days earlier than yearlings ($P = 0.02$; $n = 30$). Average hatch date for first nests was 31 May ± 1.5 SE ($n = 30$) days. Average date of renest initiation was approximately 15 days later (9 May ± 2.6 SE days; $n = 8$) than first nests, with hatch date occurring 14 June ± 2.0 SE days. Clutch size differed between nesting attempts (first nests: 8.3 ± 0.2 SE eggs; renests: 6.4 ± 0.6 SE; $P < 0.01$; $n = 64$),

but not by nest fate ($P = 0.83$), female age-class ($P = 0.98$), or year ($P = 0.10$).

One adult female in 2007 nested approximately 30.3 km from lek of capture but most females nested close to leks. In 2006, successful nests were significantly closer to an active lek ($P = 0.04$; $n = 40$) than failed nests (1.5 ± 0.3 km vs. 2.9 ± 0.5 km, $\bar{x}$ ± SE); however, there was no difference in 2007 (2.5 ± 0.5 km vs. 3.2 ± 0.7 km, $P = 0.70$; $n = 39$), or when both years were combined (2.1 ± 0.3 km vs. 3.0 ± 0.4 km, $P = 0.13$; $n = 79$). The distance that adults and yearlings nested from the nearest active lek did not differ significantly (2.2 ± 0.3 km vs. 3.3 ± 0.5 km, $P = 0.08$; $n = 79$). Sixty-eight percent of nests were within 3 km of a documented active lek, and 97% of nests were within 7 km.

Average distance between an individual's nest in 2006 to its nest in 2007 was 1.08 ± 0.40 SE km ($n = 21$). There was no difference in nest site fidelity between adults and yearlings ($P = 0.65$; $n = 21$) or between nests that either failed or were successful the first year ($P = 0.47$; $n = 21$). Mean distance between failed first nests and subsequent renests was 1.85 ± 0.55 SE km ($n = 8$). Successful renests (0.95 ± 0.36 SE km) were not significantly closer to first nests than failed renests (2.03 ± 0.91 SE km, $P = 0.17$; $n = 8$).

Resource Selection

Distribution of total cover, grass cover, grass height, visual obstruction, and sagebrush height differed between nest sites in 2006 and 2007 ($P < 0.05$; Table 8.1). In addition, all screened vegetative characteristics differed between nests and random sites (Table 8.1). The minimum AIC_c model (AIC_c weight = 0.39; Table 8.2) of nest site selection included sagebrush canopy coverage at the site level ($\beta = 0.20$, SE = 0.06) and visual obstruction at the nest ($\beta = 0.22$, SE = 0.04; Table 8.2). Increasing sagebrush cover by 5% increased the odds of use approximately 6.1 times. Increasing visual obstruction at the nest by 2.54 cm increased the odds of use 3.2 times. Predictive ability of the top model (ROC values) was excellent at 0.93 and the Hosmer–Lemeshow goodness-of-fit test was nonsignificant ($P = 0.14$), indicating acceptable model fit.

A second model including sagebrush canopy coverage, visual obstruction at the nest, and average

TABLE 8.2

Selected models from logistic regression analysis (n = 39 models) predicting Greater Sage-Grouse nest sites (n = 73) versus random sites (n = 74) in northwestern South Dakota, 2006–2007.

Model[a]	Log(L)	K[b]	Δ AICc[c]	w_i[d]
Sagebrush cover + visual obstruction 0 m	−50.80	5	0.00	0.52
Sagebrush cover + visual obstruction 0 m + max grass hgt. 0–5 m	−49.82	6	0.22	0.47
Visual obstruction 0 m	−57.50	4	11.26	0.00
Sagebrush cover	−89.14	4	74.54	0.00
Intercept only	−101.89	2	95.85	0.00
Year	−101.89	3	97.92	0.00

[a] For ease of interpretation, year variable was excluded from model column. See Kaczor (2008) for full model set.
[b] Number of habitat parameters plus intercept, SE, and year.
[c] Change in AIC_c value.
[d] Model weight.

TABLE 8.3

Selected models for daily nest survival of Greater Sage-Grouse in northwestern South Dakota, 2006–2007.

Model[a]	K[b]	AIC_c	ΔAIC_c[c]	w_i[d]
Max grass hgt. + litter	3	225.79	0.00	0.23
Max grass hgt. + litter + daily precip. + precip. lag	5	226.75	0.96	0.15
Max grass hgt. + litter + daily precip.	4	227.37	1.60	0.11
Max grass hgt. + litter + bird age	4	227.77	1.98	0.09
Constant	1	252.71	26.92	0.00

[a] See Kaczor (2008) for full model set.
[b] Number of variables plus intercept.
[c] Change in AIC_c value.
[d] Model weight.

grass height within 5 m also had strong support (AIC$_c$ weight = 0.35). Sagebrush canopy coverage and visual obstruction at the nest obtained the highest summed AIC$_c$ weights of 0.99. The combined model of sagebrush canopy cover and visual obstruction at the nest had the greatest support, but there was less support for a single-factor model, although beta estimates for the two variables were similar ($\Delta\beta$ = 0.03).

Nest Success

Most nests were located under Wyoming big sagebrush (90%) or silver sagebrush (7%; n = 79). One nest was against a large boulder, and another was in a dense stand of prairie cordgrass (*Spartina pectinata*). Egg hatchability averaged 78.3 ± 2.1 SE % (n = 513). Constant nest survival rates with no covariates were 45.6 ± 5.3 SE %, but that was a poor model of DSR. The best model for DSR (AIC$_c$ weight = 0.23) included grass height and litter cover (Table 8.3). Three other models were ΔAIC$_c$ ≤ 2 units of the top model. Grass height had a positive association with DSR (β = 0.15, SE = 0.03; Fig. 8.2), whereas percent litter cover had a negative association on DSR (β = −0.08, SE = 0.03); both factors were present in all of models with ΔAIC$_c$ < 2.0.

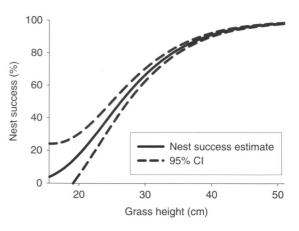

Figure 8.2. Effect of grass height on nest success of Greater Sage-Grouse in northwestern South Dakota, 2006–2007. Nest success estimates were derived from back-transformed beta estimates included in top model. Confidence intervals estimated from the delta method (Seber 1982).

The second-ranked model (AIC$_c$ weight = 0.15) included grass height, litter, daily precipitation, and a 1-day lag of precipitation. Daily precipitation had a positive association with DSR (β = 29.5, SE = 40.4) and the 1-day lag of precipitation was negatively associated with DSR (β = –1.89, SE = 0.77). These variables were only included in supported models when combined with grass height and litter. The third- and fourth-ranked models both included grass height and litter along with the variables daily precipitation and bird age, respectively. Nest success differed between years from 37.7 ± 7.3 SE % in 2006 to 52.5 ± 7.2 SE % in 2007. However, adding a year effect to the top model did not improve model fit.

DISCUSSION

Our study of Greater Sage-Grouse on the easternmost portion of their range in South Dakota identified interesting aspects of sage grouse ecology that have not previously been documented. Female body condition was above average and nesting initiation rates were also high. Similar to other studies, sagebrush cover was an important variable in nest site selection, but at a much lower density than expected. Grass structure, which far exceeded range-wide estimates, played an important role in providing increased cover for successful nests (Connelly et al. 2004). Overall, nest success was within range-wide estimates, suggesting certain features of the habitat condition in South Dakota are productive for sage grouse.

Nesting Parameters

Nest initiation rates for sage grouse are generally low compared to other prairie grouse (Bergerud

1988). However, estimates of nesting initiation based on telemetry are probably underestimated in the literature, as follicular development indicated that at least 98.2% of females laid eggs the previous spring in Idaho (Dalke et al. 1963, Schroeder et al. 1999). Nonetheless, nest initiation rates were high in this study relative to range-wide estimates (Connelly et al. 2004). Females in our study were approximately 63 g (~4%) heavier than the average for 673 individuals in eight other studies (Schroeder et al. 1999). Heavier body mass in female Wild Turkeys (*Meleagris gallopavo*) increased the likelihood of breeding (Porter et al. 1983, Hoffman et al. 1996). Sage grouse exhibit considerable temporal variation in nest initiation rates between years, which may be related to nutrition before and during the breeding season (Hungerford 1964, Barnett and Crawford 1994, Moynahan et al. 2007). High rates of initiation suggest that habitat conditions in our study site were above average.

Renesting rates in sage grouse are highly variable (0–87%), and are linked to environmental effects and habitat quality (Schroeder 1997, Moynahan et al. 2007). Low renesting rates may be related to low primary productivity in the arid and semiarid environments occupied by sage grouse (Schroeder and Robb 2003). For example, Moynahan et al. (2007) found no renesting by sage grouse in dry years with little vegetative growth. In North Dakota, Herman-Brunson et al. (2009) reported 9.5% renesting in sage grouse. The relatively high proportion of renesting females in our study and greater female mass suggest that nesting habitat in South Dakota is of higher quality than elsewhere in sage grouse range. The inverse relationship between length of incubation and renesting propensity suggests that the condition of the female may decline as incubation progresses. An inverse relationship between the duration of incubation and renesting has also been shown elsewhere (Aldridge and Brigham 2001, Herman-Brunson 2009, Martin et al., this volume, chapter 17).

Nest Success

Sage grouse in South Dakota selected nest sites with higher sagebrush cover and placed their nests beneath sagebrush plants with greater horizontal cover (VOR) than random sites. Shrub density (correlated with sagebrush cover) and nest-bowl VOR were important predictors of sage grouse nest sites in North Dakota (Herman-Brunson et al. 2009). Connelly et al. (2000) recommended 15–25% sagebrush canopy coverage for nesting sage grouse, and this recommendation has been confirmed with a range-wide meta-analysis (Hagen et al. 2007). In South Dakota, nesting sage grouse selected for sagebrush with the highest densities and protective cover, but that was less than recommended values. In contrast to sagebrush, grass structure in South Dakota exceeds both management recommendations and range-wide averages (Connelly et al. 2000, Hagen et al. 2007). Western South Dakota forms a transition zone between the northern wheatgrass-needlegrass prairie that dominates most of the Dakotas and the big sagebrush plains of Wyoming (Johnson and Larson 1999). Thus, while South Dakota had less than expected sagebrush cover for sage grouse, the grass structure likely compensated for the low sagebrush densities in providing cover for nests. Grass structure is highly correlated with annual precipitation; therefore, periodic drought may reduce nest cover for sage grouse. Poor grazing management in areas with low sagebrush cover could reduce grass structure, which may have detrimental effects on sage grouse nesting.

Sage grouse nest success varies widely across the range, from 14.5% (Gregg 1991) to 70.6% (Chi 2004), and is generally believed to be related to habitat conditions (Connelly et al. 1991, Aldridge and Brigham 2002, Hagen et al. 2007). Our estimate of nest success was similar to that of other sage grouse studies (48%; Connelly et al. 2004), despite the fact that available sagebrush canopy coverage was less than other areas. Successful nests in our study had taller grass structures than failed nests. Thus, tall grass differentiated not only suitable nest sites,

but also nesting success. Nesting cover also increased nest success in Alberta, and was suggested to provide ample nest concealment in both sagebrush and non-sagebrush overstories in Washington (Sveum et al. 1998, Aldridge and Brigham 2002). Although litter cover entered our models as being an important predictive variable for nest success, the impact litter actually has on nest success is unknown. Litter may be greater after productive growing seasons, or be lower after intensive grazing pressure (Hart et al. 1988, Naeth et al. 1991).

Our results suggest that some aspects of sage grouse habitat in our study area were conducive to maintaining sage grouse populations despite being outside of current management recommendations (Connelly et al. 2000). Although management recommendations were based on existing knowledge, our habitat also provided the necessary requirements for the nesting period, which may be an important consideration for land managers elsewhere in sage grouse ranges.

Management Implications

If sage grouse populations continue to decrease or remain listed as a sensitive species, sagebrush conservation and enhancement could be a top priority for land management agencies to enable sage grouse persistence in western South Dakota. Management for greater grass and sagebrush cover and height, and reduced conversion to tillage agriculture, could be encouraged to protect remaining habitats. Grazing by domestic sheep (*Ovis aries*) can reduce sagebrush cover (Baker et al. 1976), thereby reducing habitat quality for sage grouse. Domestic sheep grazing is not widespread in South Dakota, but was common on both private and public lands in our study area.

Range management practices that could increase sagebrush and grass cover and height include: rest-rotation grazing, where the rested pasture is not grazed until early July to allow for undisturbed nesting, or reduced grazing intensities or seasons of use to reduce impacts on sagebrush and grass growth (Adams et al. 2004). Land managers could develop grazing plans that leave or maintain grass heights ≥26 cm to try to maintain 50% nest success. In addition, we suggest annual grazing utilization not exceed 35% in order to improve rangeland conditions, particularly sagebrush cover (Holechek et al. 1999).

Wyoming big sagebrush typically recovers from a fire in 50–120 years (Baker 2006), and because of the restricted distribution and limited cover of sagebrush in South Dakota, we suggest limited use of prescribed fire or herbicides in areas with sagebrush.

ACKNOWLEDGMENTS

Funding for this study was provided by the Bureau of Land Management (ESA000013), U.S. Forest Service Rocky Mountain Research Station (05-JV-11221609-127), U.S. Forest Service Dakota Prairie National Grasslands (05-CS-11011800-022), and support from South Dakota State University. Field assistance was provided by C. Berdan, T. Berdan, B. Eastman, B. Hauser, T. Juntti, J. Nathan, S. Harrelson, and T. Zachmeier. A number of volunteers assisted during capture and radio-collaring of females and chicks. A. Apa assisted with training on trapping techniques. We also acknowledge and appreciate those land owners who granted us permission to conduct this study on their private lands. We thank D. Turner, J. Sedinger, and an anonymous reviewer for their comments on prior versions of this manuscript. Any mention of trade, product, or firm names is for descriptive purposes only and does not imply endorsement by the U.S. government.

LITERATURE CITED

Adams, B. W., J. Carlson, D. Milner, T. Hood, B. Cairns, and P. Herzog. 2004. Beneficial grazing management practices for sage-grouse (*Centrocercus urophasianus*) and ecology of silver sagebrush (*Artemisia cana*) in southeastern Alberta. Technical Report, Public Lands and Forests Division, Alberta Sustainable Resource Development. Pub. No. T/049.

Aldridge, C. L., and R. M. Brigham. 2001. Nesting and reproductive activities of Greater Sage-Grouse in a declining northern fringe population. Condor 103:537–543.

Aldridge, C. L., and R. M. Brigham. 2002. Sage-grouse nesting and brood-rearing habitat use in southern Canada. Journal of Wildlife Management 66: 433–444.

Aldridge, C. L., S. E. Nielsen, H. L. Beyer, M. S. Boyce, J. W. Connelly, S. T. Knick, and M. A. Schroeder. 2008. Range-wide patterns of Greater Sage-Grouse persistence. Diversity and Distributions 14:983–994.

Baker, M. F., R. L. Eng, J. S. Gashwiler, M. H. Schroeder, and C. E. Braun. 1976. Conservation committee report on the effects of alteration of sagebrush communities on the associated avifauna. Wilson Bulletin 88:165–170.

Baker, W. L. 2006. Fire and restoration of sagebrush ecosystems. Wildlife Society Bulletin 34:177–185.

Barnett, J. K., and J. A. Crawford. 1994. Pre-laying nutrition of sage-grouse hens in Oregon. Journal of Range Management 47:114–118.

Beck, J. L., K. P. Reese, J. W. Connelly, and M. B. Lucia. 2006. Movements and survival of juvenile Greater Sage-Grouse in southeastern Idaho. Wildlife Society Bulletin 34:1070–1078.

Benkobi, L., D. W. Uresk, G. Schenbeck, and R. M. King. 2000. Protocol for monitoring standing crop in grasslands using visual obstruction. Journal of Range Management 53:627–633.

Bergerud, A. T. 1988. Population ecology of North American grouse. Pp. 578–685 *in* A. T. Bergerud and M. W. Gratson (editors), Adaptive strategies and population ecology of northern grouse. Wildlife Management Institute, Minneapolis, MN.

Beyer, H. L. 2004. Hawth's analysis tools for ArcGIS. <http://www.spatialecology.com/htools> (10 January 2006).

Booth, D. T., S. E. Cox, T. W. Meikle, and C. Fitzgerald. 2006. The accuracy of ground-cover measurements. Rangeland Ecology and Management 59:179–188.

Boyce, M. S., P. R. Vernier, S. E. Nielson, and F. K. A. Schmiegelow. 2002. Evaluating resource selection functions. Ecological Modeling 157:281–300.

Braun, C. E. 1998. Sage grouse declines in western North America: what are the problems? Proceedings of the Western Association of Fish and Wildlife Agencies 78:139–156.

Burnham, K. P., and D. R. Anderson. 2002. Model selection and multi-model inference: a practical information theoretic approach. 2nd ed. Springer-Verlag, New York, NY.

Chi, R. Y. 2004. Greater Sage-Grouse on Parker Mountain, Utah. M.S. thesis, Utah State University, Logan, UT.

Connelly, J. W., S. T. Knick, M. A. Schroeder, and S. J. Stiver. 2004. Conservation assessment of Greater Sage-Grouse and sagebrush habitats. Western Association of Fish and Wildlife Agencies, Cheyenne, WY.

Connelly, J. W., M. A. Schroeder, A. R. Sands, and C. E. Braun. 2000. Guidelines to manage sage grouse populations and their habitats. Wildlife Society Bulletin 28:967–985.

Connelly, J. W., W. L. Wakkinen, A. D. Apa, and K. P. Reese. 1991. Sage-grouse use of nest sites in southeastern Idaho. Journal of Wildlife Management 55:521–524.

Cottam, G., and J. T. Curtis. 1956. The use of distance measures in phytosociological sampling. Ecology 37:451–460.

Crunden, C. W. 1963. Age and sex of sage grouse from wings. Journal of Wildlife Management 27:846–849.

Dalke, P. D., D. B. Pyrah, D. C. Stanton, J. E. Crawford, and E. F. Schlatterer. 1963. Ecology, productivity, and management of sage grouse in Idaho. Journal of Wildlife Management 27:811–841.

Daubenmire, R. F. 1959. A canopy-coverage method of vegetation analysis. Northwest Science 33:43–64.

Dinsmore, S. J., and D. H. Johnson. 2005. Population analysis in wildlife biology. Pp. 154–184 in C. E. Braun (editor), Techniques for wildlife investigations and management. 6th ed. The Wildlife Society, Bethesda, MD.

Dinsmore, S. J., G. C. White, and F. L. Knopf. 2002. Advanced techniques for modeling avian nest survival. Ecology 83:3476–3488.

Eng, R. L. 1955. A method for obtaining sage grouse age and sex ratios from wings. Journal of Wildlife Management 19:267–272.

Floyd, D. A., and Anderson, J. E. 1987. A comparison of three methods for estimating plant cover. Journal of Ecology 75:221–228.

Giesen, K. M., T. J. Schoenberg, and C. E. Braun. 1982. Methods for trapping sage grouse in Colorado. Wildlife Society Bulletin 10:224–231.

Gregg, M. A. 1991. Use and selection of nesting habitat by sage-grouse in Oregon. M.S. thesis, Oregon State University, Corvallis, OR.

Hagen, C. A., J. W. Connelly, and M. A. Schroeder. 2007. A meta-analysis of Greater Sage-Grouse Centrocercus urophasianus nesting and brood-rearing habitats. Wildlife Biology 13(Supplement): 42–50.

Hart, R. H., M. J. Samuel, P. S. Test, and M. A. Smith. 1988. Cattle, vegetation, and economic responses to grazing systems and grazing pressure. Journal of Range Management 41:282–286.

Herman-Brunson, K. H., K. C. Jensen, N. W. Kaczor, C. C. Swanson, M. A. Rumble, and R. W. Klaver. 2009. Nesting ecology of Greater Sage-Grouse Centrocercus urophasianus at the eastern edge of their historic distribution. Wildlife Biology 15:237–246.

Hoffman, R. W., M. P. Luttrell, and W. R. Davidson. 1996. Reproductive performance of Merriam's Wild Turkeys with suspected Mycoplasma infection. Proceedings of National Wild Turkey Symposium 7:145–151.

Holechek, J. L., H. Gomez, F. Molinar, and D. Galt. 1999. Grazing studies: what we've learned. Rangelands 21:12–16.

Hosmer, D. W., and S. Lemeshow. 2000. Applied logistic regression. 2nd ed. John Wiley and Sons Inc., New York, NY.

Hungerford, C. R. 1964. Vitamin A and productivity in Gambel's Quail. Journal of Wildlife Management 28:141–147.

Johnson, J. R., and G. E. Larson. 1999. Grassland plants of South Dakota and the northern Great Plains. South Dakota State University. Brookings, SD.

Johnson, P. R. 1976. Soil survey of Butte County, South Dakota. U.S. Department of Agriculture, South Dakota Agricultural Experiment Station, Brookings, SD.

Kaczor, N. W. 2008. Nesting and brood-rearing success and resource selection Greater Sage-Grouse in northwestern South Dakota. M.S. thesis, South Dakota State University, Brookings, SD.

Knick, S. T., D. S. Dobkin, J. T. Rotenberry, M. A. Schroeder, W. M. Vander Haegen, and C. Van Riper III. 2003. Teetering on the edge or too late: conservation and research issues for avifauna of sagebrush habitats. Condor 105:611–634.

Mielke, P. W., and K. J. Berry. 2001. Permutation methods: A distance function approach. Springer-Verlag, New York, NY.

Moynahan, B. J., M. S. Lindberg, J. J. Rotella, and J. W. Thomas. 2007. Factors affecting nest survival of Greater Sage-Grouse in northcentral Montana. Journal of Wildlife Management 71:1773–1783.

Naeth, M. A., A. W. Bailey, D. J. Pluth, D. S. Chanasyk, and R. T. Hardon. 1991. Grazing impacts on litter and soil organic matter in mixed prairie and fescue grassland ecosystems in Alberta. Journal of Range Management 44:7–12.

Pollard, J. H. 1971. On distance estimators of density in randomly distributed forests. Biometrics 27:991–1002.

Porter, W. F., G. C. Nelson, and K. Mattson. 1983. Effects of winter conditions on reproduction in a northern Wild Turkey population. Journal of Wildlife Management 47:281–290.

Robel, R. J., J. N. Briggs, A. D. Dayton, and L. C. Hulbert. 1970. Relationships between visual obstruction measurements and weight of grassland vegetation. Journal of Range Management 23:295–298.

SAS Institute Inc. 2007. JMP version 7. Cary, NC.

Schroeder, M. A. 1997. Unusually high reproductive effort by sage grouse in a fragmented habitat in north-central Washington. Condor 99:933–941.

Schroeder, M. A., and L. A. Robb. 2003. Fidelity of Greater Sage-Grouse Centrocercus urophasianus to breeding areas in a fragmented landscape. Wildlife Biology 9:291–299.

Schroeder, M. A., C. L. Aldridge, A. D. Apa, J. R. Bohne, C. E. Braun, S. D. Bunnell, J. W. Connelly, P. A. Deibert, S. C. Gardner, M. A. Hilliard, G. D. Kobriger, S. M. McAdam, C. W. McCarthy, J. J. McCarthy, D. L. Mitchell, E. V. Rickerson, and S. J. Stiver. 2004. Distribution of sage-grouse in North America. Condor 106:363–376.

Schroeder, M. A., J. R. Young, and C. E. Braun. 1999. Sage grouse (*Centrocercus urophasianus*). A. Poole and F. Gill, (editors), The Birds of North America No. 425. The Birds of North America, Inc., Philadelphia, PA.

Seber, G. A. F. 1982. The estimation of animal abundance and related parameters. 2nd ed. Charles Griffin and Company Ltd., London.

South Dakota Department of Game, Fish, and Parks. 2006. South Dakota comprehensive wildlife conservation plan. South Dakota Department of Game, Fish, and Parks, Wildlife Division Report 2006-08, Pierre, SD.

South Dakota State Climate Office. 2007. Office of the State Climatologist. <http://climate.sdstate.edu> (15 December 2007).

Stephens, P. A., S. W. Buskirk, G. D. Hayward, and C. M. Del Rio. 2005. Information theory and hypothesis testing: a call for pluralism. Journal of Applied Ecology 42:4–12.

Sveum, C. M., W. D. Edge, and J. A. Crawford. 1998. Nesting habitat selection by sage-grouse in south-central Washington. Journal of Range Management 51:265–269.

Wakkinen, W. L., K. P. Reese, J. W. Connelly, and R. A. Fischer. 1992. An improved spotlighting technique for capturing sage grouse. Wildlife Society Bulletin 20:425–426.

White, G. C., and K. P. Burnham. 1999. Program MARK: survival estimation from populations of marked animals. Bird Study 46(Supplement):120–138.

CHAPTER NINE

Use of Dwarf Sagebrush by Nesting Greater Sage-Grouse

David D. Musil

Abstract. I investigated habitat characteristics for 156 Greater Sage-Grouse (*Centrocercus urophasianus*) nests and 138 random plots among eight populations in southern Idaho during 2003–2005. In addition to traditional habitat measurement methods, I included two visual obstruction variables observed from the nesting female's perspective: effective height and horizontal cover. Independent, uncorrelated variables were obtained from principal components analysis for two subcategories of habitat: habitat throughout southern Idaho and sites dominated by dwarf species of sagebrush (little sagebrush, *Artemisia arbuscula*, and black sagebrush, *A. nova*). Height of shrubs and shrub canopy cover was the single most important factor accounting for variability among southern Idaho habitat and sites dominated by dwarf sagebrush, respectively. Shrub density and horizontal cover were the second and third factors contributing to the variability for both subcategories. Grouse used sites with greater shrub height and canopy cover than available at random. Successful nests tended to have greater shrub density and somewhat greater horizontal cover than random points. Successful yearlings had the greatest horizontal cover among the age and fate categories of nests. A gradient of shrub heights was observed, with dwarf sagebrush being the shortest and mesic sites with mixed shrubs and sites dominated by three-tip sagebrush (*A. tripartita*) being the tallest. Despite lacking height, dwarf sites maintained density and horizontal cover similar to other shrub sites. Logistic regression models were developed using variables obtained from principal components analysis. Horizontal cover 3 m from the center of the plot, alone, was the best model for separating nests from random plots in southern Idaho habitat and had the best classification rate among dwarf sagebrush habitat. Habitat measurements of dwarf sagebrush–dominated nest sites were within the habitat guidelines used for managing Greater Sage-Grouse and this is the first time this habitat type has been described. Nests ($n = 46$) in habitat dominated by dwarf sagebrush had 43% success and nests with dwarf sagebrush species concealing the nest bowl ($n = 22$) had 50% nest success. Certain grass and shrub species have similar overall heights but considerably different effective concealment cover. Grouse researchers should consider using effective height and horizontal cover habitat measurements in future research of shrub-steppe grouse.

Key Words: Artemisia, black sagebrush, *Centrocercus urophasianus,* effective height, Greater Sage-Grouse, habitat use, horizontal cover, little sagebrush, nest habitat.

Musil, D. D. 2011. Use of dwarf sagebrush by nesting Greater Sage-Grouse. Pp. 119–136 *in* B. K. Sandercock, K. Martin, and G. Segelbacher (editors). Ecology, conservation, and management of grouse. Studies in Avian Biology (no. 39), University of California Press, Berkeley, CA.

reater Sage-Grouse (*Centrocercus urophasianus*) populations have declined throughout their range (Connelly and Braun 1997, Connelly et al. 2004), and their distribution is greatly influenced by the occurrence of shrubsteppe habitat types, especially those dominated by sagebrush (Patterson 1952, Connelly and Braun 1997). Past research on nest habitat of Greater Sage-Grouse in Idaho (Connelly et al. 1991) and elsewhere (Hagen et al. 2007) has focused on landscapes dominated by taller structured sagebrush like Wyoming big sagebrush (*Artemisia tridentata wyomingensis*), mountain big sagebrush (*A. t. vaseyana*), and silver sagebrush (*A. cana*). Little is known about nesting by Greater Sage-Grouse in habitat dominated by dwarf sagebrush like little sagebrush (*A. arbuscula*) and black sagebrush (*A. nova*), which are more associated with winter cover (Dalke et al. 1963, Beck 1977). Gregg et al. (1994) sampled three successful nests in little sagebrush habitat and Popham and Gutiérrez (2003) observed four nests in little sagebrush. Current guidelines for managing Greater Sage-Grouse habitat are also based on studies conducted in areas largely dominated by taller sagebrush species (Connelly et al. 2000). Providing values for nest use in dwarf sagebrush habitat and comparing with current guidelines will assist in managing this unique habitat to benefit Greater Sage-Grouse.

Traditional habitat measurement methods have focused on droop height of the tallest structure of a grass or shrub (Connelly et al. 2003), overlooking the possible importance of visual obstruction provided by individual plants. Plants can have the same height but provide different horizontal concealment cover depending on their growth structure, soil and climate differences, and land management practices. Concealment cover is important for several aspects of grouse survival (Mussehl 1963, Bergerud and Gratson 1988), especially for nesting Greater Sage-Grouse (Hagen et al. 2007).

Nesting Greater Sage-Grouse females are known to take periodic incubation breaks to forage and defecate away from the nest (Coates and Delehanty 2008). These forays risk exposing the female and her nest to predation, especially if concealment cover is lacking. Measuring horizontal cover from the nesting female's perspective is a novel approach and may explain selection of a nest site for optimum concealment during incubation breaks.

My objective was to describe habitat use immediately around nests of Greater Sage-Grouse in southern Idaho, focusing on areas dominated by dwarf sagebrush species for comparisons to sites dominated by other shrub species. I also wanted to determine the usefulness of two new habitat measurement methods to describe cover measured from a nesting female's perspective in addition to methods previously used.

STUDY AREAS

This research was conducted on multiple study areas throughout southern Idaho (Fig. 9.1), previously described by other researchers (Klott et al. 1993, Wik 2002, Lowe 2006, Shepherd 2006), and ranging in elevation from 1,600 to 2,400 m. Most of the 20–40 cm of annual precipitation in southern Idaho occurs during the winter and spring. The landscape, in general, is flat to rolling shrubsteppe habitat on valley floors, steeper slopes along valley edges, increasing in elevation to base of mountain ranges. Mesic sites are typically on west sides of mountain ranges with deeper soils, whereas xeric sites are on east sides within the rain shadow in shallow soils. Depending on the moisture regime and soil depth, several sagebrush species dominate the cover, including Wyoming big sagebrush, mountain big sagebrush, basin big sagebrush (*A. t. tridentata*), three-tip sagebrush (*A. tripartita*), little sagebrush, and black sagebrush. Non-sagebrush shrubs include antelope bitterbrush (*Purshia tridentata*), green rabbitbrush (*Chrysothamnus viscidiflorus*), saltbush (*Atriplex* spp.), spiny hopsage (*A. spinosa*), and horsebrush (*Tetradymia canescens*). Common grasses include bluebunch wheatgrass (*Pseudoroegneria spicata*), Idaho fescue (*Festuca idahoensis*), cheatgrass (*Bromus tectorum*), western wheatgrass (*Agropyron smithii*), crested wheatgrass (*A. cristatum*), basin wildrye (*Elymus cinereus*), Sandberg bluegrass (*Poa secunda*), Indian ricegrass (*Achnatherum hymenoides*), bottlebrush squirreltail (*Elymus elymoides*), and needle-and-thread grass (*Stipa comata*). Common forbs include western yarrow (*Achillea millefolium*), wild onion (*Allium* spp.), pussytoes (*Antennaria* spp.), arrow-leaf balsamroot (*Balsamorhiza sagittata*), hawksbeard (*Crepis* spp.), buckwheat (*Eriogonum* spp.), lettuce (*Lactuca* spp.), lupine (*Lupinus* spp.), phlox (*Phlox* spp.), salsify (*Tragopogon* spp.), and mules ears (*Wyethia amplexicaulis*).

Figure 9.1. Eight populations of Greater Sage-Grouse sampled for nest habitat in southern Idaho, 2003–2005. Populations are: (1) Washington County, (2) Cow Creek, (3) Oreana/ Big Springs/Shoofly, (4) Sheep Creek, (5) Browns Bench/ Shoshone Basin, (6) Laidlaw Park, (7) Birch Creek, and (8) Little Lost Creek. Elevation is depicted with shading from low (black) to high (white) elevations.

METHODS

Habitat Sampling

This research was conducted on eight breeding populations of Greater Sage-Grouse among two moisture regimes and three shrub species categories distributed throughout southern Idaho. Xeric sites have annual precipitation <30 cm/yr, whereas mesic sites have >30 cm/yr and are dominated by a variety of sagebrush and non-sagebrush species (Hironaka et al. 1983). Sites were also categorized according to dominance of sagebrush height based on grouping by Hironaka et al. (1983): tall (*A. tridentata* spp.) and dwarf (little and black sagebrush). I also considered three-tip sagebrush as a separate category. The shrub species with the largest canopy cover at a plot was considered dominant.

Greater Sage-Grouse females were captured by night lighting (Giesen et al. 1982, Wakkinen et al. 1992), fitted with 16.5-g necklace-style radio transmitters (Riley and Fistler 1992), and monitored during laying and incubation as part of other ongoing studies. Age-class was categorized as yearling (<1 yr old) and adult (>1 yr old) and determined by

wing characteristics (Eng 1955, Dalke et al. 1963). A nest was considered successful if at least one egg hatched. Random point coordinates, independent of nest sites, were generated using ArcView Spatial Analyst (ESRI, Redlands CA 92373) and located in the field with handheld GPS units. Random points were centered on the coordinates to avoid observer bias and eliminate overestimation of available shrub canopy cover and density. Random plots were sampled within known boundaries of nesting Greater Sage-Grouse based on past and current research in Idaho.

Habitat was measured at nest sites within one week after females ceased nesting efforts or when abandoned or depredated nests would have hatched based on initiation dates and a 27-day incubation period (Patterson 1952). Random plots were measured throughout the hatching season and at least one plot was measured within a week of measuring a nest plot. Habitat measurements were sampled along four 10-m transects placed at right angles radiating from the center of the nest and oriented in a random direction similar to Wakkinen (1990), Gregg et al. (1994), and Musil et al. (1994). Droop height of the closest shrub and grass was measured for each plant species within 1 m of the transect at 1, 3, and 5 m from the center of the plot for each of the four transects. Droop height is defined as the tallest naturally growing portion of the plant (Connelly et al. 2000, 2003). Sagebrush was identified according to a dichotomous key (Hironaka et al. 1983), but subspecies and hybrids were possibly misidentified by not following more discriminating methods (Stevens and McArthur 1974, Shumar et al. 1982, Rosentreter and Kelsey 1991). Droop height of residual (previous season growth excluding flower stalks), live (current green and growing leaves), flower stalk (tallest flower structure), and number of flower stalks was measured for each grass species separately. Maximum height (tallest structure of entire plant) was generated post hoc by selecting maximum measurements from those recorded in the field. Effective height was measured by placing a meter stick behind the grass or shrub opposite the nest and estimating the tallest height concealing >50% of the 2.5-cm-wide meter stick.

Horizontal cover outside of the nest bowl was measured using a cover pole at 1, 3, and 5 m from the plot center along the transects and read from 20 cm above the ground immediately outside of the nest shrub or at the center of a random plot (Robel et al. 1970; Fig. 9.2). At least one-half of a 2.54-cm-tall segment (48 segments/pole) had to be obscured by vegetation to be counted as covered. Measurements were separated post hoc according to the following segment categories: low (0–18 cm), medium (19–61 cm), high (62–122 cm), ground (0–61 cm), and total (0–122 cm). Readings from the 1-m position were later discarded so nests and random plots could be compared.

Shrub canopy cover and shrub density was measured along the 10-m transects and segmented into 0–1, 1–3, 3–5, 5–10, and 1–10 m increments (Canfield 1941). To eliminate oversampling the center and to compare random plots to nests, the 0–1 m increment was discarded during subsequent analysis. Gaps in the canopy >5 cm were excluded (Connelly et al. 2003). Shrub density was sampled by counting the number of plants of each shrub species intersecting or within 0.5 m on both sides for the length of the transects and separated into the same segments used for canopy coverage.

Understory cover for each forb and grass species was measured with a 40 × 50 cm modified cover frame (Daubenmire 1959) at 1, 3, and 5 m from the center on each of the four transects. Cover canopies were modified from Daubenmire (1959) to include more sensitivity for lower cover values for the following cover classes: 1 (0–1%), 2 (2–5%), 3 (6–25%), 4 (26–50%), 5 (51–75%), and 6 (76–100%). Slope and aspect were measured using a clinometer and compass, respectively. Elevation was estimated by plotting locations on 7.5-minute quadrangle maps.

Statistical Analysis

Principal components analysis was used to reduce the set of correlated habitat variables into independent components using SAS (McGarigal et al. 2000, SAS Institute 2001). My goal was to test the statistical hypothesis that there is no difference in vegetation between use and non-use by nesting Greater Sage-Grouse. The biological hypothesis tested was that Greater Sage-Grouse were selecting nesting habitat for concealment cover. Varimax rotation facilitated interpretation of the variables within the components (O'Rourke et al. 2005). Meaningful factor loadings were set at ±0.40. Variables were considered complex and were removed if they significantly loaded on >1 component. Analyses were run sequentially until all complex variables were identified and eliminated. Post hoc analyses were also conducted by selecting a subset of the data to determine variables best describing nest use for habitat dominated by dwarf sagebrush.

In an attempt to avoid model dredging (Guthery et al. 2005), I identified variables from the first principal components describing at least 50% of the cumulative variability for subsequent use in developing predictive models of nest use analyzed with logistic regression (Afini and Clark 1996). One variable with the greatest loading from each principal component was selected to represent the component in the model. When loadings were tied, the variable with the greatest ease of interpretation for land management was selected. Once a list of variables was chosen, I conducted a model search using a SAS macro (All Possible Logistic Regressions, available from C. T. Moore; see White et al. 2005) and used corrected Akaike's Information Criterion (AIC$_c$) scores to rank the models (Burnham and Anderson 2002), according to methods described by White et al. (2005). The models were also ranked according to Akaike weights (w_i) and the best model picked with the

Figure 9.2. Measurement of horizontal cover at a Greater Sage-Grouse nest site, Idaho. Top: Observer reading cover pole from immediately outside the nest bowl with eye level 20 cm above ground. Pole is 3 m distance from center of nest. Bottom: View of cover pole measured in upper photo.

largest weight (Burnham and Anderson 2002). For each model, I used a SAS macro (Monte Carlo Cross-Validation for Logistic Regression, available from C. T. Moore; see White et al. 2005) to determine the predictive accuracy for each model $\leq \Delta AIC_c = 2$ of the best model and set the number of iterations at 1,200. I set the proportion of the data withheld based on the number of variables in the model using criteria described by White et al. (2005): 50%, 40%, and 35% of the data withheld for 1-, 2-, and 3-variable models, respectively.

RESULTS

I measured 156 Greater Sage-Grouse nests and 138 random plots from eight grouse breeding populations during 2003–2005 in southern Idaho, including 89 vegetation variables. Forty-four percent ($n = 68$) of the 156 nests successfully produced chicks. Adults comprised 77% of the nests sampled and of those, 43% were successful, whereas 40% of the 35 yearling nests were successful. Overall apparent success of nests sampled was 34% ($n = 62$), 49% ($n = 46$), and 44% ($n = 48$) in 2003, 2004, and 2005, respectively. For nest sites in dwarf sagebrush dominated habitat, I sampled 21 successful and 24 unsuccessful nests. Of the 22 nests with dwarf sagebrush directly over the nest bowl, 11 were successful. Analyses are separated between the data for plots sampled throughout all habitat types in southern Idaho and for the subset of plots dominated by dwarf sagebrush.

Principal Components Analysis

Southern Idaho Habitat

Eleven variables associated with horizontal cover near the ground as well as live, effective, and maximum grass heights were identified as complex variables and were removed from principal components analysis, but 78 other variables were retained. Sixteen principal components met the minimum eigenvalue of 1.0 and described 84% of the variance in the data. A majority of the variance (55%) was described by the first five principal components (Table 9.1). Eight shrub height variables were associated with the first principal component (Prin I) and accounted for 23% of the variance; Prin II had eight variables of shrub density accounting for 11% of variance; Prin III had eight variables of horizontal cover and 8% of variance; Prin IV had eight variables of shrub canopy cover and 7% of variance; and Prin V had five forb cover variables describing 6% of the variance.

Viewing the first three components three-dimensionally (Fig. 9.3A), nests were separated from random plots by having greater shrub height (Prin I) and shrub density (Prin II). Successfully hatched nests of yearling females appeared to have the greatest horizontal cover (Prin III) and shrub density. Successful adult nests had shorter shrub heights than successful yearlings, while maintaining similar shrub density. Unsuccessful nests of both adults and yearlings tended to have less horizontal cover than successful nests but values similar to random plots.

Grouping plots by sagebrush type and moisture regimes for the first three principal components showed a gradient of shrub height from dwarf-dominated plots to mesic and three-tip sagebrush plots (Fig. 9.4). There also appears to be a gradient along the Prin III axis, with mesic sites having the least and three-tip sagebrush having the most horizontal cover. Mesic random plots had the least shrub density while maintaining tall shrub height along with mesic nest sites. Dwarf sagebrush sites, though lacking the height, maintained shrub density and horizontal cover similar to the taller sites.

Dwarf Sagebrush Habitat

Analysis of 46 nests and 43 random plots dominated by dwarf sagebrush cover showed 12 variables were considered complex; these were removed, leaving 77 variables in the analysis. Complex variables were associated with the number of grass and forb species and heights of shrubs and grasses. Ten principal components met the minimum eigenvalue of 1.0 and accounted for 80% of the variability. A majority (53%) of the variance was described by the first four principal components (Table 9.2). Prin I had eight variables associated with shrub canopy cover and accounted for most (19%) of the variability. Shrub density was described by eight variables for Prin II and accounted for 15% of the variability. Prin III was associated with seven variables of horizontal cover with 10% of the variability. Five variables of grass height and one variable of horizontal cover were associated with Prin IV and described 9% of the variability.

Comparing the first three principal components three-dimensionally (Fig. 9.3B), nests of unsuccessful yearling females tended to have the least shrub density but horizontal cover similar to unsuccessful adults, whereas successful yearlings had the greatest horizontal cover and highest shrub canopy cover. Successful adult nests tended to have the greatest shrub density but shrub canopy cover was more similar to random plots than other nests. Random dwarf plots tended to have the lowest shrub canopy coverage but horizontal cover similar to adult nests.

Logistic Regression

Southern Idaho Habitat

The five variables with highest loading on the first five principal components, one from each

TABLE 9.1

Variables loading on first principal components comprising majority of variability of Greater Sage-Grouse nesting habitat in Idaho, 2003–2005.

Principal component I		Principal component II		Principal component III		Principal component IV		Principal component V	
Variables[a]	%[b]	Variables[c]	%	Variables[d]	%	Variables[e]	%	Variables[f]	%
HTSHRUBEF	89	DENSAGE110	94	HCOVTT3	89	CCSAGE110	93	CCFORBTT	94
HTSHRUBLV	87	DENSAGE35	94	HCOVMED3	88	CCSAGE35	86	CCFORB1	87
HTSHRUBEF3	86	DENSAGE510	93	HCOVTT5	88	CCSHRUB110	82	CCFORB5	87
HTSHRUBEF1	85	DENSHRUB35	92	HCOVGRD3	87	CCSAGE510	82	CCFORB3	85
HTSHRUBLV1	82	DENSHRUB110	90	HCOVHIGH5	86	CCSHRUB510	78	CCFORBSPP	59
HTSHRUBLV3	81	DENSHRUB510	89	HCOVHIGH3	80	CCSAGE13	78		
HTSHRUBEF5	80	DENSAGE13	72	HCOVMED5	79	CCSHRUB35	76		
HTSHRUBLV5	74	DENSHRUB13	68	HCOVLOW5	43	CCSHRUB13	70		
(23% of variability)		(11% of variability)		(8% of variability)		(7% of variability)		(6% of variability)	

[a] Shrub height abbreviations: EF is effective height, LV is height of live portion of tallest branch measured 3 or 5 m from center of plot.

[b] Percent loading of variable onto component; only those variables loading >40% were retained.

[c] Density of shrubs: SAGE is any sagebrush species, SHRUB is all shrub species including sagebrush measured at segments of 1–3 (13), 3–5 (35), 5–10 (510), and 1–10 (110) m from center of plot.

[d] Horizontal cover: LOW is measured from ground level to 18 cm, MED is measured from 18–61 cm above ground level, HIGH is from 61 to 122 cm, GRD is from ground level to 61 cm, and TT is entire cover pole measured 3 and 5 m from center of plot.

[e] Canopy coverage of shrubs: same designations as density.

[f] Canopy coverage of forbs for 1, 3, and 5 m from center of plot and for all distances totaled (TT).

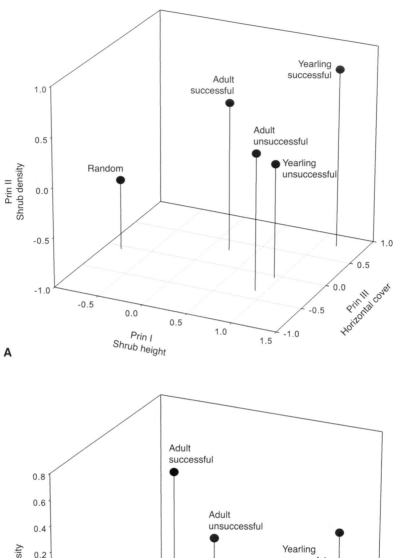

Figure 9.3. Habitat relationship of first three principal components for age-class of Greater Sage-Grouse females, nest fate, and random plots in Idaho, 2003–2005. A: Habitat throughout southern Idaho. B: Habitat dominated by dwarf sagebrush.

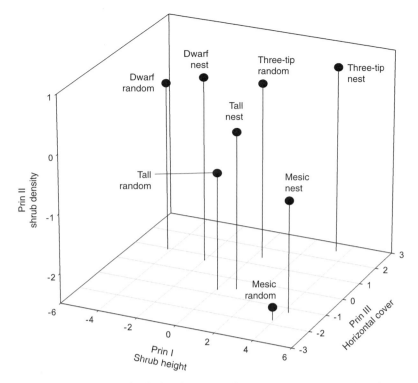

Figure 9.4. Habitat relationship for first three principal components among Greater Sage-Grouse nest sites and random plots by moisture category and shrub size in Idaho, 2003–2005.

TABLE 9.2

Variables loading on first principal components comprising majority of variability of Greater Sage-Grouse nesting habitat dominated by dwarf sagebrush in Idaho, 2003–2005.

Principal component I		Principal component II		Principal component III		Principal component IV	
Variables[a]	%[b]	Variables[c]	%	Variables[d]	%	Variables[e]	%
CCSHRUB110	95	DENSAGE110	98	HCOVTT3	95	HTGRASSFLWTT	91
CCSAGE110	94	DENSAGE510	97	HCOVHIGH5	92	HTGRASSFLW5	87
CCSHRUB510	92	DENSAGE35	96	HCOVTT5	90	HTGRASSFLW1	82
CCSAGE510	89	DENSHRUB35	95	HCOVMED3	86	HTGRASSMAX5	81
CCSAGE35	87	DENSHRUB110	94	HCOVHIGH3	85	HTGRASSFLW3	75
CCSHRUB35	86	DENSHRUB510	94	HCOVGRD3	83	HCOVLOW5	41
CCSAGE13	79	DENSAGE13	71	HCOVMED5	76		
CCSHRUB13	78	DENSHRUB13	68				
(19 % of variability)		(15 % of variability)		(10 % of variability)		(9 % of variability)	

[a] Shrub canopy cover abbreviations: SAGE is any sagebrush species, SHRUB is all shrub species including sagebrush measured at segments of 1–3 (13), 3–5 (35), 5–10 (510), and 1–10 (110) m from center of plot.

[b] Percent loading of variable onto component; only those variables loading >40% are retained.

[c] Density of shrubs: same designations as canopy coverage.

[d] Horizontal cover (HCOV): MED is measured from 18–61 cm above ground level, HIGH is from 61 to 122 cm, GRD is from ground level to 61 cm, and TT is entire cover pole measured 3 and 5 m from center of plot.

[e] Droop height (HT) of grass flowers (GRASSFLW) at 1, 3, and 5 m from center of plot, and for all distances totaled (TT). HCOVLOW5 is the horizontal cover measured from ground level to 18 cm with a cover pole at 5 m from center of plot.

Model[a]	Estimate	SE	AIC$_c$	w_i	Correctly classified (%)[b]	
					$\bar{X}$	SE
Intercept	−0.954	0.258	386.9	0.332	59.3	0.1
HCOV3	0.024	0.005				
Intercept	−0.908	0.261	387.1	0.289	59.4	0.1
HCOV3	0.026	0.006				
CCFORB	−0.016	0.012				
Intercept	−1.035	0.319	388.6	0.132	58.6	0.1
DENSAGE110	0.030	0.069				
HCOV3	0.024	0.005				
Intercept	−0.914	0.289	388.7	0.125	59.6	0.1
HTSHRUBEF	−0.002	0.006				
HCOV3	0.024	0.005				
Intercept	−0.932	0.286	388.8	0.122	59.0	0.1
HCOV3	0.024	0.005				
CCSAGE110	−0.003	0.014				

[a] Variable abbreviations: HCOV3 is horizontal cover at 3 m from center of plot, CCFORB is canopy cover of forbs throughout plot, DENSAGE110 is density of sagebrush 1–10 m from center of plot, HTSHRUBEF is the effective height of all shrubs, CCSAGE110 is canopy cover of sagebrush 1–10 m from center of plot.
[b] Cross-validation based on 1,200 iterations and withholding 50% and 40% of data to validate the 1- and 2-variable models, respectively.

component, were used to develop and compare models from all possible combinations of these variables to separate nests from random plots. Even though the two variables had identical loadings for Prin II, density of sagebrush 1–10 m from center of plot was chosen to represent Prin II instead of sagebrush density at 3–5 m due to greater ease of interpretability for landscape management. The model with the lowest AIC$_c$ score and greatest weight (w_i) had one variable (horizontal cover 3 m from center of plot) and had a classification rate of 59.3% (Table 9.3). Additional variables did not improve the classification rate.

Dwarf Sagebrush Habitat

Horizontal cover 3 m from the center of the plot and height of grass flower stalks best separated nests from random plots in dwarf sagebrush dominated habitat (Table 9.4). Horizontal cover alone had a better classification rate than the minimum AIC$_c$ model. Models with additional variables had higher AIC$_c$ and lower w_i values as well as poorer classification rates.

Dwarf Sagebrush Nest Characteristics

I sampled 46 Greater Sage-Grouse nests and 43 random plots that were dominated by dwarf sagebrush canopy cover (Table 9.5). Dwarf sagebrush tended to be shorter than other sagebrush by about 20 cm at nest sites. Nest sites tended to have greater overall shrub height than that available at random. Plant height over the nest was greater than average shrub heights 1–10 m from the nest. Females were using about 20-cm taller dwarf sagebrush to conceal the nest than the

TABLE 9.4

Models of Greater Sage-Grouse nesting habitat dominated by dwarf sagebrush in Idaho, 2003–2005.

| Model[a] | Estimate | SE | AIC_c | w_i | Correctly classified (%)[b] | |
					$\overline{X}$	SE
Intercept	−1.148	0.726	106.6	0.394	65.9	0.2
HCOV3	0.080	0.020				
HTGRASSFLW	−0.053	0.033				
Intercept	−1.937	0.560	107.3	0.285	68.4	0.2
HCOV3	0.065	0.017				
Intercept	−0.772	0.841	108.1	0.189	65.1	0.2
DENSAGE110	−0.0824	0.092				
HCOV3	0.081	0.021				
HTGRASSFLW	−0.060	0.034				
Intercept	−1.179	0.923	108.8	0.132	64.9	0.2
CCSHRUB110	0.002	0.029				
HCOV3	0.079	0.022				
HTGRASSFLW	−0.053	0.034				

[a] Variable abbreviations: HCOV3 is horizontal cover at 3 m from center of plot, HTGRASSFLW is height of grass flower, DENSAGE110 is density of sagebrush 1–10 m from center of plot, CCSHRUB110 is canopy cover of shrubs 1–10 m from center of plot.
[b] Cross-validation based on 1,200 iterations and withholding 50%, 40%, and 35% of data to validate the 1-, 2-, and 3-variable models, respectively.

average height of dwarf sagebrush in the vicinity of the nest. Nest sites had >2 cm taller effective and residual grass heights than random plots but had similar maximum grass heights. Shrub densities, canopy cover, and grass and forb cover were similar between nests and random points. Nests appeared to have greater horizontal cover 3 m from the center of the plot than was available at random. Slope and aspect were similar, but nests were located at slightly lower elevations than random plots.

Use of Effective Height

Depending on the grass species, tall grasses like bluebunch wheatgrass, crested wheatgrass, and Idaho fescue tended to have a greater proportion of the maximum height providing effective horizontal cover than shorter stature species like cheatgrass and Sandberg bluegrass (Table 9.6). Needle-and-thread and Indian ricegrass, though species of taller stature, provided less horizontal cover than bottlebrush squirreltail, a shorter species with dense foliage. Crested wheatgrass provided the tallest horizontal cover, whereas cheatgrass provided the shortest.

Effective height for shrub species was related to the overall height (Table 9.7). Little sagebrush, black sagebrush, spiny hopsage, and green rabbitbrush were low stature and tended to have shorter effective heights. The proportion of the plant providing effective cover also varied with the flower structure. Shrubs with short flower stalks, like antelope bitterbrush and spiny hopsage, which have flowers embedded within the branches, tended to have a higher proportion of the plant providing concealment cover. Species with long flower stalks providing maximum height well above the main structure of the shrub, like basin big sagebrush and Wyoming big sagebrush, tended to have a lower proportion of the overall height providing effective concealment cover. Mountain big sagebrush had a greater effective concealment cover percentage than other species of sagebrush.

TABLE 9.5
Characteristics of Greater Sage-Grouse nests in habitat dominated by dwarf sagebrush in southern Idaho, 2003–2005.

Variable	Nests (n = 46)		Random (n = 43)	
	X̄	SE	X̄	SE
Dwarf sagebrush height (cm)	30.6	1.2	25.8	1.2
Non-dwarf sagebrush height (cm)	51.2	4.6	44.2	5.7
All sagebrush height (cm)	35.7	1.8	28.4	1.6
Shrub height (cm) all shrubs combined	30.9	1.7	22.6	1.7
Effective dwarf sagebrush height (cm)	21.9	1.2	18.3	1.3
Effective non-dwarf sagebrush height (cm)	35.4	5.2	34.5	5.5
All sagebrush effective height (cm)	25.2	1.7	20.6	1.6
Effective shrub height (cm)	24.3	2.0	18.3	1.9
Plant height (cm)[a] over nest by species				
Dwarf sagebrush (22 nests)	51.5	2.3	—	—
Non-dwarf sagebrush (10 nests)	69.5	8.5	—	—
Other shrub (11 nests)	73.5	24.8	—	—
Non-shrub (2 nests)	44.0	27.0	—	—
Grass height maximum (cm)	27.1	1.3	26.9	1.5
Effective grass height (cm)	9.3	0.7	6.8	0.6
Residual grass height (cm)	11.7	1.3	9.1	0.8
Live grass height (cm)	14.4	0.6	12.8	0.7
Grass flower height (cm)	23.1	1.2	23.2	1.4
Shrub density (#/m2)	4.2	0.5	4.4	0.4
Sagebrush density (#/m2)	3.2	0.5	3.5	0.3
Dwarf sagebrush density (#/m2)	3.1	0.4	3.4	0.3
Shrub canopy cover (%)	22.8	1.3	19.3	1.3
Sagebrush canopy cover (%)	18.6	1.6	18.2	1.7
Dwarf sagebrush canopy cover (%)	17.4	1.4	16.3	1.6
Grass cover (%)	15.1	1.6	14.9	1.5
Forb cover (%)	10.5	1.6	9.2	1.5
Horizontal cover (%) 3 m from center	37.8	2.3	25.4	2.2
Slope (degrees)	3.5	0.7	3.9	0.6
Aspect (degrees)	161.7	16.4	141.5	14.2
Elevation (m)	1,678	25	1,776	28

[a]Maximum droop height of plant over nest. One nest was in open with no plants over the nest.

TABLE 9.6
Effective height (cm) compared to traditional maximum droop height for selected grass species in Greater Sage-Grouse habitat, Idaho, 2003–2005.

Species	n^a	Effective height		Maximum height		% Effective of maximum	
		$\bar{X}$	SE	$\bar{X}$	SE	$\bar{X}$	SE
Bluebunch wheatgrass	179	12.9	0.7	31.6	1.3	48.6	4.0
Bottlebrush squirreltail	175	10.6	0.5	20.6	0.8	57.8	3.2
Cheatgrass	152	1.2	0.3	12.0	0.5	13.2	3.3
Crested wheatgrass	23	16.3	1.6	34.3	2.4	50.7	5.9
Idaho fescue	52	11.6	0.7	24.5	1.8	57.8	5.1
Indian ricegrass	59	6.6	0.6	12.9	1.0	62.1	6.0
Needle-and-thread grass	30	7.5	0.7	25.9	3.6	41.2	6.9
Sandberg bluegrass	248	4.1	0.2	23.2	0.8	25.0	1.2

NOTE: Maximum number of plants measured is 12/plot; grazed plants were not measured.

[a] Number of nests and random plots combined.

TABLE 9.7
Effective height (cm) compared to traditional maximum droop height for selected shrub species in Greater Sage-Grouse habitat, Idaho, 2003–2005.

Species	n^a	Effective height		Maximum height		% Effective of maximum	
		$\bar{X}$	SE	$\bar{X}$	SE	$\bar{X}$	SE
Little sagebrush	128	22.4	0.8	29.4	0.9	79.3	5.4
Black sagebrush	25	15.6	1.1	25.2	1.9	59.9	1.7
Basin big sagebrush	47	48.8	4.0	66.5	4.3	69.7	2.6
Three-tip sagebrush	42	30.5	1.6	47.0	2.2	64.7	1.5
Mountain big sagebrush	36	63.8	4.2	69.1	4.8	88.1	3.5
Wyoming big sagebrush	130	41.9	1.8	51.1	1.6	78.2	1.9
Antelope bitterbrush	31	76.2	7.5	79.4	7.9	92.6	3.1
Spiny hopsage	35	28.0	2.6	32.7	2.9	88.3	9.0
Green rabbitbrush	162	22.6	0.9	28.4	0.9	76.3	1.6

NOTE: Maximum number of plants measured is 12/plot; grazed plants were not measured.

[a] Number of nests and random plots combined.

DISCUSSION

Nest Habitat Use

Descriptions of Greater Sage-Grouse nests in dwarf sagebrush–dominated habitat have previously been limited to a few nest sites (Gregg et al. 1994,

Popham and Gutiérrez 2003) or at the landscape scale (Shepherd 2006). Popham and Gutiérrez (2003) found Greater Sage-Grouse avoided dwarf sagebrush nesting habitat but Shepherd (2006) found almost equal proportions of nests in little and big sagebrush landscapes. In my study, nest sites in dwarf sagebrush habitat accounted for 29% of the nests sampled and were within the established guidelines for sage grouse breeding habitat characteristics (Connelly et al. 2000), despite being dominated by shorter stature species of sagebrush. Instead of shrub height, shrub canopy coverage was the factor most important in describing the variation in the data. Females also used taller sagebrush immediately over the nest for concealment, as has been found for studies in habitats dominated by taller sagebrush species (Wallestad and Pyrah 1974, Petersen 1980, Wakkinen 1990, Musil et al. 1994, Apa 1998). Females in dwarf sagebrush also tended to use taller grass, similar to results of Wakkinen (1990), Gregg et al. (1994), and Sveum et al. (1998). Use does not necessarily correlate with fitness at a landscape scale (Aldridge and Boyce 2007), but 43% of the nests we sampled in dwarf sagebrush habitat successfully hatched, as did 50% of the nests concealed by dwarf sagebrush over the nest bowl. These rates are average for nest success of Greater Sage-Grouse (Bergerud 1988, Schroeder et al. 1999), showing that dwarf sagebrush can provide productive nesting habitat.

This study is the first to compare Greater Sage-Grouse nesting habitat use across multiple habitat regimes. The models did not perform well, having a <70% correct classification rate between nest site and random plots. Also, the top variables used in the models accounted for relatively low amounts of the variability in the data, perhaps due to similarity in available vegetation within the habitat types. Klebenow (1969) had a similar problem when describing Greater Sage-Grouse nest habitat in Idaho. Despite these shortcomings, some interesting patterns were revealed with my data. It appears Greater Sage-Grouse obtain concealment cover for nests depending on the height of the dominant sagebrush species available. Within tall sagebrush, females chose sites with greater shrub density than random points, but with similar shrub heights. The same was true for mixed shrubs on mesic sites, except this habitat provided taller shrubs, probably because robust antelope bitterbrush was more common. Three-tip sagebrush nest sites had shrub heights similar to mesic sites but were among the densest shrub values across the habitat types sampled. Females selected nest sites in dwarf sagebrush habitat with more overhead cover from the shrub canopy, possibly to compensate for lack of brush height. Herman-Brunson et al. (2009)

found higher nest success on sites with shorter sagebrush but greater canopy cover.

Greater Sage-Grouse females are likely selecting sites with adequate concealment within 10 m of the nest, but I did not discern at what distance cover might open for better views of approaching predators (Götmark et al. 1995). Females require concealment while exiting from and returning to the nest during incubation breaks to avoid attracting predators. Coates and Delehanty (2008) determined that yearling female Greater Sage-Grouse take more frequent and longer incubation breaks than adults, thus exposing themselves and their nests to greater depredation. Incubation rhythms could explain why successful yearlings tended to have greater horizontal cover in my study in both overall habitat use and sites dominated by dwarf sagebrush. Coupling less incubation constancy (Coates and Delehanty 2008) with apparently less horizontal cover, unsuccessful yearling females are more exposed during their frequent movements to and from the nest and are more easily detected by predators, thus lending themselves to higher nest failure. Wiebe and Martin (1998) found a fitness trade-off, though, where concealed nests of White-tailed Ptarmigan (*Lagopus leucurus*) were less likely to be detected by predators but had higher risks of depredation on incubating females because approaching predators could not be easily seen. I had 2 (1%) females depredated on nests, similar to low values found in other studies (Moynahan et al. 2007, Coates et al. 2008), so it is probably more beneficial for Greater Sage-Grouse to find nest locations with more concealment than a view of approaching predators.

Use of Two New Habitat Measurement Variables

Effective height and horizontal cover are not necessarily new methods for measuring habitat, but they have not been previously used in conjunction to describe Greater Sage-Grouse nest habitat. Rotenberry and Wiens (1980) used effective height to measure bird habitat, some of which was in shrubsteppe cover, and it loaded significantly on a component they described as vertical cover. Similarly, Mussehl (1963) measured a form of effective height for Blue Grouse (*Dendragapus obscurus*) habitat and determined its importance for concealment during brood rearing but did not describe the protocol used to measure effective height, especially the angle of view. Visual obstruction methods have been used to describe Greater Sage-Grouse nest habitat and have been found to be important indicators of habitat use (Popham and Gutiérrez 2003, Moynahan et al. 2007, Herman-Brunson et al. 2009, Kaczor et al., this volume, chapter 8) but have not been recorded from the point of view of nesting females.

Horizontal cover, as I measured it with a cover pole, proved to be the most important metric, occurring in every top predictive model. I did record the gaps among the different levels above ground but did not discriminate among obstruction by shrubs, grasses, or forbs. In hindsight, horizontal cover should be categorized by the type of plant obscuring the pole and at what level above ground it occurs. Also, horizontal cover should be measured from different positions rather than just from the plot center, since females require concealment while traveling to and from nests during incubation breaks, not just while sitting on the nest. Cover at 5 m from the nest was typically masked by cover at 3 m, except when cover was lacking at 3 m, so observations should be no more than 2 m from the pole. Comparisons should be conducted between observing the pole from ground level and from the traditional 1-m height

above ground to see if the two metrics are correlated. Observing from the female's perspective likely includes more obstruction than from traditional measurements taken 1 m above ground.

I found differences of effective heights among different species of grasses and shrubs. Sandberg bluegrass, as with other *Poa* spp., may have tall flower stalks, but the cover they provide remains in the blades at the base, providing little concealment when compared to bottlebrush squirreltail, which has similar overall height but dense cover throughout the plant structure. Interspecific differences are illustrated in Figure 9.5, which compares two bunch grasses with similar heights but different concealment qualities. A site dominated by plants with taller effective heights would provide more concealment for nests and likely better nest success. I conclude that vegetation height alone does not adequately describe habitat for nesting Greater Sage-Grouse; quality of the plant structure is more important. Effective shrub height may not measure the true concealment by shrubs, because only cover at crown height is measured. Species like basin big sagebrush tend to have a more open structure near ground level, whereas dwarf sagebrush and

Figure 9.5. Example of two grasses with similar heights but different levels of horizontal concealment cover. Grass on left (bluebunch wheatgrass) has taller concealment cover than grass on right (needle-and-thread grass), but both species have relatively identical droop heights of residual, live, and flower structures. Meter stick is supported with a metal rod.

mountain big sagebrush have cover throughout the vertical structure, thereby providing more concealment of nests.

I recommend biologists include two techniques I used when monitoring nesting habitat for grouse populations in shrubsteppe habitat. Effective height and horizontal cover measure concealment by visual obstruction, the former for quality of individual plant species and the latter for quality of the landscape in general. A combination of the two methods provided important information for modeling the nesting habitat of Greater Sage-Grouse across a range of habitats in my study. I believe measuring habitat from the grouse's perspective adds strength to detecting relationships with the environment, especially in habitat that is naturally sparse, such as shrubsteppe, but may not be applicable to habitats with dense understories, as in prairie grasslands.

ACKNOWLEDGMENTS

This research was funded in part by the U.S. Fish and Wildlife Service Federal Aid in Wildlife Restoration Project W-160-R-34. I thank all the Idaho Department of Fish and Game personnel who participated in this project: J. Connelly and T. Hemker for supervision; C. White assisted with logistic regression analysis; and the following field technicians: B. Atkinson, P. Atwood, B. Cadwallader, A. Foley, S. Harrington, D. Lockwood, B. Lowe, R. Morris, C. Perugini, D. Plattner, T. Stirling, D. VanDoren, and R. Wilson. D. VanDoren also suggested the measurement of effective height. I thank graduate students N. Burkepile, B. Lowe, J. Shepherd, and P. Wik for the use of data from their nesting females. G. Stanford, J. Stanford, and B. Tindall were gracious hosts, allowing access to their private lands and providing campsites and hospitality for field crews. Earlier versions of this manuscript were greatly enhanced by two anonymous reviewers and the final form by associate editor B. Sandercock.

LITERATURE CITED

Afini, A. A., and V. Clark. 1996. Computer-aided multivariate analysis. 3rd ed. Chapman & Hall, New York, NY.

Aldridge, C. L., and M. S. Boyce. 2007. Linking occurrence and fitness to persistence: habitat-based approach for endangered Greater Sage-Grouse. Ecological Applications 17:508–526.

Apa, A. D. 1998. Habitat use and movements of sympatric sage and Columbian Sharp-tailed Grouse in southeastern Idaho. Ph.D. dissertation, University of Idaho, Moscow, ID.

Beck, T. D. I. 1977. Sage grouse flock characteristics and habitat selection during winter. Journal of Wildlife Management 41:18–26.

Bergerud, A. T. 1988. Population ecology of North American grouse. Pp. 578–685 in A. T. Bergerud and M. W. Gratson (editors), Adaptive strategies and population ecology of northern grouse. Wildlife Management Institute, University of Minnesota Press, Minneapolis, MN.

Bergerud, A. T., and M. W. Gratson. 1988. Survival and breeding strategies of grouse. Pp. 473–577 in A. T. Bergerud and M. W. Gratson (editors), Adaptive strategies and population ecology of northern grouse. Wildlife Management Institute, University of Minnesota Press, Minneapolis, MN.

Burnham, K. P., and D. R. Anderson. 2002. Model selection and multimodel inference: a practical information-theoretic approach. 2nd ed. Springer-Verlag, New York, NY.

Canfield, R. H. 1941. Application of the line interception method in sampling range vegetation. Journal of Forestry 39:388–394.

Coates, P. S., J. W. Connelly, and D. J. Delehanty. 2008. Predators of Greater Sage-Grouse nests identified by video monitoring. Journal of Field Ornithology 79:421–428.

Coates, P. S., and D. J. Delehanty. 2008. Effects of environmental factors on incubation patterns of Greater Sage-Grouse. Condor 110:627–638.

Connelly, J. W., and C. E. Braun. 1997. Long-term changes in sage grouse Centrocercus urophasianus populations in western North America. Wildlife Biology 3:229–234.

Connelly, J. W., S. T. Knick, M. A. Schroeder, and J. S. Stiver. 2004. Conservation assessment of Greater Sage-Grouse and sagebrush habitats. Unpublished Report. Western Association of Fish and Wildlife Agencies, Cheyenne, WY.

Connelly, J. W., K. P. Reese, and M. A. Schroeder. 2003. Monitoring of greater sage-grouse habitats and populations. Bulletin 80, University of Idaho, College of Natural Resources Experiment Station, Moscow, ID.

Connelly, J. W., M. A. Schroeder, A. R. Sands, and C. E. Braun. 2000. Guidelines to manage sage grouse populations and their habitats. Wildlife Society Bulletin 28:967–985.

Connelly, J. W., W. L. Wakkinen, A. D. Apa, and K. P. Reese. 1991. Sage-grouse use of nest sites in southeastern Idaho. Journal of Wildlife Management 55:521–524.

Dalke, P. D., D. B. Pyrah, D. C. Stanton, J. E. Crawford, and E. F. Schlatterer. 1963. Ecology, productivity, and management of sage grouse in Idaho. Journal of Wildlife Management 27:810–841.

Daubenmire, R. 1959. A canopy-coverage method of vegetational analysis. Northwest Science 33:43–64.

Eng, R. L. 1955. A method for obtaining sage grouse age and sex ratios from wings. Journal of Wildlife Management 19:267–272.

Giesen, K. M., T. J. Schoenberg, and C. E. Braun. 1982. Methods for trapping sage grouse in Colorado. Wildlife Society Bulletin 10:224–231.

Götmark, F., D. Blomqvist, O. C. Johansson, and J. Bergkvist. 1995. Nest site selection: a trade-off between concealment and view of the surroundings? Journal of Avian Biology 26:305–312.

Gregg, M. A., J. A. Crawford, M. S. Drut, and A. K. DeLong. 1994. Vegetational cover and predation of sage-grouse nests in Oregon. Journal of Wildlife Management 58:162–166.

Guthery, F. S., L. A. Brennan, M. J. Peterson, and J. J. Lusk. 2005. Information theory in wildlife science: critique and viewpoint. Journal of Wildlife Management 69:457–465.

Hagen, C. A., J. W. Connelly, and M. A. Schroeder. 2007. A meta-analysis of Greater Sage-Grouse *Centrocercus urophasianus* nesting and brood-rearing habitats. Wildlife Biology 13:42–50.

Herman-Brunson, K. M., K. C. Jensen, N. W. Kaczor, C. S. Swanson, M. A. Rumble, and R. W. Klaver. 2009. Nesting ecology of Greater Sage-Grouse *Centrocercus urophasianus* at the eastern edge of their historic distribution. Wildlife Biology 15:237–246.

Hironaka, M., M. A. Fosberg, and A. H. Winward. 1983. Sagebrush-grass habitat types of southern Idaho. College of Forestry, Wildlife, and Range Sciences Bulletin No. 35, Moscow, ID.

Klebenow, D. A. 1969. Sage-grouse nesting and brood habitat in Idaho. Journal of Wildlife Management 33:649–662.

Klott, J. H., R. B. Smith, and C. Vullo. 1993. Sage grouse habitat use in the Brown's Bench area of south-central Idaho. Technical Bulletin 93-4. United States Department of Interior, Bureau of Land Management, Idaho State Office, Boise, ID.

Lowe, B. S. 2006. Greater Sage-Grouse use of threetip sagebrush and seeded sagebrush-steppe. M.S. thesis, Idaho State Univeristy, Pocatello, ID.

McGarigal, K., S. Cushman, and S. Stafford. 2000. Multivariate statistics for wildlife and ecology research. Springer-Verlag, New York, NY.

Moynahan, B. J., M. S. Lindburg, J. J. Rotella, and J. W. Thomas. 2007. Factors affecting nest survival of Greater Sage-Grouse in north-central Montana. Journal of Wildlife Management 71:1773–1783.

Musil, D. D., K. P. Reese, and J. W. Connelly. 1994. Nesting and summer habitat use by translocated sage-grouse (*Centrocercus urophasianus*) in central Idaho. Great Basin Naturalist 54:228–233.

Mussehl, T. W. 1963. Blue Grouse brood cover selection and land-use implications. Journal of Wildlife Management 63:547–555.

O'Rourke, N., L. Hatcher, and E. J. Stepanski. 2005. A step-by-step approach to using SAS for univariate and multivariate statistics. 2nd ed. SAS Institute, Inc., Cary, NC.

Patterson, R. L. 1952. The sage grouse in Wyoming. Sage Books, Inc., Denver, CO.

Petersen, B. E. 1980. Breeding and nesting ecology of female sage grouse in North Park, Colorado. M.S. thesis, Colorado State University, Fort Collins, CO.

Popham, G. P., and R. J. Gutiérrez. 2003. Greater Sage-Grouse *Centrocercus urophasianus* nesting success and habitat use in northeastern California. Wildlife Biology 9:327–334.

Riley, T. Z., and B. A. Fistler. 1992. Necklace radio transmitter attachment for pheasants. Journal of Iowa Academy of Science 99:65–66.

Robel, R. J., J. N. Briggs, A. D. Dayton, and L. C. Hulbert. 1970. Relationships between visual obstruction measurements and weight of grassland vegetation. Journal of Range Management 23:295–298.

Rosentreter, R., and R. G. Kelsey. 1991. Xeric big sagebrush, a new subspecies in the *Artemisia tridentata* complex. Journal of Range Management 44:330–335.

Rotenberry, J. T., and J. A. Wiens. 1980. Habitat structure, patchiness, and avian communities in North American steppe vegetation: a multivariate analysis. Ecology 6:1228–1250.

SAS Institute. 2001. The SAS system for Windows, version 8.2. SAS Institute Inc., Cary, NC.

Schroeder, M. A., J. R. Young, and C. E. Braun. 1999. Sage grouse (*Centrocercus urophasianus*). A. Poole and F. Gill (editors), The birds of North America No. 425. The Birds of North America, Inc., Philadelphia, PA.

Shepherd, J. F., III. 2006. Landscape-scale habitat use by Greater Sage-Grouse (*Centrocercus urophasianus*) in southern Idaho. Ph.D. dissertation, University of Idaho, Moscow, ID.

Shumar, M. L., J. E. Anderson, and T. D. Reynolds. 1982. Identification of subspecies of big sagebrush by ultraviolet spectrophotometry. Journal of Range Management 35:60–62.

Stevens, R., and E. McArthur. 1974. A simple field technique for identification of some sagebrush taxa. Journal of Range Management 27:325–326.

Sveum, C. M., W. D. Edge, and J. A. Crawford. 1998. Nesting habitat selection by sage-grouse in south-central Washington. Journal of Range Management 51:265–269.

Wakkinen, W. L. 1990. Nest site characteristics and spring–summer movements of migratory sage-grouse in southeastern Idaho. M.S. thesis, University of Idaho, Moscow, ID.

Wakkinen, W. L., K. P. Reese, J. W. Connelly, and R. A. Fischer. 1992. An improved spotlighting technique for capturing sage grouse. Wildlife Society Bulletin 20:425–426.

Wallestad, R. O., and D. Pyrah. 1974. Movement and nesting of sage grouse hens in central Montana. Journal of Wildlife Management 38:630–633.

White, C. G., S. H. Schweitzer, C. T. Moore, I. B. Parnell, and L. A. Lewis-Weis. 2005. Evaluation of the landscape surrounding Northern Bobwhite nest sites: a multiscale analysis. Journal of Wildlife Management 69:1528–1537.

Wiebe, K. L., and K. Martin. 1998. Costs and benefits of nest cover for ptarmigan: changes within and between years. Animal Behaviour 56:1137–1144.

Wik, P. A. 2002. Ecology of Greater Sage-Grouse in south-central Owyhee County, Idaho. M.S. thesis, University of Idaho, Moscow, ID.

CHAPTER TEN

Modeling Nest and Brood Habitats of Greater Sage-Grouse

Jay F. Shepherd, John W. Connelly, and Kerry P. Reese

Abstract. We derived landscape variables and their estimates that help explain nest and brood success at multiple spatial extents for Greater Sage-Grouse (*Centrocercus urophasianus*). These variables should allow habitat managers to more accurately delineate breeding habitat in a manner that has empirical validity for effective Greater Sage-Grouse habitat conservation. We examined grouse breeding habitat at different spatial extents in southeastern Idaho to determine how well mid-scale landscape vegetation parameters might describe habitats that result in successful versus unsuccessful reproductive efforts. We extracted vegetation buffers at two spatial extents (150- and 450-m radius circular buffers) for each nest and brood site and used Akaike's Information Criterion to determine models and habitat variables that best explained nest and brood success. We defined a successful nest as a nest where one or more eggs hatch; a successful brood was defined as one in which at least one chick survives to 10 weeks of age. We used field data from 176 nests and 62 broods across five years, 1998–2002. At the 150-m spatial extent, the most explanatory and parsimonious logistic regression models determining nest success indicated that the number of sagebrush patches and grass–forb dominated cover (i.e., non-sagebrush habitat dominated by grasses and some forbs) had a negative influence on nest success, while percent land dominated by sagebrush positively influenced nest success. At the larger spatial extent (450-m buffers), edge density (m/ha) of sagebrush (a metric of sagebrush fragmentation) and grass-forb dominated cover had a negative influence on nest success, while percent land dominated by sagebrush was positively related to success. For both spatial extents, the primary difference between successful and unsuccessful nests was the amount of sagebrush and grass–forb dominated vegetation present on the landscape. At both spatial extents, percent land dominated by grass–forb cover negatively influenced brood success, while a small proportion of agricultural cover was positively related to success. Additionally, edge density of grass-forb dominated cover negatively influenced brood success at the 450-m spatial extent. Amount of grass–forb dominated cover and edge density of sagebrush had a negative influence on brood success during the early brood-rearing period

Shepherd, J. F., J. W. Connelly, and K. P. Reese. 2011. Modeling nest and brood habitats Greater Sage-Grouse. Pp. 137–150 *in* B. K. Sandercock, K. Martin, and G. Segelbacher (editors). Ecology, conservation, and management of grouse. Studies in Avian Biology (no. 39), University of California Press, Berkeley, CA.

(1–3 weeks), while sagebrush-dominated land cover had a positive effect during this period. The amount of grass–forb dominated landcover was an important variable in explaining brood success for the early and entire brood period at both spatial extents: Successful broods had less grass–forb dominated habitat in the surrounding landscape.

Key Words: brood-rearing, *Centrocercus urophasianus*, Greater Sage-Grouse, habitat models, landscape-scale, nesting, sagebrush.

Habitat variables influencing selection of nest sites by Greater Sage-Grouse (*Centrocercus urophasianus*; hereafter sage grouse) are characterized by the presence of sagebrush (*Artemisia* spp.) and herbaceous cover (Connelly et al. 1991, 2000; Gregg et al. 1994; Holloran et al. 2005; Hagen et al. 2007). Similarly, females with broods use areas characterized by both sagebrush and herbaceous cover (Klott and Lindzey 1990, Connelly et al. 2000, Hagen et al. 2007, Casazza et al., this volume, chapter 11). These investigations of nest and brood-rearing habitat have generally been conducted to understand the importance of vegetation composition and structure at relatively small spatial scales (e.g., the immediate vicinity of nests or brood locations). Recent studies have also addressed broader landscape characteristics and again demonstrated the importance of sagebrush cover to sage grouse (Aldridge and Boyce 2007, Walker et al. 2007, Aldridge et al. 2008). A negative relationship was demonstrated between mean numbers of males/lek and agricultural development on the upper Snake River plain in Idaho (Leonard et al. 2000). Additionally, in North Dakota abandoned leks had a higher percentage of tilled lands within a 4-km buffer than did active leks (Smith et al. 2005). Percent sagebrush cover on the landscape was an important predictor of winter use by sage grouse and strength of sagebrush habitat selection by sage grouse was strongest at the 4-km^2 scale in comparison to smaller scales (Doherty et al. 2008).

Sage grouse depend on relatively large blocks of intact sagebrush-dominated habitat (Connelly et al. 2000, Holloran et al. 2005, Doherty et al. 2008). Thus, the amount of sagebrush cover fragmented or broken up into smaller patches by agriculture, fires, or other disturbances may negatively affect sage grouse populations (Leonard et al. 2000, Smith et al. 2005). Studies of forest-dwelling grouse and neotropical migratory birds have demonstrated the negative effect of fragmentation on avian populations (Wilcove 1985, Wegge et al. 1992, Paton 1994, Kurki et al. 2000).

Although many land management decisions may be made at an intermediate scale (i.e., grazing pastures, energy development, vegetation treatments), few studies have attempted to assess the influence of vegetation characteristics at a mid-scale landscape extent (7–64 km^2) on sage grouse breeding habitat. Additionally, sage grouse broods use agricultural land during summer (Connelly et al. 1988, 2000), but its importance to sage grouse is not well understood.

Fragmentation and loss of sagebrush-dominated habitats have been related to the decline of Greater Sage-Grouse, an obligate of sagebrush habitats (Connelly et al. 2004). Moreover, given that Greater Sage-Grouse were recently categorized under the United States' Endangered Species Act as "warranted but precluded" (U.S. Department of Interior 2010), additional insights into the influence of habitat and landscape characteristics on population productivity are essential for improved management of the species. Thus, our overall objective is to describe mid-scale landscape-level vegetation attributes that are important in explaining reproductive success of sage grouse. We examined vegetation characteristics for sage grouse nesting habitat and brood-rearing habitat and between early (1–3 weeks) and late (3–10 weeks) brood-rearing habitats. Specifically, we attempted to develop effective models that identify important vegetation landscape metrics that are related to sage grouse nest success and brood survival at two spatial extents.

METHODS

Study Area

Our study area was located in southeast Idaho near the town of Dubois, south of the Centennial Mountains and east of the Lost River Mountain

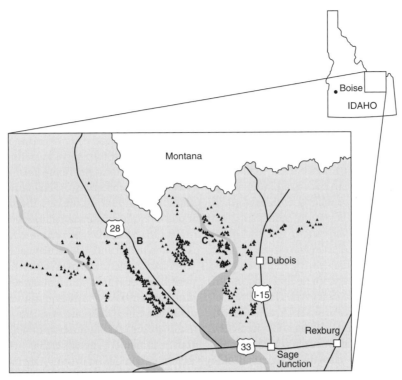

Figure 10.1. Brood locations of Greater Sage-Grouse (black triangles) in the Little Lost River (A) and Birch Creek valleys (B), and the Dubois area (C), in southeastern Idaho, 1998–2002.

Range (approximately 44°18′ N, 112°53′ W), including the Little Lost River and Birch Creek valleys and the shrubsteppe and agricultural areas within a 35–45 km radius west, south, and northeast of Dubois (Fig. 10.1). The study area was approximately 2,700 km² and ranged from 1,300 to 2,500 m in elevation. Average yearly precipitation ranged from 17.5 cm in lower elevations to 45 cm in higher elevations; average monthly temperature ranged from –10°C to 35°C (State Climate Services, Biological and Agricultural Engineering Department, University of Idaho). The study area generally supported stable, but variable, sage grouse populations and included important seasonal and year-round sage grouse habitat (Idaho Sage-Grouse Advisory Committee 2006). Cover types within the study area were dominated by subspecies of the big sagebrush complex, grasslands, and agriculture, especially alfalfa (*Medicago sativa*; Scott et al. 2002).

Greater Sage-Grouse Capture and Sampling

We used spotlighting (Wakkinen et al. 1992, Connelly et al. 2003) to capture sage grouse on or near leks during March and April from 1998 to 2002. Adult and yearling female sage grouse were leg-banded and equipped with a necklace-mounted 15-g battery-powered transmitter with a 4-hr mortality sensor. Female age-class was determined by wear patterns of outer primary feathers [<1 year old (yearling) or >1 year (adult); Crunden 1963]. We used portable receivers (Advanced Telemetry Systems, Isanti, MN) with handheld Yagi and vehicle-mounted antennas to locate radio-marked grouse, and GPS units to determine locations. Nest locations were obtained through direct observation of females on nests, and nest UTM coordinates were precisely recorded after hatching or loss occurred. Chicks were caught <48 hrs after hatching, radio-marked (Burkepile et al. 2002), and monitored to document chick survival and movements. Broods were monitored by obtaining multiple locations per week of individual females and chicks, and status of broods was determined by direct observation.

We classified nests by year and success or failure; a successful nest was defined as a nest in which at least one egg hatched. Brood locations were classified by year, female age-class, and whether the brood was successful or failed. Successful broods were defined as those in which one brood member survived to at least 10 weeks from hatch date and thus was considered recruited to the fall population. Because broods often use different habitats during the brood-rearing period (Connelly et al. 2000), we also classified brood locations by age-class of chicks (weeks 1–3 and >3–10). At 10–12 weeks of age, sage grouse juveniles approach independence, brood break-up may begin to occur, and juveniles may not be in the same location as the maternal female (Schroeder et al. 1999). Thus, brood survival cannot be reliably measured when chicks are >10 weeks of age.

Remotely Sensed Data

We obtained cover type and patch configuration and size from Landsat Thematic Mapper Imagery data classified by the 1998 Idaho GAP Analysis Project (Scott et al. 2002). The imagery was obtained from 1991–1996 spring and summer satellite passes when vegetation was actively growing (June–September). Due to cloud cover, data from several years were used to produce a statewide data layer. The imagery was updated with annual Bureau of Land Management (BLM) fire data to create a landcover layer for each year of the study, such that each year was a distinct data set (Shepherd 2006). There were 81 landcover classification categories in the original landcover map and seven new classification categories for wildfire by year from 1996–2002. Pixel size was 30 m, or 0.09 ha. Overall producer accuracy of landcover classification for nine different subsections of southern Idaho within the overall 1998 Idaho Gap Analysis Project was estimated to be 69.3% with a range of 63.6% to 79.3%. A more precise estimate of accuracy cannot be provided because an error matrix including producer and user accuracy, from which errors of omission and commission can be calculated, was not reported in the original southern Idaho classification process (Homer 1998, Scott et al. 2002).

Remotely Sensed Data Classification

We reclassified the cover type map into a simplified classification aggregating 78 of 81 landcover classifications into 12 categories. The reclassification process combined seven sagebrush types from the original Idaho GAP Analysis cover type map into one composite sagebrush category, as well as all grassland types and areas burned since the most recent satellite pass into one category, grass–forb dominated cover. The other categories were urban, disturbed areas, agriculture, other shrubs, rabbitbrush (*Chrysothamnus* spp.), deciduous trees, coniferous trees, water, riparian, and bare ground (Shepherd 2006). Field observations indicate sagebrush cover types may be misclassified and heterogeneous at a finer scale than mapped by the original Idaho GAP Analysis cover type map. Reclassification will certainly increase applicability of the results to other areas with different mixtures of sagebrush cover types.

We created cover type data files using two spatial extents: 150 and 450 m. The two buffer extents represent the maximum radiotelemetry error for hens, and therefore the highest degree of accuracy and smallest possible buffer extent (150 m) and the average daily distance moved (450 m) for adult greater sage grouse in three study areas in southern Idaho (Shepherd 2006). Moreover, the 150-m buffer approximated the stand used by grouse for nesting or brood-rearing activities. The 450-m buffer will likely contain greater amounts of fragmentation and a more heterogeneous environment for sage-grouse. Buffers were placed around individual nests and broods with ≥5 locations, and 70% least squares cross-validation fixed kernel home ranges were estimated for broods with ≥20 locations using the home range extension for Arcview (Rodgers and Carr 1998, ESRI 1999b). We used an Arc Macro Language (AML; ESRI 1999a) routine to extract vegetation buffers for each nest, creating a separate cover type data file for each nest. For brood locations, a single cover type data file was created by placing a contiguous buffer around locations of individual broods by year to avoid pseudo-replication of landscape data through sampling the same vegetation area more than once (Hurlbert 1984, Shepherd 2006).

For nests and broods, we examined data for each spatial extent separately using the program FRAGSTATS 3.3, and a suite of landscape metrics was produced (McGarigal et al. 2001). Aggregated sagebrush (>80%) or grass–forb dominated cover (>8%) types were the dominant landcover types of the 12 aggregated cover types, and therefore we derived the most relevant variables from the primary landscape metrics associated with these cover types because any relatively rare but important cover types would have been included in the sagebrush and/or grass–forb cover types. Sage grouse are dependent on comparatively large blocks of intact stands of sagebrush (Connelly et al. 2000, Holloran et al. 2005), although later brood-rearing habitat will often contain a rather high proportion of grass–forb cover or agricultural land (Connelly et al. 1988, 2000; Leonard et al. 2000; Hagen et al. 2007). Thus, the primary landscape metrics we used included percent land cover, patch density, number of patches, and edge density (m/ha) of sagebrush and grass–forb dominated cover types.

Nests were encompassed by a single uniformly sized, noncontiguous buffer, and therefore comparison of either number of patches or patch density (patches/ha) could be used in the nest analysis. The number and juxtaposition of brood locations buffered by both spatial extents differed for each brood; therefore, landscape metrics independent of size of analysis area (i.e., the 150- or 450-m buffered area) were necessary for comparison. In contrast to nest analysis, patch density (patches/ha) and number of patches could not be used in the brood analysis.

Data Analysis

We conducted univariate correlation tests for all landscape metrics to identify sets of independent variables for logistic regression that were not correlated. We used logistic regression to model nest and brood success at each spatial extent level (Zar 1999). We used Akaike's Information Criterion (AIC) and AIC differences (Δ_i), Akaike's Information Criterion for low sample size (AIC$_c$), AIC weights (w_i), and multi-model inference to assess models and associated landscape metrics (Burnham and Anderson 2002). If the ratio of n to K, or sample size to the largest number of explanatory variables in a model including the intercept, was >40, AIC was used; otherwise, AIC for small samples was used. We determined the relative importance of each variable by summing the Akaike weights (Σw_i) for each model containing the variable and used model averaging when we could not identify a top model to describe nest or brood success.

For logistic regression, we assessed global model fit using the Hosmer–Lemeshow goodness-of-fit test (Hosmer and Lemeshow 1989). Global models included all variables within the subset of best models by spatial extent and model type (nest, brood, brood age). We used the statistical software

SAS (SAS Institute, Inc. 1988), SYSTAT (SPSS, Inc. 2000), and SPSS (SPSS, Inc. 1993) for data manipulation, configuration, and analysis.

RESULTS

Nests

There was no difference in age distribution of females between successful nests (80% adult, 20% yearling) and unsuccessful nests (81% adult, 19% yearling); thus we pooled nests of all females for analysis. We used 176 nests (85 successful, 91 unsuccessful) from 1998 to 2002 for this analysis. Number of nests ranged from 14 in 1998 to 71 in 2002. Eighty-five percent of unsuccessful nests were depredated. Eighty percent ($n = 141$) of all nest attempts were by adult females, 19% ($n = 34$) by yearling females, and 1 nest attempt was by a female of unknown age. The ratio of n to K, or the number of nests (176) to the largest number of explanatory variables in a model including the intercept, for both the 150-m and 450-m spatial extents ($K = 4$), was >40.

Within both spatial extents for both successful and unsuccessful nests, approximately 95% of the area consisted of sagebrush and grass–forb cover types. The remaining 5% of the area consisted of riparian and agricultural areas, bare ground, and other shrub cover types.

Model Fit

All models derived using logistic regression fit the data. For the 150-m spatial extent, Hosmer–Lemeshow tests indicated statistical fit for the global model with three variables [percent land grass/forb (−), percent land sagebrush (+), and number of sagebrush patches (−)] for the subset of best models ($\chi^2 = 2.80$, $P = 0.423$). For the 450-m spatial extent, the global model contained four variables [percent land grass/forb (−), percent land sagebrush (+), edge density of grass/forb (−), and edge density of sagebrush (−)] for the subset of best models and also fit the data ($\chi^2 = 2.73$, $P = 0.742$).

Model Selection

At the 150-m buffer spatial extent, four models had AIC values within 3 of the lowest AIC value ($\Delta_i \leq 3$); two of the models contained only one

TABLE 10.1
Models selected to explain nest success of Greater Sage-Grouse for the 150- and 450-m spatial
extents in southeastern Idaho, 1998–2002.

Model (Estimate sign)	$-2 \ln [L]$	K	AIC	ΔAIC	w_i
150-m spatial extent					
PLGF[a] (−)	238.73	3	242.73	0.00	0.44
PLGF (−), NPSB[b] (−)	237.70	4	243.70	0.97	0.27
PLSB[c] (+)	240.72	3	244.72	1.99	0.16
PLSB (+), NPSB (−)	239.26	4	245.26	2.53	0.12
450-m spatial extent					
PLGF (−)	241.16	3	245.16	0.00	0.41
PLSB (+)	241.80	3	245.80	0.64	0.30
EDGF[d] (−)	242.93	3	246.93	1.77	0.17
EDSB[e] (−)	243.52	3	247.52	2.36	0.13

NOTE: Log likelihood ($-2 \ln [L]$), number of estimable parameters (K), Akaike's Information Criterion (AIC), difference in AIC (ΔAIC), and Akaike weights (w_i) are provided

[a] Percent of land with grass–forb dominated cover.
[b] Number of sagebrush patches.
[c] Percent land in sagebrush.
[d] Edge density of grass–forb dominated cover.
[e] Edge density of sagebrush.

explanatory variable and the other two had two variables each that were not correlated ($r \leq 0.5$; Table 10.1). For the 450-m spatial extent, all four models that had $\Delta_i \leq 3$ were single-variable models (Table 10.1). For both spatial extents, the model with the lowest AIC contained only percent land of grass–forb dominated cover, and this landscape metric was negatively associated with nest success.

Model Averaging

For the 150-m buffer extent, percent land of grass–forb dominated cover was the most important explanatory variable, with a Σw_i of 0.71 compared to $\Sigma w_i = 0.39$ and $\Sigma w_i = 0.28$ for number of patches sagebrush and percent land in sagebrush, respectively. Percent land of grass–forb dominated cover was 1.8 (0.71/0.39) times more likely an explanation for nest success (or failure) than number of patches sagebrush, and 2.5 (0.71/0.28) times more likely than percent land in sagebrush.

For the 450-m spatial extent, percent land of grass–forb dominated cover had the highest Σw_i (0.41) followed by percent land of sagebrush ($\Sigma w_i = 0.30$). Summed Akaike weights of edge density of grass–forb dominated cover and edge

density of sagebrush were not influential in explaining nest fate (Table 10.1). Scaling up resulted in a weaker predictive model for nest success.

Parameter Estimates

The most explanatory variable at each spatial extent was further analyzed with respect to nest success. At the 150-m spatial extent, the mean percent land of grass–forb dominated types was 17.50% at unsuccessful nests compared to 8.66% at successful nests (Table 10.2). At the 450-m spatial extent, the mean percent land of grass–forb dominated cover was 15.63% at unsuccessful nests compared to 9.97% at successful nests. At both spatial extents, successful nests had less grass–forb dominated land cover than unsuccessful nests. In comparison, the evidence that successful nests had more sagebrush land cover is present but weaker at both spatial extents.

Broods

We used 62 females with broods containing chicks 0–10 weeks of age (14 successful, 48 unsuccessful)

TABLE 10.2

Means of landscape metrics (SE) within two spatial extents in models of Greater Sage-Grouse nest success in southeastern Idaho, 1998–2002.

Landscape metric	Successful nests ($n = 85$)	Unsuccessful nests ($n = 91$)
150-m spatial extent		
PLSB[a]	86.97 (2.89)	78.97 (3.53)
PLGF[b]	8.66 (2.20)	17.50 (3.24)
NPSB[c]	1.00 (0.03)	1.07 (0.05)
450-m spatial extent		
PLSB	84.77 (2.79)	78.93 (3.07)
PLGF	9.97 (2.20)	15.63 (2.72)
EDSB[d]	17.88 (2.68)	19.74 (2.46)
EDGF[e]	15.47 (2.68)	19.07 (2.84)

[a] Percent of land in sagebrush.
[b] Percent of land with grass–forb dominated cover (recent fires/grasslands).
[c] Number of patches sagebrush.
[d] Edge density sagebrush (m/ha).
[e] Edge density grass–forb (m/ha).

monitored from 1998 to 2002 (number of broods ranged from 2 in 1998 to 24 in 2002) for this analysis; all females with broods had >5 locations per female. The ratio of n to K, or the number of individual broods (62) to the largest number of explanatory variables in a model including the intercept, for both the 150-m and 450-m spatial extents ($K = 3$), was <40.

Approximately 90% of the area within both spatial extents, for both successful and unsuccessful broods, consisted of sagebrush and grass–forb dominated cover types. The remaining 10% consisted of agricultural and riparian areas, bare ground, and other shrub cover types. Brood locations occurred within numerous valleys throughout the broader landscape of the upper Snake River plain, so delineation of cover types over the entire study area would have been subjective and likely misleading with respect to areas supporting sage grouse broods. Although this 90% level generally appears similar to the larger landscape, we make no inferences regarding habitat selection within the study area.

Model Fit

For the 150-m spatial extent, Hosmer–Lemeshow tests indicated statistical fit for the global model

[percent land grass/forb (−), percent land agriculture (+)] for the subset of best models ($\chi^2 = 5.47$, $P = 0.706$; Table 10.3). For the 450-m spatial extent, the global model [percent land grass/forb (−), percent land agriculture (+), and edge density of grass/forb (−)] for the subset of best models also fit the data ($\chi^2 = 4.89$, $P = 0.769$).

Model Selection and Averaging

There was one model at the 150-m spatial extent that best explained brood success with uncorrelated variables [percent land of grass–forb dominated cover (−) and percent land in agriculture (+)] and AIC$_c$ values with $\Delta_i \leq 3$ (Table 10.3). At the 150-m spatial extent, both variables had equal Σw_i of 1.00. At the 450-m spatial extent, there were two models with uncorrelated variables and AIC$_c$ values with $\Delta_i \leq 3$ (Table 10.3). The best model contained percent land in agriculture (+) and percent land of grass–forb dominated cover (−), and the second model included percent land in agriculture (+) and edge density of grass/forb (−). At this extent, both models included percent land in agriculture, which therefore had a Σw_i of 1.00, followed by percent land of grass–forb dominated cover, which had a Σw_i of 0.63. Edge density of grass/forb had a Σw_i of 0.37 (Table 10.3).

TABLE 10.3

Models selected to explain overall Greater Sage-Grouse brood success for the 150- and 450-m spatial extents in southeastern Idaho, 1998–2002.

Model (Estimate sign)	$-2 \ln [L]$	K	AIC_c	ΔAIC_c	w_i
150-m spatial extent					
PLGF[a] (−), PLAG[b] (+)	52.40	3	58.81	0.00	1.00
450-m spatial extent					
PLGF (−), PLAG (+)	54.47	3	60.88	0.00	0.63
EDGF[c] (−), PLAG (+)	55.50	3	61.92	1.04	0.37

NOTE: Log likelihood ($-2 \ln [L]$), number of estimable parameters (K), Akaike's Information Criterion (AIC_c), difference in AICC (ΔAIC_c), and Akaike weights (w_i) are provided

[a] Percent of land with grass–forb dominated cover.
[b] Percent of land in agriculture.
[c] Edge density of grass–forb dominated cover.

Parameter Estimates

At the 150-m spatial extent, the mean (SE) percent land of grass–forb dominated cover was 3.51 (1.27) for successful broods and 12.07 (2.28) for unsuccessful broods. The mean (SE) percent land in agriculture was 8.01 (3.97) for successful broods and 1.72 (0.57) for unsuccessful broods.

At the 450-m spatial extent, the mean (SE) percent land in agriculture for successful broods was 7.70 (3.24) and 2.66 (0.73) for unsuccessful broods. The mean (SE) percent land of grass–forb dominated cover was 4.16 (1.01) for successful broods and 12.57 (2.39) for unsuccessful broods, and the mean (SE) of edge density of grass/forb was 10.04 (2.47) for successful broods and 18.90 (2.70) for unsuccessful broods. These values suggest that more grass–forb dominated cover negatively affected brood success at both spatial scales while proximity to agricultural land had a positive effect. Moreover, at the broader spatial scale, increased edge density of grass–forb cover also had a negative influence on brood success.

Brood Age

We also examined habitat characteristics with respect to brood age using data from the same 62 females with broods, at the 150-m spatial extent only. We compared landscape-scale habitat use during the early brood period (weeks 1–3) between successful and unsuccessful broods, and compared habitat use by successful early broods to successful older broods (weeks >3–10). The ratio of n to K, or the number of individual broods to the largest number of explanatory variables in a model including the intercept, was <40.

Model Fit

For successful early broods (1–3 weeks old; broods had at least one chick that survived to three weeks of age), Hosmer–Lemeshow tests indicated statistical fit for the global model [percent land of grass–forb dominated cover (−), percent land in sagebrush (+), and edge density of sagebrush (−)] for the subset of best models ($\chi^2 = 6.47$, $P = 0.372$). For broods 1–3 versus 3–10 weeks old (successful broods only), the global model [percent land grass/forb (+), percent land in agriculture (−), and edge density of grass/forb (+)] for the subset of best models also fit the data ($\chi^2 = 3.74$, $P = 0.810$).

Model Selection and Averaging

There were three models with $\Delta_i \leq 3$ for each type of brood age comparison (Table 10.4). During the early brood period, models with uncorrelated variables and AIC_c values with $\Delta_i \leq 3$ indicated that percent land of grass–forb dominated cover and edge density of sagebrush had a negative influence on brood success, while percent land in sagebrush had a positive influence. Percent land of grass–forb dominated cover had the highest Σw_i (0.79) of the variables in the models explaining

TABLE 10.4

Models explaining Greater Sage-Grouse brood success during the early brood period, and between early and late brood periods for successful broods at the 150-m spatial extent in southeastern Idaho, 1998–2002.

Comparison	Model (Estimate sign)	2 ln [L]	K	AIC$_c$	ΔAIC$_c$	w_i
Success of early broods[a]	PLGF[b] (−)	31.71	3	36.75	0.00	0.56
	PLGF[b] (−), EDSB[c] (−)	30.80	4	38.62	1.86	0.22
	PLSB[d] (+)	33.66	3	38.70	1.95	0.21
Broods 1–3 vs. 3–10 weeks old[f]	PLAG[e] (−)	30.09	3	35.29	0.00	0.59
	PLGF[b] (+)	32.15	3	37.35	2.07	0.21
	EDGF[g] (+)	32.20	3	37.40	2.11	0.20

NOTE: Log likelihood (−2 ln [L]), number of estimable parameters (K), Akaike's Information Criterion (AIC$_c$), difference in AICC (ΔAIC$_c$), and Akaike weights (w_i) are provided

[a] 1–3 weeks old.
[b] Percent of land with grass–forb dominated cover.
[c] Edge density of sagebrush.
[d] Percent of land in sagebrush.
[e] Percent of land in agriculture.
[f] Successful broods only.
[g] Edge density of grass–forb dominated cover.

brood success in the early brood period, followed by edge density of sagebrush (Σw_i = 0.22), and percent land in sagebrush (Σw_i = 0.21). However, no single variable was strongly influential in explaining success.

For the comparison of successful early versus late broods, models with uncorrelated variables and AIC$_c$ values with $\Delta_i \leq 3$ contained percent land in agriculture (−), percent land of grass–forb dominated cover (+), and edge density of grass/forb (+). Percent land in agriculture had the highest Σw_i (0.59) of the variables in the models comparing habitat used by successful broods 1–3 and >3–10 weeks old, followed by percent land of grass–forb dominated cover (Σw_i = 0.21), and edge density of grass/forb (Σw_i = 0.20) but again no single variable was strongly influential.

Parameter Estimates

The mean of percent land of grass–forb dominated cover for successful and unsuccessful broods 1–3 weeks old was 5.15 and 16.08, respectively (Table 10.5), suggesting that this variable had a negative influence on survival of young broods. The mean of percent land in agriculture for successful broods 1–3 weeks and 3–10 weeks old was 2.33 and 9.06, respectively (Table 10.5), suggesting that agricultural lands were increasingly prevalent in

areas used by older broods compared to areas used by younger broods.

DISCUSSION

Our primary objective was to identify landscape metrics associated with vegetation that occurs near sage grouse nests that were successful and broods that recruited offspring to the fall population. Reclassification of the original cover type map increased the statistical validity of our sample by increasing the number of non-zero data points as opposed to extracting data from highly categorized cover type maps, which generally produce many more entries of zero.

Our approach, a map of generalized cover types relevant to sage grouse, may be better suited for addressing the basic issue of shrubsteppe fragmentation by increasing perennial grassland cover types than approaches using highly categorized cover maps.

For both spatial extents, the primary difference between successful and unsuccessful nests was the amount of sagebrush and grass–forb dominated vegetation (the combination of existing grasslands and areas affected by wildfires) present on the landscape. A lower amount of grass–forb dominated vegetation (this variable had a negative coefficient) appeared sometimes important (i.e., a relatively high summed Akaike weight)

TABLE 10.5
Means (SE) of landscape metrics in models of successful and unsuccessful Greater Sage-Grouse broods 1–3 weeks of age and successful broods 1–3 and 3–10 weeks of age in southeastern Idaho, 1998–2002.

	Successful	Unsuccessful
1–3-week-old broods		
PLGF[a]	5.15 (2.05)	16.08 (4.47)
EDSB[b]	15.94 (5.16)	28.24 (5.66)
PLSB[c]	84.07 (8.50)	68.14 (7.78)
	1–3 Weeks	**3–10 Weeks**
Successful broods		
PLAG[d]	2.33 (1.48)	9.06 (4.05)
PLGF	5.15 (2.05)	3.45 (1.66)
EDGF[e]	12.47 (5.28)	8.69 (3.57)

[a] Percent land of grass–forb dominated cover.
[b] Edge density of sagebrush.
[c] Percent land in sagebrush.
[d] Percent land in agriculture.
[e] Edge density of grass–forb dominated cover.

in explaining the success of nests at the smaller spatial extent, but less grass–forb dominated vegetation was the most important variable (highest summed Akaike weight) at the larger spatial extent. Studies of neotropical migratory birds implicate the growing amount of landscape-scale habitat fragmentation in forested areas as the ultimate cause of increasing nest predation rates (Wilcove 1985, Paton 1994); however, Trzcinski et al. (1999) reported overall habitat loss, rather than sensitivity to fragmentation, as more influential. Nonetheless, the relationship between sage grouse nest success and grass–forb dominated landcover also involved habitat loss and fragmentation, particularly fragmentation of mature sagebrush stands, that was related to increased nest failure in this study. A lower number of sagebrush patches (this variable had a negative coefficient) was a plausible variable, or had a relatively high summed Akaike weight within our suite of variables, for explaining the success of nests at the 150-m spatial extent. In comparison, less edge density, either of grass–forb dominated or sagebrush cover types, was not as useful in explaining nest failure at the 450-m spatial extent. These variables, either directly or indirectly, represent increasing levels of fragmentation and grass–forb

dominated habitat within shrubsteppe. The lack of a strong relationship between percent land in sagebrush and nest success likely is the result of our inability with remotely sensed data to differentiate between healthy sagebrush stands (with an understory of grasses and forbs) and depleted sagebrush stands (little or no herbaceous understory).

Other avian studies have found lower nest densities and higher predation near edges of non-preferred habitat and in smaller patches of preferred habitat (Albrecht 2004, Bollinger and Gavin 2004). Eighty-five percent of our unsuccessful nests were depredated, which implies that the risk of direct nest predation is higher when there is more grass–forb dominated habitat and edge associated with sagebrush in the surrounding landscape. Increased grassland and burned area, with resulting low vegetation, may increase visual nest detection typical of avian predators such as corvids (Angelstam 1985), which occurred throughout our study area. Manzer and Hannon (2005) reported higher corvid density and nest predation of Sharp-tailed Grouse (*Tympanuchus phasianellus*) in preferred grassland habitat when fragmented by degraded habitat or agriculture. Similarly, if a majority of nest predators were ravens (*Corvus corax*), artificial nests in shrubsteppe were nine times more likely to be depredated if they occurred in fragmented compared to intact landscapes (Vander Haegen et al. 2002). However, nest predation studies of Capercaillie (*Tetrao urogallus*) were inconclusive concerning the effects of fragmentation (Wegge et al. 1990).

Amount of grass–forb dominated land cover was also an important variable in explaining brood failure for the early and entire brood period at both spatial extents: successful broods had less grass–forb dominated habitat in the surrounding landscape. This appears to contradict the findings of Cassaza et al. (this volume, chapter 11), who reported that at a relatively small spatial scale (0.04 ha), grouse with broods selected areas with greater perennial forbs and higher richness of plant species. However, in our study, the reclassified category called grasslands was comprised of two main originally classified cover types (>95%), perennial grasslands and areas recently burned by wildfire, and these were considerably larger than the meadows described by Casazza et al. (this volume, chapter 11). Perennial grasslands are defined in the 1998 Idaho Gap Analysis Project as areas dominated by planted crested wheatgrass (*Agropyron cristatum*) and invasive cheatgrass (*Bromus tectorum*). Thus, scale as well as moisture help explain the difference between Casazza et al.'s (this volume, chapter 11) conclusion and our finding that increasing amounts of grass–forb dominated habitat and resulting fragmentation was related to brood failure, possibly through increased visibility to predators.

Studies of European forest-dwelling grouse have shown increased predation of broods when preferred older forest habitat becomes increasingly reduced and fragmented with deforested areas or earlier stage of forest succession (Wegge et al. 1992, Kurki et al. 2000). Sage grouse are similar to European forest grouse because they are also dependent on mature plant communities and have similar predators of chicks. The amount of sagebrush was approximately the same for both successful and unsuccessful broods at both spatial extents. Therefore, the explanation of "predator gain" or increased efficiency by searching smaller areas of brood habitat, or sagebrush patches, per brood killed is less likely for sage grouse in our study compared to brood loss of European grouse (Storaas et al. 1999). Moreover, our direct evidence

concerning the influence of edge on brood success is relatively weak.

Some studies have found that agricultural land is detrimental to grouse reproduction, possibly due to increased generalist predators (Kurki et al. 2000, Manzer and Hannon 2005). In contrast, we found a positive relationship between agricultural landcover and brood success, although agricultural land cover accounted for a relatively small portion (<10%) of the area within brood buffers at both spatial extents. The amount of agricultural land in buffers of successful broods was nearly equal to the amount of grass–forb dominated habitat found in buffers of unsuccessful broods. Although this suggests that agricultural land may have provided a benign substitute for grass–forb dominated habitat, it is more likely that agricultural lands provide a more mesic environment, and thus more food, especially during the later brood-rearing period. Sage grouse broods will use cropland (Connelly et al. 1988), particularly alfalfa (*Medicago sativa*), as a possible substitute for naturally occurring protein-rich forbs needed for growth and development (Patterson 1952, Connelly et al. 2000, Hagen et al. 2007). We found older broods (>3–10 weeks) used more agricultural land compared to younger broods. The amount of agricultural land in buffers of older broods could have been influenced by both an attraction to alfalfa and an increase in the size of areas used as the juveniles aged. The lack of agriculture in buffers of unsuccessful broods was influenced by fewer unsuccessful broods surviving to the later brood period, when agriculture was used. Moreover, if unsuccessful broods survived to the later brood period, they occurred within this period for a shorter time than successful broods since they lived less than 10 weeks, which reduced their probability of using croplands. These factors may have increased the difference in use of agricultural landcover between 14 successful and 48 unsuccessful broods, promoting it to an explanatory variable in brood success models despite its relatively low occurrence in overall brood habitat.

Irrigated alfalfa may be attractive to broods, but it is a small portion of the landscape used, and the use is confounded by mobility of older broods and the increased time that successful broods live. Given that chick survival is generally much lower during the early brood period (Dahlgren 2009) compared to the later period, modeling and eventually mapping of early brood period habitat may be more useful and relevant to future sage grouse conservation efforts than would be an assessment of later brood-rearing habitat. This would provide a more focused conservation effort on the relevant life stage of early brood-rearing and also lessen the debate as to whether agricultural lands are relevant and useful. If the early brood period is a more critical period, habitat related to the success of broods during that period is of greater conservation importance than during other life stages.

The mid-scale landscape variables we identified were empirically derived from a relatively large data set. Given the landscape-scale habitat variables related to breeding success, as potential breeding habitat is identified, prioritization and retention of habitat essential to the persistence of sage grouse populations can be mapped. Areas that are degraded at the landscape level and therefore do not have strong potential for breeding success, yet are publicly owned, could be identified as high-priority areas for restoration. This is particularly true if degraded areas are adjacent to known breeding habitat and would therefore expand the geographic area for an existing population. Moreover, areas identified as having strong potential for breeding success in terms of landscape-scale habitat can be protected from the negative effects of prescribed fire and other vegetation treatments (Klebenow 1970, Beck et al. 2009, Rhodes et al. 2010), and can also be given priority in terms of wildfire suppression. Similar conservation measures could also apply to new road construction, energy development, and agricultural conversion. Thus, landscape variables and their estimates may be used in a mapping process that has empirical validity for conserving sage grouse habitat.

ACKNOWLEDGMENTS

We thank M. L. Commons-Kemner of Idaho Department of Fish and Game and N. A. Burkepile of the University of Idaho for data assistance. This manuscript was improved by reviews provided by C. A. Hagen and an anonymous referee. We also thank Eva Strand of the College of Natural Resources, University of Idaho for GIS technical and data assistance. This is contribution 1040 from the University of Idaho Forest, Wildlife and Range Experiment Station and Idaho Federal Aid in Wildlife Restoration Project W-160-R.

LITERATURE CITED

Albrecht, T. 2004. Edge effect in wetland-arable land boundary determines nesting success of Scarlet Finches (*Carpodacus erythrinus*) in the Czech Republic. Auk 121:361–371.

Aldridge, C. L., and M.S. Boyce. 2007. Linking occurrence and fitness to persistence: habitat-based approach for endangered Greater Sage-Grouse. Ecological Applications 17:508–526.

Aldridge, C. L., S. E. Nielsen, L. B. Hawthorne, M. S. Boyce, J. W. Connelly, S. T. Knick, and M. A. Schroeder. 2008. Range-wide patterns of Greater Sage-Grouse persistence. Diversity and Distributions 14:983–994.

Angelstam, P. 1985. Predation on ground-nesting birds' nests in relation to predator densities and habitat edge. Oikos 47:365–373.

Beck, J. L., J. W. Connelly, and K. P. Reese. 2009. Recovery of Greater Sage-Grouse habitat features in Wyoming big sagebrush following prescribed fire. Restoration Ecology 17:393–403.

Bollinger, E. K., and T. A. Gavin. 2004. Responses of nesting Bobolinks (*Dolichonyx oryzivorus*) to habitat edges. Auk 121:767–776.

Burkepile, N. A., J. W. Connelly, D. W. Stanley, and K. P. Reese. 2002. Attachment of radio-transmitters to one-day-old sage grouse chicks. Wildlife Society Bulletin 30:93–96.

Burnham, K. P., and D. R. Anderson. 2002. Model selection and multimodel inference: a practical information-theoretic approach. Springer-Verlag, New York, NY.

Connelly, J. W., H. W. Browers, and R. J. Gates. 1988. Seasonal movements of sage grouse in southeastern Idaho. Journal of Wildlife Management 52:116–122.

Connelly, J. W., S. T. Knick, M. A. Schroeder, and S. J. Stiver. 2004. Conservation assessment of Greater Sage-Grouse and sagebrush habitats. Western Association of Fish and Wildlife Agencies, Cheyenne, WY.

Connelly, J. W., K. P. Reese, and M. A. Schroeder. 2003. Monitoring Greater Sage-Grouse habitats and populations. College of Natural Resources Experiment Station Bulletin 80, University of Idaho, Moscow, ID.

Connelly, J. W., M. A. Schroeder, A. R. Sands, and C. E. Braun. 2000. Guidelines to manage sage grouse and their habitats. Wildlife Society Bulletin 28:967–985.

Connelly, J. W., W. L. Wakkinen, A. D. Apa, and K. P. Reese. 1991. Sage grouse use of nest sites in southeastern Idaho. Journal of Wildlife Management 55:521–524.

Crunden, C. W. 1963. Age and sex of sage grouse from wings. Journal of Wildlife Management 27:846–849.

Dahlgren, D. K. 2009. Greater Sage-Grouse ecology, chick survival, and population dynamics, Parker Mountain, Utah. Ph.D. dissertation, Utah State University, Logan, UT.

Doherty, K. E., D. E. Naugle, B. L. Walker, and J. M. Graham. 2008. Greater Sage-Grouse winter habitat selection and energy development. Journal of Wildlife Management 72:187–195.

ESRI (Environmental Systems Research Institute, Inc,). 1999a. ARC/INFO version 8.01, Redlands, CA.

ESRI (Environmental Systems Research Institute, Inc,). 1999b. ARCVIEW version 3.3, Redlands, CA.

Gregg, M. A., J. A. Crawford, M. S. Drut, and A. K. DeLong. 1994. Vegetational cover and predation of sage grouse nests in Oregon. Journal of Wildlife Management 58:162–166.

Hagen, C. A., J. W. Connelly, and M. A. Schroeder. 2007. A meta-analysis of Greater Sage-Grouse Centrocercus urophasianus nesting and brood-rearing habitats. Wildlife Biology 13(Suppl. 1):42–50.

Holloran, M. J., B. J. Heath, A. G. Lyon, S. J. Slater, J. L. Kuipers, and S. H. Anderson. 2005. Greater Sage-Grouse nesting habitat selection and success in Wyoming. Journal of Wildlife Management 69:638–649.

Homer, C. G. 1998. Idaho/Western Wyoming landcover classification. Remote Sensing/GIS Laboratories, Utah State University, Logan, Utah.

Hosmer, D. W., and S. Lemeshow. 1989. Applied logistic regression. John Wiley and Sons, New York, NY.

Hurlbert, S. H. 1984. Pseudoreplication and the design of ecological field experiments. Ecological Monographs 54:187–211.

Idaho Sage-Grouse Advisory Committee. 2006. Conservation plan for the Greater Sage-Grouse in Idaho. Idaho Department of Fish and Game, Boise, ID.

Klebenow, D. A. 1970. Sage grouse versus sagebrush control in Idaho. Journal of Range Management 23:396–400.

Klott, J. H., and F. G. Lindzey. 1990. Brood habitats of sympatric Sage and Sharp-tailed Grouse in Wyoming. Journal of Wildlife Management 54:84–88.

Kurki, S., A. Nikula, P. Helle, and H. Linden. 2000. Landscape fragmentation and forest composition effects on grouse breeding success in boreal forests. Ecology 81:1985–1997.

Leonard, K. M., K. P. Reese, and J. W. Connelly. 2000. Distribution, movements and habitats of Sage Grouse Centrocercus urophasianus on the upper Snake River plain of Idaho. Wildlife Biology 6:265–270.

Manzer, D. L., and S. J. Hannon. 2005. Relating grouse nest success and corvid density to habitat: a multi-scale approach. Journal of Wildlife Management 69:110–123.

McGarigal, K., B. J. Marks, C. Holmes, and E. Ene. 2001. FRAGSTATS: spatial pattern analysis program for quantifying landscape structure, version 3.3.

Paton, P. W. C. 1994. The effect of edge on avian nest success: how strong is the evidence? Conservation Biology 8:17–26.

Patterson, R. L. 1952. The sage grouse in Wyoming. Sage Books Inc., Denver, CO.

Rhodes, E. C., J. D. Bates, R. N. Sharp, and K. W. Davies. 2010. Fire effects on cover and dietary resources of sage-grouse habitat. Journal of Wildlife Management 74:755–764.

Rodgers, A. R., and A. P. Carr. 1998. HRE: the home range extension for ArcView. Version 1.1. User's manual. Centre for Northern Forest Ecosystem Research, Ontario Ministry of Natural Resources, Thunder Bay, Ontario, Canada.

SAS Institute, Inc. 1988. Version 8.2. SAS Institute, Cary, NC.

Schroeder, M. A., J. R. Young, and C. E. Braun. 1999. Sage Grouse (Centrocercus urophasianus). A. Poole and F. Gill (editors), The birds of North America No. 425. Academy of Natural Sciences, Philadelphia, PA.

Scott, J. M., C. R. Peterson, J. W. Karl, E. Strand, L. K. Svancara, and N. M. Wright. 2002. A gap analysis of Idaho: Final report. Idaho Cooperative Fish and Wildlife Research Unit, Moscow, ID.

Shepherd, J. 2006. Modeling landscape-scale habitat use by Greater Sage-Grouse in southern Idaho. Ph.D. dissertation, University of Idaho, Moscow, ID.

Smith, J, T., L. D. Flake, K. F. Higgins, G. D. Kobriger, and C. G. Homer. 2005. Evaluating lek occupancy of Greater Sage-Grouse in relation to landscape cultivation in the Dakotas. Western North American Naturalist 65:310–320.

SPSS, Inc. 1993. SPSS for Windows base system user's guide, release 6.0. SPSS, Inc., Chicago, IL.

SPSS, Inc. 2000. SYSTAT user's guide, release 10.0. SPSS, Inc., Chicago, IL.

Storaas, T., L. Kastdalen, and P. Wegge. 1999. Detection of forest grouse by mammalian predators: a possible explanation for high brood losses in fragmented landscapes. Wildlife Biology 5:187–192.

Trzcinski, M. K., L. Fahrig, and G. Merriam. 1999. Independent effects of forest cover and fragmentation on the distribution of forest breeding birds. Ecological Applications 9:586–593.

U.S. Department of the Interior. 2010. Endangered and threatened wildlife and plants: 12-month finding for petitions to list the Greater Sage-Grouse as threatened or endangered. Federal Register 75:13909–14014.

Vander Haegen, W. M., M. A. Schroeder, and R. M. DeGraaf. 2002. Predation on real and artificial nests in shrubsteppe landscapes fragmented by agriculture. Condor 104:496–506.

Wakkinen, W. L., K. P. Reese, J. W. Connelly, and R. A. Fischer. 1992. An improved spotlighting technique for capturing Sage Grouse. Wildlife Society Bulletin 20:425–426.

Walker, B. L., D. E. Naugle, and K.E. Doherty. 2007. Greater Sage-Grouse population response to energy development and habitat loss. Journal of Wildlife Management 71:2644–2654.

Wegge, P., I. Gjerde, L. Kastdalen, J. Rolstad, and T. Storaas. 1990. Does forest fragmentation increase the mortality pattern of Capercaillie? Pp. 448–453 in S. Myrberget (editor), Transactions of the 19th IUGB Congress, Trondheim, Norway.

Wegge, P., J. Rolstad, and I. Gjerde. 1992. Effects of boreal forest fragmentation on Capercaillie grouse: empirical evidence and management implications. Pp. 738–749 in D. R. McCullough and R. H. Barrett (editors), Wildlife 2001: populations. Elsevier Science Publications, New York, NY.

Wilcove, D. S. 1985. Nest predation in forest tracts and the decline of migratory songbirds. Ecology 66:1211–1214.

Zar, J. H. 1999. Biostatistical analysis. 4th ed. Prentice Hall, Upper Saddle River, NJ.

CHAPTER ELEVEN

Linking Habitat Selection and Brood Success in Greater Sage-Grouse

Michael L. Casazza, Peter S. Coates, and Cory T. Overton

Abstract. Examining links between the fitness of individual organisms and their habitat-based decisions is useful to identify key resources for conservation and management of a species, especially at multiple spatial scales because selection of habitat attributes may vary with spatial scale. Decisions of habitat use by brood-rearing Greater Sage-Grouse (*Centrocercus urophasianus*) may influence the survival of chicks. We conducted radiotelemetry on 38 sage grouse broods within Mono County, California, during 2003–2005. At relocation and random sites, we measured habitat characteristics at three spatial scales using field procedures (scale, 0.03 ha) and Geographical Information System tools (scales, 7.9 ha and 226.8 ha). We then conducted three data analyses using an information-theoretic modeling approach. The purpose of these analyses was to: (1) identify habitat factors that were selected (defined as use disproportionate to availability) by sage grouse broods; (2) identify habitat factors associated with brood success (defined as ≥1 live chick at 50 days post-hatch; 24 were successful, 14 unsuccessful); and (3) evaluate brood

success as a function of habitat selection indices for brood-rearing sage grouse. At the smallest spatial scale (0.03 ha), grouse with broods selected areas with greater perennial forbs and higher richness of plant species. At larger scales (7.9 ha and 226.8 ha), areas with Utah juniper (*Juniperus osteosperma*) and singleleaf pinyon pine (*Pinus monophylla*) encroachment were avoided by grouse. Most importantly, the probability of fledging a brood increased as sage grouse females selected habitats with greater densities of perennial forbs (0.03 ha) and higher meadow edge (perimeter to edge ratio; 7.9 ha), perhaps because these areas provided a balance of food and protective cover for chicks. These results suggest that managers should discourage tree encroachment and preserve and enhance sagebrush stands interspersed with perennial forbs and a mixture of small upland meadows.

Key Words: brood success, *Centrocercus urophasianus*, forb, Greater Sage-Grouse, habitat, juniper, pinyon, selection, spatial scale, meadow.

Casazza, M. L., P. S. Coates, and C. T. Overton. 2011. Linking habitat selection and brood success in Greater Sage-Grouse. Pp. 151–167 in B. K. Sandercock, K. Martin, and G. Segelbacher (editors). Ecology, conservation, and management of grouse. Studies in Avian Biology (no. 39), University of California Press, Berkeley, CA.

P atterns in habitat selection—the disproportionate use to availability of resources or conditions by organisms—are complex, and the study of these patterns has become a priority in conserving wildlife species (Morrison 2001, Brotons et al. 2004). Organisms are thought to use resources and occupy areas that optimize their fitness (i.e., survival and reproduction; Wiens 1989, Rosenzweig 1991). Beneficial management practices are those that preserve and improve environmental factors that are selected by an individual organism for the purpose of increasing survival and reproduction (Aldridge and Boyce 2008). However, to identify these environmental factors, it is challenging and often necessary to identify links between an organism's fitness and its habitat-based decisions (Morris et al. 2008).

Greater Sage-Grouse (*Centrocercus urophasianus*; hereafter sage grouse) populations are declining throughout their range (Schroeder et al. 1999, Connelly et al. 2004), and this decline is attributed in part to low survival of broods as well as other vital rates (Schroeder et al. 1999, Aldridge and Brigham 2001). Despite the importance of this life stage, factors influencing the survival of chicks and broods are not well understood (Gregg 2006; Aldridge and Boyce 2007, 2008). Numerous studies have described habitat use and selection by female sage grouse with broods (Klebenow 1969, Wallestad 1971, Drut et al. 1994a, Sveum et al. 1998), and these studies have played important roles in management guidelines (Braun et al. 1977, Connelly et al. 2000). Although these types of studies have been largely informative they did not link habitat use or selection to an aspect of grouse fitness, such as relationships between the success of fledging chicks and habitat-related decisions by females. It is generally assumed that selection of habitat attributes is related to an aspect of fitness, but these links had not been quantified and understood for many sage grouse populations. Some recent research that explored relationships between environmental attributes that were selected by grouse and their fitness indicated that understanding these relationships further will inform grouse management decisions (Chi 2004; Aldridge and Boyce 2007, 2008).

Habitat selection and fitness research on sage grouse have been largely conducted at a single spatial scale; data analyses at multiple spatial scales are more informative. Relationships between habitat and fitness are inherently scale-sensitive (Mayor et al. 2009). Detecting informative temporal and spatial scales is essential to consider in ecological and conservation research (Allen and Hoekstra 1992), because scale can influence the strength of associations between independent and dependent variables (Boyce 2006). At relatively small spatial scales, habitat factors that are selected by brood-rearing sage grouse include forb abundance (Klebenow and Gray 1968, Drut et al. 1994a, Sveum et al. 1998), sagebrush cover (Aldridge and Brigham 2002, Thompson et al. 2006), grass cover (Thompson et al. 2006), and insect abundance (Klebenow and Gray 1968, Drut et al. 1994a, Thompson et al. 2006). Less studied are characteristics at larger spatial scales, but some authors have reported meadows and lake bottoms as important (Oakleaf 1971, Drut et al. 1994a, Aldridge and Boyce 2008). Additionally, pinyon (*Pinus* spp.) and juniper (*Juniperus* spp.) encroachment into sagebrush-steppe ecosystems is thought to negatively influence sage grouse populations (Connelly et al. 2004). However, empirical findings of large-scale effects related to this encroachment are lacking. Additionally, knowledge of specific links between habitat decisions by brood-rearing sage grouse and the success of broods at different spatial scales would benefit our understanding of sage grouse ecology and refine management strategies (Garton et al. 2005, Aldridge and Boyce 2008).

During the research design phase, the choice of scale may not be intuitive (Bowyer and Kie 2006), and a decision to choose a single scale may produce misleading conclusions (Mayor et al. 2009). For example, habitat attributes that are identified as influential to the response of an organism at one spatial scale may not be influential at another scale, or these attributes may have the reverse effect (Wiens 1989, Schneider 1994, Mahon et al. 2008). We chose to evaluate habitat attributes at multiple spatial scales to provide a more complete representation of potential habitat-related fitness associations for brood-rearing sage grouse. We employed this multiscale approach in the southwestern portion of sage grouse range because information on these geographically isolated and genetically distinct populations was limited (Benedict et al. 2003, Oyler-McCance et al. 2005). Population trends within Mono County were reported as relatively stable compared to other portions of sage grouse range (Connelly et al. 2004).

Our study consisted of three objectives. First, we examined habitat selection by females rearing broods at three spatial scales (field, 0.03 ha; GIS, 7.9 ha and 226.8 ha). Second, we identified associations between brood success (defined here as ≥1 live chick at 50 days post-hatch) and habitat factors at these three scales. Third, we developed selection indices for sage grouse (based on objective 1) and used a quantitative method to link habitat selection indices to brood success at the same three scales. Explanatory habitat factors were chosen based on factors previously reported from studies elsewhere in sage grouse range (Schroeder et al. 1999, Connelly et al. 2004, Crawford et al. 2004). These factors included riparian zones, meadows (i.e., edge vs. area), plant species richness, sagebrush cover, grasses, and forbs. To our knowledge, this study is the first to investigate the differences in annual and perennial forbs and empirically evaluate the effects of pinyon and juniper encroachment on sage grouse brood success.

STUDY AREA

We collected data in Mono County, California, at a site divided into five subareas (within 65 km of longitude 119°11′1.94″ W and latitude 38°6′30.80″ N): Sweetwater Mountains, Fales, Bodie Hills, Parker Meadows, and Long Valley (Fig. 11.1). The five subareas encompassed 481 km² and covered >59% of Mono County, which lies on the eastern side of the Sierra Nevada Mountains adjacent to the Nevada border. We defined the subareas as known concentrations of grouse that were not known to interchange with grouse in other subareas. We did not observe movements between subareas of the radio-collared grouse in this study.

Topography was highly variable, with several mountain ranges separating the northern and southern ends of the study area. Elevations ranged from 1,660–3,770 m and climate was characterized by hot, dry summers and cold winters with an average annual precipitation during the study of 36 cm. Temperatures ranged from −34°C to >32°C, with an average minimum monthly temperature of −14°C in January and an average maximum of 28°C in August (Western Regional Climate Center, Reno, NV). Vegetation types at all subareas were similar, dominated by mountain big sagebrush (*Artemisia tridentata vaseyana*), interspersed with areas of low sagebrush (*A. arbuscula*) and Wyoming big sagebrush (*A. t. wyomingensis*). Silver sagebrush (*A. cana*) and basin big sagebrush (*A. t. tridentata*) occurred

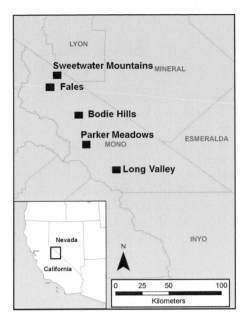

Figure 11.1. Study areas for Greater Sage-Grouse in Mono County, California, 2003–2005.

locally. Other common shrub species included snowberry (*Symphoricarpos* spp.), currant (*Ribes* spp.), bitterbrush (*Purshia tridentata*), rabbitbrush (*Chrysothamnus* spp.), and Mormon tea (*Ephedra viridis*). Primary grass species included needle grass (*Hesperostipa comata*), squirreltail (*Elymus elymoides*), and Indian ricegrass (*Achnatherum hymenoides*). Cheatgrass (*Bromus tectorum*) was present but uncommon. Dominant forbs included phlox (*Phlox* spp.), lupine (*Lupinus* spp.), buckwheat (*Eriogonum* spp.), and hawksbeard (*Crepis* spp.; Kolada et al., 2009). Singleleaf pinyon (*Pinus monophylla*) and Utah juniper (*Juniperus osteosperma*) woodlands occurred at elevations of 1,850–3,000 m.

METHODS

Field Techniques

We captured 72 female sage grouse using spotlighting techniques at night (Giesen et al. 1982, Wakkinen et al. 1992) during spring (March–April) and fall (October–November) during 2003–2005. We fitted each grouse with a 21-g necklace-mounted radio transmitter (Advanced Telemetry Systems, Isanti, MN), which included an activity sensor (Sveum et al. 1998). We located each sage grouse four times a week during the March–June breeding season using Yagi antennas and receivers

(Advanced Telemetry Systems, Isanti, MN) to within 30 m by ground. Relocation coordinates in Universal Transverse Mercator (UTM) units (datum NAD83, UTM zone 11) were recorded using hand-held Global Positioning System (GPS) devices. We assumed females were nesting when movements became localized (Connelly et al. 1993). We visually confirmed nest status but avoided flushing grouse to prevent observer-induced abandonment.

We visually checked nests every other day. We relocated 38 female sage grouse (21 adults, 17 yearlings) every 1–3 days following determination of nest fate. At relocations, we confirmed the presence or absence of chicks and then obtained micro-habitat measurements at a sample of these relocations ($n = 212$, 38 broods). If we did not detect ≥ 1 live chick, microhabitat measurements were not conducted. Therefore, we obtained habitat measurements for all female sage grouse until chicks were no longer observed. We relocated all broods at 50 days post-hatch and classified a successful brood as a female with ≥ 1 live chick at 50 days post-hatch. In instances when no chicks were detected (unsuccessful), we conducted a second search within 1–3 days to prevent scoring false negatives.

Field Explanatory Variables

In the field, we employed three methods to measure habitat characteristics that were associated with brood locations using a 0.03-ha spatial scale (centered on the location), which has been used elsewhere (Drut et al. 1994a). First, we used the line-intercept method (Canfield 1941) to estimate the percent of sagebrush canopy cover (SAC) along a 20-m transect in a random direction centered at the grouse relocation point. This technique consisted of measuring distances along each transect where shrub vegetation intersected the line, then dividing the sum of these distances by the overall transect length. Small vegetation gaps (no intersecting vegetation within a 5-cm distance) were included in the measurement as shrub canopy cover (Boyd et al. 2007). Second, we estimated percent cover of understory perennial forbs (PEF), annual forbs (ANF), and live and residual grass (GRS) using five uniformly spaced 20 × 50 cm plots at the vegetation point center and along the transect (Daubenmire 1959). Within the plots we counted the number of plant species to estimate species richness (SPR). Last, we recorded visual obstruction (VIO)

using a Robel pole (Robel et al. 1970) at the five uniformly spaced locations along each 20-m transect. Habitat measurement values were averaged across subplots to represent the relocation site at the 0.03-ha spatial scale.

GIS Explanatory Variables

We used a geographical information system (GIS) to measure multiple landscape-level covariates (ArcGIS, ESRI software, Redlands, CA). We digitized habitats into a vector coverage using digital orthophotography (1-m resolution) from the National Agriculture Imagery Program (NAIP, U. S. Department of Agriculture, Salt Lake City, UT). Mapping scale was 1:2,500 m spatial resolution. We classified landscape-level features across study areas as riparian (2.1%), meadow (7.8%), sagebrush-steppe shrubland with pinyon–juniper encroachment (5.6%), and sagebrush-steppe shrubland without pinyon–juniper (71.0%). We defined areas of pinyon–juniper encroachment by classifying areas that consisted of ≤ 40 pinyon or juniper trees per hectare in a sagebrush-dominated environment. These areas consisted of less than 5% tree canopy cover and were considered early succession, where the shrub layer was considered intact (Miller et al. 2005). Our objective was to investigate areas where pinyon and juniper were in the initial stages of encroaching (i.e., phase I). Therefore, we included these areas (≤ 40 trees/ha) in our data analyses, but excluded areas with higher tree densities (middle and late stages of succession) because these areas are not thought to be suitable for sage grouse. We categorized moist areas as meadow or riparian. Meadow consisted of seasonally wet areas vegetated primarily by non-woody plants, including succulent forbs and grasses. These areas included upland meadows (i.e., springs) and bottomland agricultural areas (i.e., fields and pastures), where the water table is at or near the surface. We classified riparian zones as streams and hydrophilic plant communities associated with the stream margins including quaking aspen (*Populus tremuloides*), Fremont cottonwood (*Populus fremontii*), and willows (*Salix* spp.). We chose to examine these factors based on *a priori* hypotheses derived from other findings and suggestions in the literature (Dunn and Braun 1986, Connelly et al. 2004). The remaining landcover types (13.5%) consisted of ponds, lakes, mountain shrub, and open canopy pine forest.

We then mapped the brood relocation sites in a GIS. We determined the spatial scales of analyses by calculating the daily mean and maximum movements of each grouse and then averaged these values across all grouse and subareas (mean = 159 m, maximum = 850 m). We used these averages as radii to calculate surface areas (7.9 and 226.8 ha, respectively) that were centered on each bird location. We used these spatial scales because they were relevant to sage grouse within Mono County. Because the same scale was required across subareas for the model analyses, we did not calculate separate averages for each subarea. We then calculated the proportion of each habitat type occurring within each radius at bird relocation sites. We developed log-ratio covariates [i.e., $\log(a_i/b)$; Aebischer et al. 1993, Kurki et al. 1998] for each habitat type of interest, where sagebrush-steppe represented the denominator (b) and the remaining classes separately represented the numerator (a; i.e., ratio meadows/sagebrush = rME, ratio mountain shrub/sagebrush = rMO, and ratio pinyon–juniper encroachment/sagebrush – rPJ; Table 11.1). This method was appropriate because it remedied a lack of independence problem that may be associated with habitat proportions (Aebischer et al. 1993, Kurki et al. 1998). Furthermore, because sage grouse are sagebrush obligates (Rowland et al. 2006), it was appropriate to interpret our results as an influence of habitat attribute a relative to sagebrush b (Kurki et al. 1998, Manzer and Hannon 2005).

We examined the influence of meadow edge density on habitat selection and brood success to evaluate the hypothesis that female grouse prefer smaller patchy meadows (Dunn and Braun 1986, Aldridge and Boyce 2007) because greater edge likely leads to successfully rearing chicks. We calculated the density of meadow edge (MEE) within each spatial scale at the relocation sites. Edge density was a ratio of meadow perimeter (m) to surface area (m²) and was used to account for differences in effects between multiple small meadows (i.e., more edge) versus fewer, larger meadows (i.e., less edge) at each scale.

We used the NEAR command in ArcGIS to measure the distance between the relocation and the nearest edge of a meadow as the variable (DIM). We then reclassified meadows and riparian zones (i.e., surface water with trees) to be the same (i.e., mesic site as a meadow or riparian) and remeasured

the distance to the nearest area as another distance variable (DIMR). The variable DIMR was compared to DIM to evaluate the hypothesis that grouse prefer moist areas regardless of differences in vegetation (i.e., trees).

We employed a used–available design to evaluate habitat selection by brood-rearing sage grouse and calculated resource selection functions (RSF; Boyce et al. 2002, Manly et al. 2002, Johnson et al. 2006). To characterize available habitat, we conducted the same habitat measurements at 175 random locations using the same field and GIS techniques as those conducted at used locations (Manly et al. 2002). The proportion of these samples of random locations for each subarea was based on the proportion of used locations. Time and logistical constraints in the field prevented sampling one random location for every used location. The range of points within each subarea was 12–39 (mean = 25), which we considered an appropriate number of random points to allow us to characterize available habitat by subarea. We calculated a minimum convex polygon (MCP) of the combined grouse relocations for each subarea. The MCPs were used to represent the boundaries of available habitat at the population level (subarea). Twenty-five random locations were removed from the analysis because these locations were not within the MCP boundaries. Thus, we used 150 remaining random locations for data analyses.

Model Development

Analysis I, Habitat Selection

We evaluated habitat selection using a design II approach (Manly et al. 2002), meaning habitat use was identified at the individual grouse level but availability assessed at the population level (i.e., subarea; Erickson et al. 2001). We classified the measured resource units as available or used. We developed generalized linear mixed models (GLMM) and specified the binomial distribution (Zuur et al. 2009). The advantage of using a binomial regression approach is that resource selection functions (RSF; Manly et al. 2002) are equivalent to the logistic discriminate function, which contrasts a sample of used and available resource units (Keating and Cherry 2004, Johnson et al. 2006). We included random effect terms in the binomial models. These terms were appropriate for representing spatial

TABLE 11.1

Explanatory variables (means ± SE) used in analyses of brood habitat use and brood survival of Greater Sage-Grouse
in Mono County, California, 2003–2005.

Measure	Abbr.	Description	Used sites		Random sites	
			Mean	SE	Mean	SE
Field[a]	SAC	Sagebrush cover (%)	27.79	0.936	23.10	1.173
	VIO	Visual obstruction (cm; i.e., Robel pole)	53.73	1.456	60.52	2.903
	SPR	Species richness of all plants	5.45	0.137	4.20	0.192
	GRS	Grass (%)	5.78	0.557	5.76	1.012
	PEF	Perennial forb (%)	4.43	0.408	2.69	0.507
	ANF	Annual forb (%)	1.67	0.298	1.94	0.443
GIS[b]	rPJ7.9	Ratio (log) pinyon-juniper encroachment to sagebrush at 7.9 ha	0.08	0.051	1.05	0.908
	rME7.9	Ratio (log) meadow to sagebrush shrub at 7.9 ha scale	0.18	0.062	0.12	0.045
	rMOS7.9	Ratio (log) mountain shrub to sagebrush shrub at 7.9 ha scale	0.03	0.005	0.01	0.004
	MEE7.9	Meadow edge as ratio of perimeter (m) to area (m²) of meadow at 7.9 ha scale	0.01	0.002	0.03	0.008
	rPJ226.8	Ratio (log) pinyon-juniper encroachment to sagebrush shrub at 226.8 ha scale	0.06	0.007	0.22	0.052
	rME226.8	Ratio (log) meadow to sagebrush shrub at 226.8 ha scale	0.15	0.039	2.10	1.853
	rMOS226.8	Ratio (log) mountain shrub to sagebrush shrub at 226.8 ha scale	0.02	0.004	0.03	0.003
	MEE226.8	Meadow edge as perimeter (m) to area (m²) of meadow at 226.8 ha scale	0.05	0.005	0.03	0.003
	DIM	Distance (km) of site to nearest meadows edge	0.46	0.028	0.65	0.063
	DIMR	Distance (km) of site to nearest meadow or riparian edge	0.44	0.023	0.55	0.061

[a] Field measurements were conducted for microhabitat covariates at 0.03 ha scale, centered on brood and random points.
[b] GIS measurements were conducted on landscape-level covariates at 7.9 and 226.8 ha scales, centered on brood and random points.

clustering (subarea), temporal correlation (year), and repeated measures for data that were gathered through time on the same individual grouse (Zuur et al. 2009). Random effects account for variation that may otherwise confound the fixed effects (e.g., forb abundance) and prevent pseudoreplication (Faraway 2006, Gillies et al. 2006, Koper and Manseau 2009). Because the variance estimate of year equaled zero, we removed it from the mixed effect models. To prevent multicollinearity in predictive models, we excluded one or two variables that covaried ($r \geq |0.65|$). No variance inflation factors were ≥ 10 (Menard 1995).

We carried out model comparisons in two steps. In step I, we developed a candidate set of models for each scale using the explanatory variables measured in the field and GIS (Table 11.1). We developed nine models using the field measurements. Three models were additive models that consisted of two covariates. One additive model included plant species richness and perennial forbs and represented the hypothesis that grouse select habitat based on food for young. A second additive model consisted of perennial forbs and annual forbs and represented the hypothesis that grouse select both types of forbs. A third model consisted of visual obstruction and sagebrush cover and represented the hypothesis that grouse selected habitats that provided vertical and horizontal cover for young. We also developed a model with a single explanatory variable for each of the six variables (Table 11.1) to compare with each other and with the additive models. Using the GIS data, we developed four models that consisted of a single explanatory variable at each spatial scale (7.9 and 226.8 ha).

We evaluated evidence of support for models at each scale using Akaike's Information Criterion (AIC) with second-order bias correction (c; Anderson 2008). We evaluated uncertainty among models using AIC_c differences (ΔAIC_c). We assigned the number of effective degrees of freedom to the individual level (i.e., individual female) and not the observation level (relocation) to prevent Type I errors. We calculated model probabilities ($w_{model\ i}$; Anderson 2008) and reported evidence ratios (ER = $w_{model\ i}/w_{model\ j}$) of the most parsimonious model compared to other models in the set (Anderson 2008). Likelihood ratio tests (Anderson 2008) were used to evaluate each model fit relative to a null model (intercept and random effects only; $\alpha = 0.05$).

In step II, we used an exploratory approach to identify the most influential covariates and spatial scales to provide information for management practices. Models were developed in step II using covariates from models that fulfilled two criteria from step I: (1) ΔAIC was ≤ 2 and (2) the model fit the data significantly better than the null model. Because we used covariates from the most parsimonious models in step I, numerous additive effects were possible. Therefore, we developed models with combinations between covariates but did not allow >2 covariates in each model. We prevented results that might be spurious by not developing more models than sampled grouse ($n = 38$; Burnham and Anderson 2002).

Analysis II, Brood Success

To estimate the effects of explanatory variables on brood success, we developed GLMM and specified the binomial distribution (Zuur et al. 2009).

Broods were scored as successful or unsuccessful. The advantage of using a binomial model in this case was to interpret the influence of explanatory variables in terms of odds ratios with 95% confidence intervals. Subarea was included as a random effect to account for spatial correlations. We did not include year in models because variance estimates equaled zero. In this case, individual grouse was not included as a random effect because measurements were averaged for each grouse. Because we were interested in the relationship between habitat selection and success, the same *a priori* models (hypotheses) that were developed for brood habitat selection were developed for brood success. Also, we used the same two-step procedure as described for habitat selection to select the most parsimonious models and identify influential covariates.

Analysis III, Linking Habitat Selection to Brood Success

We conducted a separate analysis to identify links between habitat selection and brood success. In multiple steps, we developed selection indices of each habitat factor. First, field and GIS habitat measurements at random locations were averaged at the population level (i.e., subarea). This step was necessary to characterize the available habitat per subarea. Second, we calculated averages of the explanatory variables for each individual grouse. Third, we calculated an index for each individual grouse by measuring the difference (Δ) between each averaged explanatory variable for the grouse (individual level) by the averaged explanatory variable of available habitat (population level). Fourth, we developed a candidate GLMM set, specifying the binomial distribution, by assigning brood success as a response variable (0 = unsuccessful, 1 = successful) and the selection indices for habitat factors as explanatory variables.

Models were developed for this analysis using covariates of the most parsimonious models from step II of the habitat selection and brood success analysis (described earlier). We only considered models with ΔAIC values that were <2. We used all combinations but did not include >2 covariates per model. Although this analysis was exploratory, we based our models (hypotheses) on factors that were identified as important to habitat selection and/or success. All statistical analyses were conducted using Program R ("lme4" package; Bates et al. 2008, R Development Core Team 2008).

RESULTS

The most parsimonious model for brood habitat selection (analysis I) measured in the field (0.03 ha), of the nine considered, included perennial forbs and species richness of all plants within the 20-m transect (model 1, Table 11.2). Model 2, which consisted of species richness as the only fixed effect, explained the data equally well (ΔAIC_c = 0.5; Table 11.2). However, model 1 was 1.3 ($w_{model\ 1}/w_{model\ 2}$) times more likely to be the best-approximating model than model 2 in explaining brood habitat selection. The probability that model 1 was the best of the

candidate set of models for describing brood habitat selection at this scale was 0.56 ($w_{model\ 1}$). Using the likelihood ratio test (χ^2 = 28.1, $P < 0.001$), we found that model fit was significantly improved by including these fixed covariates over a model that included no fixed effects. These analyses revealed that females with broods preferred sites with greater perennial forb and species richness of all plants than those that were randomly available (Table 11.1, Fig. 11.2). On average, using the estimated parameters, each additional plant species measured in the field appeared to increase the odds of use by approximately 33% (95% CI 19–49%). The 95% CIs for the estimated slope coefficient did not include zero. The average perennial forb cover at selected sites (4.43% ± 0.41) was nearly

TABLE 11.2

Mixed binomial regression models of Greater Sage-Grouse habitat selection models for Mono County, California, 2003–2005.

Step[a]	Analysis	No.	Model[b]	K	LL	ΔAIC_c	w	ER	χ^{2c}
I	Field, 0.03 ha scale	1	SPR (+), PEF (+)	5	−207.5	0	0.56	—	28.1*
		2	SPR (+)	4	−209.1	0.5	0.43	1.3	24.9*
	GIS, 7.9 ha scale	3	rPJ7.9 (−)	4	−215.4	0	0.87	—	12.2*
		4	MEE7.9 (−)	4	−217.9	4.9	0.07	12.4	7.3*
		5	rMO7.9 (−)	4	−218.3	5.9	0.05	17.4	6.6*
		6	rME7.9	4	−220.3	9.7	<0.01	124.3	2.5
	GIS, 226.8 ha scale	7	rPJ226.8 (−)	4	−215.1	0	0.98	—	12.9*
		8	MEE226.8 (+)	4	−219.7	9.2	0.01	98	12.2*
	GIS, distance	9	DIM	4	−217.4	0	0.95	—	3.7*
		10	DIMR	4	−220.4	6.1	0.05	19.0	2.33
II	Combined (GIS, field)	11	SPR (+), PJ226.8 (−)	5	−203.1	0	0.77	—	36.9*
		12	SPR (+), PJ7.9 (−)	5	−204.4	2.7	0.20	3.8	34.2*
		13	SPR (+), DIM (−)	5	−207.2	8.1	0.01	57.4	28.8*
		14	SPR (+), PEF (+)	5	−207.5	8.8	<0.01	81.5	28.1*
		15	SPR (+)	4	−209.1	9.3	<0.01	106.5	24.9*

[a] Step I compared models within each scale (total models, n = 19). Covariates of models that met two criteria ($\Delta AIC \leq 2$ and fit significantly better than null model) were included in step II. Step II compared models that were developed with ≤ 2 covariates of all combinations of multiple scales (total models, n = 14).
[b] All models consisted of study area and repeated measures on each individual grouse as random effects. In parentheses, signs denote positive (+) or negative (−) relationship of covariate with habitat use. Models with $\Delta AIC \geq 10$ are not presented in table. PEF = perennial forb; SPR = species richness of all plants; MEE = meadow edge; rME = ratio (log) of meadow to sagebrush; rPJ = ratio (log) of pinyon-juniper encroachment to sagebrush cover; rMO = ratio (log) mountain shrub to sagebrush; DIM = distance to nearest meadow (e.g., upland springs and dry meadows); DIMR = distance to nearest meadow or riparian area (e.g. wet area).
[c] Asterisks (*) listed in table had associated $P < 0.05$. Column abbreviations: K = number of parameters; LL = log-likelihood; ΔAIC_c = difference between model of interest and most parsimonious model with second-order bias correction; w = model probability, ER = evidence ratio (e.g., $w_{model\ 1}/w_{model\ 2}$; Anderson 2008); χ^2 = chi-square statistic to test log ratio model fit relative to null.

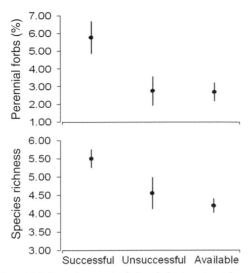

Figure 11.2. Percent perennial forb and plant species richness measured at used sites (successful and unsuccessful) of brood-rearing Greater Sage-Grouse and at random sites (available) in Mono County, California, 2003–2005.

twice as high as that of random sites (2.69% ± 0.51; Table 11.1), and plant species richness at used sites (5.45 ± 0.41) was also greater than those at random sites (4.20 ± 0.19; Table 11.1).

The most parsimonious habitat selection model measured by GIS at the 7.9 ha and 226.8 ha scales consisted of the covariate pinyon–juniper encroachment (models 3 and 7; Table 11.2). Grouse avoided areas of encroachment at both spatial scales. The probabilities of pinyon–juniper models that were best for describing the data were 0.87 ($w_{model\ 3}$) and 0.98 ($w_{model\ 7}$) within the model sets at the 7.9-ha and 226.8-ha scales.

In a separate model set we evaluated two models; one model consisted of the covariate shortest distance to mesic sites (included meadow or riparian) and the other model consisted of a covariate distance to meadow only (no riparian). We found strong evidence for a model with meadow only covariate ($w_{model\ 9}$ = 0.95), and this model was 19 times more likely to be the best-approximating model than the model including riparian areas ($w_{model\ 9}$/$w_{model\ 10}$). On average, using the model parameter estimates, for every kilometer away from a meadow, the odds of use were reduced by 52% (95% CI 42–66%). The 95% CIs for the estimated slope coefficient did not include zero.

Of the models considered in step II, we found that model 11 was the most parsimonious, consisting of the covariates plant species richness (0.03 ha) and pinyon–juniper (226.8 ha), with a model probability of 0.77 ($w_{model\ 11}$; Table 11.2). An alternative model with support included a smaller spatial scale of pinyon–juniper (7.9 ha; $w_{model\ 12}$). Model 11 was 3.8 times more likely to be the best-approximating model than model 12 ($w_{model\ 11}$/$w_{model\ 12}$), which indicated that grouse were 3.8 times more likely to avoid pinyon–juniper at the larger spatial scale.

Of the nine brood success models (analysis II) that included microhabitat covariates, the model with perennial forbs was the most parsimonious (model 1, Table 11.3). Twenty-four of 38 (63.1%) broods had ≥1 live chick at 50 days post-hatch. On average, using the model parameter estimates, a 1% increase in perennial forb coverage at the brood locations (0.03 ha) was associated with a 30% increase in the odds of success (odds ratio = 1.301, 95% CI = 1.004–1.680; Fig. 11.3A). The 95% CIs for the estimated slope coefficient did not include zero. Addition of a covariate of plant species richness did not improve model fit ($w_{model\ 2}$ = 0.23; Table 11.3) and showed similar support by the data than a forb-only model ($w_{model\ 1}$ = 0.32; Table 11.3). Probability of success was greater with an increase in plant species richness (Fig. 11.3B). We detected a greater average number of plant species in areas used by successful broods (5.5 ± 0.2) than in areas used by unsuccessful broods (4.5 ± 0.4). Additionally, the effect of perennial forbs alone was 3.6 ($w_{model\ 1}$/$w_{model\ 4}$; Table 11.3) times more likely to be the best-approximating model than a model with the additive effect of perennial and annual forbs.

We found successful females were located at areas with greater meadow edge (ratio of perimeter to area, 0.019 ± 0.004; 7.9 ha) than locations of females that were unsuccessful (0.004 ± 0.001; Fig. 11.3C). A model (5) that consisted of meadow edge at the 7.9-ha scale significantly improved model fit over a null model (χ^2 = 4.2, P = 0.03); however, a model (9) that consisted of meadow edge at the 226.8-ha scale was not supported by the data (χ^2 = 0.7, P = 0.39; Table 11.3).

In step II, the best model of the six consisted of perennial forbs (0.03 ha) and meadow edge (7.9 ha; model 15; Table 11.3). An alternative model (16) that consisted only of perennial forbs had much less support (Table 11.3). Including the additive effect of edge increased the model probability by 9.4 times ($w_{model\ 15}$/$w_{model\ 16}$).

TABLE 11.3

Mixed-effects binomial regression models of Greater Sage-Grouse brood success (≥1 live chick at 50 days post-hatch) for Mono County, California, 2003–2005.

Step[a]	Analysis	No.	Model[b]	K	LL	ΔAIC_c	w	ER	χ^{2c}
I	Field, 0.03 ha scale	1	PEF (+)	3	−22.0	0	0.32	—	5.1*
		2	PEF (+), SPR (+)	4	−21.0	0.7	0.23	1.4	7.1*
		3	SPR	3	−22.9	1.9	0.12	2.7	3.3
		4	PEF, ANF	4	−21.8	2.4	0.09	3.6	5.4
	GIS, 7.9 ha scale	5	MEE7.9 (+)	3	−22.5	0	0.58	—	4.2*
		6	rME7.9	3	−23.6	2.3	0.18	3.2	1.9
		7	rPJ7.9	3	−23.8	2.7	0.15	3.9	1.5
		8	rMO7.9	3	−24.3	3.7	0.09	6.4	0.5
	GIS, 226.8 ha scale	9	MEE226.8	3	−24.2	0	0.30	—	0.7
		10	rMO226.8	3	−23.6	0.2	0.27	1.1	0.6
		11	rME226.8	3	−24.5	0.6	0.23	1.3	0.2
		12	rPJ226.8	3	−24.6	0.8	0.20	1.5	<0.1
	GIS, distance-based	13	DIMR	3	−24.1	0	0.54	—	1.0
		14	DIM	3	−24.2	0.3	0.46	1.2	0.7
II	Combined (GIS, field)	15	PEF (+), MEE7.9 (+)	4	−18.4	0	0.75	—	12.3*
		16	PEF (+)	3	−22.0	4.5	0.08	9.4	5.1*
		17	PEF (+), SPR (+)	4	−21.0	5.2	0.06	13.5	7.1*
		18	MEE7.9 (+)	3	−22.5	5.4	0.05	15.2	4.1*
		19	SPR (+), MEE7.9 (+)	4	−21.3	5.8	0.04	18.2	6.5*
		20	SPR	3	−22.9	6.4	0.03	25.0	3.3

[a] Step I compared models within each scale (total models, $n = 19$). Covariates of models that met two criteria ($\Delta AIC \leq 2$ and fit significantly better than null model) were included in step II. Step II compared models that were developed with ≤ 2 covariates of all combinations of multiple scales (models, $N = 6$).

[b] All models consisted of study area as a random effect. In parentheses, signs denote positive (+) or negative (−) relationship of covariate with habitat use. Models with AIC value exceeding the null model were not presented in table. PEF = perennial forb; SPR = species richness of all plants; ANF = annual forb; MEE = meadow edge; rME = ratio (log) of meadow to sagebrush; rPJ = ratio (log) of pinyon-juniper encroachment to sagebrush cover; rMO = ratio (log) mountain shrub to sagebrush; DIM = distance to nearest meadow (e.g., upland springs and dry meadows); DIMR = distance to nearest meadow or riparian area (e.g. wet area).

[c] Asterisks (*) listed in table had associated $P < 0.05$. Column abbreviations: K = number of parameters; LL = log-likelihood; ΔAIC_c = difference between model of interest and most parsimonious model with second-order bias correction; w = model probability; ER = evidence ratio (e.g., $w_{model\ 1}/w_{model\ 2}$; Anderson 2008); χ^2 = Chi-square statistic to test log ratio model fit relative to null.

In our final analysis (III) to link habitat selection indices with brood success, we considered 10 models consisting of four covariates, which were plant species richness, perennial forbs, meadow edge (7.9 ha), and pinyon–juniper encroachment (226.8 ha). The model that consisted of selection indices for perennial forbs as a covariate was the most parsimonious ($w_{model\ 1} = 0.27$; Table 11.4), and the likelihood ratio test suggests it is significantly better than a null model ($\chi^2 = 4.1$, $P = 0.03$). Perennial forbs were greater at successful brood sites than at available sites, while perennial forbs at unsuccessful and random sites did not differ (Fig. 11.2). We calculated the average selection indices (difference between used and random) in percent perennial forbs as ground cover to be 3.3% ± 0.9 for successful broods and 0.5% ± 0.8 for unsuccessful broods. The 95% CIs for the estimated slope

TABLE 11.4

*Mixed-effects binomial regression models of Greater Sage-Grouse brood success (≥1 live chick at 50 days post-hatch)
as a function of indices for habitat selection in Mono County, California, 2003–2005.*

No.	Model[a]	K	LL	ΔAIC_c	w	ER	χ^{2b}
1	ΔPEF	3	−22.6	0	0.27	—	4.1*
2	ΔPEF, ΔMEE7.9	4	−21.5	0.3	0.23	1.2	6.2*
3	ΔPEF, ΔrPJ226.8	4	−22.0	1.4	0.14	2.0	5.2
4	ΔPEF, ΔSPR	4	−22.0	1.4	0.14	2.0	5.1
5	ΔSPR	3	−23.9	2.7	0.07	3.9	1.3
6	ΔMEE7.9	3	−24.2	3.3	0.05	5.2	0.7
7	ΔrPJ226.8	3	−24.5	3.8	0.04	6.8	0.2
8	ΔMEE, ΔSPR	4	−23.7	4.9	0.02	11.4	1.7
9	ΔrPJ226.8, ΔSPR	4	−23.9	5.1	0.02	13.1	1.4
10	ΔrPJ226.8, ΔMEE	4	−24.1	5.7	0.02	16.9	0.9

[a] Selection indices (Δ) were the differences in measurements of habitat attributes between used and random location for successful and unsuccessful broods. All models are listed in the table and each one included study area as a random effect. PEF = perennial forb; SPR = species richness of all plants; MEE = meadow edge (7.9 ha); rPJ = ratio (log) of pinyon-juniper encroachment to sagebrush cover (226.8 ha).
[b] Asterisks (*) listed in table had associated $P < 0.05$. Column abbreviations: K = number of parameters; LL = log-likelihood; ΔAIC_c = difference between model of interest and most parsimonious model with second-order bias correction; w = model probability; ER = evidence ratio (e.g., $w_{model\ 1}/w_{model\ 2}$; Anderson 2008); χ^2 = chi-square statistic to test log ratio model fit relative to null.

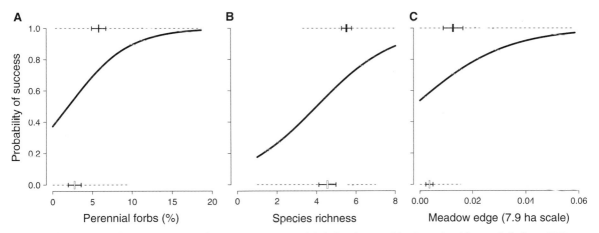

Figure 11.3. Probability of brood success in relation to (A) % perennial forb abundance (PEF), (B) species richness of all plants (SPR), and (C) meadow edge (ratio of perimeter to area) at the 7.9 ha scale (MEE) of Greater Sage-Grouse in Mono County, California, 2003–2005. Solid vertical bars represent the average of successful broods ($n = 24$) and open vertical bars represent the average of unsuccessful broods ($n = 14$). Solid horizontal bars represent standard error and dashed horizontal bars represent data range. Probability curves were derived from averaged parameters (intercept and slope) estimated from binomial models.

coefficient of selection indices for perennial forbs did not include zero. An alternative model with some support from the data ($w_{model\ 2} = 0.23$; Table 11.4) consisted of an additive effect of indices for perennial forbs and meadow edge. Model 2 also fit significantly better than a null ($\chi^2 = 6.2$, $P = 0.04$) but did not show evidence of better fit than model 1 ($\Delta AIC = 0.3$; Table 11.4). Model 1 was 1.2 times as likely as model 2 to be the best-approximating model. Two additional models were equally parsimonious ($\Delta AIC_c < 2$) and included effects of pinyon–juniper encroachment or species richness of all plants, but received less support than models 1 and 2 ($w > 1.6$).

DISCUSSION

The evidence in our study links maternal decision making in habitat use by brood-rearing grouse with their fecundity. The finding that brood-rearing grouse selected forb-rich environments supports results from other studies that have described similar clear, positive correlations (Klebenow 1969, Peterson 1970, Oakleaf 1971, Schoenberg 1982, Drut et al. 1994a). We also support earlier reported associations between forb abundance and brood success (Chi 2004; Gregg 2006; Aldridge and Boyce 2007, 2008). Here, we provide evidence of brood-rearing grouse that selected areas with greater perennial forbs (0.03-ha scale) and higher density of meadow edge (perimeter to area ratio; 7.9-ha scale) increased the success of fledging offspring.

Forbs are thought to provide a nutritional component to chick diet, which may be critical during initial stages of development (Drut et al. 1994b, Huwer et al. 2008). One likely explanation for this finding is that perennial forbs are associated with increased chick survival because increased forb abundance increases the growth rate of chicks (Huwer et al. 2008). Willow Ptarmigan (*Lagopus lagopus*) and Red Grouse (*L. l. scoticus*) chicks with higher growth rates are more likely to survive (Myrberget et al. 1977, Park et al. 2001). In a productive area (increased population growth rate) in Oregon, greater amounts of forbs were found in crops of sage grouse chicks than in a less productive area, where crop contents of chicks consisted primarily of sagebrush (Drut et al. 1994b). Similar to our findings, female grouse with ≥1 chick at 50 days post-hatch used sites with greater percent forbs than did females with no chicks at 50 days post-hatch on Parker Mountain, Utah (Chi 2004). Collectively, these findings and ours are consistent with the hypothesis that increased forb abundance promotes population growth by influencing survival of chicks during the brood-rearing stage (Drut et al. 1994b, Huwer 2004).

Increased density of meadow edge at the 7.9-ha scale was also related to brood success (i.e., analysis II) and decisions by females to select those sites near meadow edge appeared to influence the success of their broods (i.e., analysis III). These results support earlier research in Nevada

that indicated brood-rearing grouse often used smaller upland meadows with increased edge and surface water to feed on protein-rich forbs (Savage 1969, Oakleaf 1971). The effect of meadow edge on brood success weakened at the largest spatial scale in our study. Thus, the amount of meadow edge within a 7.9-ha area likely represented proximity of food resources that are relevant to the space use of brood-rearing sage grouse.

Meadow edge may have indirectly represented the important role of insects. For example, in sagebrush communities, increased moisture is associated with increased plant biomass (Whitford et al. 1995) and greater primary production is known to be associated with increased insect diversity (Lightfoot and Whitford 1991, Forbes et al. 2005). Particularly in the arid sagebrush ecosystems, moist areas are associated with increased insect distribution and diversity, perhaps by offering shade from plants, water sources, and humid microclimates (Wenninger and Inouye 2008). Insects supply essential nutrition for sage grouse chicks (Johnson and Boyce 1990, Drut et al. 1994b, Gregg 2006, Thompson et al. 2006) and appear to be critical for normal development, particularly during the first three weeks (Schroeder et al. 1999), in both wild (Gregg 2006) and captive settings (Johnson and Boyce 1990). Perhaps females select areas where sagebrush interfaces with mesic areas in search of Lepidoptera larvae, an order of insects that are thought to be the ultimate factor related to sage grouse chick survival in Oregon and Nevada (Gregg 2006). Moreover, insect species richness has been found to be greater in perennial plant communities compared to those of annuals (Lawton and Schröder 1977, Lawton and Strong 1981). We found greater support for a model with perennial forbs than a model with the additive effect of annual and perennial forbs.

It is possible that perennial forb abundance and meadow edge provided similar information in the observed pattern because of cross-scale correlation (i.e., correlations between predictor variables at different scales; Battin and Lawler 2006, Mahon et al. 2008). Because these variables were measured at different scales, one variable may positively reinforce the effect of the other. For example, perennial forbs may be an important component of meadow edge, which is a result of a hierarchical structure among habitat factors (Kristan and Scott 2006). However, the strong evidence of the single-variable forb model suggested that females are choosing forbs independent of meadow edge. Furthermore, diagnostic correlations between predictor variables did not suggest multicollinearity among variables at different scales.

In evaluating habitat selection, the distinct difference between models that consisted of distance to nearest meadow versus one with distance to mesic area (including riparian) indicated that brood-rearing grouse did not prefer mesic areas that consisted of trees. We suspected that sage grouse avoided riparian areas because mammalian and avian predator densities may be greater in areas with trees. Other authors have also reported that riparian areas appear to have higher concentrations of predators (Aldridge and Boyce 2007).

The greater importance of meadow edge compared to meadow size indicated that increasing meadow size at the expense of sagebrush loss is not beneficial. Edges of small meadows provide chicks foraging areas as well as shrub for escape cover that likely reduce predation when encountering predators. Although we did not classify meadow type (i.e., upland springs vs. agriculture), these analyses clearly indicated that small, irregularly shaped meadows were more important to sage grouse broods than large areas, such as agricultural fields. Our results support similar findings in Canada (Aldridge and Boyce 2007), where brood-rearing sage grouse avoided large cultivated cropland but selected smaller meadows with patchy cover. Some evidence suggests increased vegetation provides important structure to allow chicks to avoid predation (Thompson et al. 2006). Grouse often face tradeoffs between using protective contiguous cover for survival (McNew et al., this volume, chapter 19) and relatively open areas that are productive for foraging (Aldridge and Boyce 2007). Increased heterogeneity across a landscape was associated with increased fecundity rates (including brood survival) and with reduced adult survival of Greater Prairie-Chicken (*Tympanuchus cupido*; McNew et al., this volume, chapter 19). Landscape matrices that include small upland meadows might lessen these trade-offs by providing both cover and forage. The link between selecting areas with greater edge and success of rearing broods supports management that preserves healthy sagebrush stands around the edge of small upland meadows (Dunn and Braun 1986). Because vegetation can be managed more readily than insects, we recommend practices that increase small, irregularly shaped meadows (increased perimeter to area ratio) which

interface with sagebrush habitats, as critical brood-rearing habitat.

Strong evidence indicated that brood-rearing sage grouse avoided areas of pinyon–juniper encroachment at larger spatial scales. Despite the lack of evidence of a model that explained brood success or one that identified associations between avoidance of pinyon–juniper and brood fate, these findings should still raise conservation concern. The range of pinyon and juniper woodlands expansion into the sagebrush ecosystem has increased ten-fold since the 1800s, and is thought to adversely affect sage grouse populations (Connelly et al. 2004). This expansion is causing a replacement of sagebrush and is one of the most evident changes in vegetation within the Great Basin (Miller and Tausch 2001), largely attributed to reduced occurrence of fire (Miller and Wigand 1994, Miller and Tausch 2001). These woodlands drastically reduce understory vegetation as tree density increases (Miller et al. 2005), which has been reported for Mono County (Bi-State Local Planning Group 2004). Here, the avoidance of pinyon–juniper encroachment indicated that sage grouse spatial distribution was influenced by areas that consist of ≤40 trees/ha of pinyon or juniper. A reduction in spatial distribution with encroachment may pose a significant risk to the persistence of populations.

This study was not without sampling constraints. A larger data set would have been useful in evaluating time-dependent effects in habitat selection. For example, with increased sampling, brood age categories could contain balanced data and reduce potential temporal biases. Because most unsuccessful female sage grouse retained ≥1 chick to the later stages of the 50-day brood-rearing period, and measurements for those grouse were conducted until no chicks were found, we are confident that potential time-dependent effects did not bias our results. Although our study was limited to three years, we believe these findings are representative of the fecundity of sage grouse populations in the southwestern portion of their range.

In general, it appears that the climatic and topographic factors within Mono County are favorable to sage grouse reproductive vital rates. For example, areas have relatively more mesic sites and higher annual precipitation (average = 36 cm) where sage grouse occur in Mono County than where they occur in other areas within the Great Basin (averages = 23–33 cm; Gregg 2006, Atamian 2007,

Coates and Delehanty 2008), which is in the core of sage grouse distribution (Schroeder et al. 2004). In Mono County, success of fledging chicks at 50 days post-hatch seemed high, and nest survival rates are also higher than reported for other populations range-wide (Kolada et al. 2009). Perhaps these high elevation mesic sites provide suitable conditions for successful refugia, which may partly explain the stable population growth rates observed for sage grouse in Mono County compared to the negative trends observed in other regions (Connelly et al. 2004). Furthermore, increased precipitation levels are associated with greater growth and reproduction of perennial forbs in sagebrush steppe ecosystems (Bates et al. 2006). Thus, sage grouse fecundity responses may be particularly sensitive to variation in productivity of perennial forbs and moist meadows within relatively more arid environments than those of Mono County. Additional research that investigates relationships between habitat selection and fitness in other portions of sage grouse range would be helpful. Nevertheless, management practices range-wide that preserve and enhance a landscape matrix of sagebrush stands interspersed with small upland meadows may prove to be most beneficial to sage grouse populations.

ACKNOWLEDGMENTS

We thank M. Farinha and E. Kolada for their outstanding contribution to the study design and field data collection aspects of this study. We thank the University of Nevada, Reno, and especially J. Sedinger for supporting this research. We also thank K. Martin, J. Connelly, C. Aldridge, D. Delehanty, M. Miller, and B. Halstead for helpful manuscript review. We thank field technicians J. Felland, R. Montano, K. Nelson, B. Barbaree, K. Gagnon, S. Alofsin, and T. Skousen for field assistance. We thank D. Blankenship, S. Gardner, D. Racine, and the California Department of Fish and Game, U.S. Forest Service, and Bureau of Land Management for support. We thank D. House and the Los Angeles Department of Water and Power. We also thank P. Gore for logistical help. We extend special thanks to J. Fatooh, S. Nelson, A. Halford, and T. Taylor for providing local knowledge, help in the field, and insight into sage grouse biology. The Mono Lake Committee helped provide housing for field crews. We thank R. Haldeman and L. Fields of Quail Unlimited for financial and logistical help. Any use of trade, product, or firm names in this publication is for descriptive purposes only and does not imply endorsement by the U.S. government.

LITERATURE CITED

Aebischer, N. J., Robertson, P. A., and Kenward, R. E. 1993. Compositional analysis of habitat use from animal radiotracking data. Ecology 74:1313–1325.

Aldridge, C. L., and M. S. Boyce. 2007. Linking occurrence and fitness to persistence: habitat-based approach for endangered greater sage-grouse. Ecological Applications 17:508–526.

Aldridge, C. L., and M. S. Boyce. 2008. Accounting for fitness: combining survival and selection when assessing wildlife-habitat relationships. Israel Journal of Ecology and Evolution 54:389–419.

Aldridge, C. L., and R. M. Brigham. 2001. Nesting and reproductive activities of Greater Sage-Grouse in a declining northern fringe population. Condor 103:537–543.

Aldridge, C. L., and R. M. Brigham. 2002. Sage-grouse nesting and brood habitat use in southern Canada. Journal of Wildlife Management 66:433–444.

Allen, T. F. H., and T. W. Hoekstra. 1992. Towards a unified ecology. Columbia University Press, New York, NY.

Anderson, D. R. 2008. Model based inferences in the life sciences. Springer Science, New York, NY.

Atamian, M. T. 2007. Brood ecology and sex ratio of greater sage-grouse in east-central Nevada. Masters thesis, University of Nevada Reno, Reno, NV.

Bates, D., M. Maechler, and B. Dai. 2008. lme4: linear mixed-effects models using S4 classes. R package version 0.999375–27. <http://lme4.r-forge.r-project.org/>.

Bates, J. D., T. Svejcar, R. F. Miller, and R. A. Angell. 2006. The effects of precipitation timing on sagebrush steppe vegetation. Journal of Arid Environments 64:670–697.

Battin, J., and J. J. Lawler. 2006. Cross-scale correlations and the design and analysis of avian habitat selection studies. Condor 108:59–70.

Benedict, N. G., S. J. Oyler-McCance, S. E. Taylor, C. E. Braun, and T. W. Quinn. 2003. Evaluation of the eastern (*Centrocercus urophasianus urophasianus*) and western (*Centrocercus urophasianus phaios*) subspecies of sage-grouse using mitochondrial control-region sequence data. Conservation Genetics 4:301–310.

Bi-State Local Planning Group. 2004. Greater Sage-Grouse conservation plan for the bi-state plan area of Nevada and eastern California. 1st ed. Reno, NV.

Bowyer, R. T., and J. G. Kie. 2006. Effects of scale on interpreting life-history characteristics of ungulates and carnivores. Diversity and Distributions 12:244–257.

Boyce, M. S. 2006. Scale for resource selection functions. Diversity and Distributions 12:269–276.

Boyce, M. S., P. R. Vernier, S. E. Nielsen, and F. K. A. Schmiegelow. 2002. Evaluating resource selection functions. Ecological Modeling 157:281–300.

Boyd, C. S., J. D. Bates, and R. F. Miller. 2007. The influence of gap size on sagebrush cover estimates using line intercept technique. Rangeland Ecology and Management 60:199–202.

Braun, C. L., T. Britt, and R. O. Wallestad. 1977. Guidelines for maintenance of sage grouse habitats. Wildlife Society Bulletin 5:99–106.

Brotons, L., W. Thuiller, M. B. Araújo, and A. H. Hirzel. 2004. Presence-absence versus presence-only modeling methods for predicting bird habitat suitability. Ecography 27:437–448.

Burnham, K. P., and D. R. Anderson. 2002. Model selection and multimodel inference: a practical information-theoretic approach. 2nd ed. Springer-Verlag, New York, NY.

Canfield, R. H. 1941. Applications of the line interception method in sampling range vegetation. Journal of Forestry 39:388–394.

Chi, R. Y. 2004. Greater Sage-Grouse reproductive ecology and tebuthiuron manipulation of dense big sagebrush on Parker Mountain. Thesis, Utah State University, Logan, UT.

Coates, P. S., and D. J. Delehanty. 2008. Effects of environmental factors on incubation patterns of Greater Sage-Grouse. Condor 110:627–638.

Connelly, J. W., R. A. Fischer, A. D. Apa, K. P. Reese, and W. L. Wakkinen. 1993. Renesting of sage grouse in southeastern Idaho. Condor 95:1041–1043.

Connelly, J. W., S. T. Knick, M. A. Schroeder, and S. T. Stiver. 2004. Conservation assessment of Greater Sage-Grouse and sagebrush habitats. Unpublished report. Western Association of Fish and Wildlife Agencies, Cheyenne, WY.

Connelly, J. W., M. A. Schroeder, A. R. Sands, and C. E. Braun. 2000. Guidelines to manage sage-grouse populations and their habitats. Wildlife Society Bulletin 28:967–985.

Crawford, J. A., R. A. Olson, N. E. West, J. C. Mosley, M. A. Schroeder, T. D. Whitson, R. F. Miller, M. A. Gregg, and C. S. Boyd. 2004. Ecology and management of sage-grouse and sage-grouse habitat. Journal of Range Management 57:2–19.

Daubenmire, R. 1959. A canopy-coverage method of vegetation analysis. Northwest Science 33:43–64.

Drut, M. S., J. A. Crawford, and M. A. Gregg. 1994a. Brood habitat use by sage grouse in Oregon. Great Basin Naturalist 54:170–176.

Drut, M. S., W. H. Pyle, and J. A. Crawford. 1994b. Technical note: Diets and food selection of sage grouse chicks in Oregon. Journal of Range Management 47:90–93.

Dunn, P. O., and C. E. Braun. 1986. Summer habitat use by adult female and juvenile sage grouse. Journal of Wildlife Management 50:228–235.

Erickson, W. P., T. L. McDonald, K. G. Gerow, S. Howlin, and J. W. Kerr. 2001. Statistical issues in resource selection studies with radiotracked animals. Pp. 209–242 in J. J. Millspaugh and J. M. Marzluff (editors), Radio-tracking and animal populations. Academic Press, San Diego, CA.

Faraway, J. J. 2006. Extending the linear model with R: generalized linear, mixed effects and nonparametric regression models. Chapman and Hall, Boca Raton, FL.

Forbes, G. S., J. W. Van Zee, W. Smith, and W. G. Whitford. 2005. Desert grassland canopy arthropod species richness: temporal patterns and effects of intense, short-duration livestock grazing. Journal of Arid Environments 60:627–646.

Garton, E. O., J. T. Ratti, and J. H. Giudice. 2005. Research and experimental design. Pp. 43–71 in C. E. Braun (editor), Techniques for wildlife investigations and management. 6th ed. The Wildlife Society, Bethesda, MD.

Giesen, K. M., T. J. Schoenberg, and C. E. Braun. 1982. Methods for trapping sage grouse in Colorado. Wildlife Society Bulletin 10:224–231.

Gillies, C., M. Hebblewhite, S. E. Nielsen, M. Krawchuk, C. Aldridge, J. Frair, C. Stevens, D. J. Saher, and C. Jerde. 2006. Application of random effects to the study of resource selection by animals. Journal of Animal Ecology 75:887–898.

Gregg, M. A. 2006. Greater Sage-Grouse reproductive ecology: linkages among habitat resources, maternal nutrition, and chick survival. Ph.D. dissertation, Oregon State University, Corvallis, OR.

Huwer, S. L. 2004. Evaluating greater sage-grouse brood habitat using human-imprinted chicks. Masters thesis, Colorado State University, Fort Collins, CO.

Huwer, S. L., D. R. Anderson, T. E. Remington, and G. C. White. 2008. Using human-imprinting chicks to evaluate the importance of forbs to sage-grouse. Journal of Wildlife Management 72:1622–1627.

Johnson, C. J., S. E. Nielsen, E. H. Merrill, T. L. McDonald, and M. S. Boyce. 2006. Resource selection functions based on use-availability data: theoretical motivation and evaluation methods. Journal of Wildlife Management 70:347–357.

Johnson, G. D., and M. S. Boyce. 1990. Feeding trials with insects in the diet of sage grouse chicks. Journal of Wildlife Management 54:89–91.

Keating, K. A., and S. Cherry. 2004. Use and interpretation of logistic regression in habitat-selection studies. Journal of Wildlife Management 68:774–789.

Klebenow, D. A. 1969. Sage grouse nesting and brood habitat in Idaho. Journal of Wildlife Management 33:649–662.

Klebenow, D. A., and G. M. Gray. 1968. Food habits of juvenile sage grouse. Journal of Range Management 21:80–83.

Kolada, E. J., M. L. Casazza, and J. S. Sedinger. 2009. Ecological factors influencing nest survival of Greater Sage-Grouse in Mono County, California. Journal of Wildlife Management 73:1341–1347.

Koper, N., and M. Manseau. 2009. Generalized estimating equations and generalized linear mixed effects models for modeling resource selection. Journal of Applied Ecology 46:590–599.

Kristan, W. B., III, and J. M. Scott. 2006. Hierarchical models for avian ecologists. Condor 108:1–4.

Kurki, S., A. Nikula, P. Helle, and H. Lindén. 1998. Abundance of Red Fox and Pine Marten in relation to the composition of boreal forest landscapes. Journal of Animal Ecology 67:874–886.

Lawton, J. H., and D. Schröder. 1977. Effects of plant type, size of geographical range and taxonomic isolation on number of insect species associated with British plants. Nature 265:137–140.

Lawton, J. H., and D. R. Strong, Jr. 1981. Community patterns and competition in folivorous insects. American Naturalist 118:317–338.

Lightfoot, D. C., and W. G. Whitford. 1991. Productivity of creosotebush foliage and associated canopy arthropods along a desert roadside. American Midland Naturalist 125:310–322.

Mahon, C. L., K. Martin, and V. LeMay. 2008. Do cross-scale correlations confound analysis of nest site selection for Chestnut-backed Chickadees? Condor 110:563–568.

Manly, F. J., L. L. McDonald, D. L. Thomas, T. L. McDonald, and W. P. Erickson. 2002. Resource selection by animals: statistical design and analysis for field studies. Chapman and Hall, London, UK.

Manzer, D., and S. J. Hannon. 2005. Relating grouse nest success and corvid density to habitat: a multiscale approach. Journal of Wildlife Management 69:110–123.

Mayor, S. J., D. C. Schneider, J. A. Schaefer, and S. P. Mahoney. 2009. Habitat selection at multiple scales. Ecoscience 16:238–247.

Menard, S. 1995. Applied logistic regression analysis. Sage Publications, Thousand Oaks, CA.

Miller, R. F., J. D. Bates, T. L. Svejcar, F. B. Pierson, and L. E. Eddleman. 2005. Biology, ecology, and management of western juniper. Technical Bulletin 152, Agricultural Experiment Station. Oregon State University, Corvallis, OR.

Miller, R. F., and R. J. Tausch. 2001. The role of fire in pinyon and juniper woodlands: a descriptive analysis. Tall Timbers Research Station Miscellaneous Publication No. 11:15–30.

Miller, R. F., and P. E. Wigand. 1994. Holocene changes in semiarid pinyon–juniper woodlands: response to climate, fire and human activities in the US Great Basin. BioScience 44:465–474.

Morris, D. W., R. Clark, and M. S. Boyce. 2008. Habitat and habitat selection: theory, tests, and implications. Israel Journal of Ecology and Evolution 54:287–294.

Morrison, M. L. 2001. A proposed research emphasis to overcome the limits of wildlife-habitat relationship studies. Journal of Wildlife Management 65:613–623.

Myrberget, S., K. E. Erikstad, and T. K. Spidso. 1977. Variations from year to year in growth rates of willow grouse chicks. Astarte 10:9–14.

Oakleaf, R. J. 1971. The relationship of sage grouse to upland meadows in Nevada. Nevada Department of Fish and Game Job Completion Report W-48-2.

Oyler-McCance, S. J., S. E. Taylor, and T. W. Quinn. 2005. A multilocus population genetic survey of the Greater Sage-Grouse across their range. Molecular Ecology 14:1293–1310.

Park, K. J., P. A. Robertson, S. T. Campbell, R. Foster, Z. M. Russell, C. Newborn, and P. J. Hudson. 2001. The role of invertebrates in the diet, growth and survival of Red Grouse (*Lagopus lagopus scoticus*) chicks. Journal of Zoology 254:137–145.

Peterson, J. G. 1970. The food habits and summer distribution of juvenile sage grouse in central Montana. Journal of Wildlife Management 34:147–155.

R Development Core Team. 2008. R: a language and environment for statistical computing. R Foundation for Statistical Computing, Vienna, Austria.

Robel R. J., J. N. Briggs, A. D. Dayton, and L. C. Hulbert. 1970. Relationships between visual obstruction measurements and weight of grassland vegetation. Journal of Range Management 23:295–297.

Rosenzweig, M. L. 1991. Habitat selection and population interactions: the search for mechanism. American Naturalist 137:S5–S28.

Rowland, M. W., M. J. Wisdom, L. H. Suring, and C. W. Meinke. 2006. Greater Sage-Grouse as an umbrella species for sagebrush-associated vertebrates. Biological Conservation 129:323–335.

Savage, D. E. 1969. The relationship of sage grouse to upland meadows in Nevada. Nevada Department of Fish and Game Job Completion Report W-39-R-9.

Schneider, D. C. 1994. Quantitative ecology: spatial and temporal scaling. Academic Press, Toronto, Ontario, Canada.

Schoenberg, T. J. 1982. Sage grouse movements and habitat selection in North Park, Colorado. M.S. thesis, Colorado State University, Fort Collins, CO.

Schroeder, M. A., C. L. Aldridge, A. D. Apa, J. R. Bohne, C. B. Braun, S. D. Bunnell, J. W. Connelly, P. A. Deibert, S. C. Gardner, M. A. Hilliard, G. D. Kobridger, S. M. McAdam, C. W. McCarthy, J. J. McCarthy, D. L. Mitchell, E. V. Rickerson, and S. J. Stiver. 2004. Distribution of sage-grouse in North America. Condor 106:363–376.

Schroeder, M. A., J. A. Young, and C. E. Braun. 1999. Sage grouse (*Centrocercus urophasianus*). A. Poole and F. Gill (editors), The birds of North America No. 425. Academy of Natural Sciences, Philadelphia, PA.

Sveum, C. M., J. A. Crawford, and W. D. Edge. 1998. Use and selection of brood-rearing habitat by greater sage grouse in south central Washington. Great Basin Naturalist 58:344–351.

Thompson, K. M., M. J. Holloran, S. J. Slater, J. L. Kuipers, and S. H. Anderson. 2006. Early brood-rearing habitat use and productivity of greater sage-grouse in Wyoming. Western North American Naturalist 66:332–342.

Wakkinen, W. L., K. P. Reese, J. W. Connelly, and R. A. Fischer. 1992. An improved spotlighting technique for capturing sage grouse. Wildlife Society Bulletin 20:425–426.

Wallestad, R. O. 1971. Summer movement and habitat use by sage grouse broods in Montana. Journal of Wildlife Management 35:129–136.

Wenninger, E. J., and R. S. Inouye. 2008. Insect community response to plant diversity and productivity in a sagebrush-steppe ecosystem. Journal of Arid Environments 72:24–33.

Whitford, W. G., G. Martinez-Turanzas, E. Martinez-Meza. 1995. Persistence of desertified ecosystems: explanations and implications. Environmental Monitoring and Assessment 37:1–14.

Wiens, J. A. 1989. Spatial scaling in ecology. Functional Ecology 3:385–397.

Zuur, A. F., E. N. Ieno, N. J. Walker, A. A. Saveliev, and G. M. Smith. 2009. Mixed effects models and extensions in ecology with R. Springer-Verlag, New York, NY.

CHAPTER TWELVE

Resource Selection During Brood-Rearing by Greater Sage-Grouse

Nicholas W. Kaczor, Katie M. Herman-Brunson, Kent C. Jensen,
Mark A. Rumble, Robert W. Klaver, and Christopher C. Swanson

Abstract. Understanding population dynamics and resource selection is crucial in developing wildlife resource management plans for sensitive species such as Greater Sage-Grouse (*Centrocercus urophasianus*). Little is known about sage grouse habitats on the eastern edge of their range. We investigated resource selection of Greater Sage-Grouse during brood-rearing in North and South Dakota during 2005–2007. Resource selection models suggested sage grouse females with broods selected sites with increased vegetative cover and grass height. Composition of forbs at brood-rearing sites has been identified as important elsewhere, but we found little support for a difference in forbs between brood and random sites. Despite being sagebrush obligates, sage grouse females with broods selected areas with low sagebrush cover. Brood habitats with increased invertebrate abundance and protective cover have been shown to increase sage grouse productivity. Land managers on the eastern edge of Greater Sage-Grouse range could focus on protecting critical brood-rearing areas by maintaining at least 67% herbaceous cover and 33 cm of grass height in association with sagebrush for sage-grouse broods.

Key Words: brood-rearing habitat, *Centrocercus urophasianus*, Greater Sage-Grouse, North Dakota, resource selection, sagebrush, South Dakota.

Knowledge of seasonal habitat selection is important in developing management strategies for sensitive wildlife species. Concerns about declining populations of Greater Sage-Grouse (*Centrocercus urophasianus*; hereafter sage grouse) date back >90 years and continue today (Hornaday 1916, Aldridge et al. 2008). Sage grouse populations have declined range-wide at a rate of 2% per year since 1965 (Connelly et al. 2004). In North Dakota, populations may have declined by 67% from 1965 to 2003 and in South Dakota sage grouse populations declined steadily from 1973 to 1997, but may have recovered slightly from 1997 to 2007 (Sage and Columbian Sharp-tailed Grouse Technical Committee 2008). In the past decade, at least seven petitions have been filed to list sage

Kaczor, N. W., K. M. Herman-Brunson, K. C. Jensen, M. A. Rumble, R. W. Klaver, and C. C. Swanson. 2011. Resource selection during brood-rearing by Greater Sage-Grouse. Pp. 169–177 *in* B. K. Sandercock, K. Martin, and G. Segelbacher (editors). Ecology, conservation, and management of grouse. Studies in Avian Biology (no. 39), University of California Press, Berkeley, CA.

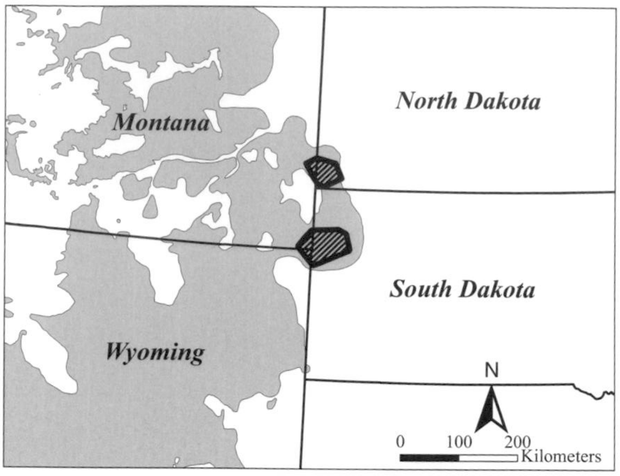

Figure 12.1. Location of study areas for Greater Sage-Grouse in northwestern South Dakota, southwestern North Dakota, southeastern Montana, and northeastern Wyoming. The dashed areas encompass all locations and the gray area is the current range of Greater Sage-Grouse (Schroeder et al. 2004).

grouse under the Endangered Species Act (ESA) of 1973, with an ESA status review recently completed by the U.S. Fish and Wildlife Service to determine the merit of the most recent petitions (Connelly et al. 2004, USFWS 2010).

Sage grouse in northwestern South Dakota and southwestern North Dakota occupy transitional vegetation communities between the northern wheatgrass (*Pascopyrum* spp.) – needlegrass (*Nassella* spp.) prairie that dominates most of the Dakotas and the big sagebrush (*Artemisia tridentata*) plains of Wyoming (Johnson and Larson 1999). In South Dakota, sage grouse are listed as a species of greatest conservation need (South Dakota Department of Game, Fish, and Parks 2006) and they are also a Priority Level 1 Species of Special Concern in North Dakota (McCarthy and Kobriger 2005). In addition, sage grouse are listed as a sensitive species for the Bureau of Land Management (BLM) and U.S. Forest Service (USFS). Despite well-understood reproductive ecology of sage grouse in the core of their range, knowledge of reproductive ecology and habitat selection by sage grouse occurring at the eastern edge of their distribution is limited. The objectives of this study were to develop an understanding of brood-rearing resource selection of sage grouse in the Dakotas. This information will be useful in developing conservation and management plans for sage grouse in eastern Montana, Wyoming, and the Dakotas.

STUDY AREA

Our study was divided into two sites, one in northwestern South Dakota and the other in southwestern North Dakota, with portions of Montana occurring in both sites and portions of Wyoming occurring in the South Dakota site (Fig. 12.1). The entire area has flat to gently rolling prairie, with a few buttes and intermittent streams. Approximately 73% of the area is privately owned. The U.S. Bureau of Land Management (BLM) manages 25% of the study area, and the remaining 2% is managed by the State School and Public Lands Divisions of North and South Dakota. Grazing is the predominant land use, with normal stocking rates between 1 and 4 hectares per animal unit month (AUM). In areas with rough terrain or less productive soils stocking rates can be as high as 6 hectares per AUM (M. Iverson, pers. comm.). In areas with large blocks of public land, livestock are rotated on a regular schedule, but may not be rotated as often in areas with less public land. Agricultural production of wheat and grass hay occurs on better soil types in both study sites. Exploration and development of oil and natural gas resources is common and has been identified as an ongoing threat to sage grouse in North Dakota (Connelly et al. 2004), with some areas having up to 16 well pads per 259 ha (square mile). Open-pit mining for bentonite occurs at the southern end of the South Dakota study site on Pierre soils (Charles Berdan, pers. comm.).

Mean annual precipitation is 35 cm, with 70% occurring during the months of April through August (South Dakota State Climate Office 2007). Temperatures in summer (May–August) average 20.1°C but can reach up to 43.3°C. Vegetation communities included mixed grass prairie with perennial and annual forbs and grasses and shrubsteppe, as described by Johnson and Larson (1999).

METHODS

Data Collection

We identified sage grouse leks within the study sites where we had land owner permission for trapping. We captured female sage grouse with large nets aided with spotlighting from all-terrain vehicles between March and mid-April of 2005–2007 (Giesen et al. 1982, Wakkinen et al. 1992). Females were weighed and equipped with 22-g necklace-style transmitters, which were approximately 1.4% of mean female sage grouse body mass and had a life expectancy of 434 days. Transmitters could be detected from approximately 2 to 5 km from the ground and were equipped with an 8-hour mortality switch. The South Dakota State University Institutional Animal Care and Use Committee approved trapping and handling techniques and study design (Protocol #07-A032).

We located radio-marked females during incubation twice per week and if the nest successfully hatched, we located females and their broods twice per week. Broods were approached cautiously to minimize the possibility of flushing or scattering the brood, with most locations taken within 50 m of actual locations. When chicks reached ~3–5 weeks of age, we flushed the brood and searched the area to obtain estimates of brood size. We discontinued telemetry of broods if no chicks were present with a female and subsequent locations of the female for two weeks showed no evidence of chicks.

We characterized vegetation at sites used by females with broods about 14 ± 2 SE days after recording the location. Two 50-m transects were established in the north–south cardinal directions, each starting at the marked brood location and terminating at their respective north–south ends. A modified Robel pole was used to quantify visual obstruction readings (VOR) and maximum grass height at 10 m intervals ($n = 11$; Robel et al. 1970, Benkobi et al. 2000). We estimated sagebrush (*Artemisia tridentata* spp. and *A. cana* spp.) density and height at 10-m intervals ($n = 11$) using the point-centered quarter method (Cottam and Curtis 1956). Canopy coverage was estimated using a 0.10-m² quadrat (Daubenmire 1959). At each 10-m interval, four 0.1 m² quadrats were placed in an H-shape with each leg of the H being 1 m long ($n = 44$ per site).

We recorded percent cover in six categories for total vegetation, grass, forb, shrub, litter, bare ground, shrub species, grass species, and forb species cover in each quadrat (Daubenmire 1959). In addition, we characterized vegetation at an equal number of random sites during the same period. Random sites were generated within a 10-km buffer of capture leks with a Geographic Information System (GIS) with Hawth's Analysis Tool (Beyer 2004). Random points were reselected if they were on a road, in a road ditch, or on private lands where we did not have access.

Data Analyses

All measurements were summarized to a value for the site. Sagebrush density was estimated from a maximum likelihood estimate (Pollard 1971). We calculated average sagebrush height for each site from the sagebrush plants that were measured to estimate density. Values recorded for canopy coverage were recoded to mid-point values of categories and summarized to an average for the site (Daubenmire 1959). To reduce non-biologically important variables, we screened canopy coverage variables and excluded any variables with canopy coverage less than 1% on sites where they were present. We then conducted a principal components analysis (PCA) to distinguish important variables that captured the variation among sites. We identified seven biologically important variables from the PCA to investigate sage grouse brood habitat resource selection (Table 12.1). Variables included sagebrush height, visual obstruction, maximum grass height, total herbaceous cover, grass cover, forb cover, and sagebrush cover. We used the multi-response permutation process to test for differences between use and random sites and between study area sites of these variables with a critical value of $\alpha < 0.05$ (Mielke and Berry 2001). We used a cluster analysis based on Euclidean distance to investigate patterns of habitat for early (0–4 weeks post-hatch), middle (4–8 weeks), and late (8–12 weeks) age-classes of broods (Ball and Hall 1967).

To investigate resource selection, we used an information-theoretic approach to estimate the importance of 173 *a priori* nominal logistic regression models including global and null models

TABLE 12.1

Observed mean values for habitat variables and associated P-values between Greater Sage-Grouse brood-rearing and random sites used in logistic regression modeling in North Dakota, 2005–2006, and South Dakota, 2006–2007; all values are reported as $\bar{x} \pm$ (SE).

Variable[a]	North Dakota			South Dakota			Pooled		
	Brood (n = 132)	Random (n = 105)	P ≤	Brood (n = 119)	Random (n = 116)	P ≤	Brood (n = 251)	Random (n = 221)	P ≤
Sagebrush height (cm)	33.2 (0.8)	32.9 (1.1)	0.51	23.8 (1.0)	22.8 (0.8)	0.14	28.8 (0.7)	27.3 (0.8)	0.09
Visual obstruction (cm)[1]	8.3 (1.3)	5.0 (0.5)	0.06	6.2 (0.4)	3.5 (0.4)	0.01	7.3 (0.7)	4.2 (0.3)	0.01
Maximum grass height (cm)[1]	35.1 (1.9)	35.2 (1.2)	0.30	30.5 (1.0)	25.7 (0.9)	0.01	32.9 (1.1)	30.2 (0.8)	0.23
Herbaceous cover (%)[2]	74.2 (1.7)	54.7 (1.8)	0.01	58.4 (1.3)	51.0 (1.5)	0.01	66.7 (1.2)	52.7 (1.2)	0.01
Grass cover (%)[2]	31.7 (1.5)	20.3 (1.4)	0.01	31.3 (1.1)	26.6 (1.3)	0.01	31.5 (0.9)	23.8 (1.0)	0.01
Forb cover (%)	10.2 (0.9)	9.1 (0.9)	0.11	7.6 (0.4)	7.1 (0.4)	0.20	9.0 (0.5)	8.1 (0.5)	0.13
Sagebrush cover (%)	4.8 (0.3)	2.9 (0.3)	0.01	4.6 (0.4)	3.6 (0.4)	0.04	4.7 (0.3)	3.3 (0.3)	0.01

[a] Paired numbers represent correlated (r > 0.70) variables that subsequently were not modeled together.

crafted from the seven habitat variables (Burnham and Anderson 2002, SAS Institute 2007). Model sets were developed to evaluate resource selection for both study sites combined and individual brood sites. To reduce variable interactions in the models, variables that were correlated ($r > 0.70$) to one another were not included in the same model (Table 12.1). Due to a small sample size with respect to the number of parameters estimated ($n/K < 40$), we used the small-sample adjustment for Akaike's Information Criterion (AIC_c) to evaluate models (Burnham and Anderson 2002). We ranked our models based on differences between AIC_c for each model and the minimum AIC_c model (ΔAIC_c), and used Akaike weights (w_i) to assess the weight of evidence in favor of each model and the sum AIC_c weight for each variable (Burnham and Anderson 2002, Beck et al. 2006). In addition, when models differed by a single parameter, we inspected the change in model deviance and investigated the slope of the coefficient estimates (β) to determine variable effects (Burnham and Anderson 2002). Model goodness-of-fit was determined using a Hosmer–Lemeshow test (Hosmer and Lemeshow 2000).

RESULTS

Capture and Monitoring

We captured and fitted 43 females with radio transmitters during spring 2005–2006 in North Dakota, and 53 females in South Dakota during spring 2006–2007, for a total of 96 individual females. We monitored 60 females in North Dakota (2005: $n = 21$, 2006: $n = 39$) and 83 females in South Dakota (2005: $n = 40$, 2006: $n = 42$). A total of 17 and 29 females were monitored in both years in North and South Dakota, respectively. After losses to nest predation, we were able to monitor 19 females with broods in North Dakota (2005: $n = 9$, 2006: $n = 10$) and 24 females with broods in South Dakota (2005: $n = 10$, 2006: $n = 14$). Average hatch date was 20 May and 31 May in North and South Dakota, respectively. Dates of brood locations ranged from 25 May to 30 August.

Resource Selection

We measured 55 and 77 brood sites and 47 and 58 random sites in North Dakota from mid-June

through August of 2005 and 2006, respectively. We also measured 59 and 60 brood sites and 56 and 60 random sites in South Dakota from mid-June through August of 2006 and 2007, respectively. All variables in both study sites, except sagebrush height, grass height, and forb cover, differed between brood and random sites ($P \leq 0.05$; Table 12.1). In addition, visual obstruction differed marginally between brood and random sites in North Dakota ($P = 0.06$). Most vegetative characteristics had higher values at brood sites compared to random sites in North Dakota ($P \leq 0.05$), with the exception of grass height, which was slightly higher at random sites. In South Dakota, vegetative characteristics at brood-rearing sites also had higher values than random sites ($P \leq 0.05$). When the two states were combined, maximum grass height did not differ between brood and random sites ($P = 0.23$). We could not distinguish sites based on brood-rearing age-classes from the cluster analysis, as the best clusters were not associated with brood age-classes.

Our best model of resource selection for sage grouse broods included sagebrush height, total herbaceous cover, and maximum grass height (AIC_c weight $= 0.46$; Table 12.2). Including variables year and state to the top model did improve model fit. Total cover and grass height were positively associated with brood-rearing site selection (total cover: $\beta = 0.05 \pm 0.01 \pm SE$, grass height: $\beta = 0.02 \pm 0.01 \pm SE$). Sagebrush height was negatively associated with brood-rearing site selection ($\beta = -0.02 \pm 0.01 \pm SE$). The Hosmer–Lemeshow test was acceptable ($P = 0.16$). Although the top model included maximum grass height, Akaike weight for all models was strongest for total cover (0.96) and sagebrush height (0.96), with total cover having a stronger effect (Fig. 12.2).

The second-ranked model (AIC_c weight $= 0.21$) included total cover, sagebrush height, and forb cover. Similar to the top model, total cover was positively associated with brood-rearing site selection ($\beta = 0.05 \pm 0.01 \pm SE$) and sagebrush height was negatively associated with brood-rearing site selection ($\beta = -0.02 \pm 0.01 \pm SE$). Forb cover was positively associated with brood-rearing site selection ($\beta = 0.04 \pm 0.02 \pm SE$). The Hosmer–Lemeshow goodness-of-fit test was nonsignificant ($P = 0.11$), indicating acceptable model fit.

TABLE 12.2

Top five ranked logistic regression models from 173 models for Greater Sage-Grouse brood-rearing sites (n = 251) and random sites (n = 221) in North Dakota, 2005–2006, and South Dakota, 2006–2007.

Model	$-2\ln(L)$	K^a	ΔAIC_c^b	w_i^c	ΔDev^d
Sagebrush hgt. + herbaceous cover + max grass hgt.	512.16	5	0.00	0.46	0.00
Sagebrush hgt. + herbaceous cover + forb cover	513.78	5	1.62	0.21	1.62
Sagebrush hgt. + herbaceous cover	516.96	4	2.76	0.12	4.80
Sagebrush hgt. + herbaceous cover + year	515.10	5	2.94	0.11	2.94
Sagebrush hgt. + herbaceous cover + state	516.28	5	4.12	0.06	4.12

[a] Number of parameters in the model including intercept and standard error.
[b] Change in AIC_c value.
[c] Model weight.
[d] Change in model deviance.

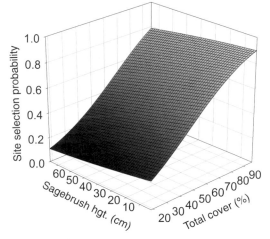

Figure 12.2. Effect of sagebrush height and total herbaceous cover on Greater Sage-Grouse brood-rearing habitat selection in North Dakota, 2005–2006, and South Dakota, 2006–2007. Probability of use derived from parameter estimates in the best approximated model.

DISCUSSION

Herbaceous Cover

Brood-rearing habitats are important for determining sage grouse productivity (Crawford et al. 1992). Our models suggested that increased herbaceous plant matter is a feature of sage grouse brood-rearing habitats in both North and South Dakota. In the Great Basin, females with chicks also selected for areas of increased herbaceous cover (Klebenow 1969, Autenrieth 1981). Although the North Dakota study site had higher values for vegetation components than the South Dakota site, resource selection by female sage

grouse with broods was invariant to the habitat conditions represented by geographic range of these two study areas. Sagebrush cover was found to be important to brood site use, but sagebrush height was identified to have a slightly negative or null effect. The apparent discrepancy may be best explained by the timing of our sampling in mid-June to mid-August, when sagebrush is not a major component of sage grouse chicks' diets (Johnson and Boyce 1990, Huwer 2004). In addition, some of the brood sites did not possess any sagebrush component, which helps explain why sagebrush height entered our models with negative coefficients. However, when females with broods did select sites with sagebrush, the sagebrush tended to be of taller stature.

Grass Cover

Taller grass provides concealment from predators and, perhaps more importantly, greater herbaceous biomass is correlated with greater invertebrate abundance (Healy 1985, Rumble and Anderson 1996, Jamison et al. 2002). Female sage grouse typically move their broods from upland nesting areas to more mesic, greener areas later in the summer (Peterson 1970, Dunn and Braun 1986, Sveum et al. 1998). Although we could not differentiate between habitats based on brood age-classes, broods may be selecting areas with higher grass cover for the increased invertebrate abundance and protection from predators that grass-dominated areas tend to provide. Females with broods in the Dakotas selected areas with greater grass cover than values typically reported in the literature for sage grouse (Drut et al. 1994, Sveum et al. 1998, Thompson et al. 2006). Our study area forms a transition zone between the northern wheatgrass–needlegrass prairie that dominates most of the Dakotas and the big sagebrush plains of Wyoming (Johnson and Larson 1999), and possesses a greater grass component compared to the shrubsteppe region (Lewis 2004). In Alberta, brood habitat was located in moist areas and drainages and was suggested to be limiting sage grouse productivity (Aldridge and Brigham 2002). In our study, brood habitat was not limiting for sage grouse populations, but grass structure, which is highly correlated with visual obstruction, provided increased protection from predators and invertebrate abundance.

Forb Cover

Availability of food resources such as forbs and insects can limit sage grouse populations through decreased recruitment of young (Peterson 1970, Wallestad 1975, Autenrieth 1981). The period from 1 to 10 days is when chick mortality is highest (Patterson 1952, Autenrieth 1981) and young need insects in close proximity to escape cover. Invertebrates are also necessary for growth, development, and survival of sage grouse chicks (Johnson and Boyce 1990). Chicks that fed in forb-rich areas gained more weight than when they fed in forb-poor habitats, and areas with greater forb cover may attract higher numbers of invertebrates (Jamison et al. 2002, Huwer 2004). Greater invertebrate abundance may explain why sage grouse tend to select areas with higher forb cover, which our second-ranked model identified as being important to sage grouse broods (Apa 1998, Sveum et al. 1998, Holloran 1999). Other studies have shown that sage grouse broods also use areas of high forb cover; however, forb cover in our study site was much lower than values reported in the literature (Schoenberg 1982, Sveum et al. 1998, Holloran 1999). Forb cover may be more important to sage grouse brood-rearing habitat in the central portion of their range, where conditions are drier, leading to more bare ground than in western North and South Dakota.

Management Implications

With possible listing under the Endangered Species Act, sage grouse conservation and preservation will be a priority for many western land management agencies. Management of sage grouse brood-rearing habitat in the Dakotas could focus on maintaining grass heights of at least 33 cm and herbaceous cover of at least 67%, which provides high visual obstruction for sage grouse broods and abundant insects for food. In addition, managers could promote and protect greener areas during mid- to late summer because these areas typically have higher production and invertebrate abundance. Programs that defer or reduce grazing and haying operations in these areas could be implemented to promote favorable conditions in brood-rearing habitats. Domestic livestock grazing may negatively influence sage grouse productivity by decreasing plant biomass and protective cover. However, light or moderate grazing in dense, grassy meadows can be beneficial to sage grouse, but overgrazing can reduce sage grouse habitats (Klebenow 1982, 1985; Oakleaf 1971).

ACKNOWLEDGMENTS

Funding for this study was provided by Federal Aid in Wildlife Restoration Act (W-67-R) through the North Dakota Game and Fish Department, Bureau of Land Management (ESA000013), U.S. Forest Service, Rocky Mountain Research Station (05-JV-11221609-127), U.S. Forest Service Dakota Prairie National Grasslands (05-CS-11011800-022), and support from South Dakota State University. Field assistance was provided by C. Berdan, T. Berdan, B. Eastman, D. Gardner, A. Geigle, B. Hauser, T. Juntti, J. Mauer, S. Harrelson, and T. Zachmeier. A number of volunteers assisted during capture and radio-collaring of females and chicks. A. Apa assisted with training on trapping techniques. We also acknowledge and appreciate those land owners who granted us permission to conduct this study on their lands. We thank D. Turner, J. Beck, and an anonymous reviewer for their comments on prior versions of this manuscript. Any mention of trade, product, or firm names is for descriptive purposes only and does not imply endorsement by the U.S. government.

LITERATURE CITED

Aldridge, C. L., and R. M. Brigham. 2002. Sage-grouse nesting and brood habitat use in southern Canada. Journal of Wildlife Management 66:433–444.

Aldridge, C. L., S. E. Nielsen, H. L. Beyer, M. S. Boyce, J. W. Connelly, S. T. Kick, and M. A. Schroeder. 2008. Range-wide patterns of Greater Sage-Grouse persistence. Diversity and Distributions 14:983–994.

Apa, A. D. 1998. Habitat use and movements of sympatric sage and Columbian Sharp-tailed Grouse in southeastern Idaho. Ph.D. dissertation, University of Idaho, Moscow, ID.

Autenrieth, R. E. 1981. Sage grouse management in Idaho. Wildlife Bulletin 9. Idaho Department Fish and Game, Boise, ID.

Ball, G. H., and D. J. Hall. 1967. A clustering technique for summarizing multivariate data. Behavioral Science 12:153–155.

Beck, J. L., K. P. Reese, J. W. Connelly, and M. B. Lucia. 2006. Movements and survival of juvenile Greater Sage-Grouse in southeastern Idaho. Wildlife Society Bulletin 34:1070–1078.

Benkobi, L., D. W. Uresk, G. Schenbeck, and R. M. King. 2000. Protocol for monitoring standing crop in grasslands using visual obstruction. Journal of Range Management 53:627–633.

Beyer, H. L. 2004. Hawth's Analysis Tools for ArcGIS. <http://www.spatialecology.com/htools> (10 January 2006).

Burnham, K. P., and D. R. Anderson. 2002. Model selection and multimodel inference: a practical information-theoretic approach. 2nd ed. Springer-Verlag, New York, NY.

Connelly, J. W., S. T. Knick, M. A. Schroeder, and S. J. Stiver. 2004. Conservation assessment of Greater Sage-Grouse and sagebrush habitats. Western Association of Fish and Wildlife Agencies, Cheyenne, WY.

Cottam, G., and J. T. Curtis. 1956. The use of distance measures in phytosociological sampling. Ecology 37:451–460.

Crawford, J. A., M. A. Gregg, M. S. Drut, and A. K. DeLong. 1992. Habitat use by female sage grouse during the breeding season in Oregon. Final report, BLM Cooperative Research Unit, Oregon State University, Corvallis, OR.

Daubenmire, R. F. 1959. A canopy-coverage method of vegetation analysis. Northwest Science 33:224–227.

Drut, M. S., W. H. Pyle, and J. A. Crawford. 1994. Diets and food selection of sage-grouse chicks in Oregon. Journal of Range Management 47:90–93.

Dunn, P. O., and C. E. Braun. 1986. Late summer–spring movements of juvenile sage grouse. Wilson Bulletin 98:83–92.

Giesen, K. M., T. J. Schoenberg, and C. E. Braun. 1982. Methods for trapping sage grouse in Colorado. Wildlife Society Bulletin 10:224–231.

Healy, W. M. 1985. Turkey poult feeding activity, invertebrate abundance and vegetation structure. Journal of Wildlife Management 49:466–472.

Holloran, M. J. 1999. Sage-grouse (*Centrocercus urophasianus*) seasonal habitat use near Casper, Wyoming. M.S. thesis, University of Wyoming, Laramie, WY.

Hornaday, W. T. 1916. Save the sage-grouse from extinction, a demand from civilization to the western states. New York Zoological Park Bulletin 5:179–219.

Hosmer, D. W., and S. Lemeshow. 2000. Applied logistic regression. 2nd ed. John Wiley & Sons, New York, NY.

Huwer, S. L. 2004. Evaluating Greater Sage-Grouse brood habitat use using human-imprinted chicks. M.S. thesis, Colorado State University, Fort Collins, CO.

Jamison, B. E., R. J. Robel, J. S. Pontius, and R. D. Applegate. 2002. Invertebrate biomass: associations with Lesser Prairie Chicken habitat use and sand sagebrush density in southwestern Kansas. Wildlife Society Bulletin 30:517–526.

Johnson, G. D., and M. S. Boyce. 1990. Feeding trials with insects in the diet of sage-grouse chicks. Journal of Wildlife Management 54:89–91.

Johnson, J. R., and G. E. Larson. 1999. Grassland plants of South Dakota and the northern Great Plains. South Dakota Agricultural Experiment Station B566 (rev). South Dakota State University, Brookings, SD.

Klebenow, D. A. 1969. Sage grouse nesting and brood habitat in Idaho. Journal of Wildlife Management 33:649–662.

Klebenow, D. A. 1982. Livestock grazing interactions with sage-grouse. Pp. 113-123 *in* J. M. Peek and P. D. Dalke (editors), Proceedings of the Wildlife-Livestock Relationships Symposium, 20–22 April 1981, Coeur d'Alene, Idaho. Proceeding 10. University of Idaho Forestry, Wildlife, and Range Experiment Station, Moscow, ID.

Klebenow, D. A. 1985. Habitat management of sage-grouse in Nevada. World Pheasant Association Journal 10:34–46.

Lewis, A. R. 2004. Sagebrush steppe habitats and their associated bird species in South Dakota, North Dakota, and Wyoming: life on the edge of the sagebrush ecosystem. Ph.D. dissertation, South Dakota State University, Brookings, SD.

McCarthy, J. J., and J. D. Kobriger. 2005. Management plan and conservation strategies for Greater Sage-Grouse in North Dakota. North Dakota Game and Fish Department, Bismarck, ND.

Mielke, P. W., Jr., and K. J. Berry. 2001. Permutation methods: a distance function approach. Springer-Verlag. New York, NY.

Oakleaf, R. J. 1971. The relationship of sage-grouse to upland meadows in Nevada. M.S. thesis, University of Nevada, Reno, NV.

Patterson, R. L. 1952. The sage grouse of Wyoming. Sage Books, Inc., Denver, CO.

Peterson, J. G. 1970. The food habits and summer distribution of juvenile sage grouse in central Wyoming. Journal of Wildlife Management 34:147–155.

Pollard, J. H. 1971. On distance estimators of density in randomly distributed forests. Biometrics 27:991–1002.

Robel R. J., J. N. Briggs, A. D. Dayton, and L. C. Hulbert. 1970. Relationships between visual obstruction measurements and weight of grassland vegetation. Journal of Range Management 23:295–297.

Rumble, M. A., and S. H. Anderson. 1996. Habitat selection of Merriam's Turkey (*Meleagris gallopavo merriami*) in the Black Hills, South Dakota. American Midland Naturalist 136:157–171.

Sage and Columbian Sharp-tailed Grouse Technical Committee. 2008. Greater Sage-Grouse population trends: an analysis of lek count databases 1965–2007. Western Association of Fish and Wildlife Agencies, Cheyenne, WY.

SAS Institute Inc. 2007. JMP version 7. Cary, NC.

Schoenberg, T. J. 1982. Sage-grouse movements and habitat selection in North Park, Colorado. M.S. thesis, Colorado State University, Fort Collins, CO.

Schroeder, M. A., C. L. Aldridge, A. D. Apa, J. R. Bohne, C. E. Braun, S. D. Bunnell, J. W. Connelly, P. A. Deibert, S. C. Gardner, M. A. Hilliard, G. D. Kobriger, S. M. McAdam, C. W. McCarthy, J. J. McCarthy, D. L. Mitchell, E. V. Rickerson, and S. J. Stiver. 2004. Distribution of sage-grouse in North America. Condor 106:363–376.

South Dakota Department of Game, Fish, and Parks. 2006. South Dakota comprehensive wildlife conservation plan. South Dakota Department of Game, Fish, and Parks, Wildlife Division Report 2006-08. Pierre, SD.

South Dakota State Climate Office. 2007. Office of the State Climatologist, Pierre, SD. <http://climate.sdstate.edu> (12 October 2007).

Sveum, C. M., J. A. Crawford, and W. D. Edge. 1998. Use and selection of brood-rearing habitat by sage-grouse in south-central Washington. Great Basin Naturalist 58:344–351.

Thompson, K. M., M. J. Holloran, S. J. Slater, J. L. Kuipers, and S. H. Anderson. 2006. Early brood-rearing habitat use and productivity of Greater Sage-Grouse in Wyoming. Western North American Naturalist 66:332–342.

U.S. Fish and Wildlife Service. 2010. 12-month findings for petitions to list the Greater Sage-Grouse (*Centrocercus urophasianus*) as threatened or endangered. Federal Register 75:13910–14014.

Wakkinen, W. L., K. P. Reese, J. W. Connelly, and R. A. Fischer. 1992. An improved spotlighting technique for capturing sage grouse. Wildlife Society Bulletin 20:425–426.

Wallestad, R. O. 1975. Male sage grouse responses to sagebrush treatment. Journal of Wildlife Management 39:482–484.

CHAPTER THIRTEEN

Habitat Selection and Brood Survival of Greater Prairie-Chickens

Ty W. Matthews, Andrew J. Tyre, J. Scott Taylor, Jeffery J. Lusk, and Larkin A. Powell

Abstract. The Greater Prairie-Chicken (*Tympanuchus cupido pinnatus*) is a species that may benefit from conversion of crop ground to grassland through the Conservation Reserve Program (CRP). CRP grasslands could provide nesting and brood-rearing habitat, an important component of population persistence. Managers and policymakers currently lack evidence of CRP's relative contribution to populations of Greater Prairie-Chicken. We used radiotelemetry to mark females (*n* = 100) in southeast Nebraska, in a landscape which had >15% of land area enrolled in CRP. We examined macrohabitat and microhabitat selection of brood-rearing females (*n* = 36) using discrete choice models, and examined the variability in brood survival using logistic exposure models. Brood-rearing females selected locations inside cool-season CRP grasslands at higher rates than rangeland, but did not select cropland. At a vegetation level,

brood-rearing locations had more bare ground and forb cover than random points. However, landcover and vegetation did not affect survival rates of broods; variation in daily brood survival was best explained by temporal effects such as hatch date and brood age. Our results suggest that CRP grasslands provide acceptable brood-rearing habitat, and managers should encourage land owners to create habitat with high forb content and an open understory. Broods in our study had low survival rates to 21 days (0.59; 95% CI: 0.41, 0.77), which may explain the low juvenile/adult ratio observed in hunter-killed birds in the region. Disturbance of CRP fields to increase bare ground and forb cover may improve their value to Greater Prairie-Chicken broods.

Key Words: brood, Conservation Reserve Program, grassland habitat, radiotelemetry, *Tympanuchus cupido*.

Matthews, T. W., A. J. Tyre, J. S. Taylor, J. J. Lusk, and L. A. Powell. 2011. Habitat selection and brood survival of Greater Prairie-Chickens. Pp. 179–191 *in* B. K. Sandercock, K. Martin, and G. Segelbacher (editors). Ecology, conservation, and management of grouse. Studies in Avian Biology (no. 39), University of California Press, Berkeley, CA.

reater Prairie-Chicken (*Tympanuchus cupido pinnatus*; hereafter prairie chicken) populations in southeast Nebraska appear to have been benefited by conversion of cropland to grassland through the U.S. Department of Agriculture's (USDA) Conservation Reserve Program (CRP). Through CRP, land owners receive an annual rental payment to remove highly erodable farm ground from production and into grassland cover. Prairie chicken populations in southeast Nebraska comprise the northernmost extension of the Flint Hills population (Vodehnal 1999, Johnsgard 2000). Unlike the Flint Hills in Kansas, southeast Nebraska's landscape was dominated by agricultural row crops. Post-settlement conversion of grasslands to croplands by European settlers caused this prairie chicken population to decline to low levels (Johnsgard 1983, Schroeder and Robb 1993). However, the population appeared to increase in the 1990s after approximately 15% of the landscape was converted to grassland through CRP (Taylor 2000). Our goal was to understand the mechanisms behind the increase in population size of prairie chickens.

The Nebraska Game and Parks Commission's (NGPC) long-term management for prairie chickens in southeast Nebraska is planned with the realization that CRP grasslands could rapidly disappear if the program was removed from the Farm Bill or if profit margins caused land owners to favor crop production over participation in CRP. Svedarsky (1988) and Westemeier et al. (1999) found that prairie chickens used grasslands similar to the low-diversity, brome-dominated CRP fields found in southeast Nebraska. However, studies of brood success in these habitats are needed to assess the ability of CRP fields to provide sufficient brood-rearing habitat needed to sustain a population.

Juvenile survival is a key demographic parameter for prairie chickens and other grouse species (Wisdom and Mills 1997, Sandercock et al. 2005, Hannon and Martin 2006), and lack of quality habitat for nesting and brood-rearing is often a limiting factor (Hamerstrom et al. 1957, Bergerud 1988). Brood habitat should provide sufficient bare ground to facilitate chick movement, adequate overhead cover to protect chicks from predators, and it should be close to nesting cover (Vodehnal and Haufler 2007). When left undisturbed for 3–4 years, CRP land can accumulate a substantial amount of vegeta-tive litter and may lack the bare ground needed by prairie chicken chicks (McCoy et al. 2001). Prairie chicken chicks also require an abundance of arthropods, the chicks' main food source during the first two weeks (Jones 1963); CRP fields may lack these essential invertebrates if vegetation diversity is low. Problems associated with low arthropod abundance may be alleviated when forb and legume species are incorporated into CRP plantings or when high-diversity grasslands are in close proximity to nesting cover.

Other types of landcover may act as important predictors of brood-rearing habitat for prairie chickens in southeast Nebraska. Pasturelands are grasslands once used for row-crop production and subsequently seeded with native or non-native grasses, whereas unplowed rangelands are native grasslands used for cattle grazing. Both habitats are thought to be used by brood-rearing females (Horak 1985, Burger et al. 1989). However, it is unclear whether short vegetation typically found in CRP provides adequate concealment from predators or protection from adverse weather to successfully produce broods.

Our goal was to examine prairie chicken use of habitat for brood-rearing in southeast Nebraska, with specific interest in habitats provided by CRP. Our objectives were to (1) examine brood habitat selection at the two scales of macrohabitat (landscape composition) and microhabitat (vegetation structure and local composition), and (2) assess the consequences of habitat use on daily survival of prairie chicken broods. Demographic information related to brood survival will be used to inform changes in land-use policy and to inform management decisions needed to support prairie chicken populations in southeast Nebraska.

METHODS

Study Area

Our study was conducted in Johnson and Pawnee counties in the tallgrass ecoregion of southeast Nebraska. The landscape consisted of rolling uplands produced mainly from glacial till and loess accumulation. Soil types in these counties were characterized by Wymore–Pawnee soil association (USDA 1986). Average annual precipitation was 84 cm, with the majority falling between the months of May and August. Average monthly temperature maximum and minimum were 32°C and −12°C, occurring in July and January, respectively. Our focus area was dominated by production of corn, soybean, and grazed grasslands, with millet and sorghum in lesser quantities. In 2007, 163.3 km² (ca. 17%) of Johnson County and 172.1 km² (ca. 15%) of Pawnee County had been enrolled in CRP (Farm Service Agency, USDA).

Trapping and Monitoring

We used walk-in traps to capture female prairie chickens from 20 March to 19 April in 2007–2008 (Schroeder and Braun 1991). We trapped birds at 13 different leks during 2007 (7 leks) and 2008 (10 leks). Male attendance ranged from 15 to 70 individuals per lek. At first capture, we fitted each female with a necklace-type radio transmitter (<20 g, Model #A3960, Advanced Telemetry Systems, Inc., Isanti, MN) and released each female immediately at the capture location. Animal capture and handling protocols were approved by the University of Nebraska–Lincoln Institutional Animal Care and Use Committee (Protocol #05-02-007).

We recorded locations of each female 5–10 times per week from time of capture to 1 August using a vehicle mounted with a null-peak dual antenna-receiver with an electronic compass (Gilsdorf et al. 2008). We randomly chose the order of fields in which we tracked the females to avoid temporal biases. We took at least three bearings within a 10-minute period for each location to minimize error caused by movements. Additional bearings were taken until we received an error polygon of less than 0.1 ha (ca. 18-m radius). All UTM coordinates and associated error polygons were calculated in the field via an onboard computer and software [Location of a Signal (LOAS), Ecological Software Solutions, Urnäsch, Switzerland, version 4.0]. We visually confirmed nest locations by approaching females with a handheld antenna and receiver, and flushed females to record the number of eggs in the nest within the first week of incubation. Once daily locations indicated the female had stopped incubating (i.e., 2–3 locations off nest), we visually inspected the nest to determine nest fate and the number of eggs that hatched. We used daily telemetry observations to locate females with broods for 21 days after hatch. We used nocturnal two-flush counts on days 10 and 21 post-hatch to determine if a brood was successful (≥1 chick at flush). We assigned failed brood fates to females that flushed long distances without returning and we performed systematic ground searches to confirm total brood failure. We used two subsequent flushes of such females to verify this classification of fate.

Landscape and Vegetation Sampling

We evaluated landscape composition by creating a year-specific, vector-based GIS by landcover layer (ArcGIS 9.0, ESRI, Redlands, CA). We used aerial photographs and extensive ground-truthing through visual inspection to classify each land cover into one of five landcover types: cropland (row crops, alfalfa), grassland, woodland, wetland, and anthropogenic (farmsteads, utility facilities). Grassland cover types were further divided in four subtypes: (1) warm-season CRP fields (predominantly switchgrass, *Panicum virgatum*; big bluestem, *Andropogon gerardii*; little bluestem, *Schizachyrium scoparium*; Indian grass, *Sorghastrum nutans*; or sideoats grama, *Bouteloua curtipendula*), (2) cool-season CRP fields (predominantly smooth brome, *Bromus inermus*; and orchard grass, *Dactylis glomerata*), (3) rangeland (unplowed native grasslands), and (4) pastureland (pastures previously plowed).

At every third brood location, we estimated percent canopy cover for cool-season grasses (COOL), warm-season grasses (WARM), forbs (FORB), standing litter (SL) and bare ground (BARE) using a 1-m diameter sampling hoop (modified from Daubenmire 1959) and a visual obstruction reading (VOR) to the nearest 0.25 dm (Robel et al. 1970). BARE was the percent ground that was not covered by residual vegetation below the vegetation canopy. We took the sample from a random location within the 0.1-ha error polygon. We also assessed the vegetation composition and VOR at five random points in the same field and habitat type. All random points were created with Hawth's Analysis Tools for ArcGIS (Beyer 2004), and we used a handheld GPS receiver to find the point in the field. Last, we classified each used brood location and random point into three topological categories (TOPO; upper, middle, or lower) relative to the maximum and minimum elevation in that particular field, using a digital elevation

TABLE 13.1

Comparison of competing discrete-choice models for macrohabitat selection of
Greater Prairie-Chicken brooding females in southeast Nebraska, 2007–2008.

Models	K	Log(L)	AIC$_c$	ΔAIC$_c$	w_i
Landcover + distance to cropland	6	−1,355.8	2,726.62	0	0.61
Landcover + distance to woodland + distance to cropland	7	−1,354.2	2,727.98	1.84	0.31
Landcover + distance to cropland + distance to edge	7	−1,355.8	2,731.24	5.43	0.06
Landcover + distance to woodland + distance to cropland + distance to edge	8	−1,354.2	2,733.54	8.59	0.02

NOTE: Models are ranked by AIC$_c$, Akaike's Information Criterion adjusted for small sample size; K is the number of parameters (note that discrete choice models have no intercept); log(L) is the log likelihood of the model; ΔAIC$_c$ is the difference of each model's AIC$_c$ value from that of the minimum AIC$_c$ model (row one); and w_i is the Akaike weight (sum of all weights = 1.00). Landcover includes five strata (cool-season CRP, warm-season CRP, pasture, rangeland, and other habitats). Twelve models with $w_i < 0.02$ are not shown.

model (DEM, UNL School of Natural Resources). All sampling was done within two days of locating brood.

Analysis and Model Selection

Discrete choice models calculate the probability of an individual selecting a resource as a function of the attributes of that resource and all other resources within the individual's available habitat (Cooper and Millspaugh 1999, McDonald et al. 2006). Discrete choice analyses estimate the probability of selection of categorical variables relative to a reference, and we used rangeland as the reference for landcover. We estimated the area of available habitat in two steps. First, we calculated the maximum displacement distance between locations collected on consecutive days for each brood. Second, we calculated available habitat as the circular area around each location with a radius equal to the minimum of these maximum displacement distances. We used Cox proportional hazards regression function (COXPH) in the survival package (Therneau and Lumley 2009) of Program R (R Development Core Team 2009) to develop our macro- and microhabitat discrete choice models. For each set of models, a correlation matrix was created to avoid any within-scale correlation.

We assessed selection of brood macrohabitat using discrete choice analysis of 16 biologically reasonable *a priori* models including a null model (Cooper and Millspaugh 1999, McDonald et al. 2006). Our macrohabitat models considered the effect of landcover type on habitat selection. We created a set of covariates for landcover (Landcover, Table 13.1) based on our classification of five habitats in our study area: cool-season CRP, warm-season CRP, pasture, rangeland, and other habitats (largely anthropogenic and water). Because woodlands and edges are known to support larger numbers of mammalian and avian predators and may be avoided by brooding prairie chicken females (Svedarsky et al. 2003), we also created covariates for linear distance to woodland (Distance to woodland) and edge (Distance to edge, Table 13.1). We defined an edge as any transition in vegetation such as fence rows, tree lines, roads, change in dominant vegetation type, or other boundaries that delineated habitat types. Finally, we created a covariate for distance to any crop field (Distance to cropland, Table 13.1), as crop fields may be an important food source for females (Svedarsky and Van Amburg 1996); brooding females could be attracted to grasslands near crop fields. We generated 20 random locations per brood location within the habitat defined as available to compare against the used locations. We used 20 random locations rather than the five suggested by McFadden (1978) to decrease the variance in covariate coefficients (Baasch 2008). We replaced any random locations that were <18 m of our brood location estimate, to match the size of our maximum allowable error polygon.

Our microhabitat analysis consisted of eight biologically reasonable *a priori* models using combinations of the following groups of covariates: vegetation composition (% Cover; WARM, COOL, FORB, SL), a quadratic model of vegetation structure (VOR + VOR^2 + Bare), and topographic position (TOPO). The seven models with covariates were compared to a null model with no covariates. We hypothesized that vegetation composition could be a key component of habitat selection for brooding prairie chicken females (Jones 1963, Kobriger 1965). We considered a non-linear effect of vegetation structure (VOR) because brooding females generally select areas less dense than nest sites (Jones 1963). Last, we hypothesized that prairie chicken females may prefer locations where topography allows better detection of potential predators or faster escape flights during predation attempts. We compared the microhabitat at brood locations with the random points taken within the field containing the nest.

We estimated daily brood survival and assessed the contribution of individual covariates in a set of eight biologically reasonable *a priori* models in a logistic-exposure structure (Shaffer 2004) using program R and logexp package (Post van der Burg 2005). Our monitoring intervals were usually 10–11 days in length, as we flushed broods at day 10 and day 21 post-hatch. We flushed at 10 days post-hatch to assess survival before any chicks had fledged. At day 21, all chicks should have fledged. Our models considered landcover (Landcover, % of observations during interval spent in: cool-season CRP, warm-season CRP, rangeland, pastureland, cropland), time (Julian day of hatch and brood age), and climate (average daily temperature during monitoring interval and average daily precipitation). We compared models with covariates to a null model. Flanders-Wanner et al. (2004) found that high temperature and precipitation events negatively impacted production of Sharp-tailed Grouse (*Tympanuchus phasianellus*). Temporal covariates, such as hatch date and brood age, have been found to affect brood survival in other galliform species (Riley et al. 1998, Fields et al. 2006, Hannon and Martin 2006). We used age as a categorical variable (1–10 or 11–21 days old) because we flushed broods in discrete time

intervals to avoid negative impacts of survival. We used female location and movement patterns to estimate date of hatch. Ten-day and 11-day period survival estimates were calculated as daily survival rates for first 10 days post-hatch ($\hat{S}_{10} = D\hat{S}R^{10}_{1-10}$ and $\hat{S}_{11} = D\hat{S}R^{10}_{11-21}$); 21-day period survival estimates were calculated as $\hat{S}_{21} = \hat{S}_{10}\hat{S}_{11}$. Variances for $\hat{S}_{10}$, $\hat{S}_{11}$ and $\hat{S}_{12}$ were approximated using the delta method (Powell 2007).

We performed model selection using an information-theoretic approach to evaluate *a priori* models for brood habitat selection and survival. We ranked each model from most to least support given the data, using Akaike's Information Criterion corrected for small sample size (AIC$_c$; Burnham and Anderson 2002). For brood survival, we used effective sample size (n = total number of days the broods survived + number of intervals that ended in failure) for the calculation of AIC$_c$ (Rotella et al. 2004). We computed Akaike weights (ω_i) for each model, where ω_i represents the probability a model being the best approximating model of those considered given the data. For each analysis, we considered a confidence set of all models with a combined model weight of $\geq$90%. We selected the top model if it was the most parsimonious of the confidence set. When the highest-ranked model was not the most parsimonious, we used conditional model averaging over the 90% confidence set to predict the covariates and associated standard errors (Burnham and Anderson 2002:152–153). We limited model-averaged predictions from continuous data to the range of data we observed.

RESULTS

We captured, radio-tagged, and monitored 100 prairie chicken females (2007: 38, 2008: 62). We monitored 36 females with broods from successful nests (2007: 11; 2008: 25). Eighteen broods (50%) survived to 10 days after hatching (2007: 2; 2008: 16). A total of 17 of 18 (95%) broods that were active at 10 days after hatching survived to day 21.

We obtained 455 locations of brooding females, which were distributed among cool-season CRP (29%), rangeland (27%), pastureland (20%), warm-season CRP (11%), cropland (7%), and other (6%) landcover classes.

TABLE 13.2

Coefficients (β), standard errors, selection ratios [exp(β)], and associated 95% confidence intervals
for covariates in the top model predicting macrohabitat selection for Greater Prairie-Chicken
females with broods in southeast Nebraska, 2007–2008.

Covariate	Coefficient	SE	Selection Ratio	95% Confidence Interval	
				Lower	Upper
Landcover					
Cool-season CRP	0.330	0.154	1.392	1.028	1.884
Warm-season CRP	0.186	0.192	1.205	0.826	1.757
Pasture	0.027	0.163	1.027	0.746	1.415
Cropland	−0.971	0.177	0.379	0.268	0.535
Rangeland	1.000	—	1.000	—	—
Distance to cropland	−0.001	0.0004	0.999	0.998	0.999

NOTE: Rangeland was set as the baseline landcover type.

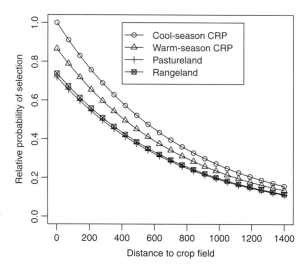

Figure 13.1. Relative probability of selection as a function of distance to cropland and landcover type in the best macrohabitat discrete choice model (Table 1) by Greater Prairie-Chicken females with broods in southeast Nebraska, 2007–2008. Relative probabilities were scaled to have a maximum value of 1.0.

Each of our females traveled >500 m at least once during a 24-hr time period while attending young. Thus, we established 500-m radii around each brood location as the available habitat for subsequent analyses. Within-scale correlations for each model set were reasonable ($r < 0.40$).

The top macrohabitat-level selection model included effects of landcover type and distance to crop field (Table 13.1). The relative probability of a brooding female selecting a location for her brood in a cool-season CRP field was 1.39 (95% CI: 1.03–1.88) times higher than rangeland (Table 13.2). A female's probability of selecting a location in a crop field for brooding is 2.64 (95% CI: 1.87–3.73) times lower than the chance of selecting rangeland. Each increase in distance of 100 m from cropland predicted a 10% decrease in the relative probability of selection, holding all other variables constant (Fig. 13.1). Selection probabilities among rangeland, pastureland, and warm-season CRP were similar.

The minimum AIC_c microhabitat-level selection model included all effects (global model)

TABLE 13.3

Coefficients (β), standard errors, selection ratios [exp(β)], and associated 95% confidence intervals
for variables in the top model predicting microhabitat selection for Greater Prairie-Chicken females
with broods in southeast Nebraska, 2007–2008.

Covariate	Coefficient	SE	Selection Ratio	95% Confidence Interval Lower	95% Confidence Interval Upper
% Cool-season grass	0.054	0.030	1.056	0.996	1.120
% Warm-season grass	0.091	0.104	1.096	0.894	1.343
% Forb	0.145	0.041	1.156	1.067	1.252
% Standing litter	−0.053	0.042	0.949	0.874	1.037
% Bare ground	0.072	0.027	1.074	1.018	1.136
VOR	5.857	1.908	—	—	—
VOR²	−1.212	0.351	—	—	—
Topography					
Bottom	−0.653	1.082	0.521	0.063	4.341
Middle	2.424	0.998	11.285	1.597	79.832
Upper	0	—	1.000	—	—

NOTE: Upper-level topography was set as the baseline topographic level. Selection ratios were not calculated for variables involved in quadratic effects due to the dependence on values of other variables.

and had an ω_i of 1.0. Prairie chicken females with a brood selected areas with mid-level topography 11.3 (95% CI: 1.6–79.8) and 21.7 (95% CI: 3.0–153) times more than bottom- or high-level topography, respectively (Table 13.3). Relative probability of selection increased as percent cover of forbs increased, with mean use of 33.6% (SE = 25.4, range: 0–85%). Selection also increased with an increase in bare ground, with a mean use of 24.8% (SE = 29.9, range: 0–100%). Probability of selection peaked at approximately 3 dm for VOR and decreased as VOR deviated from this point (Fig. 13.2). Average VOR at use points was 2.4 dm (SE = 0.8, range: 1–5 dm).

Brood survival was a function of time (brood age and Julian day of hatch; Table 13.4). Daily brood survival decreased as nests hatched later in the breeding season and increased as broods aged (Fig. 13.3). No habitat or landscape characteristics we studied accounted for variation in daily brood survival. The mean daily survival probability of a brood in the 1–10 day old age-class was 0.95 (95% CI: 0.95–0.96), and 0.99 (95% CI: 0.96–1.00) in the 11–21 day class. The mean probability of a

brood surviving to day 21 was 0.59 (95% CI: 0.41–0.77). The average number of chicks per brood surviving to day 21 was 4.50 (SE = 0.71) in 2007 (n = 2 broods) and 3.13 (SE = 2.50) in 2008 (n = 15 broods).

DISCUSSION

Brooding prairie chicken females in southeastern Nebraska selected cool-season CRP fields over any other landcover. Cool-season CRP fields provide large expanses of undisturbed grassland in Nebraska, but our finding is contrary to previous studies, which found brooding females selected disturbed areas such as cultivated pastures, recently burned grasslands, and native prairie hay fields (Jones 1963, Svedarsky 1979, Burger et al. 1989, Westemeier et al. 1995). Rangelands and pastures represented disturbed habitats in our study area, and the lower selection ratio for disturbed grassland may be related to the vertical structure of these grasslands. Large quantities of rangeland and pastureland in our study area may not have been suitable for brood-rearing due to excessive

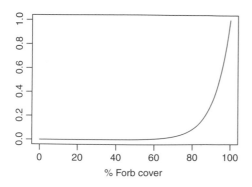

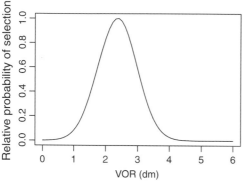

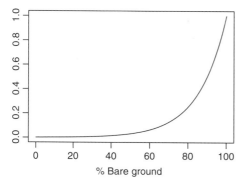

Figure 13.2. Relative probability of selection by Greater Prairie-Chicken females with broods as a function of covariates in the best microhabitat discrete choice model. All variables not plotted were held constant at their means. Relative probabilities were scaled to have a maximum value of 1.0.

grazing by cattle. Both landcover types were generally grazed year after year and provided little vegetative concealment. In contrast, grazed pastures in the Nebraska Sandhills, which is in the north-central part of the state, serve as the main brood-rearing habitat for prairie chickens (Vodehnal 1999). These grasslands are generally stocked at lower rates and are usually grazed in rotational systems, giving brooding females better overhead concealment from predators and adverse weather.

We observed that the pastures and rangelands that were used by broods were either lightly grazed grasslands or idle grasslands with no disturbance, which contrasted with the majority of available pastures and rangelands on the landscape. Brooding females also selected areas near crop fields in all landcover types, although crop fields themselves were avoided. Grassland–cropland edges may provide favorable complexes of bare ground and structurally appropriate vegetation consistent with our observations of microhabitat preferences. Females may also forage in these agricultural fields, even though there is little food for their young. Rumble et al. (1988) found that a large portion of the diets of prairie chickens in the Sheyenne National Grasslands consisted of agricultural crops, mainly corn and alfalfa. We would expect crops to be even more important to prairie chicken diets in southeastern Nebraska, because the landscape is dominated by crop fields, which increases availability and use (Horak 1985). An increase of forb and legume components, such as alfalfa and sweet clover, in grasslands may decrease dependence of crop fields by brood-rearing females (Svedarsky et al. 2003).

Within selected fields, female prairie-chickens with broods selected vegetation that was high in forb content and bare ground (Fig. 13.2). Our results are consistent with previous research (Westemeier et al. 1995, Norton 2005), and seem reasonable considering that primary brood behavior during the day consists of foraging for insects and mid-day loafing. Greater forb cover provides a higher abundance of insects, which are the primary food source of prairie chicken chicks during the critical first two weeks after hatching (Jones 1963, Schroeder and Braun 1992, Svedarsky et al. 2003). Bare ground and forb cover cannot be simultaneously maximized (Fig. 13.2), but our data suggest that habitat with a high proportion of forbs and bare ground will be preferred over brooding habitat without these features. Brooding females in our study also selected sites with intermediate topography, probably related to the vegetation densities and composition associated with each topographic strata. Newell et al. (1988) reported that broods in the Sheyenne National Grasslands in South Dakota used lowlands and midlands more than uplands due to the sparse vegetation in upland habitats. The converse may also be true in our study, with the majority of the vegetation in the lowlands

TABLE 13.4

Comparison of competing logistic-exposure models for daily brood survival of Greater
Prairie-Chickens in southeast Nebraska, 2005–2006.

Models	K	Log(L)	AIC$_c$	ΔAIC$_c$	w_i
Time[a]	3	−25.59	49.87	0.00	0.55
Time + climate[b]	5	−23.99	51.75	1.88	0.21
Landcover[c] + time + climate	9	−20.32	51.76	1.89	0.21
Landcover + time	7	−23.26	55.75	5.88	0.03

NOTE: See Table 13.1 for metrics of model selection.

[a] Includes additive effects of Julian day of hatch and brood age.
[b] Includes additive effects of temperature and precipitation.
[c] Includes effects of proportion of time spent in cool-season CRP, warm-season CRP, rangeland, pastureland, and cropland.

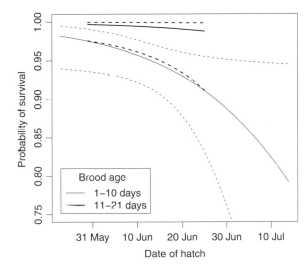

Figure 13.3. The linear effect of hatch date and brood age on daily survival of Greater Prairie-Chicken broods in southeast Nebraska, 2007–2008. Estimates are based on the best logistic-exposure model from Table 13.2 with all other variables held at their mean. Dashed lines are the associated 95% confidence intervals. No broods hatched after June 15 were alive 10 days after hatching.

being too tall and dense. Our results provide support for this assertion since broods selected sites with mid-range visual obstruction readings (1.75 to 3 dm; Fig. 13.2).

We observed low brood survival in Nebraska, especially in 2007 (21-day success: 18%) but also in 2008 (21-day success: 60%), and low chicks per brood at 21 days. Mortality resulted in a 21-day chick survival of only 14.4% for both years combined, which is less than the 37% chick survival to 24 days post-hatch reported by Newell et al. (1988) and 65% chick survival to 21 days post-hatch found by Norton (2005). Thus, it appears that low brood survival is responsible for the low juvenile:adult ratios in hunter wing surveys (J. Lusk, unpubl. data). Since hunting resumed

in 1999 in southeastern Nebraska, the average juvenile:adult ratio has been 0.91, compared to 1.77 for Nebraska's Sandhills region over the eight-year period (J. Lusk, unpubl. data).

Prairie chicken females with broods selected cool-season CRP fields. However, our survival analyses suggest that either landcover as we measured it is not an important factor in brood survival, or perhaps all the habitats used by monitored broods were equally poor. Instead, age of brood and hatch date were the only predictors of survival (Table 13.4, Fig. 13.3). One would expect selection patterns to optimize brood survival, but our data shows no evidence of a survival advantage to cool-season CRP fields despite their preferential use. It is possible that the high rate of mortality, especially in 2007, swamped any effects of habitat on brood survival. It is also possible that prairie chickens on our study site are responding to an ecological miscue associated with invasive brome in cool-season CRP (Misenhelter and Rotenberry 2000, Schlaepfer et al. 2002). Regardless, given the population growth observed on our study area since inception of the CRP program, CRP grasslands may benefit nest and/or adult survival to a degree that makes its effect on chick survival relatively unimportant to population persistence.

The relationship of brood survival with age has been well documented by Newell et al. (1988) and Norton (2005). Older chicks have greater ability to make short flights, as well as better development of thermoregulation, which decreases the need of regular brooding by the female (Svedarsky and Van Amburg 1996). Daily brood survival decreased the later in the breeding season that hatching occurred (Fig. 13.3), possibly due to declining female condition as the season progressed (Thogmartin and Johnson 1999). Svedarsky et al. (2003) suggested that prairie chickens are similar to waterfowl in that chicks have an increased survival rate when hatched by females with large fat reserves. Fields et al. (2006) also found declining brood survival in Lesser Prairie-Chickens (*Tympanuchus pallidicinctus*) with later hatch date but attributed it to confounding weather variables and insect production. Broods hatching later in the breeding season in our study may have experienced lower insect availability due to higher temperatures and lower precipitation. Higher temperature may also directly affect brood survival by inducing heat stress and water loss (Fields et al. 2006), especially when chicks are unable to thermoregulate on their own. Temperature may indirectly affect brood survival by reducing foraging time. Alhborn (1980) found that Lesser Prairie-Chickens seek shade and reduce foraging activity during periods of high temperatures. We averaged daily temperatures and precipitation during each brood's monitoring period, which tended to suppress extreme highs and lows that may cause mortality. Our monitoring intervals were set at 10 and 21 days to avoid observer bias on survival and habitat use; future studies may wish to weigh these potential biases with the need to investigate effects of short-term weather variability.

The low brood survival observed during our study was due to high levels of predation (Schole et al., this volume, chapter 18), which may be attributed to lack and distribution of quality habitat (Schroeder and Baydack 2001). Schole et al. (this volume, chapter 18) found 87% of radio-marked prairie chicken chicks died due to predation in southeastern Nebraska during our 2008 field season. The landscape was dissected with tree lines, roads, and power lines, with generally small (<32 ha) tracts of grasslands interspersed through agricultural fields. Habitat fragmentation has been linked to an increase in predation by increasing travel time in poor habitat and increasing diversity and density of predators (Schroeder and Baydack 2001). Similarly, Ryan et al. (1998) demonstrated that prairie chicken broods have smaller home ranges and higher survival in large contiguous grasslands than in a prairie–mosaic landscape.

Grasslands in our study area may have vegetative characteristics that also induce high levels of predation. CRP fields in that area are generally monocultures of brome or switchgrass with few patches of diverse vegetation and high litter accumulation from >10 years of undisturbed growth. Brooding females in CRP fields would have to make longer movements more often in search of suitable habitat, making them and their chicks more vulnerable to predation. The accumulation of litter has also been linked to an increase in small mammals, and has been suggested as attracting mammalian predators in the area (Westemeier 1988). Doxon (2005) found that an accumulation of litter impedes movement of gamebird chicks, and this may affect chicks' ability to escape predators. Undisturbed landcover like CRP fields in this area may need to be managed by rotational grazing, burning, or mowing to maintain proper height, density, and species composition (Svedarsky et al. 2003). Additionally, pastures and rangelands are commonly intensively grazed with no years of deferred grazing, which results in large expanses of areas with little to no overhead canopy cover, giving no protection from predators. Pastureland and rangeland in our area were also commonly invaded by woody vegetation, providing perches for avian predators. Frederickson (1996) stated that range conditions needed for satisfactory brood survival may be restricted by intense grazing. For optimal benefit to prairie chickens, grazed grasslands may need to adopt lower stocking rates or rotational grazing similar to the Sandhills region of north-central Nebraska.

ACKNOWLEDGMENTS

We thank M. Remund and B. Goracke, Nebraska Game and Parks Commission biologists who supported our project at the Osage Wildlife Management Area and the many land owners who allowed us access to private lands. NGPC funded our research and provided equipment, housing, and logistical support. A portion of this research was funded by a State Wildlife Grant administered through NGPC, and Nebraska Pheasants Forever provided equipment. S. Huber, A. Schole, and S. Groepper provided valuable field assistance. The School of Natural Resources provided computer and office space for TWM, AJT, and LAP. LAP was supported by Polytechnic of Namibia and a Fulbright award through the U.S. State Department during a professional development leave when this paper was written. This research was supported by Hatch Act funds through the University of Nebraska Agricultural Research Division, Lincoln, Nebraska.

LITERATURE CITED

Ahlborn, G. G. 1980. Brood-rearing habitat and fall-winter movements of Lesser Prairie Chickens in eastern New Mexico. Thesis, New Mexico State University, Las Cruces, NM.

Baasch, D. 2008. Resource selection by white-tailed deer, mule deer, and elk in Nebraska. Ph.D. dissertation, University of Nebraska, Lincoln, NE.

Bergerud, T. G. 1988. Population ecology of North America grouse. Pp. 578–777 in A. T. Bergerud and M. W. Gratson (editors), Adaptive strategies and population ecology of northern grouse. Wildlife Management Institute, University of Minnesota Press, Minneapolis, MN.

Beyer, H. L. 2004. Hawth's Analysis Tools for ArcGIS. <http://www.spatialecology.com/htools> (15 February 2009).

Burger, L. W., M. R. Ryan, and D. P. Jones. 1989. Prairie-chicken ecology in relation to landscape patterns. Missouri Prairie Journal 11:13–15.

Burnham, K. P., and D. R. Anderson. 2002. Model selection and multi-model inference: a practical information-theoretic approach. 2nd ed. Springer-Verlag, New York, NY.

Cooper, A. B., and J. J. Millspaugh. 1999. The application of discrete choice models to wildlife resource selection studies. Ecology 80:566–575.

Daubenmire, R. 1959. A canopy-coverage method of vegetational analysis. Northwest Science 33:4–63.

Doxon, E. D. 2005. Feeding ecology of Ring-necked Pheasant (*Phasianus colchicus*) and Northern Bobwhite (*Colinus virginianus*) chicks in vegetation managed using Conservation Reserve Program (CRP) practices. M.S. thesis, University of Georgia, Athens, GA.

Fields, T. L., G. C. White, W. C. Gilbert, and R. D. Rodgers. 2006. Nest and brood survival of Lesser Prairie-Chickens in west central Kansas. Journal of Wildlife Management 70:931–938.

Flanders-Wanner, B. L., G. C. White, and L. L. McDaniel. 2004. Weather and prairie grouse: dealing with effects beyond our control. Wildlife Society Bulletin 32:22–34.

Fredrickson, L. 1996. The Greater Prairie Chicken. South Dakota Conservation Digest 63:10–12.

Gilsdorf, J. M., K. C. Vercauteren, S. E. Hygnstrom, W. D. Walters, J. R. Boner, and G. M. Clements. 2008. An integrated vehicle-mounted telemetry system for VHF telemetry applications. Journal of Wildlife Management 72:1241–1246.

Hamerstrom, F. N., Jr., O. E. Mattson, and F. Hamerstrom. 1957. A guide to prairie chicken management. Wisconson Department of Natural Resources Technical Bulletin 15.

Hannon, S. J., and K. Martin. 2006. Ecology of juvenile grouse during the transition to adulthood. Journal of Zoology 269:422–433.

Horak, G. J. 1985. Kansas prairie chickens. Wildlife Bulletin No. 3. Kansas Fish and Game Commission, Pratt, KS.

Johnsgard, P. A. 1983. Grouse of the world. University of Nebraska Press, Lincoln, NE.

Johnsgard, P. A. 2000. Ecogeographic aspect of Greater Prairie-Chicken leks in southeastern Nebraska. Nebraska Bird Review 68:179–184.

Jones, R. E. 1963. Identification and analysis of Lesser and Greater Prairie-Chicken habitat. Journal of Wildlife Management 27:757–778.

Kobriger, G. D. 1965. Status, movements, habitats, and foods of prairie grouse on a Sandhills refuge. Journal of Wildlife Management 29:788–800.

McCoy, T. D., E. W. Kurzejeski, L. W. Burger, Jr., and M. R. Ryan. 2001. Effects of conservation practice, mowing, and temporal changes on vegetation structure on CRP field in northern Missouri. Wildlife Society Bulletin 29:979–987.

McDonald, T. L., B. F. J. Manly, R. M. Nielson, and L. V. Diller. 2006. Discrete-choice modeling in wildlife studies exemplified by Northern Spotted Owl nighttime habitat selection. Journal of Wildlife Management 70:375–383.

McFadden, D. 1978. Modeling the choice of residential location. Pp. 75–96 in A. Karlquist, L. Lundqvist, F. Snickars, and J. Weibull (editors), Spatial interaction theory and planning models. North Holland Publishing Company, Amsterdam, The Netherlands.

Misenhelter, M. D., and J. T. Rotenberry. 2000. Choices and consequences of habitat occupancy and nest site selection in Sage Sparrows. Ecology 81:2892–2901.

Newell, J. A., J. E. Toepfer, and M. A. Rumble. 1988. Summer brood-rearing ecology of the Greater Prairie-Chicken on the Sheyenne National Grasslands. Pp. 24–31 in A. J. Bjugstad (technical coordinator), Prairie chickens on the Sheyenne National Grasslands. U.S. Department of Agriculture Forest Service General Technical Report RM-159. Rocky Mountain Forest and Range Experiment Station, Fort Collins, CO.

Norton, M. A. 2005. Reproductive success and brood habitat use of Greater Prairie Chickens and Sharp-tailed Grouse on the Fort Pierre National Grassland of Central South Dakota. M.S. thesis, South Dakota State University, Brookings, SD.

Post van der Burg, M. 2005. Factors affecting songbird nest survival and brood parasitism in the rainwater basin region of Nebraska. M.S. thesis, University of Nebraska, Lincoln, NE.

Powell, L. A. 2007. Approximating variance of demographic parameters using the delta method: a reference for avian biologists. Condor 109:950–955.

R Development Core Team. 2009. R: a language and environment for statistical computing. R Foundation for Statistical Computing, Vienna, Austria. <http://www.R-project.org>.

Robel, R. J., J. N. Briggs, A. D. Dayton, and L. C. Hulbert. 1970. Relationships between visual obstruction measurements and weight of grassland vegetation. Journal of Range Management 23:295–297.

Rotella, J. J., S. J. Dinsmore, and T. L. Shaffer. 2004. Modeling nest-survival data: a comparison of recently developed methods that can be implemented in MARK and SAS. Animal Biodiversity and Conservation 27:187–205.

Rumble, M. A., J. A. Newell, and J. E. Toepfer. 1988. Diets of Greater Prairie-Chickens on the Sheyenne National Grasslands. Pp. 49-54 in A. J. Bjugstad (technical coordinator), Prairie chickens on the Sheyenne National Grasslands. U.S. Department of Agriculture Forest Service General Technical Report RM-159. Rocky Mountain Forest and Range Experiment Station, Fort Collins, CO.

Ryan, M. R., L. W. Burger, D. P. Jones, and A. P. Wywialowski. 1998. Breeding ecology of Greater Prairie-Chickens (Tympanuchus cupido) in relation to prairie landscape configuration. American Midland Naturalist 140:111–121.

Sandercock, B. K., K. Martin, and S. J. Hannon. 2005. Demographic consequences of age-structure in extreme environments: population models for arctic and alpine ptarmigan. Oecologia 146:13–24.

Schlaepfer, M. A., M. C. Runge, and P. W. Sherman. 2002. Ecological and evolutionary traps. Trends in Ecology & Evolution 17:474–480.

Schroeder, M. A., and R. K. Baydack. 2001. Predation and the management of prairie grouse. Wildlife Society Bulletin 29:24–32.

Schroeder, M. A., and C. E. Braun. 1991. Walk-in funnel traps for capturing Greater Prairie-Chickens on leks. Journal of Field Ornithology 62:378–385.

Schroeder, M. A., and C. E. Braun. 1992. Seasonal movement and habitat use by Greater Prairie-Chickens in northeastern Colorado. Colorado Division of Wildlife, Special Report Number 68.

Schroeder, M. A., and L. A. Robb. 1993. Greater Prairie-Chicken (Tympanuchus cupido). The birds of North America No. 36. Academy of Natural Sciences, Philadelphia, PA.

Shaffer, T. L. 2004. A unified approach to analyzing nest success. Auk 121:526–540.

Svedarsky, W. D. 1979. Spring and summer ecology of female Greater Prairie-Chickens in northwestern Minnesota. Dissertation. University of North Dakota, Grand Forks, ND.

Svedarsky, W. D. 1988. Reproductive ecology of female Greater Prarie-Chicken in Minnesota. Pp. 193–239 in A. T. Bergerud and M. W. Gratson (editors), Adaptive strategies and population ecology of northern grouse. University of Minnesota Press, Minneapolis, MN.

Svedarsky, W. D., J. E. Toepfer, R. L. Westemeier, and R. J. Robel. 2003. Effects of management practices on grassland birds: Greater Prairie-Chicken. Northern Prairie Wildlife Research Center, Jamestown, ND.

Svedarsky, W. D., and G. Van Amburg. 1996. Integrated management of the Greater Prairie Chicken and livestock on the Sheyenne National Grassland. North Dakota Game and Fish Department, Bismarck, ND. <http://www.npwrc.usgs.gov/resource/birds/sheyenne/index.htm> (16 July 1997).

Therneau, T., and T. Lumley. 2009. Survival: survival analysis, including penalised likelihood.. R package version 2.35-7. <http://CRAN.R-project.org/package=survival>

Thogmartin, W. E., and J. E. Johnson. 1999. Reproduction in a declining population of Wild Turkeys in Arkansas. Journal of Wildlife Management 63:1281–1290.

U.S. Department of Agriculture. 1986. Soil survey of Johnson County, Nebraska. U.S. Government Printing Office, Washington DC.

Vodehnal, W. L. 1999. Status and management of the Greater Prairie Chicken in Nebraska. Pp. 81–98 in W. D. Svedarsky, R. H. Hier, and N. J. Silvy (editors), The Greater Prairie Chicken: a national look. University of Minnesota Agricultural Experiment Station Miscellaneous Publication 99-1999, Saint Paul, MN.

Vodehnal, W. L., and J. B. Haufler. 2007. A grassland conservation plan for prairie grouse. North American Grouse Partnership, Fruita, CO.

Westemeier, R. L. 1988. Development of prairie pasture demonstration areas, phase II. Illinois Natural History Survey Final Report.

Westemeier, R. L., R. W. Jansen, and S. A. Simpson. 1995. Nest and brood habitat used by translocated Greater Prairie-Chickens in Illinois. P. 17 *in* G. D. Kobriger (editor), Proceedings of the Twenty-first Prairie Technical Council Conference. North Dakota Game and Fish Department, Dickinson, ND.

Westemeier, R. L., S. A. Simpson, and T. L. Esker. 1999. Status and management of Greater Prairie Chickens in Illinois. Pp. 143–152 *in* W. D. Svedarsky, R. H. Hier, and N. J. Silvy (editors), The Greater Prairie Chicken: a national look. University of Minnesota Agricultural Experiment Station Miscellaneous Publication 99-1999, Saint Paul, MN.

Wisdom, M. J., and L. S. Mills. 1997. Sensitivity analysis to guide population recovery: prairie-chickens as an example. Journal of Wildlife Management 61:302–312.

PART THREE

Population Biology

CHAPTER FOURTEEN

Testosterone Mediates Mating Success in Greater Prairie-Chickens

Jacqueline K. Augustine, Joshua J. Millspaugh,
and Brett K. Sandercock

Abstract. Testosterone plays a key role in influencing behaviors that enhance male breeding success, but elevated testosterone levels can also reduce immunocompetence and survival. In socially monogamous species, males with higher levels of circulating testosterone experience advantages during both female choice and male–male competition activities. The role of testosterone influencing male mating success in lek-mating systems is largely unknown. Over three years, we quantified natural and experimental levels of testosterone among Greater Prairie-Chickens (*Tympanuchus cupido*) at five leks. Our project had three objectives: (1) to quantify natural variation in testosterone in blood plasma and to determine whether testosterone levels are correlated with male traits and mating success; (2) to experimentally elevate testosterone levels to determine causal relationships between testosterone levels, male traits, and mating success; and (3) to analyze return rates to determine if increased levels of circulating testosterone are costly in terms of annual survival. Using a before–after control-impact (BACI) experimental design, changes in aggression, display behavior, territory size and location, mating success, and survival were compared between males with testosterone

implants (15 males) and sham implanted controls (13 males) over two years. Natural level of testosterone was a strong predictor of male mating success, but testosterone level in unmanipulated males was not related to any morphological, behavioral, or territorial traits considered. Comb area, tarsus length, and distance to the center of the lek were also significant predictors of male mating success. Similar results were also found in our field experiment: Testosterone-implanted males tended to gain more copulations than sham-implanted males, although the difference was not significant. In addition, the T- and sham-implanted males did not vary with respect to their behavior or territory attributes. Neither naturally nor experimentally elevated T levels were related to annual return rates of males. Our project is one of the first manipulations of testosterone levels in a lek-mating bird. Future work should search for links between testosterone and male attributes in lek-mating systems because females are unlikely to be able to assess testosterone levels directly.

Key Words: Galliformes, grouse, lek-mating system, mate choice, mating success, multinomial discrete choice models, physiological causes, sexual selection.

Augustine, J. K., J. J. Millspaugh, and B. K. Sandercock. 2011. Testosterone mediates mating success in Greater Prairie-Chickens. Pp. 195–208 *in* B. K. Sandercock, K. Martin, and G. Segelbacher (editors). Ecology, conservation, and management of grouse. Studies in Avian Biology (no. 39), University of California Press, Berkeley, CA.

ife history trade-offs are ubiquitous in organismal biology, but the proximate mechanisms driving variation in life history trade-offs are poorly understood (Ricklefs and Wikelski 2002). According to the immunocompetence handicap hypothesis, testosterone (T) enhances male sexual traits but has immunosuppressive effects (Folstad and Karter 1992, Mougeot et al. 2004), leading to a negative correlation between circulating T and annual survival (Ketterson and Nolan 1999, Reed et al. 2006). Nevertheless, if T increases reproductive success but decreases survival, natural selection should confer an advantage to individuals who can balance these two fitness components.

Trade-offs between mating and parental effort create a continuum among avian mating systems ranging from social monogamy, to facultative polygyny, to promiscuity. In socially monogamous species of temperate songbirds, circulating T in males peaks during pair formation and drops during brood rearing (Ketterson et al. 1992, Ketterson and Nolan 1999). In facultatively polygynous species, males have high T levels during pair formation and brood-rearing, leading to increased mating effort and reduced parental effort (Wingfield 1984). Lek mating is a type of promiscuous mating system where males perform aggregated displays at a lek site, which females visit to obtain copulations (Bradbury 1981). Only females provide parental care, so T is predicted to remain high in lek-mating males during periods of territory establishment and female visitation (Wingfield et al. 1990). However, the available data from lek-mating grouse and manakins suggest that T is elevated only during the short period of peak female visitation to the lek during the mating season (Alatalo et al. 1996, Wikelski et al. 2003). Compared to the abundance of research in monogamous and polygamous mating systems (Beletsky et al. 1989, 1990; Ketterson and Nolan 1999; Westneat et al. 2003), the role of T in lek-mating systems is largely unknown.

T may influence sexual selection through female choice or male–male competition. Because T peaks during periods of high mating activity, T-mediated traits may be used by females to select a mate. Female choice for T-dependent traits could arise via indirect benefits, such as good genes for her offspring (Andersson 1994). Studies in galliformes have found T-dependent traits are often associated with female choice, such as large fleshy head ornaments (Zuk et al. 1995a, 1995b), high-quality territories (Moss et al. 1994), and more vigorous courtship displays (Mateos and Carranza 1999). Red Jungle Fowl (*Gallus gallus*) with larger combs were more likely to attract multiple mates (Parker and Ligon 2003). T could also affect male fitness through male–male aggression, and T-mediated male ornaments or behaviors often predict the winner of paired tests (Ligon et al. 1990, Hagelin 2002).

T-implants have been used to determine experimentally how sex hormones influence male attributes, female choice, and male–male competition. T-implanted males displayed more frequently and attracted more females than sham-implanted control males (Wingfield 1984, Ketterson et al. 1992, Hill et al. 1999). T-implanted male Ring-necked Pheasants (*Phasianus colchicus*), with harem-defense polygyny, were more aggressive, had higher social rank, and increased frequency of male–male interactions (Briganti et al. 1999). T-implanted males of monogamous Red Grouse (*Lagopus lagopus scoticus*) had higher display rates and territory sizes compared to control males (Moss et al. 1994; Mougeot et al. 2003a, 2003b).

Most T-implant studies have been conducted on monogamous, territorial songbirds where T can simultaneously affect mating effort and parental care of young. Only two studies have been conducted on lek-mating males to determine the effect of T on mating effort and success while avoiding the potentially confounding effects of T on parental care. T-implanted lek-mating Golden-collared Manakins (*Manacus vitellinus*) increased rates of several display behaviors compared to sham-implanted controls during the non-breeding season (Day et al. 2006). Aggressive, display, and territory attributes of T-implanted males of the lek-mating Sharp-tailed Grouse (*Tympanuchus phasianellus*) remained unchanged relative to unmanipulated control males, but lek attendance was higher for T-implanted males (Trobec and Oring 1972).

In a three-year study, we examined how male traits, mating success, and survival vary with natural and experimentally manipulated levels of T in male Greater Prairie-Chickens (*T. cupido*), a lek-mating bird. Our objectives were: (1) to quantify natural variation in circulating

T levels during the breeding season and determine whether T levels correlate with male traits and mating success; (2) to experimentally manipulate T to determine causal relationships between T and male traits and mating success; and (3) to analyze annual return rates to determine if high levels of T are costly. Our project is one of the first experimental field studies of the interacting effects of T, male attributes, mating success, and survival in a lek-mating bird. We predicted that naturally and experimentally elevated T would have a large positive impact on male aggression, display, and mating success in prairie chickens. We expected comb size to be positively correlated with T, similar to previous studies on Red Jungle Fowl (Zuk et al. 1995a, 1995b). We did not expect T to affect other aspects of male plumage, because male plumage ornamentation is a primitive character in Galliformes, with the cryptic female plumage being a derived estrogen-dependent character (Kimball and Ligon 1999). Males in the lek center tend to obtain more copulations than those on the periphery (Hamerstrom and Hamerstrom 1973, Schroeder and Robb 1993), and we expected high T to be negatively correlated with territory distance to the center of the lek. Finally, we expected T to be negatively correlated with annual return rates according to the immunocompetence handicap hypothesis (Folstad and Karter 1992).

METHODS

We observed Greater Prairie-Chickens at lek sites during the breeding season between mid-March and mid-May, 2004–2007. All leks were located on cattle-grazed pastures in Riley and Geary Counties in northeastern Kansas (39°05′ N, 96°34′ W). Four leks were observed in 2004–2005, and we added a fifth lek in 2006. All leks were surveyed in 2007 to determine annual return rates from 2006 to 2007.

Trapping and Morphometrics

We used walk-in funnel traps and drop nets to trap males and females at lek sites during the breeding season (Hamerstrom and Hamerstrom 1973, Silvy et al. 1990, Schroeder and Braun 1991). Males were given a unique combination of colored leg bands and tail markings using nontoxic permanent markers to aid in individual identification. Seven morphometric measurements were recorded during handling: body mass (± 1 g), three linear measurements of body size (lengths of tarsus plus the longest toe, wing, tail; ± 1 mm for all), and two sexually dimorphic ornaments (comb area, measured as length $\times$ height, ± 1 mm^2; length of pinnae feathers behind head erected during display, ± 1 mm). We determined age-class as second-year (SY) or after-second-year (ASY) from the shape, coloration, and wear of the outermost two primaries (numbers 9 and 10; Schroeder and Robb 1993).

Behavioral Observations

At lek sites, male Greater Prairie-Chickens perform ritualized courtship displays and territorial behaviors (Schroeder and Robb 1993, Nooker and Sandercock 2008). A low "boom" vocalization and short "flutter-jump" flights are associated with courtship displays. Territorial behaviors include facing adjacent males and physical combat. The behavior of individual males was recorded with continuous 10-min focal observations from blinds placed ~6 m from the edge of the lek. Time spent in each of three main behavior categories (display, fighting, other) was calculated. Tallies were taken of the number of fights, males approached, boom vocalizations, and flutter-jump displays. Female presence or absence on the lek during the observation period was recorded for every focal observation. Opportunistically throughout the morning, position of males performing courtship displays or engaged in territorial disputes was recorded relative to grid stakes placed at 6-m intervals. Copulations were also recorded as they occurred and were deemed "successful" if the females shook their wings vigorously and departed the lek shortly after copulation (Schroeder and Robb 1993). To ensure repeatability among observers, pairs of observers conducted independent behavioral observations of the same male at the same time. These behavioral observations were highly correlated (PC1: $r^2 = 0.85$, $F_{1,54} = 315.8$, $P < 0.001$; PC2: $r^2 = 0.79$, $F_{1,54} = 201.4$, $P < 0.001$; see Statistical Analyses section for description of PC scores), so we did not make any adjustments for observer bias. Nevertheless, observers were rotated among the leks and males, and all observations were conducted within three hours of sunrise on days without rainfall.

Testosterone Sampling

Because handling stress can affect circulating T, leks were monitored from a blind to determine how long each male or female was in a trap. Birds were not left in the trap for more than 30 min. We collected a blood sample immediately following the bird's removal from the trap or drop net and placed the sample on wet ice. Within three hours, the blood sample was centrifuged at 14,000 rpm for 5 min to separate the blood plasma from the red blood cells. Blood plasma was collected with a pipette, transferred to a new tube, and frozen at −20°C until it could be transferred to a −70°C freezer.

We measured plasma T concentrations in each blood sample using the Salimetrics protocol and commercially available kit for salivary T enzyme immunoassay (EIA; Cat. #1-2402, Salimetrics LLC, State College, PA; Washburn et al. 2007). All plasma samples (15 μL) were diluted 1:10 with assay dilutant (135 μL). We conducted standard assay validations, including assessment of parallelism, recovery of exogenous T, intra- and inter-assay precision, and assay sensitivity (Jeffcoate 1981, Grotjan and Keel 1996, O'Fegan 2000) to confirm that T concentrations in prairie chicken plasma were measured accurately and precisely. Serial dilutions (1:1, 1:2, 1:4, 1:8, and 1:16) of two pooled plasma samples (low and high, where each pooled sample consisted of plasma from three individuals) yielded a displacement curve that was parallel (test of equal slopes, $P > 0.6$) to the standard T curve, which indicated linearity under dilution (Jeffcoate 1981). Mean recovery of exogenous T (range 38.4–240 pg/mL; levels chosen to correspond with expected plasma T levels from actual samples) was added to low (109.9 ± 1.1% ± SE, $n = 6$) and high pooled plasma samples (108.4 ± 0.7% ± SE, $n = 6$). Acceptable recovery of exogenous T (within 90–110%) verified accurate measurement throughout the working range of the assay, and demonstration of parallelism suggested no sample matrix effects (Jeffcoate 1981, Grotjan and Keel 1996, O'Fegan 2000). We used the low and high controls from the kit and analyzed them in each of the assays. Inter-assay variation was calculated from these two controls by averaging the coefficient of variation (CV) of

replicate wells from 20 randomly chosen samples. Inter-assay variation for six assays was 9.1% and average intra-assay variation was 4.7%. The sensitivity of this assay is 1.0 pg/mL. Estimates of plasma T concentration were $\log_{10}$-transformed to normalize the data.

Testosterone Implant Experiment

To test for a link between T and male behavior, we used a Before–After-Control-Impact design (BACI; Conquest 2000). Behavior, territory attributes, and mating success were compared before and after application of subcutaneous implants (35-mm lengths of silastic tubing; Dow Corning, inner diam = 1.47 mm, outer diam = 1.95 mm; sealed with silastic glue) filled with T (30 mg of testosterone propionate; Sigma-Aldrich, St. Louis, MO) or left empty (sham implants). Similar implants last up to 5–6 weeks in Red Grouse (Trobec and Oring 1972; Moss et al. 1994; Mougeot et al. 2003b, 2004; F. Mougeot, pers. comm.). The incision was closed with "liquid bandage" to prevent infection (Band-Aid brand).

The breeding season began in mid-March, when males start regularly attending the lek site. Female visitation to lek sites commenced around 25 March and peaked at 10 April, with a second, smaller peak around 5 May, presumably corresponding to renesting attempts (McNew et al., this volume, chapter 19 unpubl. data). We divided the breeding season into three periods: before the peak of female visitation (mid-March–3 April); peak female visitation and target dates for T implants (4–7 April); and post-peak female visitation (18 April–mid-May).

Only unsuccessful males received T or sham implants. Banding and behavioral observations beginning in mid-March were used to determine which males were successful and which were unsuccessful. Males were considered "successful" if they held territories that were completely surrounded by other males, fought often with many adjacent males, and had obtained copulations previously in that season or in a previous year. Males were considered "unsuccessful" if they held territories at the edge of the lek (at least 25% of territory not shared with another male), engaged in few fights, and obtained no copulations prior to implant that season or in a previous year. Unsuccessful males were captured using drop nets between 4 and 17 April and alternately assigned to the T or the sham treatment. The effects of the treatment were monitored by comparing blood serum analyses, behavioral observations, and territory size and location of T- and sham-implanted males before and after treatment. In addition, a subset of T-implanted males were recaptured 2–3 weeks after implantation, their implants were removed, dried in a desiccator at room temperature for 24 h, and weighed to determine the amount of T remaining.

Statistical Analyses

Statistical analyses were conducted in JMP IN (ver. 4.0.4, SAS Institute, 2001), except where otherwise noted. Sample sizes varied among analyses because it was not possible to measure every attribute for all males. Descriptive statistics are presented as 0 ± 1 SD unless otherwise indicated.

Eleven of 15 pairwise comparisons among the six behavioral variables (% time displaying, % time fighting, and number of boom vocalizations, flutter-jump displays, fights, and males approached) were correlated ($|r| > 0.44$; $P < 0.01$). Thus, we used principal components analysis (PCA) to obtain two principal components of behavior that were statistically independent of each other (Table 14.1), and retained two principal components (eigenvalues ≥ 1) for use in our analysis (Kaiser 1960). Female presence at the lek has a large effect on male behavior in lek-mating grouse (Höglund et al. 1997, Nooker and Sandercock

2008). To control for female presence, principal component scores were averaged separately for each male for observations when females were either present or absent from the lek.

Territorial positions of males were plotted in ArcView (ver. 3.3; Environmental Systems Research Institute, Inc., St. Charles, MO). Using the Animal Movement extension (Hooge and Eichenlaub 2000), we obtained 95% kernel estimates of territory size from positions of each male and lek size from the pooled positions of all males (± 0.1 m^2). Centroids of the 95% kernel estimates were determined using the XTools extension (Delaune 2003). Distance to lek center was defined as the distance between the centroid of a male's territory to the centroid of the entire lek (± 0.1 m).

Annual survival was estimated from return rates of male prairie chickens to lek sites. Return rates are the product of multiple probabilities, including true survival (S), site fidelity (F), and encounter rate (p; Sandercock 2006). Auxiliary data indicate that F and p are close to unity in our study population (Nooker and Sandercock 2008), and we interpret return rates as estimates of true survival for male Greater Prairie-Chickens.

Changes in behavior, territory attributes, mating success, and survival were compared before and after treatment or mean date of implant among four groups: T-implanted males, sham-implanted males, unmanipulated unsuccessful males, and unmanipulated successful males.

TABLE 14.1

Eigenvectors of a principal component analysis of reproductive behaviors from 1,332 10-min focal observations of 129 unmanipulated male Greater Prairie-Chickens at 4–5 leks/yr in northeast Kansas, 2004–2006.

Behavior	PC1 (Display)[a]	PC2 (Aggression)[a]
Percent time displaying	*0.53*	0.26
Percent time fighting	*−0.50*	0.08
Number of boom vocalizations	*0.49*	0.31
Number of flutter jump displays	*0.38*	0.02
Number of fights	−0.21	*0.65*
Number of males engaged	−0.19	*0.64*
Eigenvalue	2.79	1.69
Percent of variance explained	46.4	28.2

[a] Principal component loadings >0.4 are in *italics*.

Mean PC score for behavior, SD, and number of observations were calculated for each male before and after treatment. Effect sizes were calculated in MetaWin version 2.0 (Rosenberg et al. 2000). Differences before and after mean implant date among T-implanted, sham-implanted, unmanipulated unsuccessful, and unmanipulated successful males were analyzed using Q_B statistics from a categorical fixed effects model in MetaWin (Gurevitch and Hedges 1993). Changes in territory size and distance to lek center were analyzed using a matched pairs analysis. Mating success and return rates were analyzed using a Fisher's Exact test. Finally, the odds ratio of mating success was calculated using Proc Freq in SAS (ver. 9.1, SAS Institute 2003).

Multinomial Discrete Choice Model

We used multinomial discrete choice (MDC) models to determine the characteristics of successful males by modeling how females chose mates at unmanipulated leks (Nooker and Sandercock 2008). A limited number of copulations were observed among implanted males, and we did not model female choice at manipulated leks. Correlates of female choice were analyzed using a conditional logit multinomial discrete choice model (Proc MDC, SAS ver. 9.1). The utility of a choice (U) in benefiting the individual is modeled as:

$$U_{ij} = x_{1,ij}\beta_1 + x_{2,ij}\beta_2 + x_{n,ij}\beta_n + \varepsilon_j$$

where female i chooses among males j using characteristics of the choice j, β_n is the slope coefficient for explanatory variables, and ε is the error term. In this study, each lek represents a different set of choices (males) from which the females choose, which are represented in the model by different sets of values for the characteristics, $x_1, \ldots, x_n$.

Only copulations observed on unmanipulated leks were used in the MDC model. Each successful copulation was considered an independent choice of a female among the males at a lek. Studies of other lek-mating grouse indicate that females may copy choices of prior-mating females (Gibson et al. 1991; Gibson 1996; but see Spurrier et al. 1994), but we found no evidence of mate choice copying in Greater Prairie-Chickens (Nooker and Sandercock 2008).

In the global discrete choice model, male characteristics hypothesized to be linked to

female choice included four behavioral indices (PC scores of display and aggression with and without females), four measurements of body size (mass, tarsus, wing, and tail), two ornaments used in display (comb and pinnae), age-class (SY or ASY, coded as 0 and 1), and two territory measurements (size and distance to lek center). Morphometric and territory measurements were standardized to z-scores before analysis (mean = 0, SD = 1) so slope coefficients would be directly comparable with our PC scores for components of behavior. Backward elimination was used to simplify the model, and the resulting models were evaluated using an information-theoretic approach (Burnham and Anderson 2002). Additionally, we considered the minAIC model from Nooker and Sandercock (2008) and a minAIC model including Log_{10}T. In all, 12 models were analyzed. Akaike weights (w_i), model-averaged estimates ($\hat{\theta}_a$), and weighted unconditional standard error [SE ($\hat{\theta}_a$)] were calculated using formulae of Burnham and Anderson (2002, equations 4.1 and 4.9).

RESULTS

Natural Variation in Testosterone

Over three years, 164 blood samples were collected from 100 individuals at five leks (Table 14.2). T levels could not be quantified for six samples (3.7%) due to sample levels being below ($n = 2$) or above the detection limits ($n = 4$). An average of 1.6 ± 0.9 samples were collected from 87 males and 10 females. Log_{10}T differed by sex ($F = 23.38$, $P < 0.001$) and varied by the amount of time the bird was in the trap ($F = 4.92$, $P = 0.03$), but was not affected by lek site ($F = 1.12$, $P = 0.11$), time of day relative to sunrise ($F = 0.97$, $P = 0.34$), or either linear or quadratic seasonal trend ($F < 0.28$, $P > 0.60$; overall model $F_{10,137} = 4.01$, $P < 0.001$). T levels were four times higher in males (1.15 ± 0.91 ng/mL, $n = 82$) than females (0.29 ± 0.18 ng/mL, $n = 10$). Time in trap was negatively correlated with Log_{10}T levels ($F_{1,146} = 4.88$, $P = 0.03$), but the coefficient of determination was low ($r^2 = 0.03$). Because this correlation explained 3% of the variation, we did not control for time in trap in analyses examining associations between T, male traits, mating success, and survival.

TABLE 14.2

*Number of birds captured and uncaptured (in parentheses) per year at
Greater Prairie-Chicken leks in northeast Kansas, 2004–2006.*

Year	Lek	No of unmanipulated successful males[a]	No of unmanipulated unsuccessful males[b]	No of sham-implanted males[c]	No of T-implanted males[d]
2004	KM	4 (1)	5 (4)	0	0
	KP	3	6 (4)	0	0
	PL	1	4 (5)	0	0
	RRN	2 (3)	3 (4)	0	0
2005	KM	2 (2)	4 (5)	0	0
	KP	(1)	2	3	3
	PL	1 (1)	7 (3)	0	0
	RRN	3 (1)	4 (2)	3	3
2006	KM	0	4 (1)	0	1
	KP	(2)	1 (2)	0	0
	PL	3	1 (4)	3	2
	RRN	4 (1)	1	1	2
	HESS	0	2 (1)	3	4
Total		23 (12)	44 (35)	13	15

[a] Unmanipulated males observed receiving at least one copulation.
[b] Unmanipulated males that were not observed to receive any copulations.
[c] Implanted with an empty silastic tube.
[d] Implanted with a silastic tube containing 30 mg testosterone.

A lack of seasonal trend allowed us to average $Log_{10}T$ levels for each male each year, which were compared with male traits and mating success. A subset of males were not observed at the leks after they were captured. Non-territorial males had lower $Log_{10}T$ levels (0.86 ± 0.85 ng/mL, $n = 14$) than territorial males (1.28 ± 0.96 ng/mL, $n = 81$; $F_{1,93} = 9.72$, $P = 0.002$).

No male behavioral or morphological traits were correlated with natural variation in T ($Log_{10}T$) among territorial males at unmanipulated leks. Neither male territory size nor the territory's distance from lek center were correlated with $Log_{10}T$ (size: $F_{1,43} = 0.001$, $P = 0.97$; distance: $F_{1,43} = 0.001$, $P = 0.97$). $Log_{10}T$ did not correlate with display or aggressive behavior when females were present (display: $F_{1,40} = 0.06$, $P = 0.81$; aggression: $F_{1,40} = 0.16$, $P = 0.69$) or when females were absent (display: $F_{1,44} = 1.22$, $P = 0.28$; aggression: $F_{1,44} = 0.12$, $P = 0.73$). Of the five morphological traits considered (mass,

length of tarsus plus longest toe, keel, wing, and tail), none were correlated to T levels ($F_{1,43} < 3.10$, $P > 0.09$). Age did not affect T levels ($F_{1,43} < 0.01$, $P = 0.91$). Two sexually dimorphic traits were not associated with $Log_{10}T$ (pinnae: $F_{1,41} < 0.01$, $P = 0.99$; comb area: $F_{1,42} = 2.39$, $P = 0.13$).

The MDC model was based on 23 copulations recorded for 12 successful and 26 unsuccessful males at five unmanipulated Greater Prairie-Chicken leks over three years. Of the 15 models considered, two had ΔAIC values <2 (Table 14.3). Mating success was correlated with natural variation in $Log_{10}T$, two morphological traits and one territorial trait as indicated by the minAIC model (Table 14.3). $Log_{10}T$ had the largest slope coefficient, indicating that it was the strongest predictor of male mating success (Fig. 14.1). $Log_{10}T$ was positively correlated with mating success, such that males with higher natural levels of T obtained more copulations ($t = 2.87$, $P = 0.004$). One sexually selected trait and one measurement

TABLE 14.3

Comparison of multinomial discrete choice models examining the effects of male testosterone, morphology, behavior, and territory attributes on mating success of 23 copulations observed among 12 successful and 23 unsuccessful males at five unmanipulated Greater Prairie-Chicken leks in northeast Kansas, 2004–2006.

Model	Testosterone[a]	Morphology[b]	Behavior[c]	Territory[d]	K[e]	Dev[f]	AIC[g]	ΔAIC[h]	w_i[i]
1	T	rc	—	l	4	60.6	68.62	0.00	0.47
2	T	rc	d	l	5	60.1	70.07	1.45	0.23
3	T	rcp	d	l	6	59.2	71.17	2.56	0.13
4	T	rcp	d	ls	7	58.5	72.48	3.86	0.07
5	T	wrcp	d	ls	8	57.5	73.50	4.89	0.04
6	T	wrcpg	d	ls	9	56.4	74.43	5.81	0.03
7	T	twrcpg	d	ls	10	55.7	75.74	7.13	0.01
8	T	mrc	DAda	—	8	59.9	75.89	7.28	0.01
9	T	twrcpg	Ad	ls	11	54.9	76.89	8.27	0.01
10	—	mrc	DAda	—	7	62.9	76.92	8.30	0.01
11	T	twrcpmg	Ad	ls	12	54.4	78.37	9.75	0.00
12	T	twrcpmg	DAd	ls	13	54.1	80.13	11.51	0.00
13	T	twrcpmg	DAda	ls	14	54.0	82.00	13.38	0.00
14	—	twrcpmg	DAda	ls	13	59.7	85.71	17.09	0.00
15	T	—	—	—	1	92.3	94.33	25.71	0.00

NOTE: Model 10 is the min AIC model from Nooker and Sandercock (2008); Model 8 is model 10 + T; Model 14 is the global model without T; Model 13 is model 14 + T.

[a] Testosterone variables (T = $\log_{10}$-transformed testosterone levels, — = not included).
[b] Morphological variables (c = comb area; p = pinnae; r = tarsus; m = mass at capture; w = wing; t = tail; g = age class, — = none included).
[c] Behavioral variables (D = display with females; A = aggression with females; d = display without females; a = aggression without females, — = none included).
[d] Territory variables (l = distance to center of lek, s = territory size, — = none included).
[e] K = number of parameters.
[f] Dev = deviance.
[g] AIC = Akaike's Information Criterion.
[h] ΔAIC = difference of the AIC value in the given model compared to the minimum AIC model.
[i] w_i = Akaike weights.

of body size were positively correlated with higher mating success (comb area: $t = 2.43$, $P = 0.02$; length of tarsus plus longest toe: $t = 3.24$, $P = 0.001$). Neither display nor aggressive behavior was correlated with mating success since it was not included in the minAIC model (Table 14.3). There was weak support (ΔAIC = 1.45) for a model that included a negative, but nonsignificant, correlation between display behavior when females were absent and mating success during the breeding season ($t = -0.72$, $P = 0.47$; Table 14.3). Distance to lek center was negatively correlated with male mating success, so that males closer to the center of the lek obtained more copulations ($t = -2.26$, $P = 0.02$).

Annual survival was 55.2% ($n = 67$) among unmanipulated males and did not vary with traits that affected male mating success. Survival did not vary with natural variation in $\mathrm{Log}_{10}T$ ($\chi^2 < 0.03$, $P = 0.86$, $n = 45$), comb area ($\chi^2 = 0.23$, $P = 0.63$, $n = 50$), tarsus length ($\chi^2 = 0.06$, $P = 0.80$, $n = 50$), or distance from the center of the lek ($\chi^2 = 0.01$, $P = 0.93$, $n = 66$). Males that displayed less when females were not present had higher return rates ($\chi^2 = 6.52$, $P = 0.01$, $n = 67$), even after Bonferroni correction ($P < 0.01$). The odds of returning were 2.1 times higher (95% CI: 0.73–5.91) among males that did not receive a single copulation than among males that received one or more copulations (Fisher's Exact test: $P = 0.19$).

Experimental Implants of Testosterone

In 2005 and 2006, 15 T-implants and 13 sham implants were conducted, for a total of 28 males at five leks (Table 14.2). Male behavior before implanting did not differ between the two treatment groups (display with females: $F_{1,22} = 0.69$, $P = 0.41$; aggression with females: $F_{1,22} = 0.68$, $P = 0.42$; display without females: $F_{1,23} = 0.01$, $P = 0.94$; aggression without females: $F_{1,23} = 3.21$, $P = 0.09$). Neither territory size nor distance to lek center varied by treatment before the experiment (size: $F_{1,26} = 1.02$, $P = 0.17$; distance: $F_{1,22} = 0.20$, $P = 0.66$). Comb size, pinnae length, tarsus length, mass, wing length, and tail length did not differ between T- and sham-implanted males (all $F_{1,25} < 1.37$, $P > 0.25$). Males in the two treatments had the same proportion of yearlings and adults (T: 27% SY, $n = 15$; S: 46% SY, $n = 13$; Fisher's Exact $P = 0.43$). There was no difference in $Log_{10}T$ levels at the time of implant between T- and sham-implanted males (natural T levels of T-males: 1.33 ± 1.24 ng/mL, $n = 14$; sham-males: 1.84 ± 1.48 ng/mL, $n = 13$; $F_{1,25} = 0.94$, $P = 0.34$) or the date which the implant occurred (T-males: 10 April $\pm$ 4, $n = 15$; sham-males: 12 April $\pm$ 5, $n = 13$; $F_{1,26} = 1.67$, $P = 0.21$). After implant, the odds of T levels increasing were 2.0 times higher for of T-males than sham-implanted males (95% CI: 0.08–51.6). Three days following implantation, the T level of a single sham-implanted male decreased by 2.25 ng/mL, whereas the T level of a single T-implanted male increased by 2.36 ng/mL. Between 22 and 35 days following implant, T-implanted males (1.58 ± 0.93 ng/mL, $n = 6$) had $Log_{10}T$ levels similar to sham-implanted males (0.71 ± 0.33 ng/mL, $n = 2$; $F_{1,6} = 2.82$, $P = 0.14$), but T-implanted males had higher $Log_{10}T$ levels than successful, unmanipulated males (0.59 ± 0.34 ng/mL, $n = 6$; $F_{2,11} = 5.29$, $P = 0.02$). The T levels of sham-implanted males were intermediate between T-implanted males and unmanipulated successful males. T-implants removed from two experimental males 26 and 35 days after implant were completely empty of T. Because T-levels between sham- and T-implanted males 22 days following implant was not significant, behavioral observations >21 days post-implant were excluded from the behavioral analyses.

Experimental T treatments did not affect male morphometrics, behavior, or territory attributes.

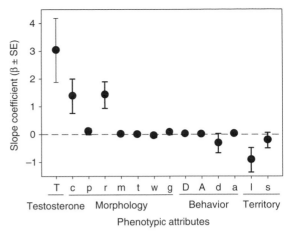

Figure 14.1. Model-averaged estimates of the slope coefficients from a multinomial discrete choice model estimating the effect of male attributes on male mating success, based on 23 copulations among 12 successful and 26 unsuccessful males at five Greater Prairie-Chicken leks in northeast Kansas, 2004–2006. Explanatory variables were z-transformed so slope estimates are directly comparable. Estimates of SE are unconditional and include uncertainty due to model selection, and confidence intervals that do not overlap zero are significant. Abbreviations for explanatory variables include: testosterone (T = log-transformed testosterone levels), morphological variables (c = comb area; p = pinnae; r = tarsus; m = mass at capture; t = tail; w = wing; g = age class); behavioral variables (D/d = display with and without females present; A/a = aggression with and without females present); and territory variables (l = distance to center of lek, s = territory size).

Comb size increased during the breeding season in both T-implanted and unmanipulated males (T males: $+25.4 \pm 69.0$ mm^2, $n = 5$; unmanipulated: $+21.5 \pm 17.7$ mm^2, $n = 4$; matched pairs test F = 0.01, df = 8, $P = 0.92$). Sham-implanted males and unsuccessful, unmanipulated males did not differ in behavior standardized for female presence (meta-analysis; display: $Q_B = 0.19$, $P = 0.66$; aggression: $Q_B = 2.36$, $P = 0.12$; $n = 23$ unsuccessful and 8 sham-implanted males) or territory attributes (matched pairs test; size: F ratio = 0.20, $P = 0.66$; distance to center: F ratio = 0.49, $P = 0.49$; $n = 15$ unsuccessful and 11 sham males, df = 25), so they were pooled in subsequent analyses and referred to as "unsuccessful" males. Changes in aggressive and display behaviors did not differ between T-implanted males, unsuccessful males, and successful males following treatment (display: $Q_B = 3.76$, $P = 0.15$; aggression: $Q_B = 5.44$,

TABLE 14.4

Comparison of effect sizes of changes in display and aggressive behavior following a testosterone implant experiment with Greater Prairie-Chickens at five leks in northeast Kansas, 2005–2006.

	Unsuccessful[a]			T-Implants			Successful		
	E+	95% CI	n	E+	95% CI	n	E+	95% CI	n
Display[b]	0.22	−0.02 to 0.47	31	0.54	0.16 to 0.92	13	0.13	−0.11 to 0.38	24
Aggression[b]	0.01	−0.24 to 0.25	31	−0.20	−0.57 to 0.16	13	0.26	0.01 to 0.51	24

[a] Sham-implanted males and unsuccessful unmanipulated males combined.
[b] Display and aggressive behavior was quantified using a principal component analysis of behavior observed during 10-min focal observations and standardized for female presence (Table 14.1).

$P = 0.07$; Table 14.4). Using a matched pairs analysis comparing territory size and distance to lek center before and after treatment, no differences were detected among T-implanted, unsuccessful and successful males (matched pairs test; size: F ratio = 1.39, $P = 0.26$; distance to center: F ratio = 1.11, $P = 0.34$; $n = 14$ T-implanted, 26 unsuccessful, 16 successful males).

T-implanted and sham-implanted males had similar mating success and survival. Four of 15 T-implanted males (14.3%) received six copulations, whereas only one of 13 sham-implanted males (3.6%) received a single copulation, but the difference was not significant (Fisher's Exact test, $P = 0.33$). The odds of returning were 4.1 times (95% CI: 0.63–26.1) higher for sham-implanted (5 of 13 returned) than T-implanted males (2 of 15 returned), but the probability of returning did not differ with respect to treatment (Fisher's Exact test, $P = 0.20$).

DISCUSSION

Although much work has investigated the role of T in monogamous and facultatively polygamous mating systems, our project is one of the few studies to demonstrate a direct link between natural variation in T and mating success in a lek-mating species. We improve on earlier studies of the natural variation of T in a lek-mating species by analyzing T levels from a larger sample of males ($n = 87$, this study; $n = 23$ males, Alatalo et al. 1996; $n = 15$–27 per species, Wikelski et al. 2003). Our experimental manipulations of T in wild birds suggest that there may be a causal link between T and mating success, but low sample sizes and high variability among T-implanted individuals limited our power for stronger inferences. Surprisingly, natural or experimental variation in T did not correspond with any male traits examined, including aggression. In addition, we did not find support for two of the predictions of the immunocompetence handicap hypothesis because natural or experimentally elevated T did not correlate with male morphological, behavioral, or territorial traits, nor did elevated T decrease annual survival of males.

T was related to male mating success, but not via any of the behavioral, morphological, or territorial traits considered. Territory size, distance from lek center, and morphology did not vary with T. A previous study found an increase in display behavior following the administration of T implants during the non-breeding season (Day et al. 2006), but we did not find any relationship between display behavior and natural or experimentally elevated T. Perhaps we did not find a relationship because our study was conducted during the breeding season, and males were already displaying at maximal levels of effort. We also did not detect a positive correlation between T and comb size or aggressive behavior, as found in previous studies (Briganti et al. 1999, Parker et al. 2002, Mougeot et al. 2005a).

A lack of a relationship between T and male behavioral, morphological, or territorial traits was surprising because females are unlikely to assess T levels directly. However, T may affect mating success via cumulative effects, unmeasured male traits, or during territory establishment in fall. First, T may have small, statistically insignificant, effects on multiple male traits that, when taken together, influence male mating success (Ligon et al. 1998), but we were unable to test for such cumulative effects. Second, T could also be related to male traits that were not measured, such as parasite load, UV reflectance, color of air sacs and combs, or attributes of vocalizations (Gibson et al. 1991; Mougeot et al. 2005b, 2006; Blas et al. 2006). Last, male grouse visit lek sites in both spring and autumn (Baines 1996, Rintamäki et al. 1999, Salter and Robel 2000, pers. obs.), but the role of lek attendance in autumn is poorly understood. Juvenile males may visit multiple leks to prospect for territories during the non-breeding season, and adult males may maintain or compete for better territories. T could enhance a male's aggressiveness outside of the breeding season and be a proximate mechanism that mediates territory establishment (Mougeot et al. 2005a).

Contrary to a prediction of the immunocompetence handicap hypothesis, T increased male mating success, but did not decrease survival. Similar to an autumn study of Red Grouse (Redpath et al. 2006), we found a nonsignificant trend for males with higher T to have decreased survival. The immunocompetence handicap hypothesis assumes that the birds live long enough that cumulative detrimental effects become apparent. The short average lifespan of Greater Prairie-Chickens (1.6 yr; Robel and Ballard 1974), and high rates of annual turnover among males at leks (Nooker 2007, Nooker and Sandercock 2008) may indicate that natural selection is a stronger determinant of male traits than sexual selection (Bleiweiss 1997, Drovetski et al. 2006). Therefore, longer-lived species of lek-mating birds, such as manakins, may be better study species for testing the immunocompetence handicap hypothesis.

The experimental T implants did not increase aggressive behavior as predicted. Changes in aggression among T-implanted males may have been difficult to detect if sham-implanted males increased their aggressive levels in response to repeated intrusions of T-implanted males (i.c., the challenge hypothesis; Wingfield et al. 1990). Alternatively, our T implants may not have had the intended physiological effects. T can be converted to estradiol in the skin of domestic chickens, producing female-like feathers on males (George et al. 1981). We were unable to examine this possibility in our field study, but estradiol also plays a role in aggression and territory defense in males (Soma et al. 2000). To separate the effects of the two sex steroids, future field studies could potentially manipulate conversion rates of T to estradiol by using implants of fadrozole, an aromatase inhibitor (Soma et al. 2000, Mougeot et al. 2005a).

The inclusion of T in the MDC model reduced the relative importance of the effects of behavior on male mating success. Model 10 in Table 14.3 included similar variables as the minAIC model presented in Nooker and Sandercock (2008; aggressive and display behavior with and without females, mass, tail or tarsus, comb). The slope coefficients of these two models were similar in magnitude and direction (data not presented), indicating that the results presented here are due to the inclusion of T in the model and not due to smaller sample sizes in this study. However, inclusion of distance to lek center in the analysis of male mating success suggested that certain specific aggressive behaviors that determine territory location, but that were not quantified in our PCA of behavior, may be important in mediating male mating success. We also found a link between comb size and mating success similar to previous studies (Rintamäki et al. 2000). Our research showed the potential importance of examining a suite of male traits since multiple traits were used by females to select a mate.

In conclusion, we found that both natural and elevated T was related to mating success of prairie chickens, but we were unable to detect any individual or suites of male traits correlated with T levels. To obtain a better understanding of the seasonal changes in natural levels of T, multiple samples from individuals throughout the breeding and non-breeding seasons are needed (Mougeot et al. 2005a, Redpath et al. 2006, Kempenaers et al. 2008). Future work should also examine parasites, plumage coloration, and vocalizations as possible intermediary steps linking T to mating success.

ACKNOWLEDGMENTS

We thank the following organizations and people for allowing access to their prairie chicken leks: Konza Prairie Biological Station, a property of The Nature Conservancy managed by the Division of Biology at Kansas State University; Rannells Flint Hills Prairie Preserve, managed by Clenton Owensby in the Department of Agronomy at Kansas State University; and private lands owned by Grant Poole and James Hess. Tom VanSlyke and Kenny Berg provided logistical support. We thank Tracey Adamson, Jeremy Baumgardt, Amanda Behnke, Jarrod Bowers,

Tara Conkling, Seth Cutright, DeVaughn Fraser, Chris Frey, Kyle Gerstner, Chod Hedinger, "Hoogy" Hoogheem, Nichole Lambrecht, and Kara Oberle for field assistance. Expert statistical assistance was provided by Thomas Loughin. Rami Woods assisted with laboratory analyses. Funding for field work included: a NSF Kansas EPSCoR Grant (KAN29509), a research grant from the American Ornithologists' Union, and an NSF Doctoral Dissertation Improvement Grant (DEB-0608477). J. K. Augustine (née Nooker) was supported by the Konza Prairie NSF Long-Term Ecological Research Grant (DEB-0218210) and by the Division of Biology at Kansas State University. The research was conducted under the following permits: Scientific, Education or Exhibition Wildlife Permit, Kansas Department of Wildlife and Parks (SC-118-2003, SC-068-2004, SC-078-2005, SC-072-2006), and Institutional Animal Care and Use Committee (Protocols 2079, 2351).

LITERATURE CITED

Alatalo, R. V., J. Höglund, A. Lundberg, P. T. Rintamäki, and B. Silverin. 1996. Testosterone and male mating success on the Black Grouse leks. Proceedings of the Royal Society of London B 263:1697–1702.

Andersson, M. B. 1994. Sexual selection. Princeton University Press, Princeton, NJ.

Baines, D. 1996. Seasonal variation in lek attendance and lekking behaviour by male Black Grouse *Tetrao tetrix*. Ibis 138:177–180.

Beletsky, L. D., G. H. Orians, and J. C. Wingfield. 1989. Relationships of steroid hormones and polygyny to territorial status, breeding experience, and reproductive success in male Red-winged Blackbirds. Auk 106:107–117.

Beletsky, L. D., G. H. Orians, and J. C. Wingfield. 1990. Steroid hormones in relation to territoriality, breeding density, and parental behavior in male Yellow-headed Blackbirds. Auk 107:60–68.

Blas, J., L. Pérez-Rodríguez, G. R. Bortolotti, J. Viñuela, and T. A. Marchant. 2006. Testosterone increases bioavailability of carotenoids: insights into the honesty of sexual signaling. Proceedings of the National Academy of Science 103:18633–18637.

Bleiweiss, R. 1997. Covariation of sexual dichromatism and plumage colours in lekking and non-lekking birds: a comparative analysis. Evolutionary Ecology 11:217–235.

Bradbury, J. W. 1981. The evolution of leks. Pp. 138–169 in R. D. Alexander and D. W. Tinkle (editors), Natural selection and social behavior. Blackwell Scientific Publications, Oxford, UK.

Briganti, F., A. Papeschi, T. Mugnai, and F. Dessí-Fulgheri. 1999. Effect of testosterone on male traits and behaviour in juvenile pheasants. Ethology, Ecology and Evolution 11:171–178.

Burnham, K. P., and D. R. Anderson. 2002. Model selection and multimodel inference. 2nd ed. Springer, New York, NY.

Conquest, L. L. 2000. Analysis and interpretation of ecological field data using BACI designs: discussion. Journal of Agricultural, Biological, and Environmental Statistics 5:293–296.

Day, L. B., J. T. McBroom, and B. A. Schlinger. 2006. Testosterone increases display behaviors but does not stimulate growth of adult plumage in male Golden-collared Manakins (*Manacus vitellinus*). Hormones and Behavior 49:223–232.

Delaune, M. 2003. XTools, a package of tools useful in vector spatial analysis. <http://arcscripts.esri.com/details.asp?dbid=11526>.

Drovetski, S. V., S. Rohwer, and N. A. Mode. 2006. Role of sexual and natural selection in evolution of body size and shape: a phylogenetic study of morphological radiation in grouse. Journal of Evolutionary Biology 19:1083–1091.

Folstad, I., and A. J. Karter. 1992. Parasites, bright males, and the immunocompetence handicap. American Naturalist 139:603–622.

George, F. W., J. F. Noble, and J. D. Wilson. 1981. Female feathering in Sebright cocks is due to the conversion of testosterone to estradiol in skin. Science 213:557–559.

Gibson, R. M. 1996. Female choice in sage grouse: the roles of attraction and active comparison. Behavioral Ecology and Sociobiology 39:55–59.

Gibson, R. M., J. W. Bradbury, and S. L. Vehrencamp. 1991. Mate choice in lekking sage grouse revisited: the roles of vocal display, female site fidelity, and copying. Behavioral Ecology 2:165–180.

Grotjan, H. E., and B. A. Keel. 1996. Data interpretation and quality control. Pp. 51–93 in E. P. Diamandis and T. K. Christopoulos (editors), Immunoassay. Academic Press, New York, NY.

Gurevitch, J., and L. V. Hedges. 1993. Meta-analysis: combining the results of independent experiments. Pp. 378–398 in S. M. Scheiner and J. Gurevitch (editors), Design and Analysis of Ecological Experiments. Chapman and Hall, New York, NY.

Hagelin, J. C. 2002. The kinds of traits involved in male-male competition: a comparison of plumage, behavior, and body size in quail. Behavioral Ecology 13:32–41.

Hamerstrom, F. N., and F. Hamerstrom. 1973. The prairie chicken in Wisconsin: highlights of a 22-year study of counts, behavior, movements, turnover and habitat. Technical Bulletin No. 64. Wisconsin Department of Natural Resources, Madison, WI. <http://digital.library.wisc.edu/1711.dl/EcoNatRes.DNRBull64>.

Hill, J. A., D. A. Enstrom, E. D. Ketterson, V. Nolan, Jr., and C. Ziegenfus. 1999. Mate choice based on static versus dynamic secondary sexual traits in the Dark-eyed Junco. Behavioral Ecology 10:91–96.

Höglund, J., T. Johansson, and C. Pelabon. 1997. Behaviourally mediated sexual selection: characteristics of successful male Black Grouse. Animal Behaviour 54:255–264.

Hooge, P. N., and W. Eichenlaub. 2000. Animal movement extension to ArcView ver 2.0. Alaska Science Center, Biological Science Office, U.S. Geological Survey, Anchorage, AK. <http://www.absc.usgs.gov/glba/gistools/index.htm>.

Jeffcoate, S. L. 1981. Efficiency and effectiveness in the endocrinology laboratory. Academic Press, San Diego, CA.

Kaiser, H. F. 1960. The application of electronic computers to factor analysis. Educational and Psychological Measurement 20:141–151.

Kempenaers, B., A. Peters, and K. Foerster. 2008. Sources of individual variation in plasma testosterone levels. Philosophical Transactions of the Royal Society B 363:1711–1723.

Ketterson, E. D., and V. Nolan, Jr. 1999. Adaptation, exaptation, and constraint: a hormonal perspective. American Naturalist 154(Supplement):S4–S25.

Ketterson, E. D., V. Nolan, Jr., L. Wolf, and C. Ziegenfus. 1992. Testosterone and avian life histories: effects of experimentally elevated testosterone on behavior and correlates of fitness in the Dark-eyed Junco (Junco hyemalis). American Naturalist 140:980–999.

Kimball, R. T., and J. D. Ligon. 1999. Evolution of avian plumage dichromatism from a proximate perspective. American Naturalist 154:182–193.

Ligon, J. D., R. Kimball, and M. Merola-Zwartjes. 1998. Mate choice by female Red Junglefowl: the issues of multiple ornaments and fluctuating asymmetry. Animal Behaviour 55:41–50.

Ligon, J. D., R. Thornhill, M. Zuk, and K. Johnson. 1990. Male–male competition, ornamentation, and the role of testosterone in sexual selection in Red Jungle Fowl. Animal Behaviour 40:367–373.

Mateos, C., and J. Carranza. 1999. Effects of male dominance and courtship display on female choice in the Ring-necked Pheasant. Behavioral Ecology and Sociobiology 45:235–244.

Moss, R., R. Parr, and X. Lambin. 1994. Effects of testosterone on breeding density, breeding success, and survival of Red Grouse. Proceedings of the Royal Society of London B 258:175–180.

Mougeot, F., A. Dawson, S. M. Redpath, and F. Leckie. 2005a. Testosterone and autumn territorial behavior in male Red Grouse Lagopus lagopus scoticus. Hormones and Behavior 47:576–584.

Mougeot, F., J. R. Irvine, L. Seivwright, S. M. Redpath, and S. Piertney. 2004. Testosterone, immunocompetence, and honest sexual signaling in male Red Grouse. Behavioral Ecology 15:930–937.

Mougeot, F., S. M. Redpath, and F. Leckie. 2005b. Ultra-violet reflectance of male and female Red Grouse, Lagopus lagopus scoticus: sexual ornaments reflect nematode parasite intensity. Journal of Avian Biology 36:203–209.

Mougeot, F., S. M. Redpath, F. Leckie, and P. J. Hudson. 2003a. The effect of aggressiveness on the population dynamics of a territorial bird. Nature 421:737–739.

Mougeot, F., S. M. Redpath, R. Moss, J. Matthiopoulos, and P. J. Hudson. 2003b. Territorial behaviour and population dynamics in Red Grouse Lagopus lagopus scoticus. I. Population experiments. Journal of Animal Ecology 72:1073–1082.

Mougeot, F., S. M. Redpath, and S. B. Piertney. 2006. Elevated spring testosterone increases parasite intensity in male Red Grouse. Behavioral Ecology 17:117–125.

Nooker, J. K. 2007. Factors affecting the demography of a lek-mating bird: the Greater Prairie-Chicken. Ph.D. dissertation, Kansas State University, Manhattan, KS.

Nooker, J. K., and B. K. Sandercock. 2008. Phenotypic correlates and survival consequences of male mating success in lek-mating Greater Prairie-Chickens (Tympanuchus cupido). Behavioral Ecology and Sociobiology 62:1377–1388.

O'Fegan, P. O. 2000. Validation. Pp. 211–238 in J. P. Gosling (editor), Immunoassays. Oxford University Press, New York, NY.

Parker, T. H., R. Knapp, and J. A. Rosenfield. 2002. Social mediation of sexually selected ornamentation and steroid hormone levels in male junglefowl. Animal Behaviour 64:291–298.

Parker, T. H., and J. D. Ligon. 2003. Female mating preferences in Red Junglefowl: a meta-analysis. Ethology Ecology & Evolution 15:63–72.

Redpath, S. M., F. Mougeot, F. M. Leckie, and S. A. Evans. 2006. The effects of autumn testosterone on survival and productivity in Red Grouse, Lagopus lagopus scoticus. Animal Behaviour 71:1297–1305.

Reed, W. L., M. E. Clark, P. G. Parker, S. A. Raouf, N. Arguedas, D. S. Monk, E. Snajdr, V. Nolan, Jr., and E. D. Ketterson. 2006. Physiological effects on demography: a long-term experimental study of testosterone's effects on fitness. American Naturalist 167:667–683.

Ricklefs, R. E., and M. Wikelski. 2002. The physiology/life history nexus. Trends in Ecology and Evolution 17:462–468.

Rintamäki, P. T., J. Höglund, E. Karvonen, R. V. Alatalo, N. Björklund, A. Lundberg, O. Rätti, and J. Vouti. 2000. Combs and sexual selection in Black Grouse (*Tetrao tetrix*). Behavioral Ecology 11:465–471.

Rintamäki, P. T., E. Karvonen, R. V. Alatalo, and A. Lundberg. 1999. Why do Black Grouse males perform on lek sites outside the breeding season? Journal of Avian Biology 30:359–366.

Robel, R. J., and W. B. Ballard, Jr. 1974. Lek social organization and reproductive success in the Greater Prairie Chicken. American Zoologist 14:121–128.

Rosenberg, M. S., D. C. Adams, and J. Gurevitch. 2000. MetaWin: statistical software for meta-analysis, version 2.0. Sinauer Assoc. Inc., Sunderland, MA.

Salter, G. C., and R. J. Robel. 2000. Capturing Lesser Prairie-Chickens on leks during fall. Transactions of the Kansas Academy of Science 103:46–47.

Sandercock, B. K. 2006. Estimation of demographic parameters from live-encounter data: a summary review. Journal of Wildlife Management 70:1504–1520.

Schroeder, M. A., and C. E. Braun. 1991. Walk-in traps for capturing Greater Prairie-Chickens on leks. Journal of Field Ornithology 62:378–385.

Schroeder, M. A., and L. A. Robb. 1993. Greater Prairie-Chicken (*Tympanuchus cupido*). A. Poole, P. Stettenheim, and F. Gill (editors), The birds of North America No. 36. Academy of Natural Sciences, Philadelphia, PA.

Silvy, N. J., M. E. Morrow, E. Shanley, Jr., and R. D. Slack. 1990. An improved drop net for capturing wildlife. Proceedings of the Annual Conference of Southeastern Fish and Wildlife Agencies 44:374–378.

Soma, K. K., A. D. Tramontin, and J. C. Wingfield. 2000. Oestrogen regulates male aggression in the non-breeding season. Proceedings of the Royal Society of London B 267:1089–1096.

Spurrier, M. F., M. S. Boyce, and B. F. Manly. 1994. Lek behaviour in captive sage grouse *Centrocercus urophasianus*. Animal Behaviour 47:303–310.

Trobec, R. J., and L. W. Oring. 1972. Effects of testosterone propionate implantation on lek behavior of Sharp-tailed Grouse. American Midland Naturalist 87:531–536.

Washburn, B. E., J. J. Millspaugh, D. L. Morris, J. H. Schulz, and J. Faaborg. 2007. Using a commercially available enzyme immunoassay to quantify testosterone in avian plasma. Condor 109:181–186.

Westneat, D. F., D. Hasselquist, and J. C. Wingfield. 2003. Tests of association between the humoral immune response of Red-winged Blackbirds (*Agelaius phoeniceus*) and male plumage, testosterone, or reproductive success. Behavioral Ecology and Sociobiology 53:315–323.

Wikelski, M., M. Hau, W. D. Robinson, and J. C. Wingfield. 2003. Reproductive seasonality of seven neotropical passerine species. Condor 105:683–695.

Wingfield, J. C. 1984. Androgens and mating systems: testosterone-induced polygyny in normally monogamous birds. Auk 101:665–671.

Wingfield, J. C., R. E. Hegner, A. M. Dufty Jr., and G. F. Ball. 1990. The "challenge hypothesis": theoretical implications for patterns of testosterone secretion, mating systems, and breeding strategies. American Naturalist 136:829–846.

Zuk, M., T. S. Johnsen, and T. Maclarty. 1995a. Endocrine-immune interactions, ornaments and mate choice in Red Jungle Fowl. Proceedings of the Royal Society of London B 260:205–210.

Zuk, M., S. L. Popma, and T. S. Johnsen. 1995b. Male courtship displays, ornaments and female mate choice in captive Red Jungle Fowl. Behaviour 132:821–836.

CHAPTER FIFTEEN

Reproductive Biology of a Southern Population of Greater Prairie-Chickens

Lance B. McNew, Andrew J. Gregory, Samantha M. Wisely, and Brett K. Sandercock

Abstract. We conducted a three-year study of the breeding chronology of Greater Prairie-Chickens (*Tympanuchus cupido*) to determine seasonal patterns of lek attendance and clutch initiation, and the duration of egg-laying and incubation for birds at the core of the species distribution. Our field study included three sites differing in landscape composition and rangeland management in the Flint Hills and Smoky Hills of Kansas. Counts of birds on leks were 30% higher when using counts from blinds compared to flush counts. Timing of lek attendance did not differ among study sites. Males attended leks from 2 March to 19 May, females were observed at leks from 20 March to 16 April, and peak lek attendance for both sexes was 9–10 April. Mean date of clutch initiation of first and renesting attempts was 26 April and 24 May, respectively, with active nests documented from 1 April to 8 July. Females delayed initiation of first nests at the most southerly study site, possibly because of a lack of suitable nesting cover early in the season due to range management practices. Although previously undocumented for prairie chickens, egg-laying rates >1 egg/day suggested that intraspecific nest parasitism occurred in 6–15% of clutches. The probability of a female renesting after first nest failure was 50%, declining with date of nest failure, but was unaffected by stage of loss or study site. On average, females initiated renests 8 days after failure of first nests. Hatch dates ranged from 18 May to 8 July, brood-rearing extended from 18 May to 22 July, and juveniles were independent by 7 September at 60 days of age. Overall, the reproductive phenology of Greater Prairie-Chickens in Kansas occurred earlier and lasted longer than in other populations. Our research results will be useful to wildlife biologists planning surveying or trapping activities, researchers conducting studies of nesting and brood ecology, and land managers concerned with minimizing the impacts of prescribed burning, cutting for hay, or other types of rangeland management.

Key Words: clutch initiation, egg flotation, incubation, lek attendance, prairie grouse, reproduction, *Tympanuchus cupido*.

McNew, L. B., A. J. Gregory, S. M. Wisely, and B. K. Sandercock. 2011. Reproductive biology of a southern population of Greater Prairie-Chickens. Pp. 209–221 in B. K. Sandercock, K. Martin, and G. Segelbacher (editors). Ecology, conservation, and management of grouse. Studies in Avian Biology (no. 39), University of California Press, Berkeley, CA.

reater Prairie-Chickens (*Tympanuchus cupido*) have shown significant population declines across their continually shrinking range over the last century. Agriculture practices have caused a drastic decline of available usable habitat since the early 20th century (>95%; Schroeder and Robb 1993, Braun et al. 1994), and prairie chicken populations declined an estimated 75–80% as a result (Johnsgard 2002). The Flint Hills region of east-central Kansas, southern Nebraska, and northeastern Oklahoma consists of intact tallgrass prairie and has been identified as a stronghold for Greater Prairie-Chickens (hereafter prairie chickens; Johnsgard 2002). This area is characterized by rocky soils that are unsuitable for cultivation and encompasses over 1.6 million ha. For this reason, many authorities consider the Flint Hills to be vital to the long-term persistence of grassland birds (Svedarsky et al. 1999, With et al. 2008). Despite large tracts of relatively intact grassland, annual lek surveys conducted by the Kansas Department of Wildlife and Parks (KDWP) show that statewide prairie chicken populations have declined annually from 4.5 birds/km² in 1980 to 1.5 birds/km² in 2008 (Applegate and Horak 1999, Rodgers 2008). The cause of population declines remains unknown, but timing of declines coincides with the introduction of the range management practice of intensive early stocking and annual spring burning (IESB; Westemeier and Gough 1999, Robbins et al. 2002). IESB benefits cattle production by increasing grass production and allowing ranchers to stock ranges with cattle early. IESB may negatively affect prairie grouse production if complete burns of large contiguous range result in significant decreases in availability of quality nesting sites (Robbins et al. 2002, Patten et al. 2007). To date, studies of the effects of rangeland management on prairie chicken breeding ecology have been limited to the selection of nest sites and relative effects on nest survival (McKee et al. 1998, Patten et al. 2007). Data are lacking regarding how these practices impact other aspects of prairie chicken breeding behavior, such as breeding phenology.

The timing of reproductive events of grassland birds is important, especially for short-lived species, whose population dynamics are sensitive to variation in reproductive success (Wisdom and Mills 1997). For prairie grouse, such as Greater Prairie-Chickens, productivity may be determined by seasonal variation in the ability of females to locate mates at mating arenas or leks, and the environmental conditions at nesting and brood-rearing habitats. For example, timing of breeding and clutch initiation should be late enough to ensure that suitable vegetative cover exists for concealment of first nesting attempts, but early enough to ensure that renesting attempts can occur if needed and that juveniles are independent before inclement winter conditions (Horak 1985, Svedarsky et al. 2003). In addition, timing of nest initiation has implications for recruitment because chick development and survival is affected by abundance and seasonal phenology of insect food items (Johnson and Boyce 1990, Park et al. 2001, Gregg and Crawford 2009). Thus, timing of reproductive events is critical for maximizing fitness of prairie chickens and may vary among areas of different habitat conditions.

Reproductive chronology of prairie chickens also has implications for population monitoring, research, and range management. Knowledge of the timing of reproductive events is necessary for wildlife biologists planning population surveys of leks or females with broods, researchers studying nesting and brood ecology, and land managers scheduling burning, grazing, or haying activities. Knowledge of reproductive chronology is particularly important for species with broad geographic ranges but regional variation in population dynamics, such as the Greater Prairie-Chicken (Rodgers 2008, McNew et al., this volume, chapter 19). Reproductive chronology has been described for isolated populations in Minnesota (Svedarsky 1983, 1988) and Wisconsin (Hamerstrom and Hamerstrom 1973), but relatively little is known about the timing of reproductive events of prairie chickens breeding in Kansas (Robel 1970, Horak 1985). Recent changes in regional land management practices over the last three decades may have altered the breeding phenology of prairie chickens in the Flint Hills, as changes in grazing and prescribed burning have affected the seasonal availability of lekking, nesting, and brood-rearing habitat (Patten et al. 2007). The landscapes of Kansas provide a unique opportunity to evaluate whether land management practices impact the breeding phenology of prairie chickens because land use and range management practices vary significantly across the species range within the state. In addition to occupying the large unfragmented grasslands of the Flint Hills, prairie chickens also occur in the more developed

Smoky Hills ecoregion (Rodgers 2008). Although grasslands in the Smoky Hills are highly fragmented by row-crop agriculture (>35% of the landscape) and improved roads (1.04 km per ha), they are not as intensively managed as grasslands in the Flint Hills and may be of better quality due to lower cattle stocking rates and less frequent burning (J. Pitman, pers. comm.).

In this paper, we describe the reproductive chronology of three declining populations of Greater Prairie-Chickens (*T. c. pinnatus*) occurring over a gradient of landscape alteration and rangeland management within the core of the species' extant range in Kansas. We expected (1) timing of breeding events to occur earlier than previous reports for northern populations due to advanced vegetation phenologies, and (2) differences in regional land use to affect the seasonal phenology and reproductive rates in our study populations. If prairie chickens require suitable cover in order to initiate nests (Pitman et al. 2005, Fields et al. 2006), clutch initiation, duration of laying and incubation, renesting propensity, and timing of brood-rearing and fledging might be delayed in areas where most residual cover is removed through extensive annual spring burning and early cattle stocking. We discuss the ecological and management implications of regional variation in the seasonal breeding chronology of prairie chickens in Kansas.

STUDY SITES

Our field study was conducted at three sites in Kansas: two sites in the Flint Hills and one site in the Smoky Hills. The three study sites differed in landscape composition and pattern, as well as rangeland management practices (Table 15.1). The southern Flint Hills site (South) was burned annually in the spring and managed with intensive early stocking (IESB, 1 head/0.8 ha for 90 days; Smith and Owensby 1978, With et al. 2008). The second study area was located in the northern Flint Hills (North). Annual spring burning is common at North and lands are managed with a mixture of IESB and season-long stock grazing and annual burning (SLSB; 1 head/1.6 ha for 180 days). The third study area (Smoky) was located in the Smoky Hills ecoregion and is more fragmented by agricultural land uses (Table 15.1). Cultivated crops include sorghum, corn, wheat, and soybeans. Native grass pastures at Smoky are burned infrequently at fire return intervals >1 year, are grazed at low intensity (1 head/>2 ha for 180 days), and cattle stocking occurs later in the season than at the Flint Hills sites. Thus, we expected the breeding phenologies of the sites to be ordered from earliest to latest: Smoky, North, and South.

TABLE 15.1

Comparison of southern Flint Hills (South), northern Flint Hills (North), and Smoky Hills (Smoky) study sites for population studies of Greater Prairie-Chickens in Kansas, 2006–2008.

	South	North	Smoky
Size (km²)	1,106	671	1,630
Prairie-chicken density index[a]	0.10	0.19	0.17
Proportion grassland	0.90	0.81	0.53
Proportion cropland	0.03	0.10	0.38
Road density (km/km²)	0.32	0.57	1.04
Mean (SE) precipitation (cm)[b]	12.3 (2.0)	11.4 (2.4)	8.2 (2.4)
Mean daily temperature (°C)[c]	15.0 (0.4)	12.9 (0.4)	12.1 (0.4)
Land management[d]	IESB	IESB, SLSB	SLSU, RG&B
No. of females radiomarked	54	77	72

[a] Males per km² = mean number of males per lek × number of leks/study site size.
[b] Mean monthly precipitation during March–May 2006–2008.
[c] Mean daily temperature during March–May 2006–2008.
[d] Dominant land management at each study site: IESB = intensive early stocking, annual burning; SLSB = season-long stock grazing, annual burning; SLSU = season-long stocking, unburned; RG&B = rotational grazing and burning (after Smith and Owensby 1978, With et al. 2008).

METHODS

Lek Attendance

During the spring lekking period (February–May), counts of birds at leks were conducted using two methods: (1) birds were flushed from untrapped leks between 0600 and 0930 hrs, and (2) prairie chickens were observed from blinds while birds were trapped at leks. We attempted to obtain counts of males and females prior to flushing by viewing leks from >100 m using binoculars when possible. For both methods, the maximum numbers of males, females, and total birds were recorded. Multiple flush counts were conducted for each lek within a breeding season but not on consecutive days. To assess whether survey method affected lek counts, we used analysis of variance (ANOVA) to compare counts of prairie chickens when leks were flushed or trapped, and among our three study sites. A Tukey–Kramer HSD was used to compare lek counts among sites at $\alpha = 0.05$ level.

We calculated the date of peak lek attendance for males and females at each study site by weighting the Julian date of lek observation (day 1 = 1 Jan) by the average number of birds attending leks:

$$\text{Day of Peak of Lek Attendance} = \frac{\sum \left(D_i \frac{A_i}{A_{i-N}} \right)}{N},$$

where D_i is the Julian day i of lek observation, A_i is the mean lek attendance by males or females for day i, $\overline{A}_{1-N}$ is the mean lek attendance for all days of observation, and N is the total number of observation days per sex. Low numbers of surveys per day at each study area precluded comparisons of peak lek attendance among sites by year. We pooled daily surveys among years of study and compared timing of peak lek attendance among study sites using ANOVA. Female lek attendance data were log-transformed to meet the normality assumption of ANOVA (Sokal and Rohlf 2000).

Egg-laying and Incubation

We captured prairie chickens with walk-in traps and drop-nets at leks during March–May of 2006–2008 (Silvy et al. 1990, Schroeder and Braun 1991). Captured birds were sexed by plumage characteristics (Henderson et al. 1967). Females were fitted with 11-g necklace-style VHF radio transmitters with an expected battery life of 12 months (Model RI-2B, Holohil Systems Ltd., Ontario, Canada). We located females ≥3 times per week during the breeding and brood-rearing seasons (March–August), and daily once females began nesting. Once a female had localized in an area for three consecutive days, we used a portable radio receiver and handheld Yagi antenna to locate and flush the bird. Nest sites were visited ≤2 times during laying and early incubation to determine clutch size and stage of incubation. Nests were not visited again until females had departed and were located away from the nest for ≥2 consecutive days. Once a female departed, we classified nest fate as successful if ≥1 egg successfully hatched chicks, or as failed if the clutch was depredated, abandoned, or destroyed for other reasons. Date of hatching was the last day the female was estimated to be incubating at a successful nest by triangulation with radiotelemetry.

To estimate duration of incubation in days, we subtracted the date of known clutch completion from the date of hatch. We assessed the influence of study site, nesting attempt, clutch size, and day of nest initiation on duration of incubation using forward stepwise regression. Alpha (α) levels of 0.05 and 0.1 were specified for entry and removal of factors from the model.

Nest and Brood Chronology

First nests were defined as the first nest discovered for an individual female within a breeding season, whereas renests were nesting attempts by radio-marked females where the first nest was known to have failed. If the clutch size increased between visits, the date of clutch initiation was determined by backdating by the number of eggs from the first visit, assuming one egg laid per day (Svedarsky 1988). If clutch size did not change between successive visits, the date of clutch initiation was determined by backdating from the hatch date, assuming an incubation period of 24 days (Schroeder and Robb 1993), or from the stage of incubation determined by egg flotation (McNew et al. 2009). We used forward stepwise regression to model dates of clutch initiation as a function of study year, study site, and nesting attempt. Alpha (α) levels of 0.05 and 0.1 were specified for entry and removal of factors from the model. We then fitted a linear model with the resulting significant predictor variables and assessed model fit.

We used logistic regression to evaluate the relationship between the probability of renesting and study site, clutch size of the first nest, day of incubation when the initial attempt failed, and the date of nest failure. Date of failure was considered to be the mid-point between the last day the nest was known to be active and the day it was identified as failed. The average interval ($\pm$ SD) between the last day a nest was known to be active and the day it was determined to have failed was 4 $\pm$ 4 d. We excluded females that were unavailable to renest if they died while incubating first nests, could not be located after first nests failed, or lost their transmitters within two weeks of failure of the first nest. We also excluded 10 nests for which explanatory data were missing. We fit 13 *a priori* models to data from 82 failed first nest attempts. We used Akaike's Information Criterion adjusted for small sample sizes (AIC_c) for model selection, and models where $\Delta AIC_c \leq 2$ were considered to be equally parsimonious (Burnham and Anderson 1998). Logistic regression analyses were conducted using the logistic procedure in SAS 9.1 (SAS Institute, Cary, NC).

We located radio-marked hens with broods daily via triangulation. Brood flushes were conducted at 14 days post-hatching to estimate pre-fledge brood survival. Prairie chickens can sustain short flights at 14 days of age (Schroeder and Robb 1993). Although juveniles can survive without the brood hen when 40 days old, they are still generally associated with the hen and brood mates until 60–80 days post-hatch (Bowman and Robel 1977; L. B. McNew, unpubl. data). Therefore, dates of fledging and independence were estimated for successful broods and compared to predicted dates for all hatched broods. Sample sizes of successful broods were too small to conduct statistical analyses, and descriptive statistics are presented. Statistics were calculated with procedures of program JMP IN (ver. 4.0.4, SAS Institute, Cary, NC).

RESULTS

Lek Attendance

During 2006–2008, we conducted 673 lek surveys at our three study sites from 2 March to 19 May. We conducted 408 lek observations from blinds during trapping activities and 265 flush counts where no traps were deployed. To assess whether our trapping activities impacted lek attendance, a random sample of 265 trapped lek observations were selected and compared to flush counts. The maximum number of prairie chickens observed was greater during lek observations of trapped leks (10.9 $\pm$ 0.4 SE birds per day) than flush counts (7.2 $\pm$ 0.4; $F_{1,522} = 56.8$, $P < 0.001$). Similarly, female lek attendance was greater for observations conducted during trapping (1.3 $\pm$ 0.9 birds per day) than during flush counts (0.4 $\pm$ 0.1; $F_{1,367} = 30.7$, $P < 0.001$), suggesting that trapping activities did not negatively impact lek attendance and that counts from lek observations of trapped leks were suitable for further analysis.

The peak of male lek attendance was 9 April across all years and study sites in Kansas, with males present on leks during the entire 79-day observation period (2 March–19 May; Fig. 15.1). Peak female attendance at leks was 10 April when data were pooled among years and sites, with 95% of female lek visitations occurring during a 28-day period between 20 March and 16 April (Fig. 15.1). Timing of peak lek attendance did not differ among study sites for males ($F_{2,172} = 0.38$, $P = 0.68$) or females ($F_{2,172} = 0.32$, $P = 0.73$), but the duration of female lek attendance appeared to be a shorter period at the South site. Copulations ($n = 13$) were observed during a 37-day period from 3 April to 9 May.

Timing of Clutch Initiation and Renesting

During 2006–2008, we located 231 nests of 155 females. A total of 167 nests were first nests, 61 nests were first renests, and three nests were third nesting attempts. Mean date of clutch initiation for first nests at all sites was 26 April (range = 1 April–22 May; $n = 162$). Mean date of clutch initiation for known renest attempts was 24 May (range = 29 April–4 July; $n = 64$). Forward stepwise regression revealed that nesting attempt and the interaction between study site and nesting attempt were significant predictors of date of clutch initiation ($r^2 = 0.45$, $P < 0.01$). Study year and site alone did not improve model fit and were removed from the model. Mean ($\pm$ SE) date of first clutch initiation differed significantly among study sites (South = 2 May $\pm$ 1.9 d, North = 30 April $\pm$ 1.5 d, Smoky = 24 April $\pm$ 1.7 d; $F_{2,150} = 3.4$,

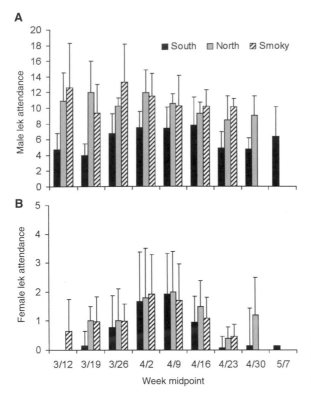

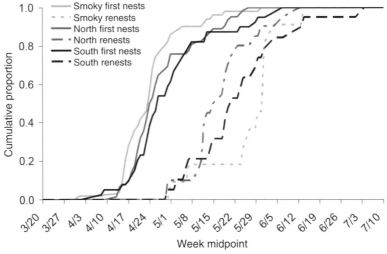

Figure 15.1. Mean daily lek attendance per week (birds per day ± SD) of male (A) and female (B) Greater Prairie-Chickens in Kansas, 2006–2008.

Figure 15.2. Cumulative clutch initiation dates for first nests and renests of female Greater Prairie-Chickens at three study sites in Kansas, 2006–2008.

$P = 0.03$), but timing of renests did not differ among study sites (Fig. 15.2). Mean date of hatching for all sites pooled was 6 June for first nests (range = 18 May–21 June) and 26 June for renests (7 June–8 July; Fig. 15.3), and date of hatching did not differ among study sites ($F_{2,40} = 2.0$, $P = 0.15$) or years ($F_{2,21} = 0.23$, $P = 0.79$).

The probability of a prairie chicken initiating a renesting attempt was influenced by the date of failure for the first nest (Fail day) and the stage of incubation at failure (First nest age). An additive model with these two factors was the minimum AIC_c model, and models that included Fail day had 98% of the relative support of the data

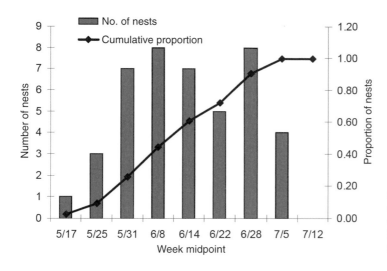

Figure 15.3. Weekly distribution of nest hatches and cumulative weekly hatch for female Greater Prairie-Chickens in Kansas, 2006–2008.

TABLE 15.2

Model selection based on minimization of AIC_c for the estimation of renesting probabilities of Greater Prairie-Chickens at three study sites in Kansas, 2006–2008.

Model[a]	K[b]	AIC_c	ΔAIC_c	w_i
Fail day + first nest age	3	91.4	0	0.41
Site + fail day + TCL + first nest age	5	92.0	0.5	0.32
Fail day + TCL + first nest age	4	92.4	1.0	0.25
Fail day	2	100.7	9.3	0.0
Site + fail day	3	101.4	10.0	0.0
Fail day + TCL	3	101.8	10.4	0.0
Site + fail day + TCL	4	105.1	11.2	0.0
TCL	2	128.4	37.0	0.0
Site	2	128.5	37.1	0.0

[a] Fail Day = date of failure for initial nesting attempt; First nest age = stage of development when initial attempt failed; Site = study site; TCL = clutch size of initial attempt.
[b] K = number of parameters; w_i = AIC_c weight or relative support for model i.

(Table 15.2). However, the regression coefficient for First nest age (β = -0.002) was not significantly different than zero (95% CI = −0.06–0.06) and was considered spurious. Females losing first nests late in the season had a lower probability of renesting (β = −0.11, 95% CI = −0.17 to −0.05; Fig. 15.4), and the odds of a female attempting a renest decreased by 11% per day during the nesting season. Prairie chickens renested with an average interval between failure of the first nest and initiation of a renesting attempt of 7.8 ± 1.1 days (range = 0–27 d, n = 45).

The fledging period, defined as the period between the dates of fledging for our first and last brood, at all study sites ranged across a 53-day period from 31 May to 22 July (mean day of fledging was 30 June). Timing of fledging did not differ for broods that successfully fledged and the dates predicted for unsuccessful broods (difference = 2 d). Prairie chicken chicks at the Smoky site tended to fledge 5–6 days earlier than the other two sites, but the difference was not significant ($F_{2,40}$ = 2.1, P = 0.13). Dates of independence for prairie chicken young at 60 days of

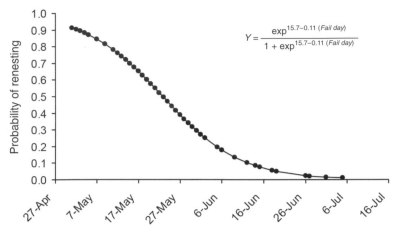

$$Y = \frac{\exp^{15.7-0.11 \, (\textit{Fail day})}}{1 + \exp^{15.7-0.11 \, (\textit{Fail day})}}$$

Figure 15.4. Probability of renesting for female Greater Prairie-Chickens as a function of date of failure for the first nesting attempt. Probability of renesting was not influenced by stage of loss, clutch size, or study site.

age would be predicted to occur from 16 July to 7 September.

Egg-Laying and Incubation

On average, prairie chickens laid an egg every 1.1 ± 0.3 days, but in 10 of 67 cases (15%), the estimated rates of egg-laying were >1 egg/day (range = 1.1–2.0). When we adjusted clutch initiation dates to account for the uncertainty of our egg flotation technique (±2 d; McNew et al. 2009), egg-laying rates at 6 of 10 nests still were >1 egg per day. Thus, 6–15% of prairie chicken clutches in our study showed evidence of intraspecific nest parasitism by other female prairie chickens. Clutch sizes of first nests (12.4 ± 2.3 eggs) were larger than renests (10.5 ± 2.4 eggs). Prairie chickens incubated nests for 25.0 ± 2.5 days on average (range = 22–29 d, $n = 38$). Forward stepwise regression indicated that duration of incubation was not affected by study site ($F_{2,34} = 0.5, P = 0.63$), date of nest initiation ($r^2 = 0.07$, $P = 0.11$, df = 1, $n = 35$), or nesting attempt ($F_{1,34} = 3.4$, $P = 0.08$; second and third nests pooled). Duration of incubation was positively related to clutch size by:

Duration of Incubation = 20.9 + 0.32 (Clutch Size)

but most of the variation was unexplained ($r^2 = 0.12$, $P = 0.03$, df = 1, $n = 35$).

DISCUSSION

Compared to populations of prairie chickens in the northern extent of their range (Hamerstrom and Hamerstrom 1973, Svedarsky 1983, 1988), the seasonal timing of lek attendance and clutch initiation was earlier in Kansas, the duration of the nesting and brood-rearing periods was longer, and rates of renesting were higher. Moreover, regional differences in landscape pattern and rangeland management resulted in differences in timing of clutch initiation among sites in the Flint Hills and Smoky Hills ecoregions of Kansas, with delayed initiation in annually burned and heavily grazed grasslands. Duration of incubation and age at fledging were similar for all populations. Egg-laying rates >1 egg per day indicate that intraspecific nest parasitism may be more common in the core range of Kansas than in relict populations elsewhere.

Timing of Lek Attendance and Nesting

Lek attendance by both male and female prairie chickens in Kansas was highest during the second week of April, with no annual variation in seasonal timing during our three-year study. Male attendance at leks was stable throughout March to May, although males were most active in display behaviors when females visited in mid-April (Nooker and Sandercock 2008). We did not observe seasonal declines in male lek attendance as previously described for prairie chickens in Kansas (Robel 1970), and our results were more consistent with the stable lek attendance reported for other populations (Hamerstrom and Hamerstrom 1973, Svedarsky 1983). Sustained male attendance may be driven by a propensity of females to initiate multiple nests or by extended nesting periods in our populations.

Clutch initiation for prairie chickens in Kansas (1 April) began earlier than populations in Minnesota and Wisconsin (20–27 April; Hamerstrom and Hamerstrom 1973, Svedarsky 1983), but later than a population of Attwater's Prairie-Chicken (*T. c. attwaterii*) in coastal Texas (12 March; Lutz et al. 1994), a latitudinal trend reported for other species of prairie grouse (Connelly et al. 1998, Schroeder et al. 1999). Latitudinal differences in the onset of clutch initiation may be due to variation in vegetation phenology across the species' range, which likely results in earlier availability of suitable resources at lower latitudes (Schoech and Hahn 2008). In addition, the nesting season was longer in Kansas than reported previously for both northern and southern populations of prairie chickens, with active nests located during a three-month period between 1 April and 4 July. Elsewhere, nests have been found during a two-month period for both interior Greater Prairie-Chickens (mid-April–early June; reviewed by Schroeder and Robb 1993) and coastal Attwater's Prairie-Chicken (mid-March–early May; Lutz et al. 1994). Early nesting and a longer breeding season may allow prairie chickens in Kansas to cope with nest failure due to initially poor nesting cover with higher rates of renesting (McNew et al., this volume, chapter 19).

We observed site differences in the timing of clutch initiation, but, unexpectedly, nests were initiated later at the most southerly study site in the Flint Hills. Differences in rangeland management and agricultural use may explain differences in timing of clutch initiation of about a week among our study sites in Kansas. Most of the native tallgrass pastures at the South (~90%) and North (~70%) sites were burned during March and April, whereas none of the native tallgrass pastures at the Smoky site were burned during our study. Burning may affect timing of nesting if female prairie chickens delay egg-laying until vegetative cover is sufficient to conceal the clutch. Although delayed nesting in response to poor habitat conditions has not previously been reported for prairie chickens, female prairie grouse tend to initiate clutches in areas with greater residual cover and visual obstruction (Pitman et al. 2005, Fields et al. 2006, L. B. McNew, unpubl. data.). Alternatively, variation in timing of clutch initiation could have

been due to site differences in food availability or weather. Prairie chickens and other grouse are income breeders that require exogenous nutritional resources for egg-laying (Meijer and Drent 1999), and site differences in rangeland quality or access to subsidies from agricultural crops could have affected variation in timing through effects on female nutritional status. Cultivated agricultural fields comprised a higher proportion of the landscape at the Smoky site, and prairie chickens will utilize grain sorghum and other crops prior to nesting (Robel et al. 1970). Females were heavier at the Smoky site (mean ± SE = 929 ± 8.8 g) than at the North (908 ± 8.7 g) or South (879 ± 7.9 g) sites. It is unclear whether females at the Smoky site were in better body condition, but larger females tended to initiate clutches earlier than smaller females (McNew et al., this volume, chapter 19). Weather can influence the timing of clutch initiation in grouse as well (Martin et al. 2000, Martin and Wiebe 2004). However, warmer average daily temperatures (~2–3°C; Table 15.1) during the period when initiation of first clutches would be impacted (March–May) did not result in advanced reproductive phenology at the South site. In contrast, clutch initiation occurred later at this site. We found no differences in mean monthly precipitation ($F_{2,18} = 0.9$, $P = 0.42$) among study sites during the clutch initiation period (Table 15.1), suggesting that weather was not responsible for the observed variation in clutch initiation timing among study sites.

The influence of landscape composition and land use on prairie chicken nesting behavior and demography is well documented. Habitat conditions like residual cover directly affect the selection of nest and brood sites and the resulting success of these vital parameters (McKee et al. 1998, Pitman et al. 2005, Fields et al. 2006). Therefore, land management practices that alter habitat conditions, such as prescribed burning, grazing, and row crop agriculture, can have significant impacts on prairie chicken habitat use, reproductive success, and survival (Patten et al. 2007, McNew et al., this volume, chapter 19). Our data suggest that the effects of landscape alteration and management can influence not only vital rates directly through impacts on availability and quality of habitat, but through impacts on the effort and timing of reproduction as well. Although it is unclear whether nest initiation was delayed at the South site due to limited nesting cover or whether initiation was advanced at the Smoky site because females were in better condition, human manipulation of prairie chicken habitats appears to be impacting the nesting phenology of prairie chickens in Kansas. Given the effects of temporal variation on nest survival and renesting propensity (Hannon et al. 1988, Sandercock et al. 2005, Martin et al., this volume, chapter 17), human activities that alter prairie chicken resources, such as range management practices, may have even greater influence on populations than previously recognized.

Egg-Laying and Incubation

The average egg-laying rate of female prairie chickens in Kansas was one egg per 1.11 days, similar to published reports from other populations (Schroeder and Robb 1993). We estimated that 6–15% of the nests in our sample had egg-laying rates of >1 egg per day, with uncertainty due to the margins of error from our egg flotation technique. Given that egg-laying rates of >1 egg per day are unknown for large-bodied birds (Welty and Baptista 1988), we conclude that a subset of our nests were affected by conspecific nest parasitism. Intraspecific nest parasitism has not been documented for prairie chickens but has been reported in a few other species of grouse (Willow Ptarmigan, *Lagopus lagopus*, Martin 1984; Sharp-tailed Grouse, *T. phasianellus*, Gratson 1989, Yom-Tov 2001).

Duration of incubation for prairie chicken nests in Kansas (25 ± 2.5 d) was similar to values reported for northern populations in Wisconsin and Minnesota (23–25 d; Hamerstrom and Hamerstrom 1973, Svedarsky 1988, Schroeder and Robb 1993). Age-specific nest mortality rates can influence patterns of nest attentiveness through effects on residual reproduction, leading to variation in duration of incubation for songbirds (Martin 2002). We found no regional variation in the duration of incubation in prairie chickens, despite pronounced differences in nest survival and adult female mortality rates among our three study sites (McNew et al., this volume, chapter 19).

Renesting Propensity

A minimum of 50% of female prairie chickens renested after failure, and the probability of renesting declined seasonally with the date of failure for first nesting attempts. Our reported estimates of renesting probability are conservative

because our method of locating nests based on tracking of radio-marked females made it difficult to find nests during the laying period, possibly resulting in many undocumented first nests that failed before discovery. Nevertheless, renesting propensity is usually lower in other species of prairie and forest grouse (<36%) and has been explained by other factors, including stage of loss during the nesting cycle and female age-class (Sopuck and Zwickel 1983, Connelly et al. 1993, Storaas et al. 2000). Prairie chickens may have had high rates of renesting for three reasons. First, they are a relatively short-lived species that make a large investment in reproduction (Bergerud and Gratson 1988), and renests can contribute to the annual fecundity of single-brooded precocial birds (Martin et al. 1989, Milonoff 1991). Second, date of first nest failure impacted the probability of renesting, and a large proportion of first nests failed early in the season (>80%; L. B. McNew, unpubl. data). Date of failure may have been more important than stage of loss because prairie chickens breed at southerly latitudes and have a longer breeding season than forest and tundra grouse (Sandercock et al. 2005). Last, differences in habitat conditions among the study sites could have influenced the probability of renesting. Because prairie chickens are income breeders, marked differences in landscape composition, fragmentation, and land management practices observed among study sites could have impacted the resources available for egg production. However, prairie chickens shared similar abilities to initiate renesting attempts among the three sites, suggesting that exogenous resources for follicle development during renesting attempts were not limiting for any of our populations. The resource availability hypothesis was also rejected for tundra grouse (Sandercock et al. 2005, Martin et al., this volume, chapter 17), suggesting alternate hypotheses may be more appropriate for explaining renesting abilities of Tetraoninae.

In summary, the reproductive chronology of prairie chickens in Kansas started earlier and lasted longer than in other populations, possibly due to a combination of longer summers at low latitudes and regional differences in landscape composition and rangeland management practices. Rangeland management practices that remove or reduce residual vegetative cover during March and April, such as annual spring burning and intensive early stocking of cattle, have the potential to negatively impact prairie chickens by delaying onset of clutch initiation and reducing nesting success. A better understanding of the breeding chronology of prairie grouse and the duration of reproductive stages will assist management efforts and provide a foundation for intensive studies of population demography in the future. For example, wildlife biologists planning lek surveys for population monitoring or live-trapping of prairie chickens for translocations or population studies would optimize field effort in Kansas by planning field work from late March to mid-April, the period of greatest lek attendance and activity. Likewise, land managers can reduce negative impacts on prairie chicken populations by managing for a shifting mosaic of burned and unburned prairie to provide patches of residual nesting cover. Landowners should also delay haying and spraying during early April–late July, as this is the primary nesting and brood-rearing period in Kansas.

ACKNOWLEDGMENTS

We thank the many field technicians who helped collect field data, especially D. Broman, T. Cikanek, L. Hunt, V. Hunter, and W. White. Funding and equipment were provided by a consortium of federal and state wildlife agencies, conservation groups, and wind energy partners under the National Wind Coordinating Collaborative, including National Renewable Energies Laboratory (DOE), U.S. Fish and Wildlife Service, Kansas Department of Wildlife and Parks, Kansas Cooperative Fish and Wildlife Research Unit, National Fish and Wildlife Foundation, Kansas and Oklahoma chapters of The Nature Conservancy, BP Alternative Energy, FPL Energy, Horizon Wind Energy, and Iberdrola Renewables. B. K. Sandercock and S. M. Wisely were supported by the Division of Biology at Kansas State University. We thank K. Martin, J. Pitman, D. Wolfe, and an anonymous reviewer for comments on the manuscript.

LITERATURE CITED

Applegate, R. D., and G. J. Horak. 1999. History and status of the Greater Prairie-Chicken in Kansas. Pp. 113–121 in W. D. Svedarsky, R. H. Hier, and N. J. Silvy (editors), The Greater Prairie-Chicken: a national look. Minnesota Agricultural Experiment Station Miscellaneous Publication 99-1999. University of Minnesota, St. Paul, MN.

Bergerud, A. T. and M. W. Gratson. 1988. Adaptive strategies and population ecology of northern grouse. University of Minnesota Press, Minneapolis, MN.

Bowman, T. J. and R. J. Robel. 1977. Brood break-up, dispersal, mobility, and mortality of juvenile prairie chickens. Journal of Wildlife Management 41:27–34.

Braun, C. E., K. Martin, T. E. Remington, and J. R. Young. 1994. North American grouse: issues and strategies for the 21st century. Proceedings of the 59th North American Wildlife and Natural Resources Conference 59:428–437.

Burnham, K. P., and D. R. Anderson. 1998. Model selection and inference: a practical information-theoretic approach. Springer New York, NY.

Connelly, J. W., R. A. Fischer, A. D. Apa, K. P. Reese, and W. L. Wakkinen. 1993. Renesting by sage grouse in southeastern Idaho. Condor 95:1041–1043.

Connelly, J. W., M. W. Gratson and K. P. Reese. 1998. Sharp-tailed Grouse (Tympanuchus phasianellus). A. Poole (editor), The birds of North America online. Ithaca: Cornell Lab of Ornithology. <http://bna. birds.cornell.edu.er.lib.k-state.edu/bna/species/354 doi:10.2173/bna.354>.

Fields, T. L., G. C. White, W. C. Gilgert, and R. D. Rodgers. 2006. Nest and brood survival of Lesser Prairie-Chickens in west central Kansas. Journal of Wildlife Management 70:931–938.

Gratson, M. W. 1989. Intraspecific nest parasitism by Sharp-tailed Grouse. Wilson Bulletin 101:126–127.

Gregg, M. A., and J. A. Crawford. 2009. Survival of Greater Sage-Grouse chicks and broods in the northern Great Basin. Journal of Wildlife Management 73:904–913.

Hamerstrom, F. N., Jr., and F. Hamerstrom. 1973. The prairie chicken in Wisconsin—highlights of a 22-year study of counts, behavior, movements, turnover, and habitat. Technical Bulletin 64. Wisconsin Department of Natural Resources, Madison, WI.

Hannon, S. J., K. Martin, and J. O. Schieck. 1988. Timing of reproduction in two populations of Willow Ptarmigan in northern Canada. Auk 105:330–338.

Henderson, F. R., F. W. Brooks, R. E. Wood, and R. B. Dahlgren. 1967. Sexing of prairie grouse by crown feather patterning. Journal of Wildlife Management 31:764–769.

Horak, G. J. 1985. Kansas prairie chickens. Kansas Fish and Game Commission, Pratt, KS.

Johnsgard, P. A. 2002. Grassland grouse and their conservation. Smithsonian Institution Press, Washington, DC.

Johnson, G. D., and M. S. Boyce. 1990. Feeding trials with insects in the diet of sage grouse chicks. Journal of Wildlife Management 54:89–91.

Launchbaugh, J. L., C. E. Owensby, J. R. Brethour, and E. F. Smith. 1983. Intensive-early stocking studies in Kansas. Kansas State University Agricultural Experiment Station Progress Report 441.

Lutz, R. S., J. S. Lawrence, and N. J. Silvy. 1994. Nesting ecology of Attwater's Prairie-Chicken. Journal of Wildlife Management 58:230–233.

Martin, K. 1984. Intraspecific nest parasitism in Willow Ptarmigan. Journal of Field Ornithology 55:250–251.

Martin, K., S. J. Hannon, and R. F. Rockwell. 1989. Clutch size variation and patterns of attrition in fecundity of Willow Ptarmigan. Ecology 70:1788–1799.

Martin, K., P. B. Stacey, and C. E. Braun. 2000. Recruitment, dispersal and demographic rescue in spatially-structured White-tailed Ptarmigan populations. Condor 102:503–516.

Martin, K. and L. Wiebe. 2004. Coping mechanisms of alpine and arctic breeding birds: extreme weather and limitations to reproductive resilience. Integrative and Comparative Biology 44:177–185.

Martin, T. E. 2002. A new view of life-history evolution tested on an incubation paradox. Proceedings of the Royal Society of London 269:309–316.

McKee, G., M. R. Ryan, and L. M. Mechlin. 1998. Predicting Greater Prairie-Chicken nest success from vegetation and landscape characteristics. Journal of Wildlife Management 62:314–321.

McNew, L. B., A. J. Gregory, S. M. Wisely, and B. K. Sandercock. 2009. Estimating the stage of incubation for nests of Greater Prairie-Chickens using egg flotation: a float curve for grousers. Grouse News 38:12–14.

Meijer, T., and R. Drent. 1999. Re-examination of the capital and income dichotomy in breeding birds. Ibis 141:399–414.

Milonoff, M. 1991. Renesting ability and clutch size in precocial birds. Oikos 62:189–194.

Nooker, J. K., and B. K. Sandercock. 2008. Correlates and consequences of male mating success in lek-mating Greater Prairie-Chickens (*Tympanuchus cupido*). Behavioral Ecology and Sociobiology 62:1377–1388.

Park, K. J., P. A. Robertson, S. T. Campbell, R. Foster, Z. M. Russell, D. Newborn, and P. J. Hudson. 2001. The role of invertebrates in the diet, growth and survival of Red Grouse (*Lagopus lagopus scoticus*) chicks. Journal of Zoology 254:137–145.

Patten, M. A., D. H. Wolfe, and S. K. Sherrod. 2007. Lekking and nesting response of the Greater Prairie-Chicken to burning of tallgrass prairie. Pp. 149–153 *in* R. E. Masters and K. E. M. Galley (editors), Proceedings of the 23rd Tall Timbers fire ecology conference: fire in grassland and shrubland ecosystems. Tall Timbers Research Station, Tallahassee, FL.

Pitman, J. C., C. A. Hagen, R. J. Robel, T. M. Loughin, and R. D. Applegate. 2005. Location and success of Lesser Prairie-Chicken nests in relation to vegetation and human disturbance. Journal of Wildlife Management 69:1259–1269.

Robbins, M. B., A. T. Peterson, and M. A. Ortega-Huerta. 2002. Major negative impacts of early intensive cattle stocking on tallgrass prairies: the case of the Greater Prairie-Chicken (*Tympanuchus cupido*). North American Birds 56:239–244.

Robel, R. J. 1970. The possible role of behavior in regulating Greater Prairie-Chicken populations. Journal of Wildlife Management 34:306–312.

Robel, R. J., J. N. Briggs, J. J. Cebula, N. J. Silvy, C. E. Viers, andf P. G. Watt. 1970. Greater Prairie-Chicken ranges, movements, and habitat usage in Kansas. Journal of Wildlife Management 34:286–306.

Rodgers, R. 2008. Prairie chicken lek surveys—2008. Performance Report, Statewide Wildlife Research and Surveys. Kansas Department of Wildlife and Parks, Pratt, KS.

Sandercock, B. K., K. Martin, and S. J. Hannon. 2005. Life history strategies in extreme environments: comparative demography of alpine and arctic ptarmigan. Ecology 86:2176–2186.

Schoech, S. J., and T. P. Hahn. 2008. Latitude affects degree of advancement in laying by birds in response to food supplementation: a meta-analysis. Oecologia 157:369–376.

Schroeder, M. A., and C. E. Braun. 1991. Walk-in traps for capturing Greater Prairie-Chickens on leks. Journal of Field Ornithology 62:378–385.

Schroeder, M. A., and L. A. Robb. 1993. Greater Prairie-Chicken (*Tympanuchus cupido*). A. Poole, P. Stettenhein and F. Gill (editors), The birds of North America No. 36. Cornell Lab of Ornithology, Ithaca, NY. <http://bna.birds.cornell.edu/bna/species/036> (18 May 2009).

Schroeder, M. A., J. R. Young, and C. E. Braun. 1999. Greater Sage-Grouse (*Centrocercus urophasianus*). A Poole (editor), The birds of North America online No. 425. Cornell Lab of Ornithology, Ithaca, NY. <http://bna.birds.cornell.edu.er.lib.k-state.edu/bna/species/425doi:10.2173/bna.425> (28 March 2006).

Silvy, N. J., M. E. Morrow, E. Shanley, and R. D. Slack. 1990. An improved drop net for capturing wildlife. Proceedings of the Annual Conference of the Southeastern Association of Fish and Wildlife Agencies 44:374–378.

Smith, E. F., and C. E. Owensby. 1978. Intensive-early stocking and season-long stocking of Kansas Flint Hills range. Journal of Range Management 31:14–17.

Sokal, R. R., and F. J. Rohlf. 2000. Biometry. 3rd ed. W. H. Freeman and Company, New York, NY.

Sopuck, L. G., and F. C. Zwickel. 1983. Renesting in adult and yearling Blue Grouse. Canadian Journal of Zoology 61:289–291.

Storaas, T., P. Wegge, and L. Kastdalen. 2000. Weight-related renesting in Capercaillie *Tetrao urogallus*. Wildlife Biology 6:299–303.

Svedarsky, W. D. 1983. Reproductive chronology of Greater Prairie-Chickens in Minnesota and recommendations for censusing and nest searching. Prairie Naturalist 15:120–124.

Svedarsky, W. D. 1988. Reproductive ecology of female Greater Prairie-Chickens in Minnesota. Pp. 193–239 *in* A. T. Bergurud and M. W. Gratson (editors), Adaptive strategies and population ecology of northern grouse. Vol. I. University of Minnesota Press, Minneapolis, MN.

Svedarsky, W. D., J. E. Toepfer, R. L. Westemeier, and R. J. Robel. 2003. Effects of management practices on grassland birds: Greater Prairie-Chicken. Northern Prairie Wildlife Research Center, Jamestown, ND. <http://www.npwrc.usgs.gov/resource/literatr/grasbird/Greater Prairie-Chicken/Greater Prairie-Chicken.htm> (28 May 2004).

Svedarsky, W. D., T. J. Wolfe, and J. E. Toepfer. 1999. Status and Management of the Greater Prairie-Chicken in Minnesota. Pp. 25–38 *in* W. D. Svedarsky, R. H. Hier, and N. J. Silvy (editors), The Greater Prairie-Chicken: a national look. Minnesota Agricultural Experiment Station Miscellaneous Publication 99-1999. University of Minnesota, St. Paul, MN.

Welty, J. C., and L. Baptista. 1988. The life of birds. 4th ed. Saunders College Publishing, Orlando, FL.

Westemeier, R. L., and S. Gough. 1999. National outlook and conservation needs for Greater Prairie-Chickens. Pp. 169–187 *in* W. D. Svedarsky, R. H. Hier, and N. J. Silvy (editors), The Greater Prairie-Chicken: a national look. Minnesota Agricultural Experiment Station Miscellaneous Publication 99-1999. University of Minnesota, St. Paul, MN.

Wisdom, M. J., and L. S. Mills. 1997. Sensitivity analysis to guide population recovery: prairie-chickens as an example. Journal of Wildlife Management 61:302–312.

With, K. A., A. W. King, and W. E. Jensen. 2008. Remaining large grasslands may not be sufficient to prevent grassland bird declines. Biological Conservation 141:3152–3167.

Yom-Tov, Y. 2001. An updated list and some comments on the occurrence of intraspecific nest parasitism in birds. Ibis 143:133–143.

CHAPTER SIXTEEN

Regional Variation in Nesting Success of Lesser Prairie-Chickens

*Eddie K. Lyons, Ryan S. Jones, John P. Leonard, Benjamin E.
Toole, Robert A. McCleery, Roel R. Lopez, Markus J. Peterson,
Stephen J. DeMaso, and Nova J. Silvy*

Abstract. Declines in Lesser Prairie-Chicken
(*Tympanuchus pallidicinctus*) populations have
been attributed to loss or fragmentation of habitat and conversion of native prairie to agricultural
cropland, and have been exacerbated by improper
grazing practices and drought. Loss of adequate
vegetation for nesting and brooding of Lesser
Prairie-Chickens have accelerated population
declines observed in the Texas Panhandle. We
monitored 114 female radio-marked Lesser
Prairie-Chickens in the Texas Panhandle from
2001 to 2007 to determine if nest success differed
in two regions (northeastern and southwestern)
of the Texas Panhandle. We used an information-
theoretic approach to test hypotheses explaining differences in nest success of Lesser Prairie-
Chickens in each region. To evaluate differences
between successful and unsuccessful nests, we
measured vegetative height, plant species at nest,
and visual obstruction readings (VOR) at each nest
and at random points. Nest success was significantly
($P = 0.040$) lower in the southwestern region
(38%) compared to the northeastern region (67%).
Evaluating factors influencing nest success, we
found that parameters examined did not explain
differences in nesting success. However, we
found nest locations had higher VOR then random sites in both the northeastern ($\bar{x} = 35$ cm,
SE = 2.3 vs. 21 cm, SE = 2.4) and southwestern
($\bar{x} = 18$ cm, SE = 2.4 vs. 10 cm, SE = 1.1) regions.
Height at nest locations ($\bar{x} = 44$ cm, SE = 1.7) was
greater than at random sites ($\bar{x} = 32$ cm, SE = 1.8)
for the southwestern region, but not the northeastern region ($\bar{x} = 52$ cm, SE = 3.9; $\bar{x} = 60$ cm,
SE = 8.2, respectively). Height and VOR at both
nest sites and random locations were higher in
the northeastern region than in the southwestern region, indicating more cover and possibly
explaining the greater nest success in the northeastern region. The effects of drought appeared to
affect nesting attempts, nest success, and renesting in both regions during our study. To increase
populations of Lesser Prairie-Chickens in Texas,
we recommend managers focus on providing vegetation with adequate height and visual structure
for successful nesting.

Key Words: Lesser Prairie-Chicken, nest success,
radiotelemetry, reproduction, Texas, *Tympanuchus
pallidicinctus*, vegetation type.

Lyons, E. K., R. S. Jones, J. P. Leonard, B. E. Toole, R. A. McCleery, R. R. Lopez, M. J. Peterson, S. J. DeMaso, and N. J. Silvy.
2011. Regional variation in nesting success of Lesser Prairie-Chickens. Pp. 223–231 *in* B. K. Sandercock, K. Martin, and
G. Segelbacher (editors). Ecology, conservation, and management of grouse. Studies in Avian Biology (no. 39), University
of California Press, Berkeley, CA.

innated grouse (*Tympanuchus* spp.) populations have declined throughout their range,
and many are considered species of concern
(Storch 2007). Declines in distribution and abundance of Sharp-tailed Grouse (*T. phasianellus*),
Greater Prairie-Chicken (*T. cupido*), and Lesser
Prairie-Chicken (*T. pallidicinctus*) populations have
been extensively documented (Taylor and Guthery
1980, Johnsgard 1983, Schroeder and Robb 1993,
Connelly et al. 1998, Silvy et al. 2004). Given their
historically small range, relatively small population
sizes, and continued declines in abundance, Lesser
Prairie-Chickens were listed as a candidate species in 1998 by the U.S. Fish and Wildlife Service
(Federal Register 1998, Hagen and Giesen 2005)
and placed on the International Union for Conservation of Nature and Natural Resources (IUCN) red
list in 2004 (IUCN 2004, Storch 2007). Declines in
Lesser Prairie-Chicken abundance have been attributed to habitat fragmentation, improper livestock
grazing, and land conversion from rangelands to
agricultural cropland (Crawford 1980, Taylor and
Guthery 1980, Hagen et al. 2004).

Historically, Lesser Prairie-Chickens occupied rangelands throughout the Texas panhandle
(Oberholser 1974, Litton et al. 1994). Changing
land-use practices have left Lesser Prairie-Chickens
in ranges dominated by woody species such as shinnery oak (*Quercus havardii*), resulting in small, isolated populations (McCleery et al. 2007). The extant
range, in Texas, consists of two disjunct metapopulations in portions of ~11 counties (Taylor and
Guthery 1980, Sullivan et al. 2000, Silvy et al. 2004).
The majority of birds are located in the northeastern

portion of the Texas Panhandle in native prairie
dominated by bunchgrasses with small amounts
of sand sagebrush (*Artemisia filifolia*), and a second
smaller population inhabiting shinnery oak rangelands of the southwestern Panhandle.

Although the mechanisms responsible for
declining Lesser Prairie-Chicken abundance are
not completely understood, previous research on
other prairie grouse has found nest success and
brood survival to be significant factors influencing grouse numbers (Bergerud and Gratson 1988,
Peterson and Silvy 1996). Numerous studies
have documented nest success of Lesser Prairie-
Chickens across their range in varying habitats (Sell 1979, Haukos 1988, Hagen et al. 2004,
Hagen and Giesen 2005), but no recent studies
have evaluated nest success in the two remaining
Lesser Prairie-Chicken populations in Texas.

Because of uncertainty surrounding recovery
efforts, we initiated field studies to determine
whether Lesser Prairie-Chicken nest success
differed between two populations in separate
regions of the Texas Panhandle. The objectives of
our study were to (1) investigate spatial variation
in the nest success of Lesser Prairie-Chickens,
and (2) determine what vegetation components
may influence nest success in two Lesser Prairie-
Chicken populations in different habitats.

STUDY AREAS

We conducted our study from April 2001 through
August 2007 in two areas in the Texas Panhandle
(Fig. 16.1). In 2001, trapping sites were located in

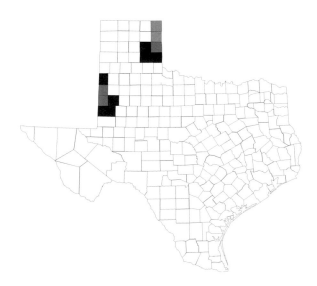

Figure 16.1. Current distribution (black) of Lesser Prairie-
Chickens in 11 counties of
Texas (after Silvy et al. 2004).
Gray areas indicate counties
where our study areas were
located, 2001–2007.

portions of Hemphill (36°01′ N, 100°11′ W) and Wheeler (35°33′ N, 100°06′ W) counties (northeastern region). In 2002, we expanded trapping to include the southern portion of Lipscomb County (36°07′ N, 100°03′ W), Texas, and added Yoakum and southern Cochran counties (33°23′ N, 102°50′ W) (southwestern region) in 2003.

The northeastern region was dominated by native prairie with sand sagebrush as the dominant woody species, with lesser amounts of Chickasaw plum (*Prunus angustifolia*) and fragrant sumac (*Rhus aromatica*). The southwestern region was dominated by shinnery oak with small amounts of sand sagebrush. Both regions contained similar grass and forb associations, as described by Jackson and DeArment (1963). However, the southwestern region had vegetative cover of 50% woody vegetation (83% shinnery oak and 17% sand sagebrush), 11% grass, 2% forbs, and 37% bare ground (Leonard 2008), whereas the northeastern region had 2% woody vegetation, 44% grass, 4% forbs, 35% bare ground, and 15% litter (Toole 2005). Common herbaceous species in both regions included little bluestem (*Schizachyrium scoparium*), big bluestem (*Andropogon gerardii*), sand bluestem (*A. hallii*), sand lovegrass (*Eragrostis tichodes*), sand dropseed (*Sporobolus cryptandrus*), and three awn (*Aristida* spp.). Common forbs included camphorweed (*Heterotheca pilosu*), Texas croton (*Croton texensis*), western ragweed (*Ambrosia psilostachya*), and queensdelight (*Stillingia sylvatica*). Taxonomic nomenclature follows Gould (1962).

Our study areas ranged from 5,000 to 18,000 ha and were bordered by center-pivot irrigated cropland, Conservation Reserve Program lands (CRP), and grazed rangelands. Primary land uses were ranching and natural gas and oil extraction. Environmental conditions were similar across both study regions. Average precipitation across the regions was approximately 48 cm/year during our study (NOAA 2009). A severe drought occurred at both sites in 2003 (NOAA 2009).

METHODS

Data Collection

We captured female Lesser Prairie-Chickens using non-explosive Silvy drop nets (Silvy et al. 1990) on leks prior to and during the breeding season from late March to 1 June during 2001 through 2007. At capture, we identified birds as yearling or adult based on shape, wear, and coloration of the ninth and tenth primaries (Amman 1944, Copelin 1963). We equipped each female with a numbered leg band, and a 12–15-g battery-powered, mortality-sensitive radio transmitter. We used two models of necklace-style radio transmitters: non-adjustable collar-style radio transmitters with fixed-loop antennas (Telemetry Solutions, Walnut Creek, CA) and adjustable collar-style transmitters with whip antennas (Wildlife Materials Inc., Carbondale, IL).

We monitored Lesser Prairie-Chickens a minimum of three days per week throughout the study using vehicle-mounted five-element Yagi antennas. Observations were increased to ≥5 times per week during the spring and early summer to allow better monitoring of nesting activity. Nests were located on foot using three-element handheld Yagi antennas after hen locations remained unchanged for approximately three days. We recorded clutch size if the hen flushed off the nest, but did not intentionally flush hens to obtain these data. We marked each nest with a handheld global positioning system unit, and nest sites were not visited again until the hen left the nest or was depredated. At that time, we relocated nests and identified their fate as abandoned, destroyed, or hatched.

After nest fate was identified, we measured nest site characteristics to evaluate differences between vegetation characteristics at successful and unsuccessful nests. At each nest site, we recorded vegetative height (of tallest material) in centimeters and species of plant providing cover to the nest bowl. We also used a range pole (Robel et al. 1970), demarked at 10-cm (1-dm) intervals, to estimate visual obstruction readings (VOR) placed in the center of the nest bowl and viewed from a height of 1 m and a distance of 4 m in four cardinal directions. We also collected VOR measurements from random points by determining a random direction (1 of 8 cardinal directions) and random distance (200–800 m in 100-m increments) to evaluate differences in vegetation at the nest compared to random locations.

Statistical Analysis

We evaluated vegetation differences between successful (incubating females with ≥1 egg hatched) and unsuccessful nests for each region separately using logistic regression (PROC GENMOD, SAS version 9.1; SAS Institute, Inc., Cary, NC; dependent variable = nest success, independent variables = candidate models). We compared candidate models using an information-theoretic approach (Burnham and Anderson 2002). We used this approach to evaluate the influence of temporal factors (Year) as well as characteristics of the nest, including VOR, height of plants at nest bowl (Height), and the species of plant providing cover to the nest bowl (Species). These four variables were combined into eight candidate models: (1) Global (all four variables considered without interactions), (2) Null (intercept only model), (3) Year, (4) Height, (5) VOR, (6) Species covering the nest, (7) VOR + Height, and (8) Species covering the nest * Height (the interactive effects of these two variables). We evaluated the fit of each model using Akaike weights (w_i) and Akaike's Information Criterion corrected for small sample size (AIC_c; Simonoff 2003), and considered models with a ΔAIC_c <2 as equally parsimonious models (Burnham and Anderson 2002).

We calculated apparent nesting rate (hens for which we found nests divided by number of hens available to nest), apparent nest success (of nests found, percent hatching at least 1 egg), and daily nest survival using a maximum likelihood estimator of the survival rate (Bart and Robson 1982) using the software routines of Krebs (1999). We used the mean laying period (11 days, assuming mean clutch size of 11 eggs with 1 egg laid/day) and mean incubation period (25 days) to estimate nest survival across the laying–incubation period (36 days). Variances were approximated for daily nest survival (36-day period) using the delta method (Powell 2007). Additionally, we used a chi-square test to determine if differences existed in frequency of successful and unsuccessful nests between regions and t-tests to determine whether differences existed in the VOR and height of vegetation between nest bowls and random sites and between regions.

RESULTS

We trapped 114 female Lesser Prairie-Chickens over the course of the study; 52 hens produced 57 nests of which 27 (47%) were successful. Of 40 females trapped in the northeastern region (2001–2003), four lost their radio transmitters before nests were located, two were killed by predators before they nested, two radios stopped transmitting (may have been destroyed during predation) before nests were located, and 12 did not nest (7 during the 2003 drought) or had nests destroyed during laying. Twenty hens incubated 21 first nests and one renest, of which 14 hatched, 2 were abandoned, 5 were destroyed by predators, and one hen was killed near presumed hatch date but the nest could not be relocated to determine nest fate. Mean hatching date for first nests ($n = 11$) in the northeastern region was 24 May (95% CI = 18–30 May). Apparent nesting rate for the northeastern Texas Panhandle was 63% and apparent nest success was 67%. Maximum likelihood daily nest survival was 0.983 (SE = 0.006) and calculated laying–incubation period (36 days) survival was 54% (95% CI = 31–78%) for the northeastern region. Of the eight females that were unsuccessful, five were killed while nesting, and of the three available to renest (number starting second nest after first nest failed), two had first nests destroyed late in incubation and only one renested during the study. All nests in the northeastern Texas Panhandle were located in clumps of little bluestem.

In the southwestern Texas Panhandle, we trapped and radio-monitored 65 female Lesser Prairie-Chickens from 2003 through 2007. Of these 65 females, four lost radios before nests were located, five were killed by predators before they nested, four radios stopped transmitting (may have been destroyed during predation) before nests were located, and 20 did not nest or had nests destroyed during laying. A total of 32 incubated first nests and three renests (one hen was captured late in breeding season with brood patch; therefore, the nest was considered a renest), of which 13 hatched, 2 were abandoned, and 19 were destroyed by predators. Mean date of hatch for first nests in the southeastern region was 25 May (95% CI = 20–30 May). Two renests in the southwestern region hatched on 5 and 7 July. Overall nesting rate for the southwestern Texas Panhandle was 62% and nest success was 38%. Maximum likelihood daily nest survival was 0.965 (SE = 0.007) and calculated laying–incubation period (36 days) survival was 29% (95% CI = 16–46%) for the southwestern region. The majority of nests (31) were established under woody shrubs. Twenty-two hens used sand sagebrush for nest cover, while nine nests were established under shinnery oak plants. Five of the

TABLE 16.1

Means and 95% confidence limits (in parentheses) for height and visual obstruction readings (VOR) at nest bowl and random
sites at successful and unsuccessful Lesser Prairie-Chicken nests by region, Texas Panhandle, 2001–2007.

Region/Fate	n	VOR at Nests	VOR at Random	Height at Nests	Height at Random
Northeastern	21	35 (31–39)	21 (16–26)	52 (44–60)	60 (44–76)
Successful	14	36 (31–41)	19 (14–25)	51 (43–59)	68 (53–84)
Unsuccessful	7	35 (24–45)	24 (23–25)	53 (34–72)	50 (38–61)
Southwestern	32	18 (13–23)	10 (8–12)	44 (41–47)	36 (32–40)
Successful	12	20 (14–26)	10 (7–13)	45 (42–48)	38 (27–48)
Unsuccessful	20	17 (10–24)	10 (7–13)	43 (38–49)	31 (26–36)

nests under woody shrubs also had bunchgrasses (little bluestem and threeawn) associated with the nest bowl. The remaining three nests were established in grass and other vegetation (one in weeping lovegrass, *Eragrostis curvula*, CRP; one in little bluestem; and one under a Buckley's yucca, *Yucca constricta*).

Apparent nest success in the northeastern region (67%, 95% CI = 43–85%) differed significantly (χ^2 = 4.199, df = 1, P = 0.040) from that in the southwestern region (38%, 95% CI = 22–56%); however, maximum likelihood daily survival estimates (0.983 and 0.965, respectively, for northeastern and southwestern regions) did not differ (χ^2 = 3.288, df = 1, P = 0.070) between regions. Evaluating factors influencing nest success for each region, we found the Null model had the lowest AIC value among the candidate models, indicating that it was the best fit for the data. Additionally, the 95% CIs of parameter estimates for all parameters in both regions contained zero. These combined findings indicated that parameters examined did not explain differences in nesting success. Models including the species of plant providing cover to the nest bowl were eliminated from the analysis of the northeastern region because all nests were covered by little bluestem.

In the northeastern region, VOR was significantly (t = 5.49, P < 0.001) higher at nest sites ($\bar{x}$ = 35 cm) than at random sites ($\bar{x}$ = 21 cm; Table 16.1). Similarly, VOR also was significantly (t = 3.55, P = 0.001) higher at nest sites ($\bar{x}$ = 18 cm) than at random sites ($\bar{x}$ = 10 cm) in the southwestern region. However, VOR at the nest bowl and at random points in the northeastern region was significantly higher than at the nest bowl and

random sites in the southwestern region (t = –5.19, P < 0.001 and t = –4.00, P < 0.001, respectively).

Vegetation height ($\bar{x}$ = 44 cm) at the nest bowl in the southwestern region (Table 16.1) was significantly (t = 4.28, P < 0.001) higher than at random sites ($\bar{x}$ = 36 cm), whereas vegetation height ($\bar{x}$ = 52 cm) at the nest bowl in the northeastern region was not significantly (t = –0.93, P = 0.360) different than at random sites ($\bar{x}$ = 60 cm). Vegetation height at the nest bowl in the northeastern region was not significantly (t = –1.77, P = 0.089) different than in the southwestern region, but vegetation height at random points in the northeastern region was significantly (t = 2.90, P = 0.009) higher than at random points in the southwestern region (Table 16.1).

DISCUSSION

Nest success of Lesser Prairie-Chickens during our study was higher (47%) than some estimates from other portions of this species' range (27%, Merchant 1982; 28%, Riley et al. 1992; 26%, Pitman et al. 2006). Riley (1978), Patten et al. (2005), and Fields et al. (2006), however, documented quite similar apparent nest success (47%, 41%, and 48%, respectively), and Copelin (1963) and Davis (2009) reported higher values (67% and 76%, respectively). Merchant (1982) documented 54% nest success during a year of average precipitation, whereas no nests were successful during a severe drought year. Hagen and Giesen (2005) estimated nest success of Lesser Prairie-Chickens at 28% based on ten studies conducted throughout this species' range, although they cautioned that these results may have been negatively influenced by observer disturbance. We did not consider disturbance to be a factor in our study, as most birds were not flushed from their nests and nests were not visited a second time until nest fate was determined.

Our apparent low nest initiation rate combined with relatively high nest success may be biased as we did not locate nests for 20 hens in the southwestern region and 12 hens in the northeastern region. These hens may have initiated laying but had their nests destroyed before incubation began. With our maximum likelihood daily nest survival estimate being lower than the apparent nest survival estimate, this could be a partial explanation. However, we also determined through back-dating from hatch date the date when incubation began and found for both regions that we located nests during the first week of incubation for both regions. Thus, it also is likely that some hens did not nest or abandoned nests early in incubation because of the droughts of 2003 and 2006. Precipitation during the five months prior to the 2003 nesting season was 23% and 88% below the 30-year normal in the northeastern (7 hens not nesting) and southwestern regions, respectively. During 2006, precipitation during the five months prior to nesting was 84% below the 30-year normal in the southwestern region, and only 10 of 24 females were known to nest. Two of 10 nests were abandoned early in incubation, and for the first time during our study, both males and females left the study area. We located radio-tagged birds up to 8 km from the display grounds where they were trapped earlier in 2006. During 2007, only four females were observed on display grounds in the southwestern region, indicating females either died (7 deaths documented) during 2006 or did not return to the study area. Patten et al. (this volume, chapter 4) found that greater movements of Greater Prairie-Chickens led to higher mortality, especially among females. Drought appeared to play a major role in lack of nesting attempts, nest success, and renesting during our study. Similarly, Wolfe et al. (2007) maintained that low Lesser Prairie-Chicken nest success, smaller clutch sizes, and no documented renesting were due primarily to severe 2006 drought conditions on their study areas in Oklahoma. They also reported that effects of the drought carried into the 2007 nesting season, when only three nests were located and no renesting was observed.

At a regional scale, we found that Lesser Prairie-Chicken nests in the northeastern Texas Panhandle were more successful than those in the southeastern Panhandle. The southwestern region was dominated by woody vegetation (shinnery oak), whereas the northeastern region was dominated by grass (bunchgrasses). The southwestern region also had less grass groundcover (11%) than did the northeastern region (44%), where all nests were located in little bluestem clumps, even though woody plants (e.g., sand sagebrush, shinnery oak, and plum) as tall or taller than the little bluestem clumps were observed in 69% of 153 (m^2) vegetation plots (Toole 2005, Leonard 2008). It appears that in areas where sufficient bunchgrasses are present, Lesser Prairie-Chickens prefer to nest in bunchgrasses (Jones 2009). Cannon and Knopf (1981) found that Lesser Prairie-Chicken densities were positively correlated with percent grass cover and negatively correlated with brush frequency and density in shinnery oak grassland. Lesser Prairie-Chicken populations in Kansas are known to inhabit grass prairies generally devoid of brush (Fields et al. 2006, Pitman et al. 2006).

Although there were no differences in VOR at successful and unsuccessful nests within either the northeastern and southwestern regions, the mean VOR from random points in the southwestern region was 10 cm, whereas the mean in the northeastern region was 21 cm. It appears that females in both regions selected nest sites in areas where VOR was maximized. However, because mean VOR was higher at nest sites in the northeastern region (35 cm) than at nest sites in the southwestern region (18 cm), it is not surprising that nest success was higher in the northeastern region. Nests in the southwestern region placed under shinnery oak or sand sagebrush had little grass cover, whereas those in the northwestern region placed within little bluestem clumps, which provided greater VOR, apparently resulted in greater nest success. Vegetation height at nest sites did not appear to contribute to the difference in nest success between regions, as it did not differ by region. Nonetheless, height at nest sites in the southwestern region was greater than height at random sites, which may have helped predators find nests by searching taller vegetation areas.

Adequate vegetation structure for nesting is probably the most important factor determining nest success of Lesser Prairie-Chickens (Kirsch

1974). Lutz and Silvy (1980), for example, found that predation of Attwater's Prairie-Chicken (*T. c. attwateri*) nests was greater in areas with lower VOR. Our findings are similar to previous research documenting the importance of vegetative structure to the success of Lesser Prairie-Chicken nests (Haukos and Smith 1989, Fields et al. 2006, Pitman et al. 2006). Improvements in habitat quality and quantity are needed to provide sufficient cover to reduce nest predation for Lesser Prairie-Chickens in Texas.

Within the northeastern region, we found Lesser Prairie-Chickens nested exclusively in little bluestem clumps, even through woody vegetation was found in 69% of all m² quadrats, with mean height of 67 cm (Toole 2005). However, within the shinnery oak vegetation type (height <0.1–2 m), we found that Lesser Prairie-Chickens nested more often under sand sagebrush plants (22 nests) than under shinnery oak plants (9 nests), where shinnery oak and sand sagebrush comprised 83% and 17%, respectively, of woody plants (Leonard 2008). Sell (1979), working in Yoakum County, Texas (shinnery oak vegetation type), found Lesser Prairie-Chickens preferred sand sagebrush for nest concealment and recommended that nesting cover in the form of sand sagebrush and residual grass cover be provided. Conversely, Crawford and Bolen (1975) reported a mix of native shinnery oak dominated rangeland and grain farming provided better habitat than 100% native rangeland. The authors suggested landscapes with <63% native rangeland were incapable of supporting Lesser Prairie-Chicken populations. More recently, Haukos and Smith (1989) reported that rangelands with <50% shinnery oak overhead cover were ideal for Lesser Prairie-Chickens. Shinnery oak also competes with food and cover plant species that are beneficial to Lesser Prairie-Chickens, and can comprise 90% of vegetation on heavily grazed rangelands (Pettit 1979).

Microhabitat use of shinnery oak rangelands by Lesser Prairie-Chickens is poorly understood; the presence of shinnery oak is cited as both beneficial (Sell 1979, Haukos and Smith 1989) and detrimental (Donaldson 1969, Martin 1990). Changes in shinnery oak age, composition, and structure may account for these conflicting results, and may also explain declining Lesser Prairie-Chicken abundance in the southwestern Texas Panhandle. When shinnery oak comprises <50% coverage of the area with a height shorter than that of the dominant grass (little bluestem), it is suitable habitat for Lesser Prairie-Chicken. However, as shinnery oak matures and concomitantly increases in density (>50% coverage) and height (>1.5 m), it often totally dominates an area. Mature oaks lead to the exclusion of important grasses and forbs through competition for space and limited moisture, especially during drought years that support invertebrates needed by Lesser Prairie-Chicken chicks.

To increase Lesser Prairie-Chicken numbers in Texas, we recommend that managers focus on providing conditions that maximize vegetative diversity and structure for successful nesting. To improve nesting success in nearby New Mexico habitats, Riley et al. (1992) recommended that managers increase grass cover at the expense of shinnery oak cover. A better understanding of how components of the maturation of shinnery oak including height, density, and structure influence dynamics of Lesser Prairie-Chicken populations is imperative to the recovery of the species.

ACKNOWLEDGMENTS

We are grateful to the land owners and land managers who allowed us access to their properties over the course of this study. We also are grateful to Duane Lucia (Texas Parks and Wildlife Department) and John Hughes (U.S. Fish and Wildlife Service) for logistical support. This project was funded by Texas Parks and Wildlife Department and Texas A&M University System.

LITERATURE CITED

Amman, G. A. 1944. Determining age of pinnated and Sharp-tailed Grouse. Journal of Wildlife Management 8:170–171.

Bart, J., and D. S. Robson. 1982. Estimating survivorship when the subjects are visited periodically. Ecology 63:1078–1090.

Bergerud, A. T., and M. W. Gratson. 1988. Adaptive strategies and population ecology of northern grouse, Vol. II: Theory and synthesis. University of Minnesota Press, St. Paul, MN.

Burnham, K. P., and D. R. Anderson. 2002. Model selection and multi-model inference: a practical information-theoretic approach. 2nd Ed. Springer Science + Business Media, New York, NY.

Cannon, R. W., and F. L. Knopf. 1981. Lek numbers as a trend index to prairie grouse populations. Journal of Wildlife Management 45:776–778.

Connelly, J. W., M. W. Gratson, and K. P. Reese. 1998. Sharp-tailed Grouse (*Tympanuchus phasianellus*).

A. Poole and F. Gill (editors), The birds of North America No. 354. Academy of Natural Science, Philadelphia, PA.

Copelin, F. F. 1963. The Lesser Prairie Chicken in Oklahoma. Technical Bulletin 6. Oklahoma Wildlife Conservation Department, Oklahoma City, OK.

Crawford, J. A. 1980. Status, problems, and research needs of the Lesser Prairie Chickens. Pp. 1–7 *in* P. A. Vohs, Jr., and F. L. Knopf (editors), Proceedings of the Prairie Grouse Symposium, Oklahoma State University, Stillwater, OK.

Crawford, J. A., and E. G. Bolen. 1975. Spring lek activity of the Lesser Prairie Chicken in west Texas. Auk 92:808–810.

Davis, D. M. 2009. Nesting ecology and reproductive success of Lesser Prairie-Chickens in shinnery oak-dominated rangelands. Wilson Journal of Ornithology 121:322–327.

Donaldson, D. D. 1969. Effect on Lesser Prairie Chickens of brush control in western Oklahoma. Ph.D. dissertation, Oklahoma State University, Stillwater, OK.

Federal Register. 1998. Endangered and threatened wildlife and plants: 12-month finding for a petition to list the Lesser Prairie-Chicken as threatened and designate critical habitat. Federal Register 63:31400–31406.

Fields, T. L., G. C. White, W. C. Gilbert, and R. D. Rodgers. 2006. Nest and brood survival of Lesser Prairie-Chickens in west central Kansas. Journal of Wildlife Management 70:931–938.

Gould, F. W. 1962. Texas plants: a checklist and ecological summary. Bulletin MS-585. Texas Agricultural Experiment Station, College Station, TX.

Hagen, C. A., and K. M. Giesen. 2005. The Lesser Prairie-Chicken (*Tympanuchus pallidicinctus*). Rev. ed. A. Poole and F. Gill (editors), The birds of North America No. 364. Academy of Natural Science, Philadelphia, PA.

Hagen, C. A., B. E. Jamison, K. M. Fiesen, and T. Z. Riley. 2004. Guidelines for managing Lesser Prairie-Chicken populations and their habitats. Wildlife Society Bulletin 32:69–82.

Haukos, D. A. 1988. Reproductive ecology of Lesser Prairie Chickens in west Texas. M. S. thesis, Texas Tech University, Lubbock, TX.

Haukos, D. A., and L. M. Smith. 1989. Lesser Prairie Chicken nest site selection and vegetation characteristics in tebuthiuron-treated and untreated sand shinnery oak in Texas. Great Basin Naturalist 49:624–626.

IUCN. 2004. 2004 Red list of threatened species. <http://www.redlist.org>.

Jackson, A. S., and R. DeArment. 1963. The Lesser Prairie Chicken in the Texas Panhandle. Journal of Wildlife Management 27:733–737.

Johnsgard, P. A. 1983. The grouse of the world. University of Nebraska, Lincoln, NE.

Jones, R. S. 2009. Seasonal survival, reproduction, and use of wildfire areas by Lesser Prairie Chickens in the northeastern Texas Panhandle. M. S. thesis, Texas A&M University, College Station, TX.

Kirsch, L. M. 1974. Habitat management considerations for prairie chickens. Wildlife Society Bulletin 2:124–129.

Krebs, C. J. 1999. Ecological methodology. 2nd ed. Benjamin/Cummings, New York, NY.

Leonard, J. P. 2008. The effects of shinnery oak removal on Lesser Prairie Chicken survival, movement, and reproduction. M. S. thesis, Texas A&M University, College Station, TX.

Litton, G. W., R. L. West, D. F. Dvorak, and G. T. Miller. 1994. The Lesser Prairie Chicken and its management in Texas. Booklet 7100-025. Texas Parks and Wildlife Department, Austin, TX.

Lutz, R. S., and N. J. Silvy. 1980. Predator response to artificial nests in Attwater's Prairie Chicken habitat. North American Prairie Grouse Conference 1:48–51.

Martin, B. H. 1990. Avian and vegetation research in the shinnery oak ecosystem of southeastern New Mexico. Thesis, New Mexico State University, Las Cruces, NM.

McCleery, R. A., R. R. Lopez, and N. J. Silvy. 2007. Transferring research to endangered species management. Journal of Wildlife Management 71:2134–2141.

Merchant, S. S. 1982. Habitat-use, reproductive success, and survival of female Lesser Prairie Chickens in two years of contrasting weather. M. S. thesis, New Mexico State University, Las Cruces, NM.

National Oceanic and Atmospheric Administration (NOAA). 2009. <http://www.noaa.gov/> (1 September 2009).

Oberholser, H. C. 1974. The bird life of Texas. University of Texas Press, Austin, TX.

Patten, M. A., D. H. Wolfe, E. Shochat, and S. K. Sherrod. 2005. Effects of microhabitat and microclimate on adult survivorship of the Lesser Prairie-Chicken. Journal of Wildlife Management 36:1270–1278.

Peterson, M. J., and N. J. Silvy. 1996. Reproductive stages limiting productivity of the endangered Attwater's Prairie Chicken. Conservation Biology 4:1264–1276.

Pettit, R. D. 1979. Effects of picloram and tebuthiuron pellets on sand shinnery oak communities. Journal of Range Management 32:196–200.

Pitman, J. C., C. A. Hagen, B. E. Jamison, R. J. Robel, T. M. Loughin, and R. D. Applegate. 2006. Nesting ecology of Lesser Prairie-Chickens in sand sagebrush prairie of southwestern Kansas. Wilson Journal of Ornithology 118:23–35.

Powell, L. A. 2007. Approximating variance of demographic parameters using the delta-method: a reference for avian biologists. Condor 109:950–955.

Riley, T. Z., 1978. Nesting and brood-rearing habitat of Lesser Prairie Chickens. M. S. thesis, New Mexico State University. Las Cruces, NM.

Riley, T. Z., C. A. Davis, M. Ortiz, and M. J. Wisdom. 1992. Vegetative characteristics of successful and unsuccessful nests of Lesser Prairie Chickens. Journal of Wildlife Management 56:383–387.

Robel, R. J., J. N. Briggs, A. D. Dayton, and L. C. Hulbert. 1970. Relationships between visual obstruction measurements and weight of grassland vegetation. Journal of Range Management 23:295–297.

Schroeder, M. A., and L. A. Robb. 1993. Greater Prairie-Chicken (*Tympanuchus cupido*). A. Poole and F. Gill (editors), The Birds of North America No. 36. Academy of Natural Science, Philadelphia, PA.

Sell, D. L. 1979. Spring and summer movements and habitat use by Lesser Prairie Chicken females in Yoakum County, Texas. M. S. thesis, Texas Tech University, Lubbock, TX.

Silvy, N. J., M. E. Morrow, E. Shanley, Jr., and R. D. Slack. 1990. An improved drop net for capturing wildlife. Proceedings of the Annual Conference of the Southeastern Association of Fish and Wildlife Agencies 44:374–378.

Silvy, N. J., M. J. Peterson, and R. R. Lopez. 2004. The cause of the decline of pinnated grouse: the Texas example. Wildlife Society Bulletin 32:16–21.

Simonoff, J. S. 2003. Analyzing categorical statistics. Springer-Verlag, New York, NY.

Storch, I. 2007. Conservation status of grouse worldwide: an update. Wildlife Biology 13(Suppl. 1):5–12.

Sullivan, R. M., J. P. Hughes, and J. E. Lionberger. 2000. Review of the historical and present status of the Lesser Prairie-Chicken (*Tympanuchus pallidicinctus*) in Texas. Prairie Naturalist 32:177–188.

Taylor, M. A., and F. S. Guthery. 1980. Status, ecology, and management of the Lesser Prairie Chicken. U.S. Department of Agriculture Forest Service General Technical Report RM77. Rocky Mountain Forest and Range Experiment Station, Fort Collins, CO.

Toole, B. E. 2005. Seasonal movements and cover use by Lesser Prairie Chickens in the Texas Panhandle. M. S. thesis, Texas A&M University, College Station, TX.

Wolfe, D. H., M. A. Patten, and S. K. Sherrod. 2007. A study of factors affecting nesting success and mortality of Lesser Prairie-Chickens in Oklahoma and removal and marking of fences to reduce collisions. Performance Reports. George M. Sutton Avian Research Center, University of Oklahoma, Bartlesville, OK.

CHAPTER SEVENTEEN

Mechanisms Underlying Variation in Renesting Ability of Willow Ptarmigan

Kathy Martin, Scott Wilson, and Susan J. Hannon

Abstract. Like many ground-nesting birds, ptarmigan are not capable of defending their clutches from most predators, relying instead on cryptic plumage and secretive behaviors to escape predation. Despite these adaptations, ptarmigan experience heavy reproductive losses (40 to 80% of first clutches) but are able to recoup some fecundity by renesting (replacing a clutch after the failure of a previous attempt). We studied two populations of Willow Ptarmigan (*Lagopus lagopus*) that varied in their probability of replacing a clutch, renest interval, and clutch size of renests. At the Chilkat Pass (CP) in subalpine tundra in northwestern British Columbia, where renests accounted for 14% of annual fecundity, renesting ability varied with female age-class (adult or yearling) and stage of loss. At La Pérouse Bay (LPB) in subarctic tundra in Manitoba, where 23% of annual fecundity resulted from renests, probability of renesting was similar for adults and yearlings, and declined when the clutch was lost later in the incubation period. When clutches were lost during laying, females

in both populations had a high probability of renesting, with LPB females able to achieve continuation laying in contrast to deterministic egg-laying for CP females. The latest stage of loss of first clutches after which females renested was day 14 of incubation at CP and day 19 at LPB. The renest interval increased with later stage of failure of first clutches at CP but not at LPB. At both sites, females that laid larger first clutches also produced larger renest clutches. Renest clutch size declined after longer renest intervals at both sites, and also with later dates of failure at LPB. Variation in renesting ability within Willow Ptarmigan populations was based mostly on stage of loss of first clutches, and not to individual traits, except for female age at CP, and with little annual variation in either population. We found no support for a bet-hedging strategy either within or between populations, as larger first clutches were correlated positively with renesting in both populations, and CP birds had lower renesting ability than LPB despite lower daily nest survival of first clutches. We

Martin, K., S. Wilson, and S. J. Hannon. 2011. Mechanisms underlying variation in renesting ability of Willow Ptarmigan. Pp. 233–246 *in* B. K. Sandercock, K. Martin, and G. Segelbacher (editors). Ecology, conservation, and management of grouse. Studies in Avian Biology (no. 39), University of California Press, Berkeley, CA.

found support for a survival–fecundity trade-off in renesting ability between populations since LPB females with the lower annual survival and higher dependence on annual fecundity were more able to replace their clutches after one or multiple failures. Thus, renesting ability may be under stronger selection at LPB than at CP, but in both populations, Willow Ptarmigan exhibited well-developed renesting abilities.

Key Words: arctic tundra, bet-hedging, clutch replacement, follicle development, gonadal recrudescence, *Lagopus lagopus*, renest clutch size, renest interval, renesting propensity, survival–fecundity trade-off.

Over half of first clutches of many bird species, particularly ground-nesting species with precocial young, fail due to predation (Ricklefs 1969, Montgomerie and Weatherhead 1988, Martin 1993). Initiating replacement clutches (renests) is a frequent response to clutch failure, and renests can contribute substantially to annual fecundity (Swanson et al. 1986, Martin et al. 1989, Milonoff 1991, but see Fletcher et al. 2006, Sandercock et al. 2008). Variation in the ability to renest has been examined in multi-brooded species, where subsequent broods greatly increase annual fecundity (Nolan 1978, Grzybowski and Pease 2005). However, there has been much less emphasis on the costs and benefits of renesting in single-brooded species of birds (Parker 1985, Martin et al. 1989, Arnold 1993, Arnold et al. 2010).

Renesting ability may vary in relation to traits of individual females (age, body mass and size, spring body condition, pair bond duration), traits of the initial clutches that failed (clutch size, initiation date, egg size), and on the temporal circumstances related to loss of the first clutches (date of loss and stage of incubation). The ability and time required to replace failed clutches may vary with female age and condition because younger females and birds in poorer condition may need more time to amass resources for a replacement clutch (Gregg et al. 2006, Devries et al. 2008). Females losing clutches later in the season or later in incubation may also require longer renest intervals to produce replacement clutches (Parker 1981, Sopuck and Zwickel 1983, Swanson et al. 1986, Arnold 1993, Fondell et al. 2006, Amat et al. 1999, Arnold et al. 2010).

Renest interval, the time in days between the loss of the first clutch and the initiation of the renest clutch, could also have a strong influence on annual fecundity. Species able to renest quickly after the loss of the first clutch may renest more than once if the second clutch is also depredated. Rapid renesting may be particularly important for populations in areas with short breeding seasons (Ricklefs 1973). In addition, since clutch size declines with advancing date of laying, shorter renest intervals will result in larger renest clutches (Klomp 1970). Rapid replacement of failed clutches also allows more time for offspring to develop before independence and fall migration (Scott et al. 1987).

We advance three hypotheses to explain variation in renesting ability between populations: bet-hedging, resource availability and survival–fecundity trade-off. If clutch predation rates are predictably high, the *bet-hedging hypothesis* predicts selection for a reduction in first clutch size so that individuals can allocate sufficient reproductive effort for a replacement clutch (Stearns 1976; Slagsvold 1982, 1984). Thus, birds in populations with higher failure should have improved renesting ability, a lower first clutch size, and/or a relatively larger renest clutch size (Milonoff 1991). Alternatively, the *resource availability hypothesis* predicts that renesting ability covaries with environmental conditions, such that populations located in more productive or resource-rich sites have higher rates of renesting (Swanson et al. 1986, Eldridge and Krapu 1988, Preston and Rotenberry 2006). Finally, the *survival–fecundity trade-off hypothesis* predicts that birds in populations with lower annual survival should have higher renesting abilities.

In this paper, we examine proximate and ultimate factors that may affect renesting ability in two populations of Willow Ptarmigan (*Lagopus lagopus*), a monogamous, ground-nesting, tundra-dwelling grouse. At Chilkat Pass (CP), British Columbia, average survival of first clutches for Willow Ptarmigan was 0.34 and renests contributed to 14% of annual fecundity (range of 2–27% annually), while at La Pérouse Bay (LPB), Manitoba, average clutch survival was 0.42 and renests contributed to 23% of annual fecundity (Martin et al. 1989, Wilson et al. 2007). In previous papers, we presented data on survival and fecundity parameters in relation to female age-class, including average probability of renesting (hereafter renesting propensity, except when referring to statistical tests) and renest clutch size for these populations. Using comparable data for both studies, the CP site had higher primary productivity and warmer spring temperatures than LPB. CP ptarmigan had stronger age-related fecundity (older females were more fecund) and lower daily nest survival. However, LPB females had greater renesting propensity and larger first and renest clutches, which resulted in greater total annual fecundity (Wiebe and Martin 1998; Sandercock et al. 2005a, 2005b; Wilson et al. 2007). LPB females had 37% annual apparent survival, while survival of females at CP was higher at 43% (Sandercock et al. 2005a). We found few effects of female body condition, type of pair bond (monogamous, polygynous, widow), pair bond duration, or age of male partner on female fecundity or survival parameters (Martin and Cooke 1987, Robb et al. 1992, Hannon and Martin 1996).

We made the following predictions. Within populations, if individual traits influence renesting ability, we expected that older and larger females and birds in better condition would have better renesting abilities than younger and smaller females and those in poor body condition. Given the strong seasonal variation in breeding phenology and first clutch attributes, we expected annual variation in renesting attributes. We expected covariation between some renesting attributes, as shorter renest intervals might result in larger renest clutches if clutch size declines seasonally, but the reverse pattern might occur if more time to acquire nutrients allowed birds to produce larger renest clutches. Between populations, if ecological variables such as predation rates of first clutches or environmental conditions influence renesting, the CP population with higher first clutch failure rates and better environmental conditions should have higher renesting ability than LPB (Sandercock et al. 2005a). Finally, if extrinsic mortality via a survival–fecundity trade-off impacts renesting rates, then females at LPB with slightly lower annual survival should have higher renesting abilities than birds at CP.

METHODS AND STUDY SITE

The study area at Chilkat Pass in northwestern British Columbia (59°50' N, 136°30' W) encompassed 4.5 km² of subalpine tundra during 1985–1988, and was described by Hannon et al. (1988). Martin conducted studies on 10 km² of subarctic tundra at La Pérouse Bay, 40 km east of Churchill,

Manitoba (58°24' N, 94°24' W), from 1981 to 1984. Martin and Cooke (1987) and Hannon et al. (1988) described the study area and general field methods. Field seasons at both sites were from mid-April to early August. There are minor differences between renesting estimates in this paper and earlier published values in Sandercock et al. (2005a, 2005b) or Hannon et al. (1998), given different data selection criteria (e.g., requirement to know exact age of female or natural failure vs. clutch removal). Here, we use the full data set for renest attributes wherever possible.

Study methods and types of data collected were similar at the two sites. More than 90% of birds were captured before incubation and uniquely color-banded. At CP, the majority of the hens were also radio-marked. Sexes were distinguished by differences in voice, plumage, and wing length (Bergerud et al. 1963). Ptarmigan were classified as yearlings (hatched the previous season) or adults (≥2 years of age) based on contrast in pigmentation between primaries 9 and 10 and the rest of the inner primary feathers. At CP, we calculated an index of female spring body condition based on the residuals of a linear regression, with date of capture and wing chord (an index of body size) as covariates. We lacked sufficient data to incorporate body condition at LPB. We did not include mate fidelity or mate age-class in our models, because in an earlier analysis these traits did not influence renesting attributes (Hannon and Martin 1996).

First and renest clutches were found by using pointing dogs or searching around roosts of territorial males (LPB), or by following radio-marked hens (CP). At both sites, we visited territories 2 to 3 times per week. We began actively searching for nests when females exhibited gravid-type behaviors related to egg formation and laying (i.e., more secretive behavior, reluctance to move, or active foraging), and increased our nest searching efforts after the territorial male was found alone. We attempted to find all nests during laying or early incubation, and we searched territories during the times of day when females were most likely to be on nests to lay an egg (i.e., 1000–1800 H; Wiebe and Martin 1995). Most males were reluctant to move in the direction of the nest when his mate was laying or incubating (K. Martin, unpubl. data). At LPB, we relied more on the behavior of territorial males (usually color-banded) to assist us in locating nests, and spent relatively more time searching for first and renests. Sometimes we visited a territory several times in sequential days before locating the nest, and in these instances, we used the behavior and location of the individual male from previous visits. Most nests were visited 2–4 times per week until hatch or clutch failure. We began checking pairs for renesting attempts 4 to 6 days after the first or previous clutch failed. The mean incubation period for Willow Ptarmigan was similar between populations (LPB: 21.2 d, range 19–25 d; CP: 21.7 d, range 19–26 d), with incubation beginning with laying of the penultimate egg (Hannon et al. 1988, Sandercock 1993). Eggs from first nests hatched in late June or early July. Renests, defined as subsequent breeding attempts after failure of first clutches, hatched from mid-July through early August. Clutches that could not be assigned as first nests or renests were deleted from the analysis. Female Willow Ptarmigan at CP and LPB lay 1 egg/d, and egg-laying gaps were rare (Sandercock 1993, Wiebe and Martin 1995). We determined date of first egg directly when we found a nest during laying. When a clutch was located during incubation, we determined the date of first egg by backdating from the stage of incubation estimated with egg flotation (accurate for the first 8 days of incubation; adapted by K. Martin after Westerskov 1956) or by backdating from the date of hatch (Hannon et al. 1988). Clutch sizes for first or renests were assigned only if a female had begun incubation.

At both sites, some renesting attempts likely failed before we found them, and therefore our estimates should be considered minimum rates of renesting. At Chilkat Pass and La Pérouse Bay, respectively, renesting attempts were found on average 5 and 12 days after onset of laying (day 1 and 6 of incubation, respectively). Nests were found at an earlier stage at CP because hens were radio-tagged at CP, therefore requiring less time to locate nests. Our earlier analyses indicated that daily nest survival averaged 0.965 at CP and 0.971 at LPB (Wilson et al. 2007). If extrapolated to the average age at which renests were found, we estimate that we may have missed up to 16% of renests at CP and 30% at LPB.

We monitored renesting after depredation of first clutches at both sites, and recorded date and stage of incubation when loss of first clutches occurred. Since it was rare to observe a predation event, we estimated date of loss of first nests by using signs indicating timing of the predation event (e.g., adult birds still defending nest sites, fresh yolk left in depredated egg, dry nest cup despite recent rain). In the absence of signs, we calculated the date of loss as the day midway between the last visit when the nest was active and the first visit to the depredated nest, if that period was 5 days or less. If the interval exceeded 5 days, a date of loss was not assigned. Stage of loss was estimated in a manner similar to date of loss, with day 1 the first day of egg-laying of the first clutch, and continuing to the end of the incubation period. Stage of loss could only be assigned for those nests where we knew date of first egg or onset of incubation. Since average first clutch size for LPB birds was 3 eggs greater than at CP, the onset of incubation (with laying of the penultimate egg) was day 7 for CP and day 10 for LPB. We measured the renest interval by calculating the time interval in days between the date the first clutch was lost to the date the first egg was laid in the renest clutch. For example, a female that lost her first nest on day 165 and laid her first egg in a renest on day 169 was assigned a renest interval of 3 days.

RENEST ANALYSES

To examine the factors affecting renesting propensity, we used a generalized linear mixed model with a binomial response variable. Data from females with failed first nests were available for 139 attempts at CP and 60 attempts at LPB. At CP, some nests failed because of experimental clutch removal; however, after controlling for age and stage of failure, there was no difference in the renesting propensity for females with natural and experimental failure (AIC_c with type of failure = 1.98 units greater than AIC_c without failure type). Therefore, all first nest failures were included in this analysis.

We initially ran an intercept-only model and then included year as a random variable. The variance accounted for by year was low at both sites (no improvement in the log-likelihood), and therefore the remaining models were run using a generalized linear model with only fixed effects. We then constructed a set of *a priori* candidate models that contained (1) female effects, including female age-class [yearling (second year) or adult (after second year)], size of the first clutch and condition (CP only), and (2) time, including date of first egg, date of failure, and stage of failure. Date of first egg and date of failure were expressed as calendar days since 1 January, and although there was some correlation between these two variables ($r = 0.45$), sufficient variation was present to include both factors in the candidate model set. Stage of failure, measured as days since the first egg was laid, was included to represent how renesting potential was influenced by reproductive effort prior to failure. Our candidate model set allowed us to examine the influence of female effects and timing variables individually as well as with other influential variables. We also included a stage by age interaction model to test whether adults and yearlings differed in their renesting characteristics depending on the stage at which the first nest was lost. We used Akaike's Information Criterion for small samples (AIC$_c$; Burnham and Anderson 2002) to rank candidate models, and the ΔAIC$_c$ and Akaike weights (w_i) to assess model likelihood and uncertainty. We used model averaging to derive parameter estimates, except in cases where one model clearly had the greatest support or if the top models contained interactions. For analysis of renesting probability, we also estimated $\hat{c}$ = residual deviance/degrees of freedom on the global model as an estimate of overdispersion; there was little evidence of overdispersion at either site (CP: $\hat{c}$ = 0.68, LPB: $\hat{c}$ = 1.17).

We used a similar approach to test what factors influenced the interval between nesting attempts (renest interval) and size of the second clutch. There was no evidence for year effects on renest interval at either CP or LPB, and therefore we used a linear model with a normal distribution in both cases. Fixed effects variables included age-class, first clutch size, condition (CP only), date of failure, and stage of failure. For analyses of second clutch size we also used a linear model with a normal distribution. Year effects were evident at CP and therefore year was treated as a random effect in many of the models. In these cases, we used a maximum likelihood rather than a restricted maximum likelihood estimator because we were comparing among models with a different fixed effects structure (Pinheiro and Bates 2004). There was no evidence for a year effect on second clutch size at LPB, and thus year was not included in the full model set. The distribution of clutch size at LPB was right-skewed, and therefore we used a square root transformation prior to analyses. All models were run using R (R Development Core Team 2009).

RESULTS

Natural clutch depredation of first nests during the study was higher at CP (mean nest success = 0.34 based on 336 nests) than at LPB (mean = 0.42, 189 nests; Wilson et al. 2007). The mean stage of loss for natural failures did not differ between the two sites [general linear model: $t = -0.97$, $P = 0.33$; CP mean stage loss: day 15.1 after onset of laying (i.e., day 9.1 of incubation), 95% CI: 14.0–16.3 d, range 1–28 d, $n = 117$ nests versus LPB mean stage loss: day 16.1 after onset of laying (i.e., day 7.1 of incubation), 95% CI: 14.6–17.7, range 1–28 d, $n = 52$ nests]. The latest stage of loss after which a female renested was day 14 of incubation at CP and day 19 at LPB. The latest initiation dates for renest clutches were 6 July at CP and 7 July at LPB.

Renesting Propensity

At CP, 54% of 138 radioed females that lost first clutches renested, and 8% of 25 females that lost second clutches attempted third clutches. At LPB, a minimum of 65% of 163 females that lost first clutches renested, and 33% of 15 females (1 of 6 yearlings and 4 of 9 adults) that lost a second clutch initiated third clutches. One of four hens at LPB completed a fourth clutch after the previous three clutches that season failed. Since the sample of females attempting second and third renest clutches was small, we restricted our analyses to the first renest clutches in both populations. Despite substantial annual variation in mean initiation dates of first clutches by up to 15 days at CP and 9 days at LPB (Hannon et al. 1988), we observed no annual variation in renesting probability for Willow Ptarmigan after loss of first clutches (Table 17.1).

Factors Affecting Renesting Propensity

The analyses for renesting propensity at Chilkat Pass used 7 variables, 23 candidate models, and 129 females whose first clutch attempt had failed (Table 17.1). Models with the greatest support contained stage of loss (females were less likely to renest with increasing days since onset of laying, $\beta_{stage} = -0.47 \pm 0.08$ SE) and female age (adults were more likely to renest than yearlings, $\beta_{age} = 2.03 \pm 0.66$; Fig. 17.1A). Date of first egg, date of failure, year, body condition, and clutch size were not included among the top models and thus had no effect on renesting rates. When the first clutch failed during laying or the first few days of incubation, almost all females renested (Fig. 17.1). The predicted renesting propensity followed a similar pattern for both age classes and declined sharply

TABLE 17.1
Model selection results of variables affecting renesting propensity for Willow Ptarmigan at Chilkat Pass and La Pérouse Bay, Canada.

Model	ΔAIC$_c$	AIC$_c$ weight	K	Deviance
A. Chilkat Pass, British Columbia				
Stage fail + age	0.00	0.37	3	86.28
Stage fail + clutch1 + age	0.90	0.24	4	85.06
Stage fail + date first egg + age	0.92	0.24	4	85.08
Stage fail*age	1.96	0.14	4	86.12
Intercept	87.77	0.00	1	178.20
B. La Pérouse Bay, Manitoba				
Stage fail	0.00	0.25	2	67.44
Stage fail + clutch1	0.81	0.16	3	66.16
Stage fail + date first egg	1.39	0.12	3	66.74
Stage fail + date fail	1.59	0.11	2	66.94
Intercept	12.00	0.00	1	81.50

NOTE: Variables included were female age-class and body condition (CP only), year, and first clutch characteristics [date of first egg (CP only), first clutch size, date of fail, and stage when failure occurred]. At both sites, year had little effect and so was not included as a random variable. Models were run using a generalized linear model and are ranked based on differences in Akaike's Criterion for small samples (AIC$_c$) and the AIC$_c$ weights. K represents the number of model parameters. In each case we report the top four models as well as the intercept-only model for comparison.

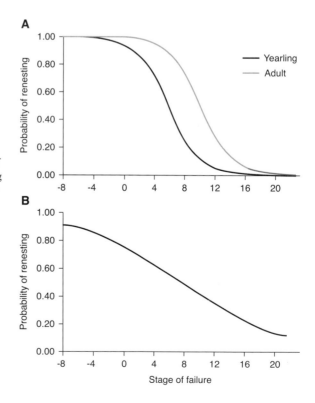

Figure 17.1. (A) Probability of renesting for Willow Ptarmigan at Chilkat Pass, British Columbia, in relation to stage of clutch loss for adult and yearling females based on the parameter estimates from the top models in Table 17.1. (B) Probability of renesting at La Pérouse Bay, Manitoba, in relation to stage of first clutch loss based on the parameter estimates for the top model in Table 17.1. Day 0 represents the onset of incubation upon laying the penultimate egg, with negative values representing the egg-laying stage. At La Pérouse Bay, adults and yearlings did not differ in their renesting probability.

4 to 5 days after onset of incubation. On day 5, adults and yearlings had a 92% and 59% probability of renesting, respectively; by day 12, those probabilities had dropped to 29% and 5%.

For La Pérouse Bay, the analyses for renesting propensity included 6 variables, 17 candidate models, and 60 females with first clutches that failed (Table 17.1B). All supported models at LPB had a strong effect of stage of loss (renesting was less likely with increasing days since onset of laying, $\beta_{stage} = -0.15 \pm 0.05$ SE). Date of first egg, year, date of failure, female age-class, and clutch size did not provide additional support to renesting propensity. At LPB, renesting propensity declined steadily with increasing stage of loss of the first clutch with comparable abilities of adult and yearling females to replace clutches (Fig. 17.1B). In fact, the female that renested after the latest incubation stage of clutch failure (day 19) was a yearling.

Between the two sites, renesting propensity after failure early in incubation was similar, but at later stages of nest loss the probability of clutch replacement declined more rapidly at CP. For instance, 82% of adult and 35% of yearling females at CP that lost clutches on day 7 of

incubation renested, while a minimum of 53% of females (both age-classes) at LPB renested. However, by day 13 of incubation, CP adults and yearlings had a 20% and 3% probability, respectively, while at LPB all females had a renesting probability of at least 31%.

Renest Interval

At CP, the mean renest interval was 6.1 days (95% CI: 5.48–6.78) with a range of 0–13 days ($n = 62$ intervals), and at LPB the mean interval was 4.8 days (95% CI: 3.0–6.5) with a range of 0–14 days ($n = 22$). If females lost their clutches after they had been incubating for 4 or more days, renest intervals in the two populations were comparable (CP: 7.1 d, 95% CI: 6.2–8.0, $n = 26$; LPB: 6.6 d, 95% CI: 3.6–9.5, $n = 7$). However, when they lost their clutches during laying or the first 3 days of incubation, LPB females were able to initiate renest clutches about 3 days earlier than females at CP.

The candidate model set for renest interval at Chilkat Pass contained 7 candidate variables, 17 models, and data for 61 renest attempts. Stage of loss had a strong influence on renesting interval (Table 17.2), with longer intervals for females that

TABLE 17.2

Model selection results for factors influencing renest interval for Willow Ptarmigan at Chilkat Pass and La Pérouse Bay, Canada.

Model	ΔAIC_c	AIC_c weight	K	Deviance
A. Chilkat Pass, British Columbia				
Stage fail	0.00	0.23	2	254.40
Stage fail*age	0.29	0.20	4	250.16
Stage fail + date fail	1.04	0.13	3	253.22
Stage fail + date first egg	1.70	0.10	3	253.88
Intercept	19.04	0.00	2	275.58
B. La Pérouse Bay, Manitoba				
Intercept	0.00	0.23	1	117.68
Age	0.54	0.18	2	115.76
Clutch1	1.74	0.10	2	116.96
Stage fail	1.78	0.10	2	117.00
Date fail	2.22	0.08	2	117.44

NOTE: The variables included were female age-class and body condition (CP only), year, first clutch size, date, and stage of failure. At both sites, year had little effect and was not included as a random variable. Variable definitions are the same as for Table 17.1. Only the top five models and the intercept-only model are shown.

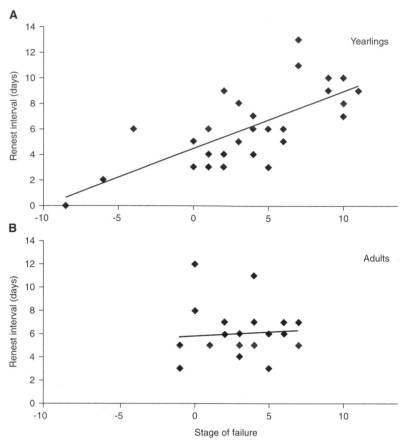

Figure 17.2. Renesting interval in relation to female age-class and stage of nesting of the first clutch at which failure occurred for Willow Ptarmigan at Chilkat Pass, British Columbia.

failed later in the nesting cycle (β_{stage} = 0.38 ± 0.08; Fig. 17.2). There was also support for an interaction effect between stage of loss and female age-class. Adults displayed a stronger positive relationship between stage of failure and renest interval, while yearlings appeared less flexible and had similar renest intervals regardless of when the first nest failed. However, range of stage of failure, which was more restricted for yearlings, may have contributed to this result.

The candidate model set for renest interval at La Pérouse Bay contained 6 variables, 14 candidate models, and data from 21 renest attempts. The intercept-only model was the top model, indicating no individual variable or set of variables best explained variation in renest interval at LPB, although there was a marginal improvement in the log-likelihood when female age-class was included. The coefficient for age-class was −2.36 (SE = 1.75), leading to a prediction of a 6-day interval for yearlings (intercept = 6.0; SE = 1.3)

and 3.6-day interval for adults. With a larger sample size, we may have detected a stronger effect of female age-class on renesting interval at LPB.

Clutch Size of First and Renests

At CP, the mean size of first clutches was 8.3 eggs (95% CI: 8.2–8.5, range 5–11 eggs, n = 252 clutches), while the first renest clutches were 6.0 eggs (95% CI: 5.9–6.1, range 4–8 eggs, n = 70 nests). Clutch sizes at LPB were significantly larger than at CP, with first clutches at LPB averaging 10.8 eggs (95% CI: 10.6–11.0, range 8–14 eggs; n = 123 nests; CP vs. LPB first clutches, general linear model: t = −19.57, P < 0.001) and first renest clutches at LPB averaging 7.6 eggs (95% CI: 7.2–8.0, range 5–11, n = 57 nests; CP vs. LPB renests, general linear model: t = −7.02, P < 0.001).

Analyses for renest clutch size at Chilkat Pass contained 7 candidate variables, 28 models,

TABLE 17.3
Model selection results for factors influencing first renest clutch size for Willow Ptarmigan
at Chilkat Pass and La Pérouse Bay, Canada.

Model	ΔAIC_c	AIC_c weight	K	Deviance
A. Chilkat Pass, British Columbia				
Clutch1 + renest int. + stage fail	0.00	0.46	4	126.08
Clutch1 + renest int.	1.30	0.24	3	129.72
Clutch1 + renest int. + stage fail + year	2.24	0.15	5	125.88
Clutch1 + stage fail	5.10	0.04	3	133.52
Intercept	16.83	0.00	1	149.66
B. La Pérouse Bay, Manitoba				
Renest int. + date fail + clutch1	0.00	0.58	4	−13.56
Renest int. + date fail	1.57	0.26	3	−8.50
Renest int. + clutch1	3.83	0.09	3	−6.24
Renest int.	5.02	0.05	2	−2.06
Intercept	12.73	0.00	1	8.24

NOTE: The variables included were female age-class, body condition (CP only), year, first clutch traits (date of first egg, clutch size, stage of failure), and renest interval. Date of failure and date of first egg of the renest clutch were too closely correlated with stage of failure of the first nest to include in the suite of models. Year was included as a random effect at Chilkat Pass. Only the top five models and the intercept-only model are shown.

and 53 nest attempts (Table 17.3). Models with the greatest support included first clutch size (females with larger first clutches had larger renest clutches, $\beta_{clutch} = 0.40 \pm 0.11$), renest interval (longer intervals resulted in smaller renest clutches, $\beta_{renest\ int} = -0.15 \pm 0.05$), and stage of failure (females with first attempts failing later in the nesting cycle laid smaller second clutches, $\beta_{stage} = -0.08 \pm 0.04$, Figs. 17.3A and 17.4A). Year was not influential when included as a random effect in models with fixed effects, accounting for only 2% of the total variance. Female age-class, condition, and date of first egg for the first attempt had little influence on renest clutch size.

Analyses for renest clutch size at La Pérouse Bay contained 7 candidate variables, 15 models, and 17 nest attempts (Table 17.3). Unlike at CP, there was no evidence for a year effect ($var_{year} \sim 0$), although sample sizes were small. Other results were similar to CP in that first clutch size ($\beta_{clutch} = 0.13 \pm 0.06$) and renest interval ($\beta_{renest\ int} = -0.04 \pm 0.01$) influenced second clutch size (Figs. 17.3B and 17.4B); however, date of failure had a greater influence than stage of failure at LPB. When first

attempts failed later in the season, females laid smaller renest clutches ($\beta_{date} = -0.02 \pm 0.006$).

DISCUSSION

Willow Ptarmigan in northern Canada were persistent and capable at producing renest clutches. Within populations, we found stage of loss during incubation to have the strongest influence on renesting propensity and renest interval, with female age-class also being important at CP. In both populations, individual variation in renest clutch size was positively correlated with the size of the first clutch and negatively correlated with length of renest interval. Compared to CP, LPB females were better able to replace clutches lost during egg-laying (including continuation egg-laying), and at later stages of incubation to renest after shorter intervals, and yearling females matched the renesting abilities of older birds.

Selection on renesting ability is likely strong in many single- and multiple-brooded avian species, and especially open-cup and ground-nesting birds with high rates of clutch failure. In many grouse species, the failure of first clutches varies

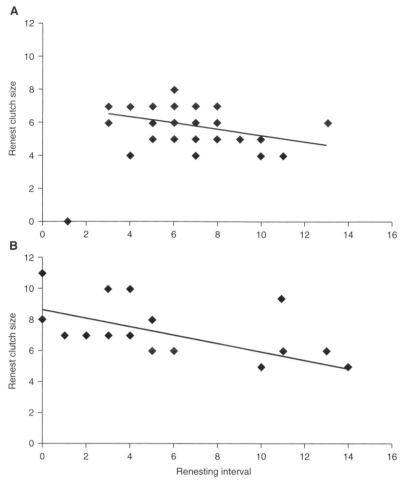

Figure 17.3. Variation in renest clutch size for Willow Ptarmigan at (A) Chilkat Pass, British Columbia, and (B) La Pérouse Bay, Manitoba, in relation to the interval between first and second nesting attempts.

from 31 to 83% (Johnsgard 1983). The efficiency with which species respond to the loss of their initial investment can be significant to annual and lifetime fecundity (Parker 1985, Martin et al. 1989, Milonoff 1991). Among the three variables we examined, renesting propensity was the most important and should be under the strongest selection because the ability to attempt a replacement clutch provides additional chances for females to have non-zero annual fecundity. In addition, nest survival for renests in most of our populations was higher than for first clutches (Wilson et al. 2007). There should also be strong selection for short renest intervals, as this time period is critical to the size of the second clutch, and possibly also will determine whether a female has time to attempt a second renest clutch should the first replacement clutch fail. Short renest intervals resulted in larger

renest clutches (this study, Arnold 1993) and perhaps an improved possibility for offspring recruitment (Martin and Hannon 1987). One source of variation in renesting ability between populations and grouse species is the ability of females to achieve continuation laying if the clutch is lost during egg-laying as observed at LPB (up to 21 eggs laid in sequential nests in 21 days) versus a determinate laying pattern recorded for CP females (Hannon et al. 1998, Sandercock 1993).

All grouse produce a single brood annually, and most can renest after failure of first nesting efforts (Sopuck and Zwickel 1983; Parker 1985; Bergerud and Gratson 1988; Sandercock et al. 2005a, 2005b). Among grouse and water birds, there is considerable variation in renest interval from ~6 days in Mallards (*Anas platyrhynchos*), American Coots (*Fulica americana*), and Willow Ptarmigan in North

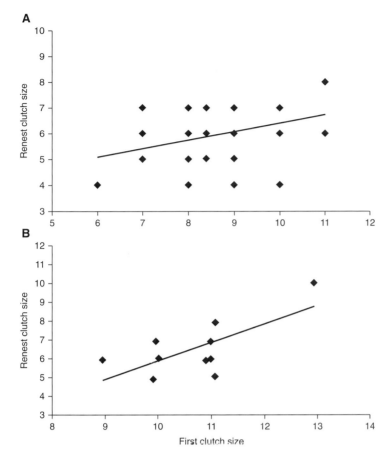

Figure 17.4. Variation in renest clutch size for Willow Ptarmigan at (A) Chilkat Pass, British Columbia, and (B) La Pérouse Bay, Manitoba, in relation to size of the first clutch. (sample size: CP = 39, LPB = 10).

America (Arnold 1993, Hannon et al. 1998), up to 9 to 14 days in Northern Pintails (*Anas acuta*), Sooty Grouse (*Dendragopus fuliginosus*), and Willow Ptarmigan in Norway (Sopuck and Zwickel 1983, Parker 1985, Grand and Flint 1996). Similar species may also differ in renesting ability even when exposed to the same environmental conditions. In the Yukon Territory, Canada, where two congeneric ptarmigan species breed sympatrically, White-tailed Ptarmigan (*L. leucura*) have a much higher renesting propensity compared to Rock Ptarmigan (*L. muta*), a trait that potentially is related to a 22% lower annual survival experienced by White-tailed Ptarmigan (Wilson and Martin 2010).

Intrapopulation Variation in Renesting

We examined a range of attributes for females, including age-class, condition, mass, and size, but only age-class at CP had a strong effect on renesting ability, with older females being better able to mobilize gonads for a renest clutch than first-time

breeders. Older female Willow Ptarmigan at CP showed improvements in potential fecundity, with 2- and 3+-year-old birds initiating clutches earlier and laying larger clutches than yearling birds (Wiebe and Martin 1998, Sandercock et al. 2005b). In most other grouse, older females were more capable at renesting and showed less variability in renesting ability than yearlings (Sopuck and Zwickel 1983, Parker 1985, Wilson and Martin 2010). The result that female traits such as body condition, mass, or size had little or no influence on renesting ability is consistent with our earlier work, where these traits had little or no influence on first clutch attributes (Robb et al. 1992). Similarly, female characteristics did not influence renesting for Northern Pintails (Grand and Flint 1996). However, the patterns differed for Mallards, where nesting effort, including renesting propensity, was higher among older and heavier females (Devries et al. 2008, Arnold et al. 2010).

One might predict that birds investing relatively more in their first clutches might have fewer body reserves for renesting, or the pattern might be reversed if first clutches were initiated earlier or indicated higher quality of females. We found no effect of first clutch characteristics (clutch size, initiation date) on renesting propensity or renest interval. We did not measure egg size in our study, but earlier at CP, hens that renested laid eggs in their first clutches that were 4.6% larger than females that did not renest, suggesting an effect of female quality on renesting propensity (Sandercock and Pederson 1994). Eggs in Willow Ptarmigan renests were about 1.6% and 5% larger than for first clutches in CP and in Norway, respectively (Parker 1981, Sandercock and Pederson 1994). Since we found a positive relationship between the size of first and renest clutches in both of our populations, it appears that, similar to American Coots (Arnold 1993), renesting Willow Ptarmigan females may not be constrained by food or nutrient reserves available for egg production.

Stage of Loss and Mobilization of Gonads

For renesting propensity and renest interval, the stage of loss during the breeding effort was the most important attribute in both populations. Thus, variation in renesting may be driven by the capacity of females to mobilize their gonads after clutch failure. The phase of rapid yolk synthesis begins in the ovum of large birds 7–10 days before ovulation and 3–4 days before ovulation in small birds (King 1973). When gonads are mobilized in domestic chickens (*Gallus domesticus*), the minimum possible renest interval is 8 days from the day when a nest was last active and preceding the day the first replacement egg is laid (7 d of rapid yolk deposition plus 1 d in the oviduct; Scott et al. 1987). Willow Ptarmigan in Canada with a full renest interval (6 d) are able to achieve rapid yolk synthesis in 5 days.

Little study has been done on the underlying mechanisms and the likely influences of physiological and environmental components on breeding capacity and gonad mobilization for renesting. However, Donham et al. (1976), using captive wild and game farm Mallards, found that within 1 day after clutch removal, three of eight females had distinctly enlarged follicles and others had increased plasma luteinizing hormone levels, and within 3 days, seven of eight females had yellow follicles one-third to one-half the size of fully developed follicles. Donham et al. (1976) suggested several stages where variation in renesting

ability could occur: Females could delay recrudescence of ovarian follicles longer into incubation, or they could mobilize endocrine functions and follicle development within one day of clutch loss. Willow Ptarmigan likely have some capacity to delay recrudescence of follicles after onset of incubation because some females at LPB that had completed or almost completed laying their first clutch had renest intervals of only 1 or 2 days. They were able to initiate renests quickly because they develop many more follicles than they use in their first clutch. At LPB, several females that were killed by predators near the end of laying or day 1 of incubation of the first clutch had 10 or more sequentially developed follicles (several full or almost full size) that likely could have been deployed to immediately initiate a second clutch (K. Martin, unpubl. data). It is unknown whether there are physiological costs to developing these extra follicles and maintaining them for a few days after the onset of incubation. In Sooty Grouse and Lesser Snow Geese (*Anser caerulescens caerulescens*), extra follicles were reabsorbed and presumably used as energetic reserves during incubation (Hannon 1981, Hamann et al. 1986).

Renest interval and first clutch size were the most important variables explaining variation in renest clutch size for both populations. Renest interval integrates both the stage of loss and initiation dates for the renest clutch variables. Females that laid larger first clutches may have developed larger numbers of additional follicles and therefore had more follicles for their renest clutch. It is also possible that the first clutch size is related to female quality or the quality of territories or environmental conditions. Regardless, in both of our populations, and in other studies of renesting, there were positive correlations between the size of first and renest clutches, suggesting that Willow Ptarmigan and American Coot females were not bet-hedging investment in their first clutches to allow for renesting opportunities (Parker 1981, Arnold 1993).

Variation in Renesting Ability Between Populations

We considered three hypotheses (bet-hedging, resource availability, survival–fecundity trade-off) to explore the interpopulation variation in renesting abilities, which may be driven by reproductive failure, environmental variation, or annual mortality of adults. If nest survival rates of first clutches are a selective force for renesting ability, then populations with the lower nest survival of first clutches, such as at CP, should be under stronger selection to renest. We can reject a bet-hedging strategy as an explanation for differences between populations because birds at CP with higher daily nest failure did not have higher renesting abilities. We also did not find support for the resource availability hypothesis, as females in the CP population, the site with higher primary productivity and more precipitation, did not renest at higher rates than females at LPB (Sandercock et al. 2005a). Overall, as income breeders, Willow Ptarmigan appear able to amass and replace resources for egg-laying readily, and thus it appears that females may be more constrained by temporal factors than by food or nutrient reserves available for egg production.

If survival–fecundity trade-offs are driven by extrinsic mortality, as found for several ptarmigan species in northern Canada (Sandercock et al. 2005a, Wilson and Martin 2010), then renesting ability may be under stronger selection for Willow Ptarmigan at LPB than at CP, given that the strongest fitness benefits would be derived in populations with lower adult survival and higher renesting. The larger body size of Willow Ptarmigan at LPB possibly facilitated renesting or might be a consequence of this ability. Although renesting abilities may be under stronger selection at LPB, Willow Ptarmigan in both North American populations exhibited well-developed and persistent renesting abilities.

LITERATURE CITED

Amat, J. A., R. M. Fraga, and G. M. Arroyo. 1999. Replacement clutches by Kentish Plovers. Condor 101:746–751.

Arnold, T. 1993. Factors affecting renesting in American Coots. Condor 95:273–281.

Arnold, T. W., J. H. Devries, and D. W. Howerter. 2010. Factors that affect renesting in Mallards (Anas platyrhynchos). Auk 127:212–221.

Bergerud, A. T., and M. W. Gratson. 1988. Adaptive strategies and population ecology of northern grouse. University of Minnesota Press, Minneapolis, MN.

Bergerud, A. T., S. S. Peters, and R. McGrath. 1963. Determining sex and age of Willow Ptarmigan in Newfoundland. Journal of Wildlife Management 27:700–711.

Burnham, K. P., and D. R. Anderson. 2002. Model selection and multimodel inference: a practical information theoretic approach. 2nd ed. Springer-Verlag, New York, NY.

Devries, J. H., R. W. Brook, D. W. Howerter, and M. G. Anderson. 2008. Effects of spring body condition and age on reproduction in Mallards (Anas platyrhynchos). Auk 125:618–628.

Donham, R. S., C. W. Dane, and D. S. Farner. 1976. Plasma luteinizing hormone and the development of ovarian follicles after loss of clutch in female Mallards (Anas platyrhynchos). General and Comparative Endocrinology 29:152–155.

Eldridge, J. L., and G. L. Krapu. 1988. The influence of diet quality on clutch size and laying pattern in Mallards. Auk 105:102–110.

Fletcher, R. J., R. R. Koford, and D. A. Seaman. 2006. Critical demographic parameters for declining songbirds breeding in restored grasslands. Journal of Wildlife Management 70:145–157.

Fondell, T. F., J. B. Grand, D. A. Miller, and R. M. Anthony. 2006. Renesting by Dusky Canada Geese on the Copper River Delta, Alaska. Journal of Wildlife Management 70:955–964.

Grand, J. B., and P. L. Flint. 1996. Renesting ecology of Northern Pintails on the Yukon–Kuskokwim Delta, Alaska. Condor 98:820–824.

Gregg, M. A., M. R. Dunbar, J. A. Crawford, and M. D. Pope. 2006. Total plasma protein and renesting by Greater Sage-Grouse. Journal of Wildlife Management 70:472–478.

Grzybowski, J. A., and C. M. Pease. 2005. Renesting determines seasonal fecundity in songbirds: what do we know? What should we assume? Auk 122:280–291.

Hamann, J., B. Andrews, and F. Cooke. 1986. The role of follicular atresia in inter- and intra-seasonal clutch size variation in Lesser Snow Geese (Anser caerulscens caerulscens). Journal of Animal Ecology 55:481–489.

Hannon, S. J. 1981. Postovulatory follicles as indicators of egg production in Blue Grouse. Journal of Wildlife Management 45:1045–1047.

Hannon, S. J., and K. Martin. 1996. Mate fidelity and divorce in ptarmigan: polygyny on the tundra. Pp. 192–210 in J. M. Black (editor), Partnerships in birds, the study of monogamy. Oxford University Press, Oxford, UK.

Hannon, S. J., K. Martin, and J. O. Schieck. 1988. Timing of reproduction in two populations of Willow Ptarmigan in northern Canada. Auk 105:330–338.

Hannon, S. J., P. K. Eason, and K. Martin. 1998. Willow Ptarmigan. A. Poole and F. Gill (editors), The birds of North America, No. 369. Academy of Natural Sciences, Philadelphia, PA.

Johnsgard, P. A. 1983. The grouse of the World. University of Nebraska Press, Lincoln, NE.

King, J. R. 1973. Energetics of reproduction in birds. Pp. 78–120 in D. S. Farmer (editor), Breeding biology of birds. National Academy of Sciences, Washington, DC.

Klomp, H. 1970. The determination of clutch-size in birds, a review. Ardea 58:1–124.

Martin, K., and F. Cooke. 1987. Bi-parental care in Willow Ptarmigan: a luxury? Animal Behaviour 35:369–379.

Martin, K., and S. J. Hannon. 1987. Natal philopatry and recruitment of Willow Ptarmigan in north central and northwestern Canada. Oecologia 71:518–525.

Martin, K., S. J. Hannon, and R. F. Rockwell. 1989. Clutch size variation and patterns of attrition in fecundity of Willow Ptarmigan. Ecology 70:1788–1799.

Martin, K., and K. L. Wiebe. 2004. Coping mechanisms of alpine and arctic breeding birds: extreme weather and limitations to reproductive resilience. Integrative Comparative Biology 44:177–185.

Martin, T. E. 1993. Nest predation among vegetation layers and habitat types: revising the dogmas. American Naturalist 141:897–913.

Milonoff, M. 1991. Renesting ability and clutch size in precocial birds. Oikos 62:189–194.

Montgomerie, R. D., and P. J. Weatherhead. 1988. Risks and rewards of nest defence by parent birds. Quarterly Review of Biology 63:167–187.

Nolan, V., Jr. 1978. The ecology and behavior of the Prairie Warbler Dendroica discolor. Ornithological Monographs 26.

Parker, H. 1981. Renesting biology of Norwegian Willow Ptarmigan. Journal of Wildlife Management 45:858–864.

Parker, H. 1985. Compensatory reproduction through renesting in Willow Ptarmigan. Journal of Wildlife Management 49:599–604.

Pinheiro, J. C., and D. M. Bates. 2004. Mixed effects models in S and S-plus. Springer, New York, NY.

Preston, K. L., and J. T. Rotenberry. 2006. The role of food, nest predation, and climate in timing of Wrentit reproductive activities. Condor 108:832–841.

R Development Core Team. 2009. R: a language and environment for statistical computing, reference index version 2.9.1. R Foundation for Statistical Computing, Vienna Austria. <http://www.R-project.org>.

Ricklefs, R. E. 1969. An analysis of nesting mortality in birds. Smithsonian Contributions to Zoology 9:1–48.

Ricklefs 1973. Fecundity, mortality, and avian demography. Pp. 366–435 in D. S. Farner (editor), Breeding biology of birds. National Academy of Science, Washington, DC.

Robb, L. A., K. Martin, and S. J. Hannon. 1992. Spring body condition, fecundity and survival in female Willow Ptarmigan. Journal of Animal Ecology 61:215–223.

Sandercock, B. K. 1993. Free-living Willow Ptarmigan are determinate egg-layers. Condor 95:554–558.

Sandercock, B. K., W. E. Jensen, C. K. Williams, and R. D. Applegate. 2008. Demographic sensitivity of population change in Northern Bobwhite. Journal of Wildlife Management 72:970–982.

Sandercock, B. K., K. Martin, and S. J. Hannon. 2005a. Life history strategies in extreme environments: comparative demography of alpine and arctic ptarmigan. Ecology 86:2176–2186.

Sandercock, B. K., K. Martin, and S. J. Hannon. 2005b. Demographic consequences of age-structure in extreme environments: population models for arctic and alpine ptarmigan. Oecologia 146:13–24.

Sandercock, B. K., and H. C. Pederson. 1994. The effect of renesting ability and nesting attempt on egg-size variation in Willow Ptarmigan. Canadian Journal of Zoology 72:2252–2255.

Scott, D. M., R. E. Lemon, and J. A. Darley. 1987. Relaying interval after nest failure in Gray Catbirds and Northern Cardinals. Wilson Bulletin 99:708–712.

Slagsvold, T. 1982. Clutch size, nest size, and hatching asynchrony in birds: experiments with the Fieldfare (Turdus pilaris). Ecology 63:1389–1399.

Slagsvold, T. 1984. Clutch size variation of birds in relation to nest predation: on the cost of reproduction. Journal of Animal Ecology 53:945–953.

Sopuck, L. G., and F. C. Zwickel. 1983. Renesting in adult and yearling Blue Grouse. Canadian Journal of Zoology 61:289–291.

Stearns, S. C. 1976. Life-history tactics: a review of the ideas. Quarterly Review of Biology 51:1–46.

Swanson, G. A., T. L. Shaffer, J. F. Wolf, and F. B. Lee. 1986. Renesting characteristics of captive Mallards on experimental ponds. Journal of Wildlife Management 50:32–38.

Westerskov, K. 1956. Age determination and dating nesting events in the Willow Ptarmigan. Journal of Wildlife Management 20:274–279.

Wiebe, K. L., and K. Martin. 1995. Ecological and physiological effects on egg laying intervals in ptarmigan. Condor 97:708–717.

Wiebe, K. L., and K. Martin. 1998. Age-specific patterns of reproduction in White-tailed and Willow Ptarmigan Lagopus leucurus and L. lagopus. Ibis 140:14–24.

Wilson, S., and K. Martin. 2010. Variable reproductive effort for two sympatric ptarmigan in response to spring weather conditions in a northern alpine ecosystem. Journal of Avian Biology, 41:319–326.

Wilson, S., K. Martin, and S. J. Hannon. 2007. Nest survival patterns in Willow Ptarmigan: influence of time, nesting stage, and female characteristics. Condor 109:377–388.

CHAPTER EIGHTEEN

Chick Survival of Greater Prairie-Chickens

Adam C. Schole, Ty W. Matthews, Larkin A. Powell,
Jeffrey J. Lusk, and J. Scott Taylor

Abstract. Chick survival during the first three weeks of life is a critical stage in the demography of Greater Prairie-Chickens (*Tympanuchus cupido*), but little information is available. Biologists often estimate brood success using periodic flushes of radio-marked females, but it is impossible to determine mortality factors if chicks are not radio-marked. We used sutures to attach 0.5-g transmitters to 1- to 2-day-old chicks in Johnson County, Nebraska, during 2008. Our objectives were to (1) assess causes of mortality of 0- to 21-day-old chicks, (2) estimate daily survival probability for 0- to 21-day-old chicks, and (3) evaluate the effect of applying transmitters with suture attachment to chicks. We monitored a total of 221 prairie chicken chicks from 20 broods. We radio-marked 27 chicks from 10 broods of radio-marked females (one to five chicks per brood). The chicks were located twice per day to ensure that they were within a 10-m radius of the female. Our limited sample showed a weak effect of radio-marking on the survival of prairie-chicken chicks ($\beta = -0.54$; SE $= 0.33$). Forty-two (19%; 95% CI: ±5%) of the 221 chicks in our sample survived to day 21, confirming low rates of productivity observed in hunter wing surveys and brood flushes of radio-marked females in a concurrent study. All radio-marked chicks in our sample died (13% exposure; 87% predators) before 21 days of age. Survival of chicks increased with age, and survival decreased during periods with high precipitation. Daily and 21-day survival rate estimates for all chicks in our sample were 0.926 (95% CI: 0.915–0.937) and 0.193 (95% CI; 0.155–0.255), respectively. Predation appeared to be the most critical factor for chick survival, so management of landscapes to reduce risk from predators may have a positive effect on Greater Prairie-Chicken populations.

Key Words: brood, chick, Greater Prairie-Chicken, radiotelemetry, survival, *Tympanuchus cupido*.

Schole, A. C., T. W. Matthews, L. A. Powell, J. J. Lusk, and J. S. Taylor. 2011. Chick survival of greater prairie-chickens. Pp. 247–254 *in* B. K. Sandercock, K. Martin, and G. Segelbacher (editors). Ecology, conservation, and management of grouse. Studies in Avian Biology (no. 39), University of California Press, Berkeley, CA.

Chick survival is a critical phase for Greater Prairie-Chicken (*Tympanuchus cupido*; hereafter prairie chicken) population dynamics; Wisdom and Mills (1997) reported that finite rates of population growth of prairie chickens were highly sensitive to juvenile survival rates. No data on cause-specific chick survival exists for prairie chickens; such information is critical for species of conservation concern. Biologists often use periodic flushes of radio-marked females to estimate brood success for grouse species, but this method does not provide information about the cause of mortality of chicks. Radio-marked chicks can be used to efficiently identify mortality events, suitable brood rearing habitat, and movements (Burkepile et al. 2002). However, radio-marking chicks requires the proper size transmitter and an effective and unobtrusive attachment method to avoid increasing mortality (Millspaugh and Marzluff 2001).

Hunter wing surveys in southeast Nebraska have indicated low productivity (0.92 chicks/adult) during 2001–2007 compared with north-central Nebraska's Sandhills population of prairie chickens in the same period (1.77 chicks/adult; J. Lusk, unpubl. data). Empirical data from a sample of radio-marked females in southeast Nebraska during 2007–2008 suggested that brood survival was low ($S_{21\text{-day}}$ = 0.59; Matthews et al., this volume, chapter 13). However, Matthews et al. (this volume, chapter 13) monitored unmarked broods and could not determine the cause of chick mortality. Our goal was to radio-mark chicks to more precisely assess variation in chick survival of prairie chickens in southeastern Nebraska. Our three objectives were to (1) assess the causes of mortality of 0- to 21-day-old chicks, (2) estimate daily survival probability of 0- to 21-day-old chicks, and (3) evaluate effects of handling and applying radio markers with suture transmitters on survival of chicks.

METHODS

Study Area

Johnson and Pawnee counties (average precipitation: 840 mm; University of Nebraska–Lincoln High Plains Climate Center) in Nebraska contain a population of Greater Prairie-Chickens, thought to be the northernmost extension of the Flint Hills population. The topography of these counties is rolling uplands, and the landscape of our study site was dominated by corn, soybean, and alfalfa production with significant areas of pasture and rangeland. In 2007, 163.3 km² (40,345 acres; ca. 17%) of Johnson County and 172.1 km² (42,533 acres; ca. 15%) of Pawnee County was enrolled in the Conservation Reserve Program (Farm Service Agency, USDA).

Field Methods

We randomly selected broods from radio-marked females in a concurrent study (Matthews et al., this volume, chapter 13) during 2008. Depending on brood size, one to five chicks in each brood were fitted with a 0.5-g (<3% chick mass) transmitter (Advanced Telemetry Systems, Isanti, MN, model A2415). We radio-marked 27 chicks from ten broods with a suture attachment method (Burkepile et al. 2002). The suture method was used for attachment because minimal training was needed, the transmitters could be attached at the nest site, and it was less invasive than prong-and-suture attachment (Mauser and Jarvis 1991) or subcutaneous implants (Korshgen et al. 1996).

We monitored female movements to ascertain nest hatch date, and we located each brood 1–2 days post-hatch to capture and mark chicks. The brood was caught by hand shortly after sunset using spotlights to maximize the chance of capturing the entire brood; potential brood numbers were determined by comparing number captured with number of eggshells when chicks departed the nest. We placed the chicks in an insulated box containing a warm bottle of water to maintain the chicks' body heat during transmitter application. We randomly selected chicks for radio-marking, and followed methods of Burkepile et al. (2002) for suture attachment. We inserted monofilament suture into a 12-ga syringe needle. The transmitters were sutured in the mid-dorsal region directly between the wings with the transmitter antennae positioned toward the tail. We inserted the needle subcutaneously, ensuring about 5 mm of skin was between the insertion and exit hole. We pushed the monofilament suture through the needle and removed the needle, leaving the monofilament in the epidermal tissue. We positioned both free ends of the monofilament through the transmitter's anterior backpack attachment once and tied a square knot. We repeated the same process for the posterior attachment of the transmitter. To ensure room for tissue growth the transmitters were sutured to leave a ca.

TABLE 18.1

Comparison of competing models to explain variation in survival of radio-marked Greater Prairie-Chicken chicks in southeast Nebraska, 2008.

Model	DIC	ΔDIC	w_{DIC}	K
Handled + marked + age + precipitation	567.9	0.00	1.0	5
Handled + precipitation	592.0	24.1	0.0	3
Handled + marked + age	808.1	240.2	0.0	4
Handled + age	916.1	348.2	0.0	3
Handled + age + precipitation	917.9	350.0	0.0	4
Handle + marked	922.1	354.2	0.0	3
Handled + marked + precipitation	922.8	354.9	0.0	4
Age	1,012.8	444.9	0.0	2
Precipitation + age	1,013.1	445.2	0.0	3
Handled	1,211.2	643.3	0.0	2
Precipitation	1,270.2	702.3	0.0	2
Null	1,273.3	705.4	0.0	1

NOTE: Models are ranked by Deviance Information Criterion (DIC) score. Differences between the top model and all other models are shown by ΔDIC; K is the number of model parameters. The model weight (w_{DIC}) is the certainty that each model is the best model of the models compared.

2-mm suture gap (Burkepile et al. 2002). We placed the entire brood, consisting of radio-marked and handled-only chicks, at the capture location to allow the female to relocate them.

Following marking, we determined location of chicks twice each day to ensure chicks were alive and within a 10-m radius of the female. If the radio-marked chick was not within 10-m of the female, we conducted an immediate, extensive search for the transmitter to determine chick fate. We performed brood flushes at 10 and 21 days post-hatch to determine chick survival of unmarked young in the same brood.

We randomly selected 10 broods from the 2008 sample of the concurrent study on the same site by Matthews et al. (this volume, chapter 13) to serve as a control group to compare survival with the handled-only and radio-marked chicks in our 10 study broods. Control chicks were never captured and never handled. Like the handled-only chicks, the control broods were flushed at 10 and 21 days post-hatch to determine brood size. Animal capture and handling protocols were approved by the University of Nebraska–Lincoln Institutional Animal Care and Use Committee (Protocol #05-02-007).

Statistical Methods

We developed an *a priori* set of 12 models, which included main effects models of chick age, precipitation, and handling, a global model with all effects, and six other additive models with biologically reasonable combinations of the effects (Table 18.1). We compared all models to a null model, with constant survival through time and space. Our age model allowed survival to vary in a linear fashion as a function of the number of days since hatch. We used the average daily precipitation as a covariate for each monitoring interval in our precipitation model. Our final models incorporated the type of handling and marking each chick received. First, a "handled" model assessed the effect of handling chicks during capture; chicks that were handled, as well as radio-marked chicks, were considered handled; chicks in broods without captures were used as controls. Second, we used an additional model to assess the effect of radio-marking and included the nested effects of handling and radio-marking in a two-factor, additive model of handling and radio-marking, which allowed us to separate the effects of radio-marking from handling.

TABLE 18.2

Parameter estimates of slope coefficients (SE and 95% confidence interval) from the best model (Table 18.1) for effects of handling, radio-marking, age, and precipitation on survival of Greater Prairie-Chicken chicks in southeast Nebraska, 2008.

Parameter	β(SE)	95% Confidence Interval
Intercept	1.30 (0.22)	$0.88 < \beta < 1.72$
Handling	0.15 (0.21)	$-0.24 < \beta < 0.54$
Radio-marking	-0.54 (0.33)	$-1.18 < \beta < 0.10$
Age	0.12 (0.02)	$0.08 < \beta < 0.16$
Precipitation	-0.54 (0.04)	$-0.62 < \beta < -0.46$

NOTE: Control young not handled or radio-marked serve as the baseline ($\beta = 0.00$) for the comparison with discrete effects of handling and radio-marking.

We used a logistic exposure structure to estimate daily chick survival ($\hat{S}_D$; Shaffer 2004). We combined two known fate data structures in the same model; radio-marked chicks had monitoring intervals corresponding with telemetry observations, while non–radio-marked chicks had 10- or 11-day monitoring intervals corresponding with flush counts at 10 and 21 days after hatch. The logistic exposure structure allowed us to include data with unequal intervals (Shaffer 2004). We encountered convergence difficulties with standard methods based on iterated weighted least squared method because of the survival patterns in control birds. Thus, we used a Markov Chain Monte Carlo (MCMC) framework using WinBUGS (version 1.4.2) and program R with R2WinBUGS package (R package version 2.1-8). We used three replicated chains with 100,000 iterations, each sampling with a starting value from a normal distribution with a mean of 0 and a standard error of 0.2. We had a burn-in of the first 50,000 samples and set our thinning at 150 for the subsequent samples. We then calculated a Deviance Information Criterion (DIC) score for each model (Spiegelhalter et al. 2002). DIC is used in the MCMC framework and is similar to Akaike's Information Criterion, so we selected the model with the lowest DIC score as the best model (Burnham and Anderson 2002). We used the 95% confidence interval (CI) surrounding the covariate estimate (β) to evaluate the strength of the parameter's effect on chick survival (Table 18.2). We calculated a mean daily survival rate using our top model (Table 18.1), setting parameters at their mean. We estimated a 21-day success rate ($\hat{S}_{21}$) as $\hat{S}_{21} = (\hat{S}_D)^{21}$, and we used the delta method to approximate the variance of $\hat{S}_{21}$ and calculate the 95% CIs (Powell 2007).

RESULTS

We monitored 221 chicks from 20 broods; 27 chicks from 10 broods were radio-marked, and 56 chicks from the same broods were handled but not radio-marked. Our control sample consisted of 138 chicks (not handled or radio-marked) from 10 broods. The suture procedure for each chick took approximately 3 minutes, with brood handling time <20 minutes. No chicks died during the suture process. We did not observe infection or inflammation of the area of suture attachment on recaptured chicks, nor did we observe abnormal movement of radio-marked chicks relative to their unmarked brood-mates. Female abandonment of broods after radio-marking did not occur, and females usually remained <20 m from us while we attached chick transmitters.

Data from three of 27 (11%) radio-marked chicks were censored when the sutures failed prior to 21 days post-hatch; two failed at day 7, one at day 9. When we recovered these transmitters, we believed the chicks were still alive because the brood was still in the vicinity, in contrast to broods which left the vicinity after partial losses due to predators. In addition, our subsequent flushes of these broods showed no mortality of young. Because we lost radio contact with these chicks before our 21-day monitoring interval ended, we were only able to assess the fate of 24 chicks.

Three of 24 (13%) radio-marked chicks died from apparent exposure. The intact remains of one chick were found shortly after a heavy rain, which

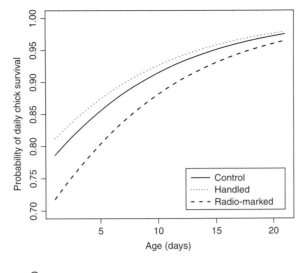

Figure 18.1. Relationship between daily chick survival and age of young (0–21 days) for control (not handled or radio-marked), handled, and radio-marked Greater Prairie-Chicken chicks in southeast Nebraska, 2008. Confidence intervals overlapped and are omitted for clarity.

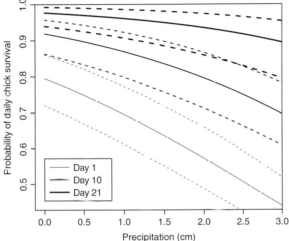

Figure 18.2. Daily survival probability for 1-, 10-, and 20-day-old Greater Prairie-Chicken chicks as a function of daily precipitation in southeast Nebraska, 2008.

suggested hypothermia as the cause of death. Two other dead chicks were found intact (day 2 and 4 post-hatch) with no visible signs of cause of disease or mortality, suggesting other exposure causes. Two of 24 (8%) radio-marked chicks died from known predation events. One chick's transmitter condition included a curled antenna and abrasions, which suggested predation by a raptor, and another transmitter was found in a pile of plucked chick and adult prairie chicken feathers, also suggesting raptor predation. Nineteen of 24 (79%) radio-marked chicks' fate was uncertain, as transmitters disappeared and were never recovered. We observed >300-m movements by radio-marked females immediately after disappearance of radio-marked young. Because the movement of broods during periods absent of chick mortality usually was localized, we believe the missing transmitters were ingested or destroyed by predators. Hence, we

recorded these chicks' fates as mortalities caused by predation. On five occasions, >1 radio-tagged chick disappeared from the same brood simultaneously, also indicative of mortality rather than radio failure. All radio-marked chicks died before 21 days of age, and 84% (n = 24) of mortalities occurred 6–13 days after hatch, 4% (n = 24) occurred 1–5 days after hatch, and 12% (n = 24) occurred during days 14–21. Twenty-seven of 56 (48%) handled-only chicks died before the first flush at day 10, and 36 of 56 (64%) died before day 21. One hundred thirteen of 138 (83%) control chicks died before day 10 and 116 of 138 (84%) died before day 21. Forty-two (19%; 95% CI: ±5%) of the 221 chicks in our sample survived to day 21.

Daily chick survival varied with age (β = 0.12, SE = 0.02; Fig. 18.1) and precipitation (β = −0.54, SE = 0.04; Fig. 18.2). Our estimate of daily survival of chicks was 0.926 (95% CI: 0.915–0.937);

probability of survival to 21 days post-hatch was 0.193 (95% CI: 0.155–0.255). Our best model ($w_{DIC} = 1.00$) included effects of precipitation, age, handling, and radio-marking. However, the estimates indicated no negative effect of handling ($\beta = 0.15$, SE = 0.21) and weak evidence for an effect of radio-marking ($\beta = -0.54$, SE = 0.33) on chick survival (Table 18.2).

DISCUSSION

Our data suggested that 80% of chicks on our study site in 2008 died prior to 21 days after hatch. The high rate of chick mortality we observed may explain the low juvenile-to-adult ratios observed by NGPC in hunter wing surveys from 2001–07 in southeast Nebraska. Matthews et al. (this volume, chapter 13) reported 21-day success rates of 7.4% for non–radio-marked broods. Our study confirms that the brood survival estimates of Matthews et al. (this volume, chapter 13) were not negatively biased because chicks were missed during flushes; our combined results suggest that a well-designed effort to monitor broods using radio-marked females can provide unbiased estimates of brood survival. Chick mortality events were highest within 14 days of hatch, similar to research reviewed by Hannon and Martin (2006). Our daily survival rate was very similar to the rate from the Flint Hills reported by McNew et al. (this volume, chapter 19). The majority of mortalities (88%) were apparently due to predation, which was similar to other grouse studies (Riley et al. 1998, Gregg et al. 2006, Manzer and Hannon 2008). Our study site has a complex predator community, so chicks could have been depredated by mammals, raptors, or reptiles. Riley and Schulz (2001) suggested that the majority of Ring-necked Pheasant (*Phasianus colchicus*) chick mortalities in the central U.S. were caused by mammals.

Precipitation decreased brood survival of Lesser Prairie-Chickens (*Tympanuchus pallidicinctus*; Fields et al. 1998), and our data suggested the same trend for Greater Prairie-Chicken chicks. Precipitation reduces arthropod numbers during above normal rainfall (Riley et al. 1998), and chicks that ingested large amounts of arthropods had 50% higher survival than chicks that ingested a diet lower in arthropods (Hill 1985). Cool, wet weather also reduces chicks' ability to thermoregulate (Flanders-Wanner et al. 2004), and most exposure mortalities of chicks occur when precipitation is >109% above average (Riley et al. 1998).

Fields et al. (1998) also documented the effect of age on chick survival of Lesser Prairie-Chickens. As chicks age, their mobility improves, which may allow them to more effectively catch arthropods and avoid predators. The primary food source during early development for grouse chicks is insects (Hannon and Martin 2006); Hill (1985) reported that Ring-necked Pheasant chicks increased in body weight as arthropod food intake increased.

The transmitter attachment method worked well, but three (11%) of our transmitters were known to have fallen off prematurely. Burkepile et al. (2002) reported that 11% of posterior sutures on Greater Sage-Grouse chicks failed, and they suggested that suture failure was caused when the anterior suture failed as a result of the sutures restricting tissue growth and expansion. By using suture attachment, we reduced the risk of infection associated with subcutaneous implanted transmitters (Gaunt et al. 1997). The application process required little training and minimal brood handling, and could be performed in the field, reducing risk of female abandonment due to chick translocation off site.

Our highest-ranked chick survival model was well supported ($w_{DIC} = 1.00$) and included effects of handling and radio-marking, which provides some evidence that radio-marking with a suture method and handling accounted for variation in survival of our sample of prairie chicken chicks (Burnham and Anderson 2002). The direction of the effect for radio-marking was negative, although the CI suggested there was, at best, a weak effect of radio-marking on survival. Burkepile et al. (2002) suggested that Greater Sage-Grouse chick survival was not affected by 1-g suture-attached radio transmitters. Our study does not provide strong evidence that chick survival is negatively affected by radio-marking, but it is possible that our sample size was inadequate to make definitive conclusions. For this reason, we encourage future field research to test for effects of transmitters on chicks. Our simultaneous assessment of radio-marked, handled, and control chicks may serve as a useful framework for future investigations.

Transmitter size is a critical consideration for effective radiotelemetry studies. We often found that our 0.5-g transmitters could be not located from our truck-mounted antenna system at distances >300 m, which made it impossible to find 19 (90%) of 21 predated chicks. We were thus unable to determine the factors responsible for mortality. We selected the 0.5-g transmitters to minimize the potential negative effects of transmitters on chicks, but at a trade-off of transmitter performance versus radio mass. Burkepile et al. (2002) lost contact with <10% of 1-g transmitters in a similar study, and radio transmitters weighing 7% of the chick body weight did not reduce survival or weight gain in Ring-necked Pheasant chicks (Ewing et al. 1994) or Wood Duck (*Axis sponsa*) ducklings (Davis et al. 1999). Based on results with gamebirds, biologists may want to consider using 1-g transmitters on prairie chicken chicks when study objectives require relocation of chicks from long distances.

Our radiotelemetry study of Greater Prairie-Chicken chicks provided valuable information on survival rates, and we continue to investigate the unique population dynamics of this stable population with low rates of productivity. Predation was apparently the largest cause of chick mortality, but management of predators is complex (Riley and Schulz 2001). Previous management plans for prairie chickens in agricultural landscapes have focused on providing suitable nesting cover for females. Our data suggest that predation of broods may be a limiting factor for prairie chicken populations, and we encourage landscape-level research efforts to evaluate factors that may contribute to high predation rates of prairie chicken chicks (Schmitz and Clark 1999).

ACKNOWLEDGMENTS

We thank M. Remund and B. Goracke, NGPC biologists who supported our project at the Osage Wildlife Management Area. NGPC funded our research and provided equipment, housing, and logistical support. A portion of this research was funded by a State Wildlife Grant administered through NGPC, and Nebraska Pheasants Forever provided equipment. S. Groepper provided valuable field assistance. ACS was supported by an Undergraduate Creative Activities and Research Experience grant from the University of Nebraska–Lincoln. The School of Natural Resources provided computer and office space for ACS, TWM, and LAP. LAP was supported by Polytechnic of Namibia and a Fulbright award through the U.S. State Department during a professional development leave when this paper was written. This research was supported by Hatch Act funds through the University of Nebraska Agricultural Research Division, Lincoln, Nebraska.

LITERATURE CITED

Burkepile, N. A., J. W. Connelly, D. W. Stanley, and K. P. Reese. 2002. Attachment of radio transmitters to one-day-old sage grouse chicks. Wildlife Society Bulletin 30:93–96.

Burnham, K. P., and D. R. Anderson. 2002. Model selection and multi-model inference: a practical information-theoretic approach. 2nd ed. Springer-Verlag, New York, NY.

Davis, J. B., D. L. Miller, R. M. Kaminski, and M. P. Vtriska. 1999. Evaluation of a radio transmitter for Wood Duck ducklings. Journal of Field Ornithology 70:107–113.

Ewing, D. E., W. R. Clark, and P. A. Vohs. 1994. Evaluation of implanted radio transmitters in pheasant chicks. Journal of the Iowa Academy of Sciences 101:86–90.

Fields, T. L., G. C. White, W. C. Gilbert, and R. D. Rodgers. 1998. Nest and brood survival of Lesser Prairie-Chickens in western Kansas. Journal of Wildlife Management 70: 931–938.

Flanders-Wanner, B. L., G. C. White, and L. L. McDaniel. 2004. Weather and prairie grouse: dealing with effects beyond our control. Wildlife Society Bulletin 32:22–34.

Gaunt, A. S., L. W. Oring, K. P. Able, D. W. Anderson, L. F. Baptista, J. C. Barlow, and J. C. Wingfield. 1997. Guidelines for the use of wild birds in research. The Ornithological Council, Washington, DC.

Gregg, M. A., M. Dunbar, and J. A. Crawford. 2006. Use of implanted radio-transmitters to estimate survival of Greater Sage-Grouse chicks. Journal of Wildlife Management 71:646–651.

Hannon, S. J., and K. Martin. 2006. Ecology of juvenile grouse during the transition to adulthood. Journal of Zoology 269:422–433.

Hill, D. A. 1985. The feeding ecology and survival of pheasant chicks on arable farmland. Journal of Applied Ecology 22:645–654.

Korschgen, C. F., K. P. Kenow, W. L. Green, M. D. Samuel, and L. Sileo. 1996. Technique for implanting radio transmitters subcutaneously in day-old ducklings. Journal of Field Ornithology 67:392–397.

Manzer, D. L., and S. J. Hannon. 2008. Survival of Sharp-tailed Grouse chicks and hens in a fragmented prairie landscape. Wildlife Biology 14:16–25.

Mauser, D. M., and R. L. Jarvis. 1991. Attaching radio transmitters to 1-day-old Mallard ducklings. Journal of Wildlife Management 55:488–491.

Millspaugh, J. J., and J. M. Marzluff. 2001. Radio tracking and animal populations. Academic Press, London, UK.

Powell, L. A. 2007. Approximating variance of demographic parameters using the delta method: a reference for avian biologists. Condor 109:950–955.

Riley, T. Z., W. R. Clark, E. Ewing, and P. A. Vohs. 1998. Survival of Ring-necked Pheasant chicks during brood rearing. Journal of Wildlife Management 62:36–44.

Riley, T. Z., and J. H. Schulz. 2001. Predation and Ring-necked Pheasant population dynamics. Wildlife Society Bulletin 29:33–38.

Schmitz, R. A., and W. R. Clark. 1999. Survival of Ring-necked Pheasant hens during spring in relation to landscape features. Journal of Wildlife Management 63:147–154.

Shaffer, T. L. 2004. A unified approach to analyzing nest success. Auk 121:526–540.

Spiegelhalter, D. J., N. G. Best, B. P. Carlin, and A. Van der Linde. 2002. Bayesian measures of model complexity and fit (with discussion). Journal of the Royal Statistical Society, Series B 64:583–616.

Wisdom, M. J., and L. S. Mills. 1997. Sensitivity analysis to guide population recovery: prairie-chickens as an example. Journal of Wildlife Management 61:302–312.

CHAPTER NINETEEN

Human-Mediated Selection on Life-History Traits of Greater Prairie-Chickens

Lance B. McNew, Andrew J. Gregory, Samantha M. Wisely, and Brett K. Sandercock

Abstract. Predation, food, climate, and other environmental factors have a significant influence on selection processes and evolution of vertebrate life-history traits. Growing evidence indicates that human activities can also affect evolutionary processes by a range of mechanisms, including impacts on life-history traits mediated by the effects of habitat management on survival of nests and adults. We tested for anthropogenic effects on the life-history evolution of Greater Prairie-Chickens (*Tympanuchus cupido*) breeding at three sites across a gradient of landscape alteration in eastern Kansas. Female prairie chickens breeding in an area heavily fragmented by row-crop agriculture and roads had low annual survival probabilities (0.32 ± 0.001 SE) and higher survival of nests (0.16 ± 0.04) and broods (0.48 ± 0.12) than the other two study areas. In contrast, two populations breeding in areas with large tracts of contiguous heavily grazed tallgrass prairie had higher annual survival (0.47 ± 0.002 and 0.68 ± 0.01) and lower survival of nests (0.07 ± 0.02 and 0.12 ± 0.03) and broods (0.29 ± 0.09 and 0.38 ± 0.09, respectively). Consistent with life-history theory predictions, the population in the fragmented area with higher adult mortality also had greater reproductive effort, and egg and clutch volumes were 5% and 9% larger than at the other study areas. Reproductive effort was not influenced by other explanatory variables, including residual female body mass. Overall, variation in the life-history traits of prairie chickens was most consistent with site differences in nest predation rates and mortality of adult females. Predation on breeding females was positively associated with the anthropogenic effects of road development and conversion of grasslands to cropland. Our results indicate that land use and land cover change can influence selection on life-history traits for a short-lived species at small spatial and short temporal scales, even after adjusting for potential phenotypic plasticity.

Key Words: anthropogenic impacts, demography, evolution, grouse, reproduction, survival.

McNew, L. B., A. J. Gregory, S. M. Wisely, and B. K. Sandercock. 2011. Human-mediated selection on life-history traits of Greater Prairie-Chickens. Pp. 255–266 *in* B. K. Sandercock, K. Martin, and G. Segelbacher (editors). Ecology, conservation, and management of grouse. Studies in Avian Biology (no. 39), University of California Press, Berkeley, CA.

life-history theory predicts that the diversity of life-history strategies in vertebrates can be explained by trade-offs among demographic traits that maximize lifetime reproductive success and fitness. Species with low adult survival should invest heavily in components of reproduction, whereas longer-lived organisms should invest less in current reproduction, at least early in their lives, to maximize benefits from residual reproductive value in future breeding attempts (Roff 1992, Martin 2002). Interspecific comparisons of variation in avian life-history traits have provided evidence for trade-offs between annual survival and the components of reproductive effort, including the probabilities of breeding and renesting, clutch size, and egg mass (Martin 1995, Ricklefs 2000, Sæther and Bakke 2000, Martin et al. 2006). Studies seeking ecological correlates of patterns of avian life-history variation have usually focused on four major factors: predation, food limitation, climatic conditions, and duration of the breeding season (Badyaev 1997, Conway and Martin 2000, Sandercock et al. 2005). Of these four factors, predation may be most important for explaining life-history variation within and among different species of birds because most demographic losses are caused by predator activity (Ricklefs 1969, 2008; Martin 1995).

High rates of nest predation are predicted to favor reductions in reproductive effort (Martin 2004). In songbirds, high levels of nest predation are associated with reductions in egg size, clutch mass, and nest attentiveness, and increases in nestling growth rates (Conway and Martin 2000, Remeš and Martin 2002, Fontaine and Martin 2006, Martin et al. 2006). Reductions in reproductive effort may be mediated by trade-offs among the different components of fecundity if finite resources must be partitioned between the number and size of offspring (Smith and Fretwell 1974, Winkler and Wallin 1987). Juvenile survival may place thresholds on the minimum size of offspring, and large eggs tend to produce large chicks that have higher survival rates in birds with precocial young (Myrberget 1977, Moss et al. 1981). Intraspecific trade-offs between clutch and egg size are rarely observed in birds, in part because egg size is highly heritable (Christians 2002). Nevertheless, egg mass decreased with increased clutch size in an interspecific comparison of songbird demography (Martin et al. 2006), and egg mass

increased in response to removals of nest predators (Fontaine and Martin 2006).

In contrast to the effects of nest predation, low rates of adult mortality are predicted to favor reduced reproductive effort (Martin 2004). In songbirds, species with low adult mortality exhibit reduced rates of nest attendance, and lower attentiveness is associated with longer incubation periods (Martin 2002). Trade-offs between survival and reproductive effort have been documented for precocial species as well, with females that have lower annual survival laying larger clutches (Patten et al. 2007) or exhibiting a higher propensity to renest (Martin et al., this volume, chapter 17). The effects of predators on juvenile survival may also play a critical role in shaping avian life histories, with low rates of juvenile mortality favoring increased reproductive effort (Russell 2000, Martin 2002). Life-history studies that address juvenile survival are fairly limited, primarily because of logistical difficulties in tracking and monitoring mobile young during natal dispersal (Hannon and Martin 2006).

Differences in resource acquisition among females can confound the detection of life-history trade-offs if life-history traits are phenotypically plastic (van Noordwijk and de Jong 1986). Trade-offs between realized fecundity and annual survival can be produced by resource limitations (Ricklefs 2000). For example, clutch size, nesting propensity, and the interval between nesting attempts were associated with the spring body condition of female Mallards (*Anas platyrhynchos*; Devries et al. 2008), and plasma protein and female age were significant predictors of renesting probability in Greater Sage-Grouse (*Centrocercus urophasianus*; Gregg et al. 2006). In addition, egg size has been found to vary among species in relation to residual body mass, an index of condition (Rahn et al. 1985, Sæther 1987). Indeed, the positive relationships between food resources and clutch and egg size have been invoked often to explain observations that do not support the clutch size:egg mass trade-off (Lack 1968, Sæther 1987, Martin et al., this volume, chapter 17).

Comparative studies of grouse (Tetraoninae) have played an important role in the development and testing of life-history theory. Interspecific studies have demonstrated that grouse exhibit the same fast–slow continuum in life-history strategies that is found in other groups of vertebrates, including trade-offs between clutch size and adult survival (Zammuto 1986, Arnold 1988, Jönsson et al. 1991). Demographic studies of ptarmigan (*Lagopus* spp.) have shown that alpine populations at southern latitudes have lower fecundity and higher adult survival than arctic populations at northern latitudes, and that predation is important as an environmental factor (Sandercock et al. 2005, Novoa et al. 2008). To date, most studies of life-history variation in birds have focused on the impacts of environmental factors under relatively undisturbed or natural conditions (Bears et al. 2009, Martin et al. 2009, Martin et al., this volume, chapter 17). However, mounting evidence now indicates that human activities can affect evolutionary processes through a variety of mechanisms, including habitat modification, selective harvest, captive breeding, and translocations (Carroll et al. 2007, Smith and Bernatchez 2008). Anthropogenic effects on land use and habitat fragmentation may have led to the observed changes in the demographic traits of Lesser Prairie-Chickens (*Tympanuchus pallidicinctus*; Patten et al. 2005). Historic differences in land tenure created major differences in the extent of fencing, power lines, and roads in rural areas of Oklahoma and New Mexico. Collisions with fences are a major cause of mortality of female prairie chickens in Oklahoma (Wolfe et al. 2007), and higher adult mortality due to collisions was correlated with larger clutch sizes and higher renesting rates in Oklahoma as compared to New Mexico (Patten et al. 2005). However, the indirect impacts of nest failure and adult mortality due to human-caused habitat alteration on the selection for demographic traits have not been assessed.

In this study, we compare the demographic traits of three independent populations of Greater Prairie-Chickens across a gradient of human landscape alteration. The landscapes of Kansas provide a unique opportunity to evaluate whether alteration of habitats impacts the selection of life-history traits of Greater Prairie-Chickens (*T. cupido*) because land use and range management practices vary significantly within the state. In the Flint Hills, large contiguous tracts of grassland are intensively managed for cattle production, whereas in the Smoky Hills, smaller tracts of less heavily grazed grassland are fragmented by row-crop agriculture (McNew et al., this volume, chapter 15). Habitat conditions impact the seasonal availability of lekking, nesting, and brood-rearing habitat (Patten et al. 2007), the phenology of breeding events (McNew et al., this volume, chapter 15), and variation in reproductive success and survival (McKee et al. 1998, Matthews et al., this volume, chapter 13). If anthropogenic changes lead to rapid selection for avian life-history traits, we expected that Greater Prairie-Chickens might be good candidates to investigate these effects because this species has large clutch sizes, low adult survival, and presumably shorter generation times than tundra or forest grouse (Patten 2009). We also expected that changes in vital rates might be mediated by nest predation because Greater Prairie-Chickens experience considerable variation in nest survival among different populations (0–72%), and nest predation is the primary cause of reproductive losses (Schroeder and Robb 1993, Peterson and Silvy 1996). If large variations in habitat conditions influence demographic rates, we expected greater reproductive effort in populations experiencing higher reproductive success or lower adult survival. Finally, our analyses were strengthened by use of standardized field protocols to investigate a suite of demographic traits among multiple populations of a single species. Our approach controls for differences in methodology and phylogenetic relationships that can be an issue for interspecific comparisons of life-history traits (Martin 1995, Sandercock et al. 2005, Martin et al. 2006).

STUDY SPECIES AND STUDY SITES

Greater Prairie-Chickens (hereafter prairie chickens) are endemic to the native grasslands of the central United States. Prior to European settlement, prairie chickens were distributed across all areas occupied by tallgrass prairie in North America (Schroeder and Robb 1993). Large-scale conversions of native prairies to row-crop agriculture during the last century are thought to be the major cause of declines in both the distribution and number of prairie chickens, which have led to population bottlenecks (Westemeier et al. 1998, Johnson and Dunn 2008). The core of the extant range of prairie chickens occurs in Kansas and adjacent states (Schroeder and Robb 1993). In Kansas, prairie chickens primarily occur in areas that are dominated by native grasslands, such as the Flint Hills ecoregion. Nevertheless, prairie chickens can tolerate moderate amounts of cultivated agriculture (<40% of total area), and populations of prairie chickens are also found in more developed regions of Kansas. Elsewhere, cultivation, grazing, and other types of human land use have reduced the population viability of prairie chickens, but the potential role of land use and land cover change as drivers of natural selection have not been investigated (Svedarsky et al. 2003).

Our study occurred at three discrete study sites: two sites located in the southern and northern Flint Hills (South and North, respectively) and one site in the Smoky Hills (Smoky). The three study areas were ≥112 km apart and differed in landscape composition and pattern as well as rangeland management practices (McNew et al., this volume, chapter 15). The South site (635 km^2) had landcover of 90% grassland and 3% cropland, a mean grassland patch size of 185 ha, and a road density of 0.32 km of roads per km^2. The majority of the site was managed with range management practice of intensive early stocking and burned annually each spring (IESB, 1 head/0.8 ha for 90 days; Smith and Owensby 1978, With et al. 2008). The North site (533 km^2) had landcover of 81% grassland and 10% cropland, a mean grassland patch size of 51 ha, and a road density of 0.57 km per km^2. Annual spring burning was common and lands were managed with a mixture of IESB and season-long stock grazing and annual burning (SLSB; 1 head/1.6 ha for 180 days). The Smoky site (1,295 km^2) was more fragmented, with landcover of 53% grassland and 38% cropland, a mean grassland patch size of 15 ha, and a higher road density of 1.4 km per km^2. Cultivated crops include sorghum, corn, wheat, and soybeans. Native grass pastures at study area 3 were burned infrequently at fire return intervals >1 year, grazed at low intensity (1 head/>2 ha for 90 days), and cattle stocking occurred later in the season than at the other two study sites. Indices of prairie chicken densities for years of study, calculated as: mean number of prairie-chickens per lek × number of leks per study area size, were 0.10, 0.19, and 0.17 birds/km^2 for the South, North, and Smoky sites, respectively.

METHODS

Field Methods

Prairie-chickens were captured at lek sites during the spring with walk-in traps and dropnets (Silvy et al. 1990, Schroeder and Braun 1991). Captured birds were sexed by plumage characteristics (Henderson et al. 1967). We determined age-class as yearling or adult from the shape, coloration, and wear of the outermost two primaries (numbers 9 and 10; Schroeder and Robb 1993). Morphometrics of adults, including total mass and length of the tarsus–metatarsus, were measured at the time of capture. All birds were individually marked with color leg bands and females were fitted with 11-g necklace-style VHF radio transmitters, equipped with mortality switches and an expected battery life of 12 months (Model RI-2B, Holohil Systems Ltd., Ontario, Canada). Radio-marked females were monitored ≥3 times per week from vehicles during the nesting and brood-rearing period (April–August) and ≥1 time per week during the rest of the year (September–March). Once a female localized in an area for three successive days, we used a portable radio receiver and handheld Yagi antenna to locate the nest. We flushed the female once in early incubation to count the eggs, to determine the stage of incubation, and to record the nest location. Females with nests were monitored daily at a distance ≥100 m by triangulation of the radio signal. Once it was determined that the female was no longer tending the nest, we classified nest fate as successful (≥1 chick produced) or failed.

Body Mass of Females

Reproductive effort of female prairie chickens at the different study sites could be influenced by site differences in food resources if females with heavier body mass were in better nutritional condition and laid larger eggs. Alternatively, site differences in body mass could be a result of seasonal differences in ovarian development among females at capture. We evaluated the first possibility by regressing female mass at capture on length of the tarsus–metatarsus as an index of body size. Residual body mass of females was used as an index of spring body condition before egg-laying. Assessment of ovarian development was difficult because we were unable to determine if females were gravid at capture. We used the interval between the day of capture and the day of nest initiation as a covariate (McNew et al., this volume, chapter 15). Mass of a female grouse increases before the onset of egg-laying (Hannon and Roland 1984), and we expected that females with shorter intervals between capture and nest initiation were more likely to be gravid. We used analysis of covariance to test whether regional differences in female mass at capture were influenced by the length of time between capture and nest initiation. We tested factorial models with main effects and interaction terms, and all parametric statistics were calculated using procedures of program SAS (ver. 9.1, SAS Institute, Cary, NC).

Clutch Size and Egg Volume

Clutch size was calculated as the maximum number of eggs recorded per clutch once egg-laying was completed and a female had started incubation. We floated all eggs from clutches determined to be in incubation to assess stage of development from egg buoyancy, adjusting for cases where egg-laying rates exceeded one egg per day (McNew et al. 2009, McNew et al., this volume, chapter 15). We measured egg volume only once during incubation to minimize the impacts of nest visits. Egg length (L) and breadth (B) were measured to the nearest 0.1 mm using calipers, and linear measurements were converted to an estimated egg volume (V) with the following equation (Narushin 2005):

$$V = (0.6057 - 0.0018B)LB^2$$

Mean clutch size and egg volumes were compared among study areas using analysis of variance. We also compared egg volume and clutch size relative to residual body mass of females. Analysis of covariance was used to test whether site differences in clutch size and egg volume could be explained by potential variation in the nutritional condition or the age of females.

Nest and Brood Survival

Nest survival was the probability of a nest producing ≥1 hatched chick, whereas brood survival was defined as the probability that ≥1 chick survives to fledging at 14 days after hatching. We calculated daily rates of nest and brood survival for each study area with the nest survival model of program MARK (ver. 4.3; White and Burnham 1999, Dinsmore et al. 2002). Multiple model selection and inference was used to evaluate the importance of three factors on daily nest survival (Burnham and Anderson 1998). The three factors included in the global model for nest survival included nesting attempt (first or renest), female age, and study site. We estimated a corrected probability of nest survival by raising the daily nest survival probabilities to a power equal to the duration of the nest exposure period (37 days; Dinsmore et al., 2002, Sandercock et al. 2005, McNew et al., this volume, chapter 15). This method assumes that daily nest survival is similar across the nest exposure period within a study site. Duration of the nesting cycle was calculated assuming an egg-laying rate of one egg per day and an average incubation period of 25 days (Nooker 2007, McNew et al., this volume, chapter 15). To estimate brood survival prior to fledging, we conducted early-morning flush counts of females attending broods at 14 days post hatch. (Hubbard et al. 1999, Fields et al. 2006). If no chicks were counted, we used subsequent flush counts at 10-day intervals to confirm presence or absence of chicks. We updated 14-day flush counts for 5% of cases from zero to the maximum number of observed chicks at later flush counts. The probability of brood survival to fledging was calculated as the product of the estimates of daily brood survival from the top model for a 14-day period from hatching until fledging. Variances of derived parameters were calculated using the delta method (Powell 2007).

Survival of Females

We estimated monthly survival of female prairie chickens during a two-year period between March 2007 and February 2009 with the nest survival procedure of program MARK. The nest survival model is a general procedure for known-fate data and is useful for estimating survival from "ragged" telemetry data from radio-marked birds (Hartke et al. 2006, Mong and Sandercock 2007). Multiple model selection and inference was used to evaluate the importance several factors on monthly adult survival (Burnham and Anderson 1998), including female age, study area, residual body mass adjusted for tarsus–metatarsus length, and linear and quadratic time trends. We used the most parsimonious model to derive monthly survival probabilities, and then extrapolated annual survival rates at each study area as the product of monthly survival rates during the entire study period. Variances of derived parameters were calculated using the delta method (Powell 2007).

RESULTS

Body Mass of Females

A total of 203 individual female prairie chickens were captured before egg-laying at our three study areas in Kansas. Reproductive data were available for 159 females. We excluded females for which the capture to clutch initiation interval was less than zero ($n = 8$). Analysis of covariance showed there was no interaction between the effects of study site and the interval between capture and egg-laying ($F_{2,124} = 0.93$, $P = 0.40$). Body mass of female prairie chickens differed significantly among the three study sites ($F_{2,124} = 7.7$, $P < 0.001$), and females were heaviest at the Smoky site (929 ± 8.8 g), intermediate at the North site (908 ± 8.7 g), and lightest at the South site (879 ± 7.9 g; Table 19.1). The interval between capture and egg-laying was unrelated to female mass ($F_{1,124} = 1.7$, $P = 0.20$). Mass did not differ between age-classes ($F_{1,124} = 0.39$, $P = 0.54$), and the interaction between female age and study site was not significant ($F_{1,124} = 0.88$, $P = 0.42$). In addition, the interval between female capture and clutch initiation did not differ among age classes ($F_{1,136} = 1.6$, $P = 0.21$). We found no significant relationship between female mass and tarsus–metatarsus length ($r^2 < 0.01$, df $= 1$, $P = 0.64$).

Clutch Size and Egg Volume

Analysis of covariance showed there was no interaction between the effects of study area or nesting attempt on clutch size ($F_{2,151} = 0.13$, $P = 0.88$). First nests were consistently larger than renests ($F_{1,151} = 39.1$, $P < 0.001$); the number of eggs per clutch averaged 12.5 to 13.1 eggs for first nesting attempts and 10.2 to 10.9 eggs for renests (Table 19.1). First nests at the Smoky site tended to be larger by about 0.5 eggs per clutch, but overall, clutch size did not differ significantly among study areas ($F_{1,151} = 0.44$, $P = 0.65$). Clutch size did not differ between female age classes (mean $\pm$ SE $= 12.7 \pm 0.25$ for both groups; $F_{1,141} = 0.98$, $P = 0.32$), and there was no interaction between age-class and study site ($F_{2,141} = 0.28$, $P = 0.75$). Analysis of covariance showed that there was no interaction between the effects of residual female mass and study area on egg volume ($F_{2,143} = 1.07$, $P = 0.35$). Egg volume differed among the three study areas ($F_{2,142} = 3.2$, $P = 0.04$), with the largest eggs laid at the Smoky site (24.7 ± 0.2 ml) and the smallest eggs at the South site (23.7 ± 0.2 ml; Table 19.1). Egg volume did not differ between female age classes ($F_{1,140} = 2.8$, $P = 0.09$), and there was no interaction between age-class and study site ($F_{2,140} = 2.5$, $P = 0.08$). Egg volume was not related to clutch size ($r^2 = 0.01$, $P = 0.20$).

Nest Survival and Brood Survival

During the breeding seasons of 2006–2008, 231 nests of 155 female prairie chickens were located and monitored, of which 44 were successful, for an apparent nest success rate of 19%. Daily nest survival was modeled for a 37-day exposure period during a 103-day nesting season from 23 April to 19 July. The most parsimonious model ($\Delta AIC_c = 0$) included a group effect for study area. Models where nest survival varied among study areas were 9.9 times more likely than models where nest survival was constant ($w_i/w_j = 0.79/0.08$). Variation in survival among study areas accounted for 79% of the relative support of the data. Nest survival was lower at the South site (0.07 ± 0.02) compared to the North (0.12 ± 0.03) and Smoky sites (0.16 ± 0.04; Table 19.1). Overall nest survival for all sites and nesting attempts combined was 0.12 ± 0.04 SE. Evidence at failed nests indicated that predation was the primary cause of nest mortality, accounting for 94% of all losses.

Forty-three broods were monitored from hatch until fledging at 14 days of age. Daily brood survival during this period was modeled for a 69-day brood-rearing period from 17 May to 24 July. A model that contained an effect of study area was considered parsimonious ($\Delta AIC_c = 0.37$). Models where brood survival varied among study areas had 44% of the relative support. Site differences in brood survival were similar to patterns of nest survival: Survival of broods was highest at the Smoky site (0.45 ± 0.11), intermediate at the North site (0.32 ± 0.12), and lowest at the South site (0.24 ± 0.10; Table 19.1). Overall, the model-averaged estimate of brood survival until fledging across all study areas was 0.35 ± 0.07.

TABLE 19.1

Mean estimates (± SE) for body mass and demographic traits of female Greater Prairie-Chickens breeding at three study areas in eastern Kansas, 2006–2008.

Parameter	South	n	North	n	Smoky	n	Statistics[a]
Body mass of females (g)	879 (7.9)	61	908 (8.7)	51	929 (8.8)	50	$F = 6.8$, $P < 0.01$
Clutch size of first nests	12.5 (0.3)	41	12.6 (0.3)	43	13.1 (0.3)	40	$F = 1.6$, $P = 0.21$
Clutch size of renests	10.4 (0.4)	21	10.9 (0.5)	14	10.2 (0.5)	10	$F = 0.5$, $P = 0.62$
Egg volume (ml)	23.7 (0.2)	62	24.2 (0.2)	58	24.7 (0.2)	51	$F = 2.3$, $P = 0.05$
Clutch volume (ml)	278 (6.8)	62	290 (7.1)	58	304 (7.5)	51	$F = 2.8$, $P = 0.06$
Nest survival	0.07 (0.02)	83	0.12 (0.03)	85	0.16 (0.04)	63	$\Delta AIC_c = 0.0$, $w_i/w_j = 7.3$
Brood survival	0.29 (0.09)	15	0.38 (0.09)	12	0.48 (0.12)	16	$\Delta AIC_c = 0.37$, $w_i/w_j = 1.2$
Annual survival of females	0.68 (0.01)	55	0.47 (0.002)	84	0.32 (0.001)	69	$\Delta AIC_c = 0.0$, $w_i/w_j = 99.0$

[a] Parametric statistics were based on analysis of variance. Analyses of survival were based on model selection with AIC_c, where ΔAIC_c = difference in AIC_c between a model where survival differs among the three study areas and the minimum AIC_c model, and w_i/w_j = evidence ratios calculated as the ratio of relative support for the pooled weights of models where survival rates differed among the three study areas versus models where survival did not differ among areas.

Female Annual Survival

Monthly survival probabilities were estimated for 203 females. Model selection based on AIC_c indicated that variation in survival among study sites was strongly supported by the data, accounting for more than 99% of the relative support. Estimates of annual survival extrapolated from monthly rates were greater at the South site (0.68 ± 0.01) than at the North (0.47 ± 0.002) and Smoky sites (0.32 ± 0.001; Table 19.1). Overall annual survival of females during the 12-month period from March to February for all sites pooled was 0.48 (±0.001).

DISCUSSION

Female Greater Prairie-Chickens breeding at three sites across a gradient of human landscape alteration and use in the Flint Hills and Smoky Hills of Kansas exhibited variation in a suite of eight life-history traits. Females breeding at a study site consisting of large, contiguous blocks of heavily grazed native prairie (South) had the lightest body mass, laid the smallest eggs, and had the lowest clutch volume. Nest and brood survival were low but annual survival was high for prairie chickens breeding in large tracts of heavily grazed and intensively burned prairie. In contrast, females breeding at a highly fragmented, moderately grazed, and infrequently burned site (Smoky) had the heaviest body mass, laid the largest eggs, and had the greatest clutch volume. The Smoky site had the highest rates of nest and brood survival, although our estimates were depressed compared to values compiled for other populations (Peterson and Silvy 1996). In fact, our estimates of annual survival for females at the fragmented Smoky site are among the lowest values ever reported for a field study of prairie chickens. The study site in the northern Flint Hills (North) had intermediate amounts of habitat fragmentation and grazing intensity, and the vital rates of female prairie chickens were intermediate as well. We thus evaluate the potential roles of phenotypic plasticity and evolutionary processes as potential explanations for the results of our demographic analyses.

Trade-offs between realized fecundity and annual survival are often interpreted as resulting from evolutionary processes, but trade-offs can also be produced by phenotypic plasticity and resource limitations (Ricklefs 2000). For example, site differences in female mass in our study could have been an artifact of differences in date of capture and the degree of gravidity among females before egg-laying. Timing of lek attendance did not differ among the three study areas, but clutch initiation was delayed at the South site, and females at Smoky could have been closer to egg-laying at capture (McNew et al., this volume, chapter 15). Alternatively, variation in female mass could have been the result of site differences in female age structure provided there are differences in mass between yearling and adult females. We reject differences in seasonal phenology as an explanation for variation in female mass at capture, because body mass was not related to the interval between capture and date of nest initiation, and reproductive effort still differed among areas after adjustment for the covariate. Likewise, we reject the latter explanation because the age structure of captured females was similar among sites (~50% yearlings: 50% adults) and female mass did not differ between the age-classes.

Phenotypic plasticity (i.e., the ability of females to alter their reproductive effort based on body condition) could also be relevant if site differences in body mass, clutch size, and egg volume were due to regional differences in food availability that impacted the body condition of egg-laying females. Females had the highest body mass and laid the largest clutches and eggs at the Smoky site, a site fragmented by agricultural development. Cultivated agricultural fields comprised a higher proportion of the landscape at the Smoky site, and prairie chickens will utilize grain sorghum and other crops during winter and early spring (Robel et al. 1970). Two lines of evidence suggest that body condition cannot explain regional variation in reproductive effort of prairie chickens in Kansas. First, residual female mass did not explain variation in egg volume among our three study areas. Food supplementation usually has little impact on egg size of birds but can have larger effects on timing of laying and clutch size (Christians 2002). Estimates of heritability for egg size are often high in birds, suggesting that egg size may be under selective pressures unrelated to the nutritional status of laying females. Second, egg volume of prairie chickens was not related to clutch size. Life-history theory predicts a negative relationship between egg size and clutch size if female resources must be partitioned (Roff 1992), but a positive association would be expected if both traits are impacted by nutritional condition, which we did not observe.

Lower reproductive effort among prairie chickens breeding in heavily grazed contiguous grasslands and higher reproductive effort among prairie chickens in moderately grazed and fragmented grasslands was consistent with life-history theory, which predicts that high nest predation and high adult survival should select for reduced reproductive effort (Roff 1992). Mortality of female prairie chickens was almost entirely the result of predation (90%; L. B. McNew, unpubl. data). Thus, the most important environmental factor leading to divergence in the life-history traits of prairie chickens appears to be the impacts of predators on the survival of adults and nests. We lacked estimates of predator abundance for our three study areas, but fragmentation by agricultural development and road density were ranked Smoky > North > South. Known predators of prairie chickens, such as coyotes (*Canis latrans*), use edge habitats and roads for travel and foraging (Kuehl and Clark 2002, Tigas et al. 2002). Higher-quality nesting and brood-rearing habitat as a result of greater residual cover due to infrequent burning and lower cattle grazing intensity (McNew et al., this volume, chapter 15) could explain greater reproductive success at the Smoky site. Thus, anthropogenic changes in land use and habitat fragmentation may have led to differential rates of exposure to predators. Limited data from prior to large-scale implementation of IESB suggest that nest success of prairie chickens in the Flint Hills was similar (35%) to our estimates from the Smoky Hills (Robel 1970). Therefore, it appears that the direct effects of human activities on grassland ecosystems and the indirect impacts of habitat modification upon predator–prey interactions have influenced the selection of life-history traits of Greater Prairie-Chickens in Kansas over a relatively short time period. Notwithstanding, our results should be viewed in the context of a relatively short-term field study.

There is mounting evidence that human activities have led to ecologically significant evolutionary change in a variety of taxa, and at a range of temporal and spatial scales, contributing to growing interest in the study of contemporary evolution (Carroll et al. 2007, Smith and Bernatchez 2008). Relatively few studies have evaluated the impacts of habitat loss and degradation on the life-history evolution of terrestrial vertebrates. Cutting of grasslands for hay production destroys nests of grassland songbirds, including Savannah Sparrows (*Passerculus sandwichensis*) breeding in dairy pastures in Vermont. Perlut et al. (2008) showed that timing of hay cutting altered mating strategies and the occurrence of extra-pair copulations, as well as the strength of selection on morphological traits. Fencing of pastures for livestock is a landscape modification that poses a risk of collision mortality for female Lesser Prairie-Chickens (Wolfe et al. 2007), and Patten et al. (2005) presented evidence that female prairie chickens subject to higher fence collision mortality laid larger clutches and had a greater probability of renesting than birds at less heavily fenced sites. Our study extends these previous results by showing that landscape modification by humans may lead to differential rates of predation that affect the life-history traits of Greater Prairie-Chickens. Mammalian predators play an important role in structuring terrestrial ecosystems (Pace et al. 1999), but previous studies investigating trophic dynamics have primarily focused on the ecological consequences of the removal of top predators and mesopredator release (Elmhagen and Rushton 2007, Berger et al. 2008). Changes in predator abundance and diversity can also drive evolutionary change in the life-history strategies of lower trophic levels. For example, predators can determine the life-history evolution of guppies (*Poecilia reticulata*) in captivity and natural environments (Reznick et al. 2008). Selective removal of top predators is one way that humans influence life-history evolution, but our results suggest that indirect effects of landscape modification on predation risk can also be important.

Our analysis is one of the first studies to assess the influence of human landscape alteration on the life-history evolution of grouse, and our work could be extended in two ways. First, we observed the impacts of predation on the demographic parameters of prairie chickens but were unable to determine whether variation in predation rates was due to a numerical or a functional response. We lacked estimates of predator abundance, and the identity of major predators was surmised by inspecting the remains of depredated nests and carcasses. Our analyses would be informed by a better understanding of predator abundance and activity in relation to land use and landcover changes. Second, our analyses were based on retrospective comparisons of demographic data for prairie chickens at three study sites over a short time, and life-history traits could have covaried with an environmental factor that we failed to consider (Ricklefs 2000). Experimental protocols are a stronger approach to testing for local adaptation but would require raising birds in a common environment or reciprocal transplants among different populations (James 1983, Rhymer 1992, Bears et al. 2008). Experimental tests will be logistically difficult for prairie chickens because of their large home range requirements, vagility, and conservation status. Wildlife management activities are rarely considered from an evolutionary perspective but could have potential for analyses of contemporary life-history evolution in prairie chickens. For example, comparisons of performance between wild prairie chickens and pen-reared Attwater's Prairie-Chickens (*T.c. attwateri*) might yield insights into the selection conditions of captive-rearing environments (Peterson and Silvy 1996, Hess et al. 2005). Finally, ongoing translocations of prairie chickens from Kansas to relict populations in Illinois and Missouri (Westemeier et al. 1998, J. C. Pitman, pers. comm.) will provide future opportunities for investigating adaptation in wild populations in new environments.

ACKNOWLEDGMENTS

We thank the many field technicians who helped collect field data, especially D. Broman, T. Cikanek, L. Hunt, V. Hunter, and W. White. Funding and equipment were provided by a consortium of federal and state wildlife agencies, conservation groups, and wind energy partners under the National Wind Coordinating Collaborative including National Renewable Energies Laboratory (DOE), U.S. Fish and Wildlife Service, Kansas Department of Wildlife and Parks, Kansas Cooperative Fish and Wildlife Research Unit, National Fish and Wildlife Foundation, Kansas and Oklahoma chapters of The Nature Conservancy, BP Alternative Energy, FPL Energy, Horizon Wind Energy, and Iberdrola Renewables. B. K. Sandercock and S. M. Wisely were supported by the Division of Biology at Kansas State University. A. J. Gregory was supported by a research fellowship from the NSF-funded GK-12 Program (DFE-0841414). We thank J. F. Cully, Jr., K. Martin, and two anonymous reviewers for comments on the manuscript.

LITERATURE CITED

Arnold, T. W. 1988. Life histories of North American game birds: a reanalysis. Canadian Journal of Zoology 66:1906–1912.

Badyaev, A. V. 1997. Avian life history variation along altitudinal gradients: an example with cardueline finches. Oecologia 111:365–374.

Bears, H., M. C. Drever, and K. Martin. 2008. Comparative morphology of Dark-eyed Juncos *Junco hyemalis* breeding at two elevations: a common aviary study. Journal of Avian Biology 39:152–1621.

Bears, H., K. Martin, and G. C. White. 2009. Breeding in high-elevation results in shifts to slower life-history strategy within a single species. Journal of Animal Ecology 78:365–375.

Berger, K. M., E. M. Gese, and J. Berger. 2008. Indirect effects and traditional trophic cascades: a test involving wolves, coyotes, and pronghorn. Ecology 89:818–828.

Burnham, K. P. and D. R. Anderson. 1998. Model selection and inference: a practical information-theoretic approach. Springer. New York, NY.

Carroll, S. P., A. P. Hendry, D. N. Reznick, and C. W. Fox. 2007. Evolution on ecological time-scales. Functional Ecology 21:387–393.

Christians, J. K. 2002. Avian egg size: variation within species and inflexibility within individuals. Biological Reviews 77:1–26.

Conway, C. J., and T. E. Martin. 2000. Evolution of passerine incubation behavior: influence of food, temperature, and nest predation. Evolution 54:670–685.

Devries, J. H., R. W. Brook, D. W. Howerter, and M. G. Anderson. 2008. Effects of spring body condition and age on reproduction in Mallards (*Anas platyrhynchos*). Auk 125:618–628.

Dinsmore, S. J., G. C. White, and F. C. Knopf. 2002. Advanced techniques for modeling avian nest survival. Ecology 83:3476–3488.

Elmhagen, B., and S. P. Rushton. 2007. Trophic control of mesopredators in terrestrial ecosystems: top-down or bottom-up? Ecology Letters 10:197–206.

Fields, T. L., G. C. White, W. C. Gilgert, and R. D. Rodgers. 2006. Nest and brood survival of Lesser Prairie-Chickens in west-central Kansas. Journal of Wildlife Management 70:931–938.

Fontaine, J. J., and T. E. Martin. 2006. Parent birds assess nest predation risk and adjust their reproductive strategies. Ecology Letters 9:428–434.

Gregg, M. A., M. R. Dunbar, J. A. Crawford, and M. D. Pope. 2006. Total plasma protein and renesting by Greater Sage-Grouse. Journal of Wildlife Management 70:472–478.

Hannon, S. J., and K. Martin. 2006. Ecology of juvenile grouse during the transition to adulthood. Journal of Zoology 269:422–433.

Hannon, S. J., and J. Roland. 1984. Morphology and territory acquisition of Willow Ptarmigan. Canadian Journal of Zoology 62:1502–1506.

Hartke, K. M., J. B. Grand, G. R. Hepp, and T. H. Folk. 2006. Sources of variation in survival of breeding female Wood Ducks. Condor 108:201–210.

Henderson, F. R., F. W. Brooks, R. E. Wood, and R. B. Dahlgren. 1967. Sexing of prairie grouse by crown feather patterning. Journal of Wildlife Management 31:764–769.

Hess, M. F., N. J. Silvy, C. R. Griffin, R. R. Lopez, and D. S. Davis. 2005. Differences in flight characteristics of pen-reared and wild prairie-chickens. Journal of Wildlife Management 69:650–654.

Hubbard, M. W., D. L. Garner, and E. E. Klaas. 1999. Wild Turkey poult survival in south-central Iowa. Journal of Wildlife Management 63:199–203.

James, F. C. 1983. Environmental component of morphological differentiation in birds. Science 221:184–186.

Johnson, J. A., and P. O. Dunn. 2008. Low genetic variation in the Heath Hen prior to extinction and implications for the conservation of prairie-chicken populations. Conservation Genetics 7:37–48.

Jönsson, K. I., P. K. Angelstam, and J. E. Swenson. 1991. Patterns of life-history and habitat in Palaearctic and Nearctic forest grouse. Ornis Scandinavica 22:275–281.

Kuehl, A. K., and W. R. Clark. 2002. Predator activity related to landscape features in northern Iowa. Journal of Wildlife Management 66:1224–1234.

Lack, D. 1968. Ecological adaptations for breeding in birds. Methuen, London, UK.

Martin, M., A. F. Camfield, and K. Martin. 2009. Demography of an alpine population of Savannah Sparrows. Journal of Field Ornithology 80:253–264.

Martin, T. E. 1995. Avian life history evolution in relation to nest sites, nest predation, and food. Ecological Monographs 65:101–127.

Martin, T. E. 2002. A new view of avian life-history evolution tested on an incubation paradox. Proceedings of the Royal Society of London 269:309–316.

Martin, T. E. 2004. Avian life-history evolution has an eminent past: does it have a bright future? Auk 121:289–301.

Martin, T. E., R. D. Bassar, S. K. Bassar, J. J. Fontaine, P. Lloyd, H. A. Mathewson, A. M. Niklison, and A. Chalfoun. 2006. Life-history and ecological correlates of geographic variation in egg and clutch mass among passerine species. Evolution 60:390–398.

McKee, G., M. R. Ryan, and L. M. Mechlin. 1998. Predicting Greater Prairie-Chicken nest success from vegetation and landscape characteristics. Journal of Wildlife Management 62:314–321.

McNew, L. B., A. J. Gregory, S. M. Wisely, and B. K. Sandercock. 2009. Estimating the stage of incubation for nests of Greater Prairie-Chickens using egg flotation: a float curve for grousers. Grouse News 38:12–14.

Mong, T. W., and B. K. Sandercock. 2007. Optimizing radio retention and minimizing radio impacts in a field study of Upland Sandpipers. Journal of Wildlife Management 71:971–980.

Moss, R., A. Watson, P., Rothery, and W. W. Glennie. 1981. Clutch size, egg size, hatch weight and laying date in relation to early egg mortality in Red Grouse, *Lagopus lagopus scoticus* chicks. Ibis 123:450–462.

Myrberget, S. 1977. Size and shape of eggs of Willow Grouse *Lagopus lagopus*. Ornis Scandinavica 8:39–46.

Narushin, V. G. 2005. Egg geometry calculation using the measurements of length and breadth. Poultry Science 84:482–484.

Nooker, J. K. 2007. Factors affecting the demography of a lek-mating bird: the Greater Prairie-Chicken. Ph.D. dissertation, Kansas State University, Manhattan, KS.

Novoa, C., A. Besnard, J. F. Brenot, and L. N. Ellison. 2008. Effect of weather on the reproductive rate of Rock Ptarmigan *Lagopus muta* in the eastern Pyrenees. Ibis 150:270–278.

Pace, M. L., J. J. Cole, S. R. Carpenter, and J. F. Kitchell. 1999. Trophic cascades revealed in diverse ecosystems. Trends in Ecology and Evolution 14:483–488.

Patten, M. A. 2009. Are forest and grassland grouse on different life history tracks? Grouse News 37:15–19.

Patten, M. A., D. H. Wolfe, and S. K. Sherrod. 2007. Lekking and nesting response of the Greater Prairie-Chicken to burning of tallgrass prairie. Pp. 149–153 *in* R. E. Masters and K. E. M. Galley (editors), Proceedings of the 23rd Tall Timbers fire ecology conference: fire in grassland and shrubland ecosystems. Tall Timbers Research Station, Tallahassee, FL.

Patten, M. A., D. H. Wolfe, E. Shochat, and S. K. Sherrod. 2005. Habitat fragmentation, rapid evolution and population persistence. Evolutionary Ecology Research 7:235–249.

Perlut, N. G., C. R. Freeman-Gallant, A. M. Strong, T. M. Donovan, C. W. Kilpatrick, and N. J. Zaliks. 2008. Agricultural management affects evolutionary processes in a migratory songbird. Molecular Ecology 17:1248–1255.

Peterson, M. J., and N. J. Silvy. 1996. Reproductive stages limiting productivity of endangered Attwater's Prairie-Chicken. Conservation Biology 10:1264–1276.

Powell, L. A. 2007. Approximating variance of demographic parameters using the delta method: a reference for avian biologists. Condor 109:949–954.

Rahn, H., P. R. Sotherland, and C. V. Paganelli. 1985. Interrelationships between egg mass and adult body mass and metabolism among passerine birds. Journal of Ornithology 126:263–271.

Remeš, V., and T. E. Martin. 2002. Environmental influences on the evolution of growth and developmental rates in passerines. Evolution 56:2505–2518.

Reznick, D. N., C. K. Ghalambor, and K. Crooks. 2008. Experimental studies of evolution in guppies: a model for understanding the evolutionary consequences of predator removal in natural communities. Molecular Ecology 17:97–107.

Rhymer, J. M. 1992. An experimental study of geographic variation in avian growth and development. Journal of Evolutionary Biology 5:298–306.

Ricklefs, R. E. 1969. An analysis of nesting mortality in birds. Smithsonian Contributions to Zoology 9:1–48.

Ricklefs, R. E. 2000. Density dependence, evolutionary optimization, and the diversification of avian life histories. Condor 102:9–22.

Ricklefs, R. E. 2008. The evolution of senescence from a comparative perspective. Functional Ecology 22:379–392.

Robel, R. J. 1970. Possible role of behavior in regulating Greater Prairie-Chicken populations. Journal of Wildlife Management 34:306–312.

Robel, R. J., J. N. Briggs, J. J. Cebula, N. J. Silvy, C. E. Viers, and P. G. Watt. 1970. Greater Prairie-Chicken ranges, movements, and habitat usage in Kansas. Journal of Wildlife Management 34:286–306.

Roff, D. A. 1992. The evolution of life-histories. Chapman and Hall, New York, NY.

Russell, E. M. 2000. Avian life histories: is extended parental care the southern secret? Emu 100:377–399.

Sæther, B.-E. 1987. The influence of body weight on the covariation between reproductive traits in European birds. Oikos 48:79–88.

Sæther, B.-E., and Ø. Bakke. 2000. Avian life history variation and contribution of demographic traits to the population growth rate. Ecology 81:642–653.

Sandercock, B. K., K. Martin, and S. J. Hannon. 2005. Life history strategies in extreme environments: comparative demography of arctic and alpine ptarmigan. Ecology 86:2176–2186.

Schroeder, M. A., and C. E. Braun. 1991. Walk-in traps for capturing Greater Prairie-Chickens on leks. Journal of Field Ornithology 62:378–385.

Schroeder, M. A., and L. A. Robb. 1993. Greater Prairie-Chickens. A. Poole, P. Stettenheim, and F. Gill (editors), The birds of North America No. 36, ed. Academy of Natural Sciences, Philadelphia, PA.

Silvy, N. J., M. E. Morrow, E. Shanley, and R. D. Slack. 1990. An improved drop net for capturing wildlife. Proceedings of the Annual Conference of the Southeastern Association of Fish and Wildlife Agencies 44:374–378.

Smith, C. C., and S. D. Fretwell. 1974. The ultimate balance between size and number of offspring. American Naturalist 108:499–506.

Smith, E. F., and C. E. Owensby. 1978. Intensive early stocking and season-long stocking of Kansas Flint Hills range. Journal of Range Management 31:14–17.

Smith, T. B., and L. Bernatchez. 2008. Evolutionary change in human-altered environments. Molecular Ecology 17:1–8.

Svedarsky, W. D., J. E. Toepfer, R. L. Westemeier, and R. J. Robel. 2003. Effects of management practices on grassland birds: Greater Prairie-Chicken. Northern Prairie Wildlife Research Center, Jamestown, ND. <http://www.npwrc.usgs.gov/resource/literatr/grasbird/gpch/gpch.htm> (28 May 2004).

Tigas, L. A., D. H. VanVuren, and R. M. Sauvajot. 2002. Behavioral responses of bobcats and coyotes to habitat fragmentation and corridors in an urban environment. Biological Conservation 108:299–306.

van Noordwijk, A. J., and G. de Jong. 1986. Acquisition and allocation of resources: their influence on variation in life history tactics. American Naturalist 128:137–142.

Westemeier, R. L., J. D. Brawn, S. A. Simpson, T. L. Esker, R. W. Jansen, J. W. Walk, E. L. Kershner, J. L. Bouzat, and K. N. Paige. 1998. Tracking the long-term decline and recovery of an isolated population. Science 282:1695–1698.

White, G. C., and K. P. Burnham. 1999. Program Mark: survival estimation from populations of marked animals. Bird Study 46(Suppl.):S120–139.

Winkler, D. W., and K. Wallin. 1987. Offspring size and number: a life history model linking effort per offspring and total effort. American Naturalist 129:708–720.

With, K. A., A. W. King, and W. E. Jensen. 2008. Remaining large grasslands may not be sufficient to prevent grassland bird declines. Biological Conservation 141:3152–3167.

Wolfe, D. H., M. A. Patten, E. Shochat, C. L. Pruett, and S. K. Sherrod. 2007. Causes and patterns of mortality in Lesser Prairie-Chickens *Tympanuchus pallidicinctus* and implications for management. Wildlife Biology 13(Suppl. 1):95–104.

Zammuto, R. M. 1986. Life histories of birds: clutch size, longevity, and body mass among North American game birds. Canadian Journal of Zoology 64:2739–2749.

CHAPTER TWENTY

Demographic Traits of Two Alpine Populations of Rock Ptarmigan

*Claude Novoa, Jean-François Desmet, Jean-François Brenot,
Bertrand Muffat-Joly, Marc Arvin-Bérod, Jean Resseguier,
and Bastien Tran*

Abstract. We examined survival and reproduction of Rock Ptarmigan (*Lagopus muta*) on two study areas located in the northern French Alps (Haut Giffre: HG) and in the eastern Pyrenees (Canigou Massif: CM). Part of HG included an intensively frequented ski resort. Data on annual reproductive success of Rock Ptarmigan were obtained from counts of broods and adults made with pointing dogs in early August. Additional data on fecundity parameters (nest success, age-dependent reproduction) were obtained from radio-tagged females. From 1998 to 2007, we radio-tagged 50 Rock Ptarmigan at HG and 55 at CM. The proportion of radio-tagged females that successfully hatched a clutch was lower at HG (0.40; 95% CI = 0.23–0.59) than at CM (0.68; 95% CI = 0.44–0.86), but this difference decreased when we considered the proportion of successful nesting females that reared a brood (0.55; 95% CI = 0.39–0.70 vs. 0.66; 95% CI = 0.51–0.78). Over all years, the mean number of fledglings per adult in early August was greater at CM (0.38; 95% CI = 0.35–0.41) than at HG (0.27; 95% CI = 0.25–0.30). Predation was the main cause of death for adult birds at both sites, but the primary predators differed between sites. Raptors accounted for 85% of predator kills at CM but only 40% at HG, where mammalian predators were more common. The annual survival rate of adult females tended to be lower at HG (0.61; 95% CI = 0.44–0.75) than at CM (0.70; 95% CI = 0.43–0.86), but the difference was not significant. A population viability analysis showed that with the recurrent low rates of reproduction recorded in these southern latitude populations, survival rates of juveniles or adults would have to be higher to maintain populations.

Key Words: causes of death, *Lagopus muta*, life history strategies, population model, radiotelemetry, reproductive success, survival.

Novoa, C., J.-F. Desmet, J.-F. Brenot, B. Muffat-Joly, M. Arvin-Bérod, J. Resseguier, and B. Tran. 2011. Demographic traits of two alpine populations of Rock Ptarmigan. Pp. 267–280 *in* B. K. Sandercock, K. Martin, and G. Segelbacher (editors). Ecology, conservation, and management of grouse. Studies in Avian Biology (no. 39), University of California Press, Berkeley, CA.

Rock Ptarmigan (*Lagopus muta*) are small-bodied grouse occurring over a wide latitudinal range, from 83° N in northern Greenland to less than 40° N in Japan (Johnsgard 1983, Storch 2000). Although most populations inhabit subarctic or arctic lands above 60° N, the species is also found in alpine areas of North America and in the southern mountains of Europe and Japan. Like other species of ptarmigan living all year round in arctic–alpine habitats, Rock Ptarmigan have developed particular life history strategies and physiological adaptations to cope with extreme environmental conditions (Martin 2001, Martin and Wiebe 2004). In this respect, ptarmigan are considered to be some of the most effective sentinel species for evaluating responses to environmental change (Sandercock et al. 2005a). Given both large numbers and wide distributional range, Rock Ptarmigan are not considered to be a threatened species at a continental scale (Storch 2000, 2004). However, the situation of southern populations of Rock Ptarmigan may be viewed differently because of their small population size and geographical isolation, which may increase their vulnerability to demographic and environmental stochasticity (Shaffer 1987, Lande 1993, Storch 2000). Moreover, the loss of genetic diversity recently found in Pyrenean Rock Ptarmigan (Caizergues et al. 2003, Bech et al. 2009) may be an additional constraint for the long-term persistence of this population. Another substantial difference between northern and southern populations of Rock Ptarmigan is the potential effect of human activities. Indeed, unlike arctic areas, most of the Rock Ptarmigan habitats in the Alps and the Pyrenees are affected by livestock grazing, by infrastructure associated with ski resorts, and by disturbance from wilderness sports in alpine areas (Ingold et al. 1996).

Previous studies of population dynamics of Rock Ptarmigan have concerned cyclic northern populations in Scotland (Watson 1965, Watson and Rae 1998), Iceland (Gardarsson 1988), Alaska (Weeden 1965, Weeden and Theberge 1972), and Canada (Cotter et al. 1992, Cotter and Gratto 1995). Reproductive success has often been studied, but rarely survival of adults or fledged young. The same is true of most previous investigations of southern alpine populations (Glutz von Blotzheim et al. 1973, Morscheidt 1994, Miquet 1995), although data on demographic rates of radio-marked adults have been recently recorded in Italy (Scherini et al. 2003, in press). In the current study, we report on reproductive and survival rates recorded in two populations of Rock Ptarmigan living at the southern limit of the range of the species. Our demographic data allowed us to estimate population growth rates and the probabilities of long-term persistence.

METHODS

Study Areas

The study was conducted from 1998 to 2007 on two study areas approximately 520 km apart, the Haut Giffre (HG) in the northern French Alps and the Canigou Massif (CM) in the eastern Pyrenees. The latter represents the eastern limit of the distribution of Rock Ptarmigan in the French Pyrenees and one of the southernmost populations in Western Europe.

In both areas, Rock Ptarmigan occur in alpine habitats (2,000–2,900 m a.s.l) dominated by ericaceous shrubs mixed with other dwarf alpine plants. At CM, open woodlands of mountain pine (*Pinus uncinata*) occur at the lowest elevations of Rock Ptarmigan habitat. Substrate and climate differ markedly between the two sites. HG is a limestone plateau with abundant crevasses and small cliffs, whereas CM is a succession of eroded plateaus and U-shaped valleys with large extents of gneiss screes. HG is one of the wettest regions of the French Alps (mean annual precipitation 1,666 mm at 700 m a.s.l.), while the climate at CM is relatively dry (850 mm at 1,550 m a.s.l.).

Cattle and sheep grazing were moderate, with stocking rates of 0.05–0.10 animal unit month/ha, on both areas. Shooting of Rock Ptarmigan was not allowed at CM and only occasionally at some sites surrounding HG. Both areas were intensively frequented by hikers in July–August. The main difference in recreational activities between CM and HG concerned winter sport activities. While CM was sparsely visited by people in winter, HG included a renowned popular ski resort (Flaine-Grand Massif), which covered a third of the study area.

Field Methods

From 1998 to 2007, we radio-tagged 50 Rock Ptarmigan at HG (3 juveniles, 8 adult males, and 39 adult females) and 55 birds at CM (10 juveniles, 6 adult males, and 39 adult females). The most effective capture method was to lure brood-rearing females toward a net with a tape-recorded chick distress call (Brenot et al. 2002), which explains the female-biased sex ratio of our captures.

To study age-dependent reproduction, we determined the age of females by the amount of black pigment on the second and third outermost wing primaries (Weeden and Watson 1967, Ellison and Léonard 1996). Females were classified as yearlings (12–24 months) if the ninth wing primary (P9) was more darkly pigmented than the eighth (P8), and adults (≥24 months) when P9 had the same amount or less pigment than P8. However, unless we make a distinction between adults and yearlings, the term adult includes yearlings. Chicks were considered fledged at the age of 4–5 weeks. Juveniles were young birds independent of parental care and less than 12 months old.

In 1998–1999, birds were fitted with 11–13-g necklace radio tags (Holohil System Ltd.) with an expected lifespan of 12 months, and after 2000, with 7–9-g necklace tags with an expected lifespan of 12 to 24 months. All tags included a mortality sensor that activated after 12 hours of inactivity. Birds were located at least once per month from the ground using a portable receiver (Yaesu or Custom Electronics) and a handheld Yagi antenna. Aerial surveys by fixed-wing aircraft equipped with an antenna were undertaken if any transmitter signals were lost.

Cause of death was classified as predation, accident (ski-lift collision, HG only), natural event (birds trapped in snow burrow by avalanche), or unknown cause. Differences between predation by mammalian predators or raptors was determined by evidence at kills, including: tags buried or not, marks on tags, feathers plucked clean from the body (raptors) or broken (mammal), and scat or other sign in the vicinity of the kills. In both study areas, potential predators of Rock Ptarmigan included Golden Eagle (*Aquila chrysaëtos*), Goshawk (*Accipiter gentilis*), Peregrine Falcon (*Falco peregrinus*), red fox (*Vulpes vulpes*), pine marten (*Martes martes*), stone marten (*Martes foina*), and stoat (*Mustela erminea*). Red fox, pine marten, stoat, and occasionally alpine marmot (*Marmota marmota*) were the potential nest predators.

Estimation of Demographic Parameters

Data on reproductive parameters such as the mean clutch size, the proportion of females that successfully hatched a clutch or reared a brood, and the mean brood size in late August were obtained from radio-tagged females. Clutch size was determined at the onset of incubation, either by flushing the females off the nest or by counting the eggs during an incubation recess. We considered that a female had successfully hatched a clutch or reared a brood when at least one egg hatched or one chick fledged. The 95% confidence intervals for both nesting and rearing success were calculated following the Wilson procedure as modified by Newcombe (1998). We analyzed differences in hatching and rearing success between sites or age-classes using chi-square tests. We also compared nest success between sites by modeling daily nest survival rates using the nest survival procedure in program MARK, version 4.3 (White and Burnham 1999, Dinsmore et al. 2002).

An index of annual reproductive success of Rock Ptarmigan was obtained from counts made in early August using pointing dogs. We systematically searched for adult males and females, including hens both with and without broods, on a reference area of 440 ha at HG and 410 ha at CM. These areas included different habitat types and elevations, and we assumed that both broods and adults had the same probability of being detected. We could not always determine the sex of adults within the groups because birds often flushed at long distances during these counts. For this reason, the number of broodless females was unknown and the index of reproductive success was defined as the number of fledglings per adult (males plus females). We used estimators taken from Gart and Zweifel (1967) to calculate the 95% confidence interval of the reproductive success index, assuming a random sampling of birds and taking into account the finite population correction for a sampling fraction of 80% (Cochran 1977).

We arranged data on reproductive success from summer censuses in a three-way table (10 years × 2 sites × 2 age-classes) and used a log-linear model for analysis (Sokal and Rohlf 1981). Effects of the variables year, site, and their interaction were tested using likelihood ratio tests. All analyses were conducted using program R (ver. 2.7.1; R Development Core Team 2006).

Survival rates of birds were estimated by modeling individual encounter histories from the date of capture to death or end of radio signal. Because numbers of adult males were small, we limited adult survival analysis to females. Females fitted with the heaviest tags (>11 g) during the first two years of the study (1998–1999) were excluded from the analysis because these tags resulted in premature radio failures and high rates of early predation. For our survival analyses, we used 34 adult females (2001–2007) at HG and 30 adult females (2000–2007) at CM. Some birds (10/30 at CM and 11/34 at HG) were monitored over several successive years, owing to recapture. We pooled the 13 juveniles marked at the two sites for estimating the survival rate of juvenile birds from August to July. Survival histories were analyzed using program MARK and the known-fate model (White and Burnham 1999). For each month (encounter occasion) we entered the bird's status (alive or dead). The encounter history files totaled 84 encounter occasions at HG (7 years × 12 months) and 96 at CM (8 years × 12 months). We tested first for between-year differences in monthly survival rates, before testing for seasonal (October–March vs. April–September) or site (HG vs. CM) effects. For this analysis, we created a set of *a priori* candidate models, each involving one of the explanatory variables mentioned above. Because few birds were tracked each year, data from different years were combined to obtain acceptable precision of survival estimates.

For each candidate model, program MARK provides Aikake's Information Criterion corrected for small sample sizes (AIC$_c$), which allows comparisons between non-nested models and represents a good compromise between the model fit and the number of parameters (Anderson and Burnham 1999). Because the variance inflation factor ($\hat{c}$), which allows a test of the goodness of fit of the general model, cannot be estimated for nest survival or known-fate procedures of program MARK, we used AIC$_c$ for inference instead of quasi-AIC$_c$.

We considered the model with the lowest AIC$_c$ to be best supported by the data, and we used differences in AIC$_c$ between the best model and the other candidate models (ΔAIC$_c$) to determine the relative ranking of each model. We followed Burnham and Anderson (2002) in selecting equally parsimonious models, where ΔAIC$_c \leq 2$ exhibits strong support for a given model. Program MARK also provided the AIC$_c$ weights for each set of candidate models. Weights summed to 1.0 and can be interpreted as the weight of evidence in favor of each model, where w_i/w_j is used to quantify degree of support of one model over the other (Burnham and Anderson 2002). When two models were equally parsimonious (ΔAIC$_c \leq 2$), we used the model averaging procedure of program MARK to estimate the survival parameter. Seasonal or annual survival rates were then calculated as monthly survival rate raised to the 6th or 12th power.

Population Model

We incorporated annual survival and reproduction estimates from the present study into an age-structured stochastic model (Caswell 1989, Johnson and Braun 1999). We simulated a female population with two age-classes (yearlings and adults) without including gains or losses from immigration or emigration (closed population model). To take into account demographic stochasticity, we based our model on a matrix-type design where each transition (survival, fecundity) was defined as a binomial distribution (Fig. 20.1). Fecundity, defined as the number of female fledglings reared per adult female, was extrapolated from the index of reproductive success (number of fledglings per adult), assuming an equal sex ratio for adults and fledglings (Ellison and Léonard 1996).

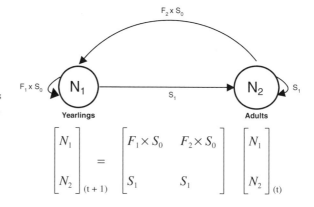

Figure 20.1. Life cycle diagram for Rock Ptarmigan and associated Leslie matrix. Females are distributed in two age classes, yearlings (1 year, N_1) and adults ($\geq$ 2 years, N_2), which produce, respectively, F_1 and F_2 juvenile females. Juveniles survive to one year of age with a survival rate of S_0. S_1 is the annual survival rate of yearling and adult ptarmigan older than one year.

$$\begin{bmatrix} N_1 \\ N_2 \end{bmatrix}_{(t+1)} = \begin{bmatrix} F_1 \times S_0 & F_2 \times S_0 \\ S_1 & S_1 \end{bmatrix} \begin{bmatrix} N_1 \\ N_2 \end{bmatrix}_{(t)}$$

Models were run using ULM software (Legendre and Clobert 1995, Ferrière et al. 1996), which calculated the asymptotic growth rate and the probability of extinction. For each site we defined different models, using initial population sizes of 100 or 200 females, with 70% adult females and 30% yearling females in each case. For each model, we ran 1,000 Monte Carlo simulations over 25 years, the time recommended by Beissinger and Westphal (1998) to minimize propagation errors. The probability of extinction was the percentage of 1,000 random simulations leading to extinction, defined as $n \leq 1$. The limits of the 95% confidence interval of population growth rate (λ) were calculated as $\lambda \pm 1.96 * SE(\lambda)$ (Caswell 1989).

RESULTS

Reproductive Success

From 1998 to 2007, data on fecundity parameters (clutch size, nest success, age-dependent reproduction) were obtained from 29 radio-marked females (6 yearlings, 23 adults) at HG and 38 radio-marked females (8 yearlings, 30 adults) at CM. At both areas, a few females (see Methods) were monitored over successive breeding seasons (from 2 to 5). In total, we monitored 30 nesting attempts at HG and 19 at CM, and 42 brood rearing attempts at HG and 50 at CM.

Mean clutch size of first nests and renests combined was higher at HG (6.5 eggs, SE = 0.25, $n = 21$) than at CM (5.9, SE = 0.28, $n = 17$) (one-sided t-test: $t = -1.799$, df = 36, $P = 0.04$),

but the apparent nest success (i.e., the proportion of radio-tagged females that successfully hatched a clutch, was lower at HG (0.40; 95% CI = 0.23–0.59) than at CM (0.68; 95% CI = 0.44–0.86) ($\chi^2 = 3.76$, df = 1, $P = 0.05$). Likewise, the daily survival rate of nests tended to be lower at HG (0.961, 95% CI = 0.933–0.978) than at CM (0.983, 95% CI = 0.960–0.993), but the small number of nests monitored at both sites resulted in large confidence intervals of the estimates of daily nest survival rates and therefore provide only a small improvement of the model likelihood ($\Delta AIC_c = 0.65$). Predation, often associated with adverse weather conditions, was the main cause of nesting failures. At CM, the fates of 19 nesting attempts were: 13 successful nests, 5 nests depredated, and 1 unknown case, while at HG the fates of 30 nesting attempts were: 12 successful nests, 6 nests depredated with 4 incubating females killed, 1 nest destroyed by hailstorm, 1 nest abandoned, and 6 unknown fates.

As most females were caught and radio-marked after the clutch hatched, our estimates of breeding success for these hens was restricted to the brood-rearing period. The apparent rate of brood survival (i.e., the proportion of radio-tagged females that fledged one or more chicks) was 0.55 (23/42) at HG and 0.66 (33/50) at CM. Brood size in late August averaged 3.0 chicks (5–6 weeks old) at HG (SE = 0.40, $n = 23$) and 2.7 at CM (SE = 0.30, $n = 33$), and the distributions of brood size did not differ between sites (chi-square exact test, df = 6, $P = 0.67$; Fig. 20.2). Overall, the number of fledglings per radio-tagged hen in late August

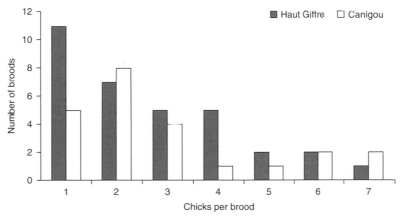

Figure 20.2. Number of chicks in radio-monitored Rock Ptarmigan broods in the Haut Giffre (HG–northern French Alps) and the Canigou (CM–eastern French Pyrenees). Chicks were 5–6 weeks old.

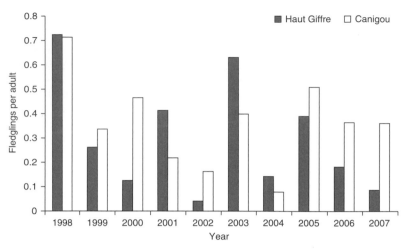

Figure 20.3. Index of reproductive success of Rock Ptarmigan determined by summer counts with pointing dogs in the Haut Giffre (northern French Alps) and the Canigou (eastern French Pyrenees), 1998–2007. Counts include broods (chicks) and adults (females with and without broods and males).

TABLE 20.1

Model selection procedure for estimation of monthly survival rates of female Rock Ptarmigan at Canigou in the eastern French Pyrenees (n = 30), 2000–2007 and Haut Giffre in the northern French Alps (n = 34), 2001–2007.

Sites	Model	AIC_c	ΔAIC_c	AIC_c Weight	k	Deviance
Canigou	Season	100.41	0.00	0.515	2	43.65
	Constant	100.54	0.13	0.482	1	45.80
	Year	110.54	10.14	0.003	8	41.41
Haut Giffre	Constant	160.90	0.00	0.515	1	57.82
	Year	162.30	1.41	0.255	7	46.99
	Season	162.51	1.61	0.230	2	57.41
Canigou vs. Haut Giffre	Constant	259.60	0.00	0.674	1	103.80
	Sites	261.05	1.45	0.326	2	103.24

did not differ between sites (1.6 at HG vs. 1.8 at CM; $\chi^2 = 0.13$, df = 1, $P = 0.72$). At both sites, yearling females tended to fledge fewer chicks than did adult females (1.0 vs. 1.7 at HG and 1.5 vs. 1.8 at CM), but these differences were not significant ($\chi^2 = 0.80$, df = 1, $P = 0.37$; $\chi^2 = 0.17$, df = 1, $P = 0.68$).

From 1998 to 2007, counts with pointing dogs in August indicated that reproductive success (fledglings per adult) varied greatly among years at both sites ($\chi^2 = 55.07$, df = 9, $P < 0.001$). The between-site differences in reproductive success also varied annually (age-ratio * site: $\chi^2 = 19.57$,

df = 9, $P = 0.02$), but on average the number of fledglings per adult in early August was greater at CM (0.38; 95% CI = 0.35–0.41) than at HG (0.27; 95% CI = 0.25–0.30) ($\chi^2 = 3.75$, df = 1, $P = 0.05$; Fig. 20.3).

Survival Rates and Causes of Death

Survival rates did not vary among years at CM ($\Delta AIC_c > 10$), but we could not completely reject a year effect at HG ($\Delta AIC_c < 2$; Table 20.1). This effect was associated with the year 2004, when Rock Ptarmigan suffered a high rate of mortality,

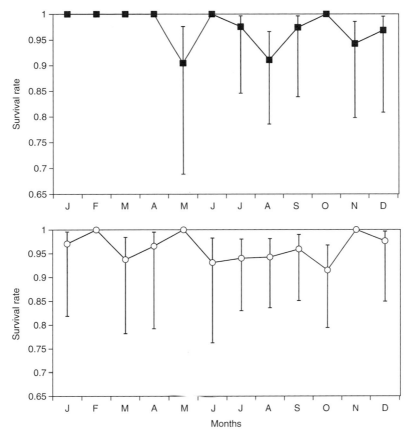

Figure 20.4. Monthly survival rate (95% CI) of radio-marked female Rock Ptarmigan in the Canigou (top; n = 30 females) and the Haut Giffre (bottom; n = 34 females).

with 7 of 13 radio-tagged birds being killed that year.

Estimates of monthly survival rates, derived from model "constant," were 0.959 (SE = 0.009) at HG and 0.970 (SE = 0.009) at CM and were not significantly different (ΔAIC_c < 2, Table 20.1). We found strong evidence at CM, but also to a lesser extent at HG, that survival varied between seasons, with a higher survivorship during the six-month period of autumn and winter (October–March) than during the six-month breeding season (April–September; Fig. 20.4). Out of 30 females radio-marked at CM, only 2 were killed during autumn–winter versus 7 during the breeding season. At HG, out of 34 radio-marked females, 7 died during autumn–winter versus 12 during the breeding season. The two models "season" and "constant" had equal support (ΔAIC_c < 2; Table 20.1), and the parameter estimates were calculated using the model-averaging procedure. Derived survival estimates for the autumn–winter

and breeding survival periods were 0.87 (95% CI = 0.69–0.95) versus 0.80 (95 % CI = 0.62–0.90) at CM and 0.79 (95% CI = 0.67–0.87) versus 0.77 (95% CI = 0.65–0.86) at HG (Table 20.2).

Predation was the main cause of death at both areas, but the number of unknown causes of death at HG prevented us from testing for a difference in mortality patterns between the two areas. At CM, the causes of death of 15 full-grown Rock Ptarmigan were raptor (n = 11), mammal (n = 2), and unknown predator (n = 2), while at HG the causes of deaths of 21 birds were raptor (n = 4), mammal (n = 6), unknown predator (n = 7), collision with ski-lift cables (n = 1), death in snow burrow (n = 1), and unknown causes (n = 2).

Demographic Models

Female-only population models were computed using annual survival rates of 0.70 at CM and

TABLE 20.2

Estimates of monthly survival rates of female Rock Ptarmigan at Canigou in the eastern French Pyrenees (n = 30) and Haut Giffre in the northern French Alps (n = 34), 2000–2007.

Sites	Period	Monthly survival (SE)	Six-month survival rate (95% CI)
Canigou	Oct.–Mar.	0.977 (0.009)	0.87 (0.69–0.95)
	Apr.–Sep.	0.963 (0.012)	0.80 (0.62–0.90)
Haut Giffre	Oct.–Mar.	0.961 (0.010)	0.79 (0.67–0.87)
	Apr.–Sep.	0.958 (0.011)	0.77 (0.65–0.86)

NOTE: Parameter estimates were derived from model averaging between models "constant" and "season." Six-month survival rates were extrapolated as the monthly survival raised to the sixth power.

0.61 at HG, estimates calculated as the product of the seasonal rates during autumn–winter and the breeding periods. At both areas, we used a juvenile survival rate of 0.62, the estimate calculated from September to July for 13 radio-marked juveniles. The data from radio-monitored females suggested that reproductive success varied by age-class. Therefore, at CM we set the number of female fledglings reared at 0.33 per yearling hen and at 0.40 per adult hen (0.38 female fledglings per breeding female for both age-classes combined). At HG, we set the reproductive success at 0.18 and 0.31 female fledgling per yearling and adult hen, respectively (0.27 young female per breeding female for both age-classes combined).

These parameters led to values of asymptotic rates of population growth of λ = 0.934 (95% CI = 0.933–0.936) at CM and 0.804 (95% CI = 0.799–0.810) at HG (Table 20.3), which meant that the current demographic parameters observed would not result in long-term persistence of either population if closed. Models gave a mean population age structure of 27% yearlings and 73% adults at CM versus 16% and 84% at HG. At CM, these age structures were similar to the age ratio of captures (21% yearlings). Without immigration from surrounding areas, the probability of extinction in the short term of the HG population would be quite high (Table 20.3; Fig. 20.5).

The minimum survival rates of adult females needed for population growth rates (λ) to be equal to 1.0 (stationary) were 0.77 at CM and 0.79 at HG. Only at CM was this theoretical survival rate included in the 95% confidence interval of our annual survival rate estimates (CM: 0.43–0.86; HG: 0.44–0.75).

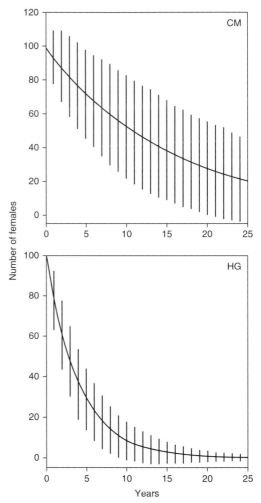

Figure 20.5. Mean trends of Rock Ptarmigan over a 25-year period as predicted from a female-based stochastic model for Canigou in the eastern French Pyrenees, and Haut Giffre in the northern French Alps. Each point represents the number of females (95% CI) resulting from Monte Carlo simulations with 1,000 iterations and an initial population size of 100 females.

TABLE 20.3

Finite rate of population growth (λ) and probability of extinction for two Rock Ptarmigan populations, predicted from a female-based stochastic model without immigration.

| Sites | Initial population size | Annual survival | | Fecundity (Female fledg-lings per female) | | Population growth (λ, 95% CI) | Probability of extinction (25 yrs.) |
		Adult	Juvenile	Adult	Yearling		
Canigou	100	0.70	0.62	0.40	0.33	0.934 (0.933–0.936)	0.019
Canigou	200	0.70	0.62	0.40	0.33	0.935 (0.934–0.936)	0.000
Haut Giffre	100	0.61	0.62	0.31	0.18	0.804 (0.799–0.810)	0.922
Haut Giffre	200	0.61	0.62	0.31	0.18	0.792 (0.790–0.795)	0.811

DISCUSSION

This 10-year study provides comprehensive demographic data for two southern alpine populations of Rock Ptarmigan in France. Both populations have low reproductive rates and high survival rates of adults, demographic traits similar to alpine populations of White-tailed Ptarmigan (*Lagopus leucura*) in North America (Sandercock et al. 2005a). Despite the high rates of survival that we recorded, demographic modeling suggested that long-term trends in numbers on both of our study areas were negative. Below we discuss the demographic traits that we recorded and the limits of this first trial of modeling population viability.

Reproductive Success

Average clutch size, calculated from first nests and renests combined, was greater in the Alps (6.5 eggs per clutch at HG) than in the Pyrenees (5.9 at CM). Overall, these results agree with those of other southern Rock Ptarmigan populations: clutch size of 5.5 in Japan (Sakanakura, pers. comm.), 6.5 in the Italian Alps (Scherini et al. 2003), and 6.9 in the French Alps (Miquet and Deana 2004) and confirm that Rock Ptarmigan lay small clutches at high-elevation sites. We could not determine the renesting rate at HG or CM, but replacement clutches were rare at both study areas, particularly CM. Our results differ from other southern populations, where studies have reported a renesting rate of 17% (3/18) in the Italian Alps (Scherini et al. 2003) and 27% (4/15) in the French Alps (Miquet and Deana 2004). In Alaska, late-hatching nests, hence presumably replacement clutches, comprised only 3% of the 228 records of nests or known-age broods (Weeden 1965).

In our study, the percentage of radio-tagged females that successfully hatched a clutch differed between study areas in a pattern opposite that of clutch size: Apparent nesting success was lower at HG (0.41) than at CM (0.68). Apparent nesting success at HG was similar to the 50% reported in the Italian Alps (Scherini et al. 2003) but much greater than the 22% observed in the National Park of Vanoise (Miquet and Deana 2004). Neither the average brood size nor the number of fledglings per radio-tagged hen in August differed between sites, suggesting that overall brood-rearing success was equal at HG and CM.

At HG and in Vanoise National Park, the high rate of nest failure may be associated with abundant mammalian predators, mainly red fox. Indeed, these two study areas in the Alps are heavily impacted by tourism, which likely favor mammalian nest predators (Storch and Leidenberger 2003, Watson and Moss 2004). However, as suggested by Scherini et al. (2003), the intensity of nest predation may be related to variation in spring weather conditions affecting the body condition of females. Elsewhere, predation on nests has little influence on overall breeding success of Rock Ptarmigan. Instead, number of fledglings is influenced more by viability of young, as determined by body condition of females (Watson et al. 1998, Novoa et al. 2008). It is possible that the frequent disturbance of Rock Ptarmigan females wintering in ski resorts at HG (during late November to early April) may affect their body condition before the onset of breeding season and hence their reproductive performance.

Reproductive effort in Rock Ptarmigan increases from southern to northern populations, the latter laying larger clutches and fledging more young per hen in August (Ellison and Léonard 1996, Novoa et al. 2008). Geographic variation in reproductive parameters associated with a latitudinal gradient have also been reported for closely related species in northern Canada, where Willow Ptarmigan (*Lagopus lagopus*) and White-tailed Ptarmigan breeding at southern subalpine and alpine sites had smaller clutches and lower nesting, fledging, and renesting success than Willow Ptarmigan breeding at a low-elevation arctic site (Sandercock et al. 2005b). However, in contrast with this general trend, Kaler et al. (2010) found that annual fecundity of Evermann's Rock Ptarmigan (*Lagopus muta evermanni*) in the Aleutian Islands was relatively low, at <1 female fledgling per breeding female.

Survival Rates and Causes of Death

The annual survival rates of adult females found in this study varied from 0.61 (95% CI: 0.44–0.75) at HG to 0.70 (95% CI: 0.43–0.86) at CM, with no significant difference between sites. The annual survival rates calculated from 13 adult female Rock Ptarmigan radio-monitored in the Italian Alps (Scherini et al. in press) was 0.68 (95% CI = 0.42–0.86), an estimate close to ours. At all sites in France and Italy, seasonal mortality of females was highest during the breeding season (April to September), similar to Willow Ptarmigan (Hannon et al. 2003). The relatively high survival rates that we recorded are in agreement with our additional data on longevity. At HG, an adult hen radio-monitored since 2003 was at least 7 years old in 2008. At CM an adult hen caught in 1999 and re-sighted in 2006 was at least 9 years old, while another adult hen caught in 2002 was regularly radio-monitored until late 2006, that is, until at least 6 years old. In Japan, Sakanakura (pers. comm.) reported for Rock Ptarmigan a maximum longevity of 11 years for males and 8 years for females.

Estimates of survival from radio-tracking data offer the advantage that mortality is not confounded by immigration/emigration, but assume that radio transmitters do not influence demographic rates (Sandercock 2006). Recently, the detrimental effect of radio transmitters on the survival of game birds has been questioned for Northern Bobwhite (*Colinus virginianus*) (Guthery and Lusk 2004), but the subject is still controversial (Folk et al. 2007). Recent studies on Red Grouse (*Lagopus lagopus scoticus*) (Thirgood et al. 1995), Willow Ptarmigan (Hannon et al. 2003), and Lesser Prairie-Chicken (*Tympanuchus pallidicinctus*) (Hagen et al. 2006) have reported equal survival rates between radio-marked birds and control birds marked with wing tags or leg bands, but Cotter and Gratto (1995) found a higher survival in June of color-banded males of Rock Ptarmigan versus radio-marked males. In our study, we could not completely discard a negative effect of capture and tagging on the survival of females, because several females were depredated in the month following their capture. Thus, the survival rates estimated at CM and HG should be considered as minimum rates.

Estimates of apparent survival rates from counts or re-sightings of banded birds in northern populations suggest that adult mortality is higher in the arctic and subarctic. In Alaska, the average total losses from late August to May (~9 months) over a five-year study varied from 0.40 to 0.67, and apparent survival rates during these nine months ranged from 0.33 to 0.60 (Weeden 1965). In Scotland, Watson (1965) found an average rate of total losses of 50% between August populations and subsequent spring populations, for an apparent survival rate of 0.50 over a 10-month period. In the Canadian arctic, Cotter et al. (1992) reported a survival rate from June through July

of only 0.82 for radio-monitored adult Rock Ptarmigan. Combining radio-tracking data and re-sighting of banded birds, these authors estimated that the annual survival rate would average 0.43, with 0.50 for males and 0.39 for females. Even though these latter estimates may be biased by different sources of error, comparison of survival rates between northern and southern populations shows a trend opposite to the latitudinal variation in annual fecundity.

Predation was the major cause of death at both areas, but the pattern of mortality varied between sites, with raptors accounting for 85% of identified predation events at CM, but only 40% at HG. Predation by raptors has been identified as the major cause of non-hunting mortality in ptarmigan (Bergerud 1988, Cotter et al. 1992, Nielsen 1999, Smith and Willebrand 1999). Surprisingly, only one radioed bird died by collision with ski lift cables despite many females staying throughout the year in the core of Flaine station. A recent survey carried out at 252 French ski resorts showed that collision mortality is a frequent cause of death (OGM 2006).

Demographic Models

Demographic models have been used to make decisions for managing grouse populations (Caizergues and Ellison 1997, Johnson and Braun 1998, Grimm and Storch 2000). Here, we used a female-based model including demographic stochasticity for modeling the short-term trajectories of two populations of Rock Ptarmigan. Our model with demographic stochasticity is appropriate for small, low-density populations of Rock Ptarmigan such as those found in the Alps and Pyrenees. Indeed, demographic stochasticity is inherent to any demographic process, regardless of the environment, and its strength increases as population size declines (Legendre et al. 1999). Even though the precision and the accuracy of our estimates of survival and reproduction may be affected by the small sample sizes of radioed females, and by possible bias due to capture and tags, the estimated population growth rates indicated in both areas a negative trend ($\lambda < 1$). Whatever the uncertainty associated with these first attempts to model Rock Ptarmigan populations, our results suggest that the two populations under study are not viable without immigration. Despite a high probability of extinction, the HG population is interconnected with surrounding populations and immigration likely compensates for part of its demographic unbalance, as demonstrated in spatially structured populations of ptarmigan (Smith and Willebrand 1999, Martin et al. 2000). In contrast with HG, the CM population represents an eastern isolate likely unconnected with the main part of Pyrenean Rock Ptarmigan range. Thus at CM, the effects of demographic stochasticity, combined with loss of genetic diversity (Caizergues et al. 2003, Bech et al. 2009), may increase the risk of extinction of this small population.

The life-history strategies of two species of North American ptarmigan form a continuum between the high reproductive strategy of arctic populations of Willow Ptarmigan and the survivorship strategy of alpine populations of White-tailed Ptarmigan (Sandercock et al. 2005a). Generation times reported for all these populations were less than 2.9 years (Sandercock et al. 2005a). With a generation time between 3.3 (Alps) and 3.5 years (Pyrenees) (Tran 2007), the demographic traits of southern populations of Rock Ptarmigan represent an extreme case of survivorship strategy. As a consequence, southern populations of Rock Ptarmigan will be more affected by perturbations of survival than reproduction. Hence, management to reduce the mortality rates of adult birds by means of shooting plans and making ski-lift cables more visible (Ellison et al. 2003, OGM 2006) may be important in maintaining these Rock Ptarmigan populations. Predator control could be justified in situations where human activities have led to artificially high predator densities, such as corvids or foxes with access to supplemental food around tourist sites (Storch and Leidenberger 2003, Watson and Moss 2004).

ACKNOWLEDGMENTS

Jérôme Sentilles, the agents of the Eastern Pyrenees Departmental Service of the Office National de la Chasse et de la Faune Sauvage, and many students helped with collecting the field data. Philippe Aubry, François Sarrazin, and Scott Wilson kindly helped with data analysis. Laurence Ellison greatly encouraged the initiation of the study and provided constructive comments on successive versions of the manuscript. Brett Sandercock and an anonymous referee greatly improved a later draft of this manuscript. We are extremely grateful to all of the above for their time and assistance.

LITERATURE CITED

Anderson, D. R., and K. P. Burnham. 1999. Understanding information criteria for selection among capture-recapture or ring recovery models. Bird Study 46(Suppl.):14–21.

Bech, N., J. Boissier, S. Drovetski, and C. Novoa. 2009. Population genetic structure of Rock Ptarmigan (*Lagopus muta pyrenaica*) in the "sky island" of French Pyrenees: implications for conservation. Animal Conservation 12:138–146.

Beissinger, S. R., and I. M. Westphal. 1998. On the use of demographic models of population viability in endangered species management. Journal of Wildlife Management 62:821–841.

Bergerud, A. T. 1988. Population ecology of North American grouse. Pp. 578–685 *in* A. T. Bergerud and M. W. Gratson (editors), Adaptive strategies and population ecology of northern grouse. University of Minnesota Press, Minneapolis, MN.

Brenot, J.-F., J.-F. Desmet, and J. Morscheidt. 2002. Mise au point d'une méthode de capture des poules de lagopède alpin *Lagopus mutus* accompagnées de jeunes. Alauda 70:190–191.

Burnham, K. P., and D. R. Anderson. 2002. Model selection and multimodel inference: a practical information theoretic approach. Springer-Verlag, New York, NY.

Caizergues, A., A. Bernard-Laurent, J.-F. Brenot, L. N. Ellison, and J. Y. Rasplus. 2003. Population genetic structure of Rock Ptarmigan *Lagopus mutus* in northern and western Europe. Molecular Ecology 12:2267–2274.

Caizergues, A., and L. N. Ellison. 1997. Survival of Black Grouse *Tetrao tetrix* in the French Alps. Wildlife Biology 3:177–186.

Caswell, H. 1989. Matrix population models. Sinauer, Sunderland, MA.

Cochran, W. G. 1977. Sampling techniques. 3rd ed. Wiley & Sons, New York, NY.

Cotter, R. C., D. A. Boag, and C. C. Shank. 1992. Raptor predation on Rock Ptarmigan (*Lagopus mutus*) in the central Canadian arctic. Journal of Raptor Research 26:146–151.

Cotter, R. C., and C. J. Gratto. 1995. Effects of nest and brood visits and radio transmitters on Rock Ptarmigan. Journal of Wildlife Management 59:93–98.

Dinsmore, S. J., G. C. White, and F. L. Knopf. 2002. Advanced techniques for modeling avian nest survival. Ecology 83:3476–3488.

Ellison, L., and P. Léonard. 1996. Validation d'un critère d'âge chez le lagopède alpin (*Lagopus mutus*) et sexe et âge ratios dans des tableaux de chasse des Alpes et des Pyrénées. Gibier Faune Sauvage [Game and Wildlife] 13:1495–1510.

Ellison, L., Y. Magnani, and C. Novoa. 2003. Management of Black Grouse and Rock Ptarmigan in France. Pp. 5–11 *in* A. Brugnoli and U. Zamboni (editors), Atti del seminario "esperienze di gestione dei galliformi di montagna con particolare riferimento alla programmazione venatoria," Trento, Associazione Cacciatori della Provincia di Trento.

Ferrière, R. F., F. Sarrazin, S. Legendre, and J. P. Baron. 1996. Matrix population models applied to viability analysis and conservation: theory and practice using the ULM software. Acta Oecologica 17:629–656.

Folk, T. H., J. B. Grand, W. E. Palmer, J. P. Carroll, D. C. Sisson, T. M. Terhune, S. D. Wellendorf, and H. L. Stribling. 2007. Estimates of survival from radiotelemetry: a response to Guthery and Lusk. Journal of Wildlife Management 71:1027–1033.

Gardarsson, A. 1988. Cyclic population changes and some related events in Rock Ptarmigan in Iceland. Pp. 300–329 *in* A. T. Bergerud and M. W. Gratson (editors), Adaptive strategies and population ecology of northern grouse. University of Minnesota Press, Minneapolis, MN.

Gart, J. J., and J. R. Zweifel. 1967. On the bias of various estimators of the logit and its variance with application to quantal bioassay. Biometrika 54:181–187.

Glutz von Blotzheim, U. N., K. M. Bauer, and E. Bezzel. 1973. Handbuch der Vögel Mitteleuropas, Vol. 5: Galliformes und Gruiformes, Akademische Verlagsgesellschaft, Francfort/Main, Germany.

Grimm, V., and I. Storch. 2000. Minimum viable population size of Capercaillie *Tetrao urogallus*: results from a stochastic model. Wildlife Biology 6:219–225.

Guthery, F. S., and J. J. Lusk. 2004. Radiotelemetry studies: are we radio-handicapping Northern Bobwhites? Wildlife Society Bulletin 32:194–201.

Hagen, C. A., B. K. Sandercock, J. C. Pitman, R. J. Robel, and R. D. Applegate. 2006. Radiotelemetry survival estimates of Lesser Prairie-Chickens in Kansas: are there transmitter biases? Wildlife Society Bulletin 34:1064–1069.

Hannon, S. J., R. C. Gruys, and J. O. Schieck. 2003. Differential mortality of the sexes in Willow Ptarmigan *Lagopus lagopus* in northern British Columbia, Canada. Wildlife Biology 9:317–326.

Ingold, P., R. Schnidrig-Petrig, H. Mabercher, U. Pfister, and R. Zeller. 1996. Tourisme/sports de loisir et faune sauvage dans la région alpine suisse. Cahier de l'environnement No. 62. Office fédéral de l'environnement, des forêts et du paysage (OFEFP), Berne, Switzerland.

Johnsgard, P. A. 1983. The grouse of the world. Croom Helm, London, UK.

Johnson, K. H., and C. E. Braun. 1999. Viability and conservation of an exploited sage grouse population. Conservation Biology 13:77–84.

Kaler, R. S. A., S. E. Ebbert, C. E. Braun, and B. K. Sandercock. 2010. Demography of a reintroduced population of Evermann's Rock Ptarmigan in the Aleutian Islands. Wilson Journal of Ornithology 122:1–14.

Lande, R. 1993. Risks of population extinction from demographic and environmental stochasticity, and random catastrophes. American Naturalist 142:911–927.

Legendre, S., and J. Clobert. 1995. ULM, a software for conservation and evolutionary biologists. Journal of Applied Statistics 22:817–834.

Legendre, S., J. Clobert, A. P. Møller, and G. Sorci. 1999. Demographic stochasticity and social mating system in the process of extinction of small populations: the case of passerines introduced to New Zealand. American Naturalist 153:449–463.

Martin, K. 2001. Wildlife communities in alpine and sub-alpine habitats. Pp. 285–310 *in* D. H. Johnson and T. A. O'Neil (managing directors), Wildlife-habitat relationships in Oregon and Washington. Oregon State University Press, Corvallis, OR.

Martin, K., P. B. Stacey, and C. E. Braun. 2000. Recruitment, dispersal and demographic rescue in spatially-structured White-tailed Ptarmigan populations. Condor 102:503–516.

Martin, K., and L. K. Wiebe. 2004. Coping mechanisms of alpine and arctic breeding birds: extreme weather conditions and limitations to reproductive resilience. Integrative Comparative Biology 44:177–185.

Miquet, A. 1995. Le lagopède alpin, *Lagopus mutus*, dans le Parc National de la Vanoise. Parc National de la Vanoise, Chambery, France.

Miquet, A., and T. Deana. 2004. Etude du lagopède alpin *Lagopus mutus helveticus* dans le Parc National de la Vanoise: résultats préliminaires. Travaux Scientifiques du Parc National de la Vanoise 22:137–154.

Morscheidt, J. 1994. Densités au printemps et succès de la reproduction chez le lagopède alpin *Lagopus mutus* dans la réserve domaniale du Montvallier (Ariège, France). Alauda 62:123–132.

Newcombe, R. G. 1998. Two-sided confidence intervals for the single proportion: comparison of seven methods. Statistics in Medicine 17:857–872.

Nielsen, O. K. 1999. Gyrfalcon predation on ptarmigan: numerical and functional responses. Journal of Animal Ecology 68:1034–1050.

Novoa, C., A. Besnard, J.-F. Brenot, and L. N. Ellison. 2008. Effect of weather on the reproductive rate of Rock Ptarmigan *Lagopus muta* in the eastern Pyrenees. Ibis 150:270–278.

Observatoire des Galliformes de Montagne (OGM). 2006. Percussion des oiseaux dans les câbles aériens des domaines skiables. Zoom No. 4, Sevrier, France.

R Development Core Team. 2006. R: a language and environment for statistical computing. R Foundation for Statistical Computing, Vienna, Austria.

Sandercock, B. K. 2006. Estimation of demographic parameters from live-encounter data: a summary review. Journal of Wildlife Management 70:1504–1520.

Sandercock, B. K., K. Martin, and S. J. Hannon. 2005a. Demographic consequences of age-structure in extreme environments: population models for arctic and alpine ptarmigan. Oecologia 146:13–24.

Sandercock, B. K., K. Martin, and S. J. Hannon. 2005b. Life history strategies in extreme environments: comparative demography of arctic and alpine ptarmigan. Ecology 86:2176–2186.

Scherini, G. C., M. Favaron, and G. Tosi. (in press). Sopravvivienza della pernice bianca alpina (*Lagopus mutus helveticus*) e prelievo sostenibile. Atti del IV Convegno Nazionale dei Biologi della Selvaggina, INFS, Bologna, Italy.

Scherini, G. C., G. Tosi, and L. A. Wauters. 2003. Social behaviour, reproductive biology and breeding success of alpine Rock Ptarmigan *Lagopus mutus* in northern Italy. Ardea 91:11–23.

Shaffer, M. 1987. Minimum viable populations: coping with uncertainty. Pp. 69–86 *in* M. E. Soulé (editor), Viable populations for conservation. Cambridge University Press, New York, NY.

Smith, A., and T. Willebrand. 1999. Mortality causes and survival rates of hunted and unhunted Willow Grouse. Journal of Wildlife Management 63:722–730.

Sokal, R. R., and F. J. Rohlf. 1981. Biometry. 2nd ed. Freeman and Company, New York, NY.

Storch, I. 2000. Grouse: status survey and conservation action plan 2000–2004. Cambridge, IUCN/World Pheasant Association, Reading, UK.

Storch, I. 2004. Conservation status of grouse worldwide: an update. Wildlife Biology 13:5–12.

Storch, I., and C. Leidenberger. 2003. Tourism, mountain huts and distribution of corvids in the Bavarian Alps, Germany. Wildlife Biology 9:301–308.

Thirgood, S. J., S. M. Redpath, P. J. Hudson, M. M. Hurtley, and N. J. Aebischer. 1995. Effects of necklace radio transmitters on survival and breeding of Red Grouse *Lagopus lagopus scoticus*. Wildlife Biology 1:121–126.

Tran, B. 2007. Analyse de viabilité d'une population chassée de lagopède alpin. Mémoire de Master II, Université Paris VI, Paris, France.

Watson, A. 1965. A population study of ptarmigan (*Lagopus mutus*) in Scotland. Journal of Animal Ecology 34:135–172.

Watson, A., and R. Moss. 2004. Impacts of ski-development on ptarmigan (*Lagopus mutus*) at Cairn Gorm, Scotland. Biological Conservation 116:267–275.

Watson, A., R. Moss, and S. Rae. 1998. Population dynamics of Scottish Rock Ptarmigan cycles. Ecology 79:1174–1192.

Weeden, R. B. 1965. Breeding densities, reproductive success, and mortality of Rock Ptarmigan at Eagle Creek, Central Alaska, from 1960 to 1964. Proceedings of North American Wildlife Conference 30:336–348.

Weeden, R. B., and J. B. Theberge. 1972. The dynamics of a fluctuating population of Rock Ptarmigan in Alaska. International Ornithological Congress 15:90–106.

Weeden, R. B., and A. Watson. 1967. Determining the age of Rock Ptarmigan in Alaska and Scotland. Journal of Wildlife Management 31:825–826.

White, G. C., and K. P. Burnham. 1999. Program Mark: survival estimation from populations of marked animals. Bird Study 46:120–136.

PART FOUR

Conservation and Management

CHAPTER TWENTY-ONE

Effects of Climate Change on Nutrition and Genetics of White-tailed Ptarmigan

Sara J. Oyler-McCance, Craig A. Stricker, Judy St. John,
Clait E. Braun, Gregory T. Wann, Michael S. O'Donnell,
and Cameron L. Aldridge

Abstract. White-tailed Ptarmigan (*Lagopus leucura*) are well suited as a focal species for the study of climate change because they are adapted to cool, alpine environments that are expected to undergo unusually rapid climate change. We compared samples collected in the late 1930s, the late 1960s, and the late 2000s using molecular genetic and stable isotope methods in an effort to determine whether White-tailed Ptarmigan on Mt. Evans, Colorado, have experienced recent environmental changes resulting in shifts in genetic diversity, gene frequency, and nutritional ecology. We genotyped 115 individuals spanning the three time periods, using nine polymorphic microsatellite loci in our genetic analysis. These samples were also analyzed for stable carbon and nitrogen isotopic composition. We found a slight trend of lower heterozygosity through time, and allelic richness values were lower in more recent times, but not significantly using an alpha of 0.05 ($P < 0.1$). We found no changes in allele frequencies across time periods, suggesting that population sizes have not changed dramatically. Feather $\delta^{13}C$ and $\delta^{15}N$ values decreased significantly across time periods, whereas the range in isotope values increased consistently from the late 1930s to the later time periods. Inferred changes in the nutritional ecology of White-tailed Ptarmigan on Mt. Evans relate primarily to increased atmospheric deposition of nutrients that likely influenced foraging habits and tundra plant composition and nutritional quality. Future work seeks to integrate genetic and isotopic data with long-term demographics to develop a detailed understanding of the interaction among environmental stressors on the long-term viability of ptarmigan populations.

Key Words: climate change, genetics, *Lagopus leucura*, stable isotopes, temporal variation.

Oyler-McCance, S. J., C. A. Stricker, J. St. John, C. E. Braun, G. T. Wann, M. S. O'Donnell, and C. L. Aldridge. 2011. Effects of climate change on nutrition and genetics of White-tailed Ptarmigan. Pp. 283–294 *in* B. K. Sandercock, K. Martin, and G. Segelbacher (editors). Ecology, conservation, and management of grouse. Studies in Avian Biology (no. 39), University of California Press, Berkeley, CA.

All major ecosystems are predicted to experience alterations resulting from climate change (Intergovernmental Panel on Climate Change 2007). Alpine ecosystems may be particularly susceptible to warming because their existence is partially affected by low temperature conditions, which are expected to rise (Armstrong and Halfpenny 2001). Projected changes in alpine systems will likely be detrimental to vertebrate species inhabiting these ecosystems since alpine habitats are analogous to islands, separated by expanses of low-lying, warmer habitats. Alpine vertebrates are highly specialized in their habitat requirements (Armstrong and Halfpenny 2001); as treeline is expected to advance in elevation and plants from lower altitudes invade, concomitant changes to alpine environments may influence the viability of endemic species (Price 1997).

Changes in climate over the past century have been shown to influence many aspects of avian population biology (Crick 2004, Wormworth and Mallon 2006). Advancing trends in breeding phenology (e.g., egg-laying and hatching dates) are the most widely observed response to increasing spring temperatures (Crick 2004, Lyon et al. 2008); however, few studies have investigated how climate-induced responses in avian breeding phenology have impacted species at the population level. This is due in part to a general paucity of long-term data sets that include demographic information in addition to data on nesting activities. Ecosystem changes mediated through anthropogenic impacts or climate change have the potential to directly affect demographic parameters, but studies are needed that address other potentially relevant responses, such as genetic and nutritional responses, to gain a more comprehensive understanding of alterations at the population and individual levels.

The White-tailed Ptarmigan (*Lagopus leucura*) is endemic to alpine habitats at or above timberline throughout the mountain west of the U.S. and Canada (Braun et al. 1993). Unlike other avian species breeding in alpine habitats, White-tailed Ptarmigan remain at high altitudes throughout the year and, with few exceptions, spend almost their entire life history above treeline (Hoffman 2006). The species is well adapted to harsh environments found in the alpine and has developed several behavioral and physiological traits that allow it to survive under such conditions. In cold weather, for example, ptarmigan choose microclimates that are sometimes 7°C above ambient temperatures, and they have a tendency to walk rather than fly, both traits that help ptarmigan maximize metabolic efficiency (Braun et al. 1993, Martin et al. 1993). The thermoneutral zone of this species is broad, allowing it to survive in a wide range of temperatures without having to expend excess energy (Johnson 1968). In Colorado, the primary food of this species throughout the non-breeding season (September through April) includes the buds, twigs, and leaves of *Salix* species, with forbs and berries becoming highly important during the breeding season (May through August) (May and Braun 1972). Insects are vital components of chick diets the first three weeks after hatching (May 1975).

White-tailed Ptarmigan on Mt. Evans, Colorado, have been studied annually since 1966 (major findings reviewed by Hoffman 2006), representing the longest-running data set known for the species. The overall focus of this research was to investigate demographic patterns and included annual observations of individually marked birds, allowing estimation of survival and breeding success, various morphometric measurements (i.e., mass, primary feather length, tarsus length, etc.), and locations and occupancy of territories. Such long-term data sets are extremely valuable for investigating a species' response to environmental change. While the alpine ecosystem may be less likely to be directly impacted by human disturbance, it is thought to be particularly vulnerable to environmental change due to climate warming (Price 1997, Baron et al. 2000a) and atmospheric deposition of airborne pollutants (Baron et al. 2000b).

Investigation into the interaction between ecological and evolutionary responses to environmental change is an important aspect of ecosystem-based studies. Such responses can be examined using molecular genetic and stable isotope techniques (Kelly 2000, Kelly et al. 2005, Reusch and Wood 2007, Valenzuela et al. 2009). Documenting shifts in genetic allele frequencies and changes in levels of genetic diversity can provide clues as to past demographic patterns, movement among populations, and the ability of a species to adapt (Avise 1994). Similarly, stable isotope analysis of consumer tissues can document shifts in foraging ecology, which may be associated with changes in landcover or landscape biogeochemistry (Hilton et al. 2006, Inger and Bearhop 2008). We conducted a temporal comparison spanning the last 70 years to begin to investigate whether White-tailed

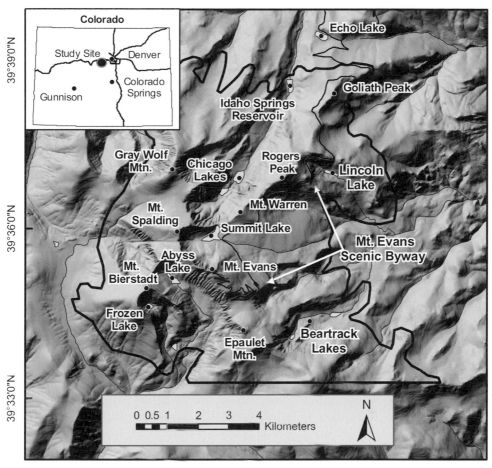

Figure 21.1. Study area for White-tailed Ptarmigan at Mt. Evans, Colorado.

Ptarmigan on Mt. Evans have experienced environmental changes resulting in shifts in genetic diversity, gene frequency, and nutritional ecology. This study represents a preliminary examination into whether selected responses of ptarmigan to environmental change could be detected. Our specific objectives were to (1) assess whether microsatellite allele frequencies have changed, (2) document changes in genetic diversity, and (3) detail potential shifts in the nutritional ecology of White-tailed Ptarmigan on Mt. Evans.

METHODS

Study Site

A population of White-tailed Ptarmigan was studied in the Mt. Evans Wilderness Area from 1966 to 2008 (herein Mt. Evans, 39°35' N, 105°38' W;

Fig. 21.1). The study area is ~16 km southwest of Idaho Springs, Colorado, and ranges in elevation from 3,500 to 4,350 m at the summit, covering 13.2 km² of alpine habitat. Vegetation is typical of alpine habitat in the southern Rocky Mountains and includes communities of cushion plants, *Kobresia* sedge meadows, and sedge–grass wet meadow and willow–sedge hummock stands (Braun and Rogers 1971). Ptarmigan are typically found in *Trifolium* cushion fellfields, *Carex* meadows, *Carex–Geum* rock meadows, and near receding snowfields in mid- to late summer (Braun et al. 1993). Maximum daily air temperature measured at the nearby Niwot Ridge Long-Term Ecological Research site (50 km north of Mt. Evans; 1964–2006) has averaged 12.6°C and –9.4°C for July and January, respectively (http://culter.colorado.edu/NWT/data/datmanaccess.html; accessed 4 October 2007). Daily precipitation for July and January has averaged 4.6 and 2.4 mm, respectively, with annual precipitation averaging 1,153.4 mm. These data also suggest an increase in the number of days when maximum temperatures exceed 15°C from 1966 to 2008 ($r^2 = 0.19$, $P = 0.003$).

Tissue Collection

Ptarmigan were captured at the Mt. Evans study site from 1966 to 2008 using a noose design modified from Zwickel and Bendell (1967). In a typical year, captures occurred during two intensive field sessions in May–June, and again in August–September. In mid-May, ptarmigan were located by playing tape recordings of male territorial calls (Braun et al. 1973). Males on breeding territories were typically accompanied by hens, both of which were subsequently captured. Ptarmigan were located during the second capture session by playing tape recordings of chick distress calls. Feather samples were collected from birds during the 1966–1968 period (herein the late 1960s). Additionally, blood and feather samples were collected from birds captured in 2007 and 2008 (herein the late 2000s). Blood samples were collected by clipping a toenail and collecting 2–3 drops in a microfuge tube previously coated with EDTA (Brinkmann). Retrix and/or contour feathers were taken from individual birds. Contour feathers were collected from the body of captured birds, primarily on the upper breast, scapulars, and upper tail coverts. During both capture sessions, only contour feathers that were dark, and hence newly grown on breeding areas, were plucked from birds. Ptarmigan begin molting from their all-white basic plumage in April to darker, cryptic nuptial plumages, which persist throughout the summer and early fall (Braun et al. 1993). Hence, all contour feather samples were grown while on breeding areas. Blood samples and feathers were frozen at –20°C.

Additionally, we located nine White-tailed Ptarmigan collected specifically from Mt. Evans in 1937 (herein the late 1930s). These specimens are preserved as study skins at the Denver Museum of Nature and Science. Skin and feather samples from each of the nine specimens were obtained using clean techniques for use in this study.

Molecular Genetics

DNA was extracted from skin and feathers of nine museum specimens collected on Mt. Evans in the late 1930s using the GENECLEAN kit for Ancient DNA (Bio 101) following manufacturer's protocol using buffer dehybridization solution B, and including the proteinase K step. Extractions from museum samples were conducted in an isolated ancient DNA facility to avoid contamination from modern samples. The 42 feather samples collected in the late 1960s were extracted using the Promega Wizard DNA Purification Kit following the manufacturer's instructions (Promega Corporation) with the following modifications: The DNA precipitation step was allowed to incubate for 3 hr on ice with a subsequent 15 min centrifugation step. The DNA was rehydrated in 30 μL of the kit supplied DNA rehydration solution. DNA was extracted from 64 blood samples collected during the late 2000s with the GenomicPrep Blood DNA Isolation Kit (Amersham Biosciences Corp.) using the manufacturer's instructions with modifications following Oyler-McCance et al. (2005).

All 115 White-tailed Ptarmigan samples were screened using nine nuclear microsatellite loci. Primer pairs for three of these loci (LLST1, LLSD4, and LLSD8) were originally designed for Red Grouse, *Lagopus lagopus scoticus* (Piertney and Dallas 1997). The forward primer for microsatellite locus LLSD4 was redesigned to make the PCR product smaller (LLSD4.2F CATGGTT-GTCTTATCCTCTGAGAAAACTG). An additional microsatellite locus (SGCA11) was targeted using primers designed for Greater Sage-Grouse (*Centrocercus urophasianus*) by Taylor et al. (2003). Primers for this locus were also redesigned to shorten the locus (SGCA11.2F GAATATCTT-TCTTTAACAGAATCC, SGCA11.2R CTACTGT-TCTGTTGTGCAAGAC). Five new microsatellite loci that were isolated from Gunnison Sage-Grouse (Oyler-McCance and St. John 2010) were used in this study (SGMS06-8, MSP 7, MSP 18, SGMS06-2, and SGMS06-6). The Polymerase Chain Reaction (PCR) was used to amplify each microsatellite locus with a fluorescently labeled forward primer (Beckman Coulter). Thermocycler conditions for conventional one-step PCRs were as follows: Preheat at 94°C for 2 min, followed by 35 cycles of 94°C for 40 sec, anneal for 1 min (temperatures in Table 21.1), and 1-min extension at 72°C. The reaction concluded with a 5-min final extension at 72°C.

Additionally, a two-step or multiplex PCR procedure (Piggott et al. 2004) was performed to amplify samples that were homozygous for

TABLE 21.1

*Nine polymorphic microsatellite loci assessed in a population of
White-tailed Ptarmigan at Mt. Evans, Colorado, 1937–2008.*

Microsatellite Locus	Annealing Temperature (C)	Allele Size Range
SGCA11.2	52	154–161
SGMS06.8	52	121–135
MSP7	55	137–139
MSP18	55	93–95
LLSD4.2	63	154–156
SGMS06.2	55	107–143
SGMS06 .6	58	127–154
LLST1	55	134–159
LLSD8	55	141–151

eight of the nine loci, for samples that would not amplify with a standard single PCR reaction consisting of 35 cycles, and for all the museum specimens. All two-step PCRs were repeated completely (both steps) at least twice. Additionally, museum specimens that were genotyped as homozygotes were amplified multiple times to minimize the chance for allelic dropout. In the multiplex procedure, step one involved the amplification of each sample with two separate pooled primer sets each consisting of six individual primer pairs. A second PCR was used to amplify each locus individually by using 2 μL from the first PCR in a 20-μL reaction. This second PCR consisted of 40 cycles with the parameters previously mentioned (Piggott et al. 2004). All PCR amplifications were performed with GoTaq Flexi (Promega) DNA polymerase. PCR products were run on the CEQ8000 XL DNA Analysis System following manufacturer's protocol (Beckman-Coulter). All samples included the size standard S400 (Beckman-Coulter) and were analyzed using the Frag 3 method of the CEQ Genetic Analysis Software Package (version 6.0). The resulting size fragment chromatograms were scored independently by two individuals.

Stable Isotopes

Feathers were cleaned in a 2:1 chloroform–methanol solution and allowed to air dry for 24–48 hours. Approximately 1 mg of vane material was clipped from the distal end of individual contour feathers and transferred into 5 × 9 mm tin capsules for stable isotope analyses. Only retrix feathers were available for all but three birds sampled in 1968; subsamples were obtained similarly to contour feathers. Samples were analyzed for stable carbon (C) and nitrogen (N) isotopic composition using an elemental analyzer (Carlo Erba, NC1500) interfaced to a micromass Optima mass spectrometer operated in continuous flow mode (Fry et al. 1992). Isotope values are expressed in delta (δ) notation:

$$\delta X = (R_{sample}/R_{standard}) - 1$$

where X represents ^{13}C and ^{15}N in parts per thousand (‰) deviation relative to a standard (monitoring) gas and R represents ^{13}C/^{12}C and ^{15}N/^{14}N for samples and the standard, respectively. Isotopic data were normalized to V-PDB and Air using the internationally distributed primary standards USGS 40 (–26.24‰ and –4.52‰ for δ^{13}C and δ^{15}N, respectively) and USGS 41 (37.76‰ and 47.57‰ for δ^{13}C and δ^{15}N, respectively). Analytical error was assessed through replicate measures of primary standards, which was less than 0.2‰ measured across all analytical sequences. A secondary standard (reagent grade keratin) was analyzed in duplicate within each analytical sequence and used as a quality control check; reproducibility was better than 0.2‰ for δ^{13}C and δ^{15}N. Accuracy was assessed by analyzing primary standards as unknowns and was within 0.2‰ for δ^{13}C and δ^{15}N.

Data Analysis

For all analyses, we assessed genetic and isotopic data for differences between the three sampling periods (late 1930s, late 1960s, and late 2000s). Microsatellite genotypes were tested for departures from Hardy–Weinberg equilibrium within each time period, using a Bonferroni corrected P value of 0.002, in the computer program ARLEQUIN 2.00 (Schneider et al. 2000). ARLEQUIN uses a Markov-chain random walk algorithm (Guo and Thompson 1992) that is analogous to Fisher's Exact test but extends it to an arbitrarily sized contingency table. We used 300,000 as the forecasted chain length and 5,000 dememorization steps for this analysis. Linkage disequilibrium for each pair of loci was evaluated in each time period in GENE-POP (Raymond and Rousset 1995; Markov chain parameters: 10,000 dememorization steps, 1,000 batches, 10,000 iterations per batch). Significance was determined using a Bonferroni corrected P value of 0.002.

The amount of genetic diversity per time period was documented several different ways. We calculated mean observed and expected heterozygosity levels per time period using ARLEQUIN. Allelic richness, which adjusts for discrepancies in sample size by incorporating a rarefaction method, was estimated in FSTAT 2.9.3.2 (Goudet 1995). We tested whether the allelic richness differed significantly between time periods using a Wilcoxon matched-pairs signed-ranks test and an alpha level of 0.05. In addition, mean number of alleles per locus per time period was calculated.

We used pairwise population F_{ST} significance tests with a Bonferroni corrected alpha level of 0.02 between each pair of time periods, using ARLEQUIN to examine whether allele frequencies had changed over time. We also tested for differences in allele frequencies based on gender using a pairwise population F_{ST} test with an alpha value of 0.05. Additionally, we used the software program STRUCTURE 2.00 (Pritchard et al. 2000) as an alternative approach to explore changes in allele frequencies. STRUCTURE uses a model-based clustering analysis that groups individuals into genetic clusters without regard to their original sampling locale. We estimated the number of genetic clusters (K) by conducting 10 independent runs each for $K = 1 - 5$ with 500,000 Markov chain Monte Carlo repetitions with a 500,000 burn-in period using the model with admixture, correlated allele frequencies, and no prior information on sampling locales (popinfo = 0).

We tested for gender and capture date effects on stable isotope values using two sample t-tests. Captures in the 1960s and 2000s were conducted in both spring (May–June) and summer (July–September). We used ANOVA to test for differences among the three time periods. Systat (version 8) was used for all statistical tests related to stable isotope data using an alpha of 0.05.

RESULTS

Molecular Genetics

One of the nine museum samples from 1937 failed to amplify in any microsatellite locus. The number of microsatellite alleles per locus across time periods ranged from two to nine. There were no significant departures from Hardy–Weinberg equilibrium and no two loci were found to be linked. The mean number of alleles per time period ranged from 3.1 in the late 1930s samples to 4.6 in the late 2000s samples (Table 21.2). Mean observed heterozygosity was highest for ptarmigan in the late 1930s and lowest in those

TABLE 21.2
Levels of genetic diversity among three different time periods for a population of White-tailed Ptarmigan at Mt. Evans, Colorado.

Time period	Sample size	Mean no. of alleles per locus (SE)	Mean observed heterozygosity (SE)	Mean expected heterozygosity (SE)	Allelic richness
1930s	8	3.1 (0.43)	0.52 (0.09)	0.60 (0.04)	2.5
1960s	42	4.4 (0.84)	0.51 (0.08)	0.51 (0.09)	2.1
2000s	64	4.6 (0.85)	0.49 (0.08)	0.52 (0.08)	2.1

sampled in the late 2000s. Allelic richness, which corrects for unequal sample sizes, was highest in the late 1930s (2.5) and similar in the late 1960s and 2000s (2.1 in both; Table 21.2), although not significantly different among any pair of time periods.

There were no significant differences in allele frequencies among any of the time periods (pairwise F_{ST} tests). However, the small sample size for the late 1930s limited our power to detect differences. We found no differences by gender when all genetic data were pooled across years ($n = 110$). The most appropriate value of K in the STRUCTURE analysis given our data was 1, suggesting that all birds from all time periods represent one genetically distinct population with no detectable differences in allele frequencies through time.

Stable Isotopes

Gender differences were not statistically significant when all isotope data were pooled across years ($n = 113$). Isotopic differences between retrix and contour feathers were not obvious, and sample size was too limiting ($n = 3$ for contour feathers in 1968) to evaluate using parametric statistical procedures. Additionally, systematic differences between the two feather types were unlikely, based on the narrow 95% confidence intervals; gender and feather types were pooled to evaluate isotopic shifts over time. Statistically significant differences among capture seasons ($n = 49$) were observed for both $\delta^{13}C$ and $\delta^{15}N$. However, mean differences were approximately 0.3 and 0.7‰ for C and N, respectively, which were smaller than overall differences observed among the three time periods, so these data were pooled for subsequent analyses.

Carbon isotope values in feathers averaged −20.7‰ during the late 1930s and late 1960s, but declined significantly ($F_{2,114} = 151.9$, $P < 0.001$) to −22.2‰ during the late 2000s (Fig. 21.2A). The range in $\delta^{13}C$ within time periods increased from 0.7‰ in the late 1930s ($n = 9$) to 1.4‰ and 2.4‰ during the late 1960s ($n = 42$) and 2000s ($n = 66$). Nitrogen isotope values in feathers averaged 5.4‰ during the late 1930s and declined significantly to 4.8‰ and 3.7‰ in the late 1960s and 2000s, respectively ($F_{2,114} = 33.0$, $P < 0.001$; Fig. 21.2B). Similar to $\delta^{13}C$, the range in measured $\delta^{15}N$ values increased across time periods from 2.3‰ in the late 1930s to 3.9‰ and 4.5‰ in the late 1960s and 2000s.

DISCUSSION

All genetic diversity levels except mean number of alleles per locus were highest in the samples from the late 1930s (Table 21.2). Due to the low sample size from this time period ($n = 8$), it is not surprising that the mean number of alleles per locus was lower. When we corrected for differences in sample size (allelic richness), the samples from the late 1930s had the highest diversity (2.5 compared to 2.1 in both the 1960s and the 2000s), although not statistically significant using an alpha of 0.05 ($P < 0.1$). Both mean observed and mean expected heterozygosity were highest in the late 1930s and lower in the more modern sample periods. Loss of genetic diversity is typically associated with population decline (Cornuet and Luikart 1996). Although we documented a slight trend of lower diversity and heterozygosity, these changes are minimal compared to those associated with severe population bottlenecks in other species (Bellinger et al. 2002). Maintenance of genetic diversity is important, as it is relevant to the health and viability of a population and preserving its ability to adapt to and survive future environmental challenges (O'Brien and Evermann 1988, Quattro and Vrijenhoek 1989). While our data may suggest a trend toward loss of diversity, analysis of additional samples from the 1930s is needed to reliably verify this trend.

Our analysis also revealed that there have been no significant shifts in allele frequencies between the late 1960s and 2000s. We failed to detect significant differences between the modern samples and the late 1930s, but sample size was likely inadequate in the latter to make a robust assessment. Results from the STRUCTURE analysis also did not reflect a shift in allele frequencies, as all birds were considered to have come from one population, suggesting there have not been any detectable shifts in allele frequencies for this population over the last 70 years. Shifts in allele frequencies within a population over time can suggest significant changes in population size (Glenn et al. 1999, Bellinger et al. 2002) or that a population is small, isolated, and strongly affected by genetic drift (Hilfiker et al. 2004, Oyler-McCance et al. 2005). Either the Mt. Evans population is sufficiently large that genetic drift

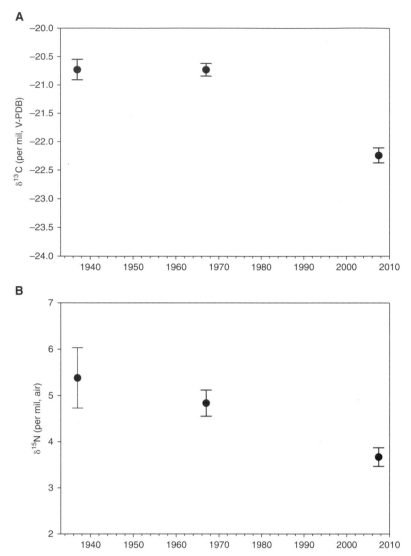

Figure 21.2. Average values for stable isotopes of (A) carbon and (B) nitrogen for feathers of White-tailed Ptarmigan collected at Mt. Evans, Colorado, during three discrete time periods (late 1930s, late 1960s, and late 2000s). Error bars indicate 95% CI.

is not a strong factor or this population may be loosely linked to other nearby populations (e.g., Hall Valley, Kenosha Pass, Square Tops, Argentine Pass, Montezuma Basin, Waldorf, and Loveland Pass). While birds captured at Mt. Evans appear to return to their breeding territories regularly (C. E. Braun, unpubl. data), this population moves to wintering areas comprised of birds from multiple breeding areas. At this point, the extent of mixing between populations remains unknown (but see Hoffman and Braun 1975).

A significant shift in carbon isotope values was observed only since the late 1960s, which is best

explained by altered foraging ecology or changes in landcover during recent times. However, the carbon isotopic composition of atmospheric CO_2 has changed significantly since at least the late 1950s, with ~1‰ decrease in $\delta^{13}C_{CO2}$ for the period 1957 to 1995 (Keeling et al. 1979, Francey et al. 1995). Additionally, isotopic data for atmospheric CO_2 at Niwot Ridge, Colorado (http://gaw. kishou.go.jp/wdcgg/; accessed 14 March 2009) suggest ~0.2–0.3‰ decline since the mid-1990s. The apparent shift in feather $\delta^{13}C$ of approximately 1.5‰ since the late 1960s is in line with isotopic changes in atmospheric CO_2, suggesting that alpine animals may be excellent sentinels for changes in atmospheric chemistry. However, changes in atmospheric chemistry do not fully explain the isotopic shift and range in feather data since the late 1960s, suggesting that altered CO_2 concentrations and increased deposition of nutrients such as nitrogen may be influencing the productivity, nutritional quality, and foraging habitats of White-tailed Ptarmigan in the Mt. Evans study area. For example, Baron et al. (2000a) predicted increased photosynthetic rates and enhanced water use efficiency of alpine tundra following a 2°C warming and doubling of atmospheric CO_2. While such conditions have not yet been expressed, tundra plant productivity may have changed since the late 1960s, and tundra responses are thought to be more tightly linked to winter and spring conditions than to summer (Baron et al. 2000a). Evidence for altered foraging habits is also supported by the increased range of $\delta^{13}C$ and $\delta^{15}N$ feather values across time periods. Isotopic variability within populations has been used as a proxy for niche width, with increased variability equating to a greater diversity of foraging habits (Bearhop et al. 2004, Newsome et al. 2007). It would appear the trophic niche breadth of Mt. Evans ptarmigan has expanded in recent times.

A linear decline in feather nitrogen isotope values was evident across all time periods. Historical records of N deposition in alpine ecosystems suggest an increase during the last century due to anthropogenic sources (Williams et al. 1996, Baron et al. 2000b, Wolfe et al. 2001). An increase in soil N stores due to atmospheric deposition would likely shift the isotopic composition of this nutrient pool, with concomitant shifts in primary consumer tissues. Recent work by Nanus et al. (2008) has shown that the $\delta^{15}N$ of atmospherically derived nitrate ranges from −6.6 to 4.6‰, with precipitation generally having lower values compared to surface waters. Similarly, declining $\delta^{15}N$ values of sedimentary organic matter since at least the 1950s has been demonstrated in alpine lakes of the southern Rocky Mountains (Wolfe et al. 2001). Collectively, these accounts are in agreement with changes in $\delta^{15}N$ of ptarmigan feathers over the last 70 years. Additionally, the increase in N deposition during the 20th century has enhanced the nutritional quality of alpine and subalpine plant species (Rueth and Baron 2002, Rueth et al. 2003). Bowman et al. (2006), through field experiments, showed changes in plant community composition and tissue nutrient concentrations, suggesting that alpine vegetation can potentially serve as a net sink for atmospherically derived N. Increased nutritional quality of available ptarmigan forage may lead to a positive feedback on population demographics, but appears to be at odds with observed declines in recruitment (C. E. Braun, unpubl. data). Currently, we are unable to resolve potential shifts in foraging ecology with changes in plant community composition and nutritional quality. Energetic costs to ptarmigan associated with a rapidly warming alpine may offset benefits associated with enhanced nutritional quality. More detailed studies are required to address these complex relationships.

Studies investigating the impacts of climate on alpine avian species are rare, although a recent study by Wang et al. (2002) of White-tailed Ptarmigan at nearby Rocky Mountain National Park suggests this species has been affected by warming temperature trends. Wang et al. (2002) reported the nesting phenology of this species advanced by ~15 days from 1975 to 1999. Hatch dates were found to be negatively related to average temperatures in May and June. A preliminary analysis of ptarmigan hatch dates from the Mt. Evans study area suggests similar trends, with advanced hatching of roughly 11 days from 1980 to 2008 ($P < 0.05$; G. T. Wann and C. E. Braun, unpubl. data). In addition, there is strong evidence that the Mt. Evans population is changing in its population–age structure. From 1966 to 2008 the age-class ratio of juveniles to adults has sharply declined ($P < 0.001$; G. T. Wann and C. E. Braun, unpubl. data), an indication that juvenile recruitment into the population has decreased. We speculate this may be a result of increased susceptibility of nests and young chicks to weather events that is expected to occur with earlier nesting, or to increased survival of adults because of decreased harvest rates over this period, but more work is needed.

Alpine species may be ideal sentinels for monitoring change in sensitive ecosystems, but long-term perspectives are generally lacking and difficult to unravel due to a suite of co-interacting factors. However, focused studies that seek to set demographic, genetic, and isotopic baselines are desperately needed to successfully identify future change/stressors and implement adaptive management strategies for the conservation of focal species and habitats. Here we have shown subtle shifts in genetic diversity through time, concomitant with significant shifts in feather isotope values. Importantly, the genetic data collected in this study provide an excellent control for inferences regarding long-term changes in the nutritional ecology of Mt. Evans White-tailed Ptarmigan. Substantial long-term changes in allele frequencies could indicate movement of ptarmigan from neighboring populations that may have different isotopic signatures. Because such changes were not observed, we are confident that shifts in feather isotope values reflect environmental change specific to the Mt. Evans study area. Further, studies employing combined molecular and stable isotope techniques offer clear advantages in species-based investigations of long-term ecosystem change, since diverse aspects of life history and population ecology are explored simultaneously (Valenzuela et al. 2009).

Future research in the Mt. Evans alpine ecosystem will focus on assessing relationships between genetic and dietary shifts and long-term population, climate, and landcover data. Additionally, a better understanding of the genetic basis of phenotypes under selection may allow prediction and mitigation of the effects of climate change on population viability (Reusch and Wood 2007). Identifying genetic markers under selection and examining whether these markers can be correlated with environmental changes associated with climate change and population-level demographic changes should be a focal point. Moreover, there is great synergistic potential if ecologists work more closely with biogeochemists that have traditionally focused on changes in landscape nutrient cycling. In particular, a more detailed analysis of controls on plant species distributions in the alpine will be helpful in assessing potential changes in the foraging ecology of the Mt. Evans ptarmigan population and other alpine species. Finally, the role of ungulate grazing on plant productivity and nutrient cycling remains tenuous (Singer and Schoenecker 2003), although the seasonal abundance of elk (*Cervus elaphus*) and mountain goats (*Oreamnos americanus*) has increased substantially in the Mt. Evans area since the late 1960s (C. E. Braun, pers. obs.). Competition for seasonally nutritious forbs may be an important interaction among alpine species that may ultimately be expressed negatively on ptarmigan demography. Further, there is some evidence that increased N

deposition may indirectly influence plant–insect interactions (Throop and Lerdau 2004), and herbivorous insects are known to be an important dietary resource for ptarmigan chicks (May and Braun 1972). Focused studies on sentinel animals in sensitive ecosystems, like the alpine, are certain to advance our understanding of climate change effects on ecosystems.

ACKNOWLEDGMENTS

Long-term research and data collection were supported by the Colorado Division of Wildlife. Contemporary data collection was financially supported by the U.S. Geological Survey, Colorado State University, Grouse Inc., and the Rocky Mountain Center for Conservation and Systematics at the University of Denver. The Denver Museum of Nature and Science generously granted us access to their collection of White-tailed Ptarmigan specimens. The University of Denver High Altitude Laboratory provided accommodations for field crews during the spring and summer capture periods. Climate data were provided by the NSF-supported Niwot Ridge Long-Term Ecological Research project and the University of Colorado Mountain Research Station. Finally, we thank Ernest Valdez, Bradley Fedy, William Iko, and three anonymous reviewers for their suggestions and help with manuscript preparation. The use of any trade, product, or firm names is for descriptive purposes only and does not imply endorsement by the U.S. Government.

LITERATURE CITED

Armstrong, D. M., and J. C. Halfpenny. 2001. Vertebrates. Pp. 128–156 in W. D. Bowman and T. R. Seastedt (editors), Structure and function of an alpine ecosystem: Niwot Ridge, Colorado. Oxford University Press, New York, NY.

Avise, J. C. 1994. Molecular markers, natural history and evolution. Chapman and Hall, New York, NY.

Baron, J. S., M. D. Hartman, L. E. Band, and R. B. Lammers. 2000a. Sensitivity of a high-elevation Rocky Mountain watershed to altered climate and CO_2. Water Resources Research 36:89–99.

Baron, J. S., H. M. Rueth, A. M. Wolfe, K. R. Nydick, E. J. Alstott, J. T. Minear, and B. Moraska. 2000b. Ecosystem responses to nitrogen deposition in the Colorado Front Range. Ecosystems 3:352–368.

Bearhop, S., C. E. Adams, S. Waldron, R. A. Fuller, and H. Macleod. 2004. Determining trophic niche width: a novel approach using stable isotope analysis. Journal of Animal Ecology 73:1007–1012.

Bellinger, M. R., J. A. Johnson, J. Toepfer, and P. Dunn. 2002. Loss of genetic variation in Greater Prairie

Chickens following a population bottleneck in Wisconsin, USA. Conservation Biology 17:717–724.

Bowman, W. D., J. R. Gartner, K. Holland, and M. Wiedermann. 2006. Nitrogen critical loads for alpine vegetation and terrestrial ecosystem response: are we there yet? Ecological Applications 16:1183–1193.

Braun, C. E., and G. E. Rogers. 1971. The White-tailed Ptarmigan in Colorado. Technical Publication 27. Colorado Division of Game, Fish and Parks, Fort Collins, CO.

Braun, C. E., R. K. Schmidt, Jr., and G. E. Rogers. 1973. Census of Colorado White-tailed Ptarmigan with tape-recorded calls. Journal of Wildlife Management 37:90–93.

Braun, C. E., K. Martin, and L. A. Robb. 1993. White-tailed Ptarmigan (*Lagopus leucurus*). A. Poole and F. Gill (editors), The birds of North America No. 68. Academy of Natural Sciences, Philadelphia, PA.

Cornuet J. M., and G. Luikart. 1996. Description and power analysis of two tests for detecting recent population bottlenecks from allele frequency data. Genetics 144:2001–2014.

Crick, H. Q. P. 2004. The impact of climate change on birds. Ibis 146(Supplement 1): 48–56.

Francey, R. J., C. E. Allison, and E. D. Welch. 1995. The 11-year high precision in situ CO_2 stable isotope record from Cape Grim, 1982–1992. Pp. 16–25 in C. Dick and P. J. Fraser (editors), Baseline Atmospheric Program (Australia) 1992. CSIRO Division of Atmospheric Research, Melbourne, Australia.

Fry, B., W. Brand, F. J. Mersch, K. Tholke, and R. Garritt. 1992. Automated analysis system for coupled $\delta^{13}C$ and $\delta^{15}N$ measurements. Analytical Chemistry 64:288–291.

Glenn, T. C., W. Stephan, and M. J. Braun. 1999. Effects of a population bottleneck on Whooping Crane mitochondrial DNA variation. Conservation Biology 13:1097–1107.

Goudet, J. 1995. Fstat version 1.2: a computer program to calculate F statistics. Journal of Heredity 86:485–486.

Guo, S. W., and E. A. Thompson. 1992. Performing the exact test of Hardy–Weinberg proportions for multiple alleles. Biometrics 48:361–372.

Hilfiker, K., F. Gugerli, J. P. Schütz, P. Rotach, and R. Holderegger. 2004. Low RAPD variation and female-biased sex ratio indicate genetic drift in small populations of the dioecious conifer *Taxus baccata* in Switzerland. Conservation Genetics 5:357–365.

Hilton, G. M., D. R. Thompson, P. M. Sagar, R. J. Cuthbert, Y. Cherel, and S. J. Bury. 2006. A stable isotopic investigation into the causes of decline in a sub-Antarctic predator, the Rockhopper Penguin *Eudyptes chrysocome*. Global Change Biology 12:611–625.

Hoffman, R. W. 2006. White-tailed Ptarmigan (*Lagopus leucura*): a technical conservation assessment. USDA Forest Service, Rocky Mountain Region. <http://www.fs.fed.us/r2/projects/scp/assessments/whitetailedptarmigan.pdf> (1 March 2009).

Hoffman, R. W., and C. E. Braun. 1975. Migration of a wintering population of White-tailed Ptarmigan in Colorado. Journal of Wildlife Management 39:485–490.

Inger, R., and S. Bearhop. 2008. Applications of stable isotope analyses to avian ecology. Ibis 150: 447–461.

Intergovernmental Panel on Climate Change. 2007. Climate change 2007: the physical science basis. Contribution of Working Group I to the Fourth Assessment Report of the Intergovernmental Panel on Climate Change (S. Solomon, D. Qin, M. Manning, Z. Chen, M. Marquis, K. B. Averyt, M. Tignor, and H. L. Miller, editors). Cambridge University Press, Cambridge, UK.

Johnson, R. E. 1968. Temperature regulation in the White-tailed Ptarmigan, *Lagopus leucurus*. Comparative Biochemistry and Physiology 24:1004–1014.

Keeling, C. D., W. G. Mook, and P. P. Tans. 1979. Recent trends in the $^{13}C/^{12}C$ ratio of atmospheric carbon dioxide. Nature 277:121–123.

Kelly, J. F. 2000. Stable isotopes of carbon and nitrogen in the study of avian and mammalian trophic ecology. Canadian Journal of Zoology 78:1–27.

Kelly, J. F., K. C. Ruegg, and T. B. Smith. 2005. Combining isotopic and genetic markers to identify breeding origins of migrant birds. Ecological Applications 15:1487–1494.

Lyon, B. E., A. S. Chaine, and D. W. Winkler. 2008. A matter of timing. Science 321:1051–1052.

Martin, K., R. F. Holt, and D. W. Thomas. 1993. Getting by on high: ecological energetics of arctic and alpine grouse. Pp. 33–41 in C. Carey, G. L. Florant, B. A. Wunder, and B. Horwitz (editors), Life in the cold III: ecological, physiological, and molecular mechanisms. Westview Press, Boulder, CO.

May, T. A. 1975. Physiological ecology of White-tailed Ptarmigan in Colorado. Ph.D. dissertation, University of Colorado, Boulder, CO.

May, T. A., and C. E. Braun. 1972. Seasonal foods of adult White-tailed Ptarmigan in Colorado. Journal of Wildlife Management 36:1180–1186.

Nanus, L., M. W. Williams, D. H. Campbell, E. M. Elliott, and C. Kendall. 2008. Evaluating regional patterns in nitrate sources to watersheds in national parks of the Rocky Mountains using nitrate isotopes. Environmental Science and Technology 42:6487–6493.

Newsome, S. D., C. M. Del Rio, S. Bearhop, and D. L. Phillips. 2007. A niche for isotopic ecology. Frontiers in Ecology and the Environment 5:429–436.

O'Brien, S. J., and J. F. Evermann. 1988. Interactive influence of infectious disease and genetic diversity in natural populations. Trends in Ecology and Evolution 3:254–259.

Oyler-McCance, S. J., and J. St. John. 2010. Characterization of small microsatellite loci for use in non invasive sampling studies of Gunnison Sage-Grouse (*Centrocercus minimus*). Conservation Genetics Resources 2:17–20.

Oyler-McCance, S. J., J. St. John, S. E. Taylor, and T. W. Quinn. 2005. Population genetics of Gunnison Sage-Grouse: implications for management. Journal of Wildlife Management 69:630–637.

Piertney, S. B., and J. F. Dallas. 1997. Isolation and characterization of hypervariable microsatellites in the Red Grouse *Lagopus lagopus scoticus*. Molecular Ecology 6:93–95.

Piggott, M.P., E. Bellemain, P. Taberlet, and A. C. Taylor. 2004. A multiplex pre-amplification method that significantly improves microsatellite amplification and error rates for faecal DNA in limiting conditions. Conservation Genetics 5:417–420.

Price, M. 1997. Global change in the mountains. Parthenon Publishing, London, UK.

Pritchard, J. K., M. Stephens, and P. J. Donnelly. 2000. Inference of population structure using multilocus genotype data. Genetics 155:945–959.

Quattro, J. M., and R. C. Vrijenhoek. 1989. Fitness differences among remnant populations of the endangered Sonoran topminnow. Science 245:976–978.

Raymond, M., and F. Rousset. 1995. GENEPOP (version 1.2): population genetics software for exact tests and ecumenicism. Journal of Heredity 86:248–249.

Reusch. T. B., and T. E. Wood. 2007. Molecular ecology of global change. Molecular Ecology 16:3973–3992.

Rueth, H. M., and J. S. Baron. 2002. Differences in Englemann spruce forest biogeochemistry east and west of the Continental Divide in Colorado, USA. Ecosystems 5:45–57.

Rueth, H. M., J. S. Baron, and E. J. Allstott. 2003. Responses of Engelmann spruce forests to nitrogen fertilization in the Colorado Rocky Mountains. Ecological Applications 13:664–673.

Schneider, S., D. Roessli, and L. Excoffier. 2000. ARLEQUIN, version 2.000: a software for population genetics data analysis. Genetics and Biometry Laboratory, University of Geneva, Switzerland.

Singer, F. J., and K. A. Schoenecker. 2003. Do ungulates accelerate or decelerate nitrogen cycling? Forest Ecology and Management 181:189–204.

Taylor, S. E., S. J. Oyler-McCance, and T. W. Quinn. 2003. Isolation and characterization of microsatellite loci in Greater Sage-Grouse (*Centrocercus urophasianus*). Molecular Ecology Notes 3:262–264.

Throop, H. L., and M. T. Lerdau. 2004. Effects of nitrogen deposition on insect herbivory: implications for community and ecosystem processes. Ecosystems 7:109–133.

Valenzuela, L. O., M. Sironi, V. J. Rowntree, and J. Seger. 2009. Isotopic and genetic evidence for culturally inherited site fidelity to feeding grounds in southern right whales (*Eubalaena australis*). Molecular Ecology 18:782–791.

Wang, G., N. T. Hobbs, K. M. Giesen, H. Galbraith, D. S. Ojima, and C. E. Braun. 2002. Relationships between climate and population dynamics of White-tailed Ptarmigan *Lagopus lecurus* in Rocky Mountain National Park, Colorado, USA. Climate Research 23:81–87.

Williams, M. W., J. S. Baron, N. Caine, R. Sommerfeld, and R. Sanford, Jr. 1996. Nitrogen saturation in the Rocky Mountains. Environmental Science and Technology 30:640–646.

Wolfe, A. P., J. S. Baron, and R. J. Cornett. 2001. Anthropogenic nitrogen deposition induces rapid ecological changes in alpine lakes of the Colorado Front Range (USA). Journal of Paleolimnology 25:1–7.

Wormworth, J., and K. Mallon. 2006. Bird species and climate change: the global status report, version 1.1. <www.climaterisk.com.au/wpcontent/uploads/2006/CR_Report_BirdSpeciesClimateChange.pdf> (15 March 2009).

Zwickel, F. C., and J. F. Bendell. 1967. A snare for capturing Blue Grouse. Journal of Wildlife Management 31:202–204.

CHAPTER TWENTY-TWO

Effects of Translocation on the Behavior of Island Ptarmigan

Robb S. A. Kaler and Brett K. Sandercock

Abstract. Evermann's Rock Ptarmigan (*Lagopus muta evermanni*) are endemic to the Near Islands of the western Aleutian Archipelago, Alaska, but introductions of nonnative arctic foxes extirpated ptarmigan from all islands except Attu Island. Fox removals were completed at Agattu Island in 1979 but natural recolonization did not occur, and 75 ptarmigan were translocated from Attu to Agattu during 2003–2006 to reestablish a breeding population. We used radiotelemetry to assess the impacts of translocation on the post-release movements, nest site selection, and brood movements of 28 females. Behavior of 11 translocated birds was compared to 17 established birds resulting from translocations completed in previous years. Nest sites of translocated females were not different from nest locations of established females with regard to topographical features or nest cover. Nest site selection was influenced by aspect and percent cover of rocks and forbs, but not by slope or general habitat features. After hatching, translocated females moved their broods greater distances from nest sites than did established females (845 vs. 190 m), and all females moved their broods to higher elevations above the nest site (62 to 108 m). The size of the brood home range was similar for established (3.6 ha, SE = 1.6) and translocated females (6.7 ha, SE = 2.4). Overall, translocated and established ptarmigan had similar movements, nest site selection, and reproductive performance at Agattu. Our results are encouraging for future efforts to reestablish populations of endemic ptarmigan and land birds elsewhere in the Aleutian Islands.

Key Words: Agattu Island, Aleutian Archipelago, colonization, Evermann's Rock Ptarmigan, *Lagopus muta evermanni*, post-release movements, radiotelemetry, reintroduction.

Kaler, R. S. A., and B. K. Sandercock. 2011. Effects of translocation on the behavior of island ptarmigan. Pp. 295–306 *in* B. K. Sandercock, K. Martin, and G. Segelbacher (editors). Ecology, conservation, and management of grouse. Studies in Avian Biology (no. 39), University of California Press, Berkeley, CA.

Oceanic island systems are highly suscepti- ble to extinction, largely owing to reduced diversification, simplified food webs, and high rates of endemism, which make island spe- cies highly sensitive to ecological perturbations and disturbance (Moors 1993, Courchamp et al. 2003). Introduction of nonnative mammals has been a major cause of biodiversity loss on islands, due to the small size of island populations and a lack of adaptive behavioral responses among island species which have evolved without preda- tors or competitors (Savage 1984, Atkinson 1985, King 1985, Moors and Atkinson 1984). Less than 20% of the world's bird species are restricted to islands, but over 90% of bird extinctions during historic times were island forms, largely attrib- uted to the effects of introduced species (Johnson and Stattersfield 1990). Restoration of island eco- systems is an urgent need in conservation biology, but may also be one of the most feasible goals. If biodiversity loss has been due to impacts of intro- duced species, restoration via predator removal may be easier to achieve in closed island systems than mainland sites (Courchamp et al. 2003). Predator removals can restore island habitats, but direct intervention may still be needed to reestab- lish taxa with limited dispersal ability. Transloca- tions of wild-caught animals are an important tool in conservation biology for reintroductions and supplementation of declining populations (Scott and Carpenter 1987, Griffith et al. 1989).

Translocations have been widely used to estab- lish or reintroduce populations of ptarmigan (Braun et al. 1978, Hoffman and Giesen 1983), forest grouse (Woolf et al. 1984), and prairie grouse (Toepfer et al. 1990, Hoffman et al. 1992, Musil et al. 1993). Post-release monitoring of habitat selection and movements can provide insights into the behavioral mechanisms that contribute to successful settlement and establish- ment of new populations. Monitoring of recently released individuals can also help to determine the ecological requirements of a species, particu- larly during the early stages of colonization. If translocated individuals disperse from release sites before settlement, the new home ranges may indicate preferred environmental condi- tions (Hirzel et al. 2004). Field studies conducted during multi-year reintroduction projects have the added benefit of measuring the impacts of translocation by comparing performance of newly released birds with established individuals and their offspring (Saltz and Rubenstein 1995, Sarrazin and Barbault 1996).

The Aleutian Islands are an archipelago of more than 200 islands extending from the Alaska Penin- sula west toward Asia and geographically separat- ing the North Pacific Ocean from the Bering Sea (Fig. 22.1). Historically, the Aleutian Islands had no native terrestrial mammals west of Umnak Island (Murie 1959, Gibson and Byrd 2007). Many island populations of native birds were negatively impacted by deliberate introductions of arctic foxes (*Alopex lagopus*) by fur trappers between 1750 and 1940 (Bailey 1993, Williams et al. 2003, Maron et al. 2006). Depredation of eggs, young, and breeding birds led to population declines and local extirpation of waterfowl, seabirds, and land birds from many islands (Murie 1959). In 1949, as part of the recovery efforts for the Aleutian Cackling Geese (*Branta hutchinsii leucopareia*), the U.S. Fish and Wildlife Service (USFWS) began systematic removal of foxes from Amchitka and Agattu Islands.

The Rock Ptarmigan (*Lagopus muta*, formerly *L. mutus*) is an arctic breeding bird with a Nearc- tic distribution (Holder and Montgomerie 1993). Within the Aleutian Archipelago, the species exhibits considerable phenotypic differentiation with a range of plumage coloration. Eight subspe- cies have been described from different groups of Aleutian Islands (Holder et al. 2000, 2004). In four subspecies, males have pale nuptial plumage, three subspecies have dark plumage, and only one subspecies has black plumage: Evermann's Rock Ptarmigan (*L. m. evermanni*). Genetic analy- ses based on mitochondrial DNA have shown that Evermann's Rock Ptarmigan are markedly differ- ent from all other Aleutian Rock Ptarmigan, and the origins of this isolated population remain unclear (Holder et al. 2000, Pruett et al. 2010). Evermann's Rock Ptarmigan once occurred throughout the Near Islands group (Fig. 22.1) in the western range of the Aleutians but were extir- pated from all islands except Attu by 1936 (Turner 1886, Murie 1959). Following completion of fox removals in 1999, the extant population on Attu was estimated to be about 1,000 birds (Ebbert and Byrd 2002). Due to a small population size and limited geographic range, Evermann's Rock Ptarmigan was designated as a species of special management concern by the USFWS.

Regional migratory movements of Rock Ptarmigan are common in mainland Alaska and

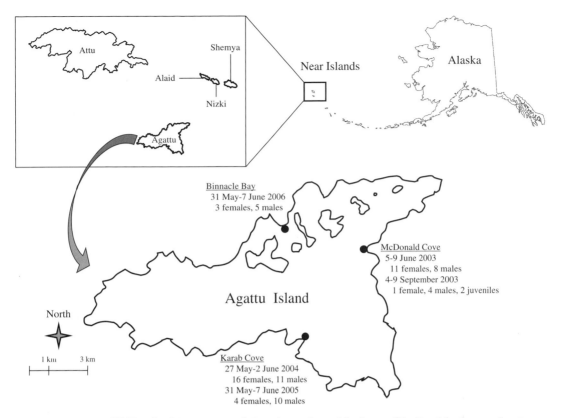

Figure 22.1. Location of field study of Evermann's Rock Ptarmigan at Agattu Island, part of the Near Islands group (inset), Alaska, 2005–2006. Evermann's Rock Ptarmigan once occupied the Near Islands (Attu, Agattu, Shemya, Nizki, and Alaid Islands), but predation by introduced foxes restricted their range to Attu by 1936. Ptarmigan were translocated from Attu to Agattu Island from 2003 to 2006 to restore birds to their former range. Outline of mountains in northeast portion of Agattu Island represents 300-m contour lines of montane habitats where ptarmigan nested. Points mark three coastal sites where ptarmigan were released after translocation from Attu to Agattu Island.

northern Canada, but ptarmigan in the Aleutian Islands are more sedentary. Long-distance movements have been reported for other species of ptarmigan (Hannon et al. 1998, Martin et al. 2000), including dispersal across marine waters (Zimmerman et al. 2005), but natural recolonization across the 28-km strait between the islands of Attu and Agattu did not occur during the 25-year period after foxes were successfully eradicated from Agattu. As part of a regional effort by the Alaska Maritime National Wildlife Refuge (NWR) to restore wildlife populations after the removal of nonnative predators, translocations were used to reestablish Evermann's Rock Ptarmigan at Agattu Island over a four-year period from 2003 to 2006. Previously, we compared the demographic performance of newly translocated birds with established birds from earlier translocations, and found that translocated birds delayed clutch initiation and laid smaller clutches, but that seasonal

productivity and survival rates of the two groups did not differ (Kaler et al. 2010). Here, we compare behavior of translocated and established birds to examine the settlement decisions, nest site selection, and brood movements of female ptarmigan during the early stages of establishment of a new population.

Our general objectives were to determine the ecological requirements of island ptarmigan and to develop translocation methods for restoration of endemic birds in the Aleutian Islands. We tested three hypotheses regarding habitat preferences of island ptarmigan: (1) If newly translocated females are unconstrained in their settlement decisions, they may select new habitats at Agattu compared to their capture sites at Attu; (2) if settlement decisions are constrained by presence of territorial birds, established females at Agattu might occupy preferred nesting habitats and recently translocated females might be limited to nest sites in suboptimal sites; and (3) if habitat use is affected by timing of breeding or prior breeding experience, recently translocated females might have larger brood home ranges than established females with familiar nest and brood-rearing areas.

STUDY AREA

Attu Island (52.85° N, 173.19° E; 89,279 ha) and Agattu Island (52.43° N, 173.60° E; 22,474 ha) are in the Near Islands group in the western range of the Aleutian archipelago and are part of the Alaska Maritime National Wildlife Refuge (Fig. 22.1). Attu is a mountainous island, composed of steep hillsides rising from sea level to elevations up to 861 m. In contrast, Agattu is an island that is primarily maritime tundra <200 m in elevation, with a single mountain range covering the northern third of the island and a maximum elevation of 634 m. Climatic conditions in the Near Islands are characteristic of a maritime environment in the north Pacific and exhibit limited daily and annual variation. A weather station at Shemya Island, approximately 30 km to the northeast, reported a mean annual temperature of approximately 3.9°C, with precipitation occurring on >200 days and averaging 80.6 cm per year (1949–1995; Western Regional Climate Center). Wind velocities averaged 42 km per hour on Shemya and gusts of 165–200 km per hour are common.

Plant communities on the two islands are similar. At coastal areas, the upper beach strand plant community forms a narrow fringe around each island, dominated by beach rye (Leymus arenarius), beach fleabane (Senecio pseudo-arnica), and beach greens (Honckenya peploides). Wet meadow communities in sheltered valley bottoms are dominated by grasses (Poa eminens, Calamagrostis canadensis) and sedges (Carex spp., Eriophorum russeolum). Sloping hillsides are dominated by two plant community types: tall forbs (Geranium erianthum, Anemone narcissiflora, Geum calthifolium) or dwarf shrub meadows with a mixture of crowberry (Empetrum nigrum) and lichens (Cladina spp.). Woody plants (Salix spp., Sorbus sambucifolia) are sparse but occur in protected low-lying areas where they may grow to 0.5–1.0 m in height. Higher elevations (>250 m) show a transition to upland dwarf shrub mats dominated by dwarf willow (Salix arctica), heaths (Phyllodoce aleutica,

Cassiope lycopodioides), and forbs (Saxifaga spp., Geum calthifolium). Glaucous-winged Gulls (Larus glaucescens) and Common Ravens (Corvus corax) were potential predators of eggs and chicks, whereas Peregrine Falcons (Falco peregrinus) and Snowy Owls (Nyctea scandiaca) were a threat to adult ptarmigan (Gibson and Byrd 2007). Agattu Island does not have introduced populations of rats (Rattus spp.), which are a problem elsewhere in the Aleutian Islands.

METHODS

Evermann's Rock Ptarmigan were captured with noose poles, noose carpets, and ground nets in the area of Massacre Bay, Attu Island. Five sets of translocations from Attu to Agattu Island were conducted: four between 27 May and 9 June of 2003–2006 and one during 4–8 September 2003 (Kaler et al. 2010). Ptarmigan were transported from Attu Island by ship and released at one of three coastal beach sites at Agattu Island: MacDonald Cove (2003), Karab Cove (2004–2005), or Binnacle Bay (2006; Fig. 22.1). During June–August of 2005 and 2006, we remained at Agattu during the breeding season to conduct an intensive radiotelemetry study of ptarmigan movements and behavior. We compared behavior between two groups of birds: ptarmigan that had been previously translocated or hatched at Agattu (hereafter, established birds) versus ptarmigan that were newly translocated and released at Agattu during 2005 or 2006 (hereafter, translocated birds).

All ptarmigan were uniquely color banded and each female was fitted with a bib-style radio collar (2005: 15 g, Telemetry Solutions, Concord, CA), or a necklace radio collar (2006: 6 g, Holohil Ltd., Carp, ON). Radios had an expected battery life of 12–18 months and were equipped with mortality switches to facilitate detection of mortality events and dropped collars. Radio-tracking began immediately after release using a three-element Yagi antenna and portable radio receivers (R2000, Advanced Telemetry Systems, Isanti, MN). For the first two weeks after release, we located each bird daily using standard triangulation techniques. A compass bearing was recorded in the direction of a radio-marked bird from each of 3–4 georeferenced points spaced ≥100 m apart. Positions of females attending nests or broods were triangulated from distances of 30–50 m to reduce location errors (Garrott et al. 1986). All locations were recorded in Universal Transverse Mercator (UTM) coordinates using a handheld Global Positioning System receiver (Garmin GPSmap 76, Garmin International, Olathe, KS). Locations were downloaded using DNR Garmin version 5.1.1 software (Minnesota DNR, 2001) and entered into a Geographic Information System (GIS) using ArcView GIS 3.2a software (Environmental Systems Research Institute, Redlands, CA).

Nest site characteristics were measured after completion of each nesting attempt. Vegetation data were collected at each nest plot and four dependent non-nest plots placed at 50 m from the nest in each of the four cardinal directions. Using a 25-m-radius circular plot (0.2 ha) at each nest, or non-nest plot, percentage classes of each general vegetation type present were estimated using the classification system of Viereck et al. (1992), based on cover of exposed soil and rocks, and major plant functional groups such as lichens, mosses, grasses, forbs, and woody plants (Kaler 2007). Elevation, slope, aspect, and topography were also recorded for each 25-m plot. Using a 5-m-radius plot nested within the 25-m plot and centered at the nest site or middle of the non-nest plot center, we classified ground cover into 13 categories after Frederick and Gutiérrez (1992). Data were converted to the median point of each group. To maintain consistency among study plots, a single field observer conducted all measurements (RSAK).

Home Range and Habitat Use

We used the Animal Movement extension for ArcView GIS software to estimate movement rates and home range size based on minimum convex polygons (Hooge and Eichenlaub 1997). We quantified home range size of broods from locations collected during a five-week period (30 June–9 August). We excluded points collected within the first three days after hatching because most females remained close to the nest site. All locations receive equal weight in estimates of home range based on minimum convex polygons and thus our estimates of home range size may be biased low (Swihart and Slade 1985, Barg et al. 2005). We estimated an arithmetic center for home ranges of each brood and calculated linear distances between the nest site and the center of each home range. Change in elevation was calculated as the difference between the average

elevation of brood locations and the elevation of the nest.

Statistical Analysis

Statistical analyses were conducted with procedures of Program SAS (ver. 8, SAS Institute, Cary, NC). To determine habitat characteristics associated with the nest sites of Rock Ptarmigan, we used discriminant function analysis (DFA) to compare nest plots and non-nest plots. A stepwise discriminant analysis with 18 habitat characteristics (8 and 10 variables for 5-m and 25-m radius plots, respectively) was used to determine which variables best discriminated between nest sites and non-nest sites (Johnson 1998). A significance level of $\alpha = 0.5$ was used for parameter entry into the analysis while an $\alpha = 0.2$ significance level was used for parameter retention. We then conducted a DFA on the remaining subset of habitat characteristics to discriminate between the two groups. A cross-validation procedure was used to determine misclassification rates for nest and non-nest sites. DFA and cross-validation are usually conducted with a subset of the data to identify significant variables, and then tested on an independent set of the remaining data. Due to small sample sizes, we opted to use all data for both steps of the analysis. Analyses were conducted using Proc STEPDISC and Proc DSCRIM. Circular statistics and the Watson–Williams test were calculated with program Oriana (ver. 3.0, Kovach Computing Services, Anglesey, Wales). Prior to analysis, all non-normal data were log$_e$-transformed to meet the assumptions of normality. Differences in brood home range size between established and translocated ptarmigan were compared using t-tests (Proc TTEST). Fisher Exact tests were used for the analysis of 2×2 contingency tables. All means are presented with standard errors (SE) unless otherwise noted. All tests were two-tailed and considered significant at $\alpha \leq 0.05$.

RESULTS

During our two-year radiotelemetry study of Evermann's Rock Ptarmigan, we radio-tracked movements of 17 established females (6 in 2005, 11 in 2006) and 11 newly translocated females (9 in 2005, 2 in 2006). Nest failure ($n = 5$) and total brood loss ($n = 10$) reduced the number of broods available for monitoring, and our analyses of brood home range and distances moved between nest sites and brood home ranges were based on six established and eight translocated females. Three females were followed in both years, of which two were followed as translocated birds during the first year of the radiotelemetry study. One established female was right-censored due to harness failure and loss of the radio transmitter.

Post-release Movements and Nest-site Selection

We calculated straight-line distances from release locations to nest sites for all 11 translocated females. Nest sites of translocated females averaged 4.2 km from their respective release location (SE = 0.7, range 0.7–7.6 km, $n = 11$). Average distance between the release location and the nest site was greater for the nine females released at the south side of the island in 2005 (mean = 5.1 km, SE = 1.9, range = 0.7–7.6 km, $n = 9$) than for two females released at the north side of the island in 2006 (mean = 2.7 km, SE = 1.6, range = 1.1–4.3 km, $n = 2$).

We detected no differences between nests of established and translocated females based on elevation, slope, aspect, and nest cover (Table 22.1), and we pooled information to characterize habitat requirements of the island population. Nests were simple 3–5-cm-deep scrapes with eggs usually laid on a thin layer of vegetation and a few ptarmigan contour feathers. Nests were well concealed; 50% (14 of 28 nests) were placed beneath a large rock or boulder (>30 cm diameter) and 46.4% (13 of 28 nests) were found among dense vegetation. Nest sites of established females were not more likely to be associated with either rocks or vegetation than those of translocated females (Fisher's Exact test: $P = 0.70$). Nest sites were

affected by slope aspect in the alpine nesting habitats, and had a nonrandom orientation with a preference for south-facing slopes (mean vector $\mu = 166°$, SE = 12.4°, $r = 0.562$, SE = 0.08, $n = 28$, Rayleigh test: $z = 8.845$, $P < 0.001$).

We compared 28 nest sites with 112 non-nest plots in a discriminant function analysis (DFA) at two scales: eight variables in 5-m-radius plots to quantify ground cover composition, and 10 variables in 25-m-radius plots to describe general habitat types and topographic features. At the 5-m scale, 3 of 8 habitat characteristics were selected in the stepwise procedure that best discriminated between nest sites and non-nest sites. The significant parameters retained in the analysis ($P < 0.2$) included cover of rocks >20 cm ($F_{1,138} = 8.33$, $P = 0.005$), forbs ($F_{1,138} = 8.48$, $P = 0.058$), and rocks <20 cm ($F_{1,138} = 1.91$, $P = 0.169$). A DFA with all eight habitat characteristics correctly classified nest plots 46.4% (13 of 28) and non-nest plots 68.8% (77 of 112) of the time. When the three key habitat characteristics were used alone, the ability to discriminate between plot types increased marginally. A DFA based on the three significant variables correctly classified nest plots 67.9% (19 of 28) of the time; non-nest plots were correctly classified 65.2% (73 of 112) of the time. The proportion of rocks >20 cm in diameter ($F_{1,138} = 8.33$, $P = 0.005$) and forb cover ($F_{1,138} = 8.48$, $P = 0.058$) were the most important variables differentiating between nest and non-nest plots.

Nest plots could not be differentiated from non-nest plots at the 25-m-radius scale. The stepwise procedure selected two variables that best discriminated between nest plots and non-nest plots— open low scrub ($F_{1,138} = 20.2$, $P < 0.001$) and cover of mesic forbs ($F_{1,138} = 2.0$, $P = 0.17$)—but

TABLE 22.1
Topographic and nest cover measurements [means ± SE (N)] for nests of established and translocated female Evermann's Rock Ptarmigan at Agattu Island, Alaska, 2005–2006.

Parameter	Established	Translocated	Pooled	Test[a]	df	Statistic	P
Elevation (m)	237 ± 28 (18)	251 ± 13.6 (10)	242 ± 14 (28)	t	26	−0.46	0.65
Slope (degree)	35 ± 57 (18)	27 ± 10 (10)	32 ± 6 (28)	t	25	−0.55	0.59
Aspect (degree)	154 ± 19 (18)	179 ± 16 (10)	166 ± 12 (28)	F	1,26	1.04	0.32
Nest cover (%)	90 ± 6 (18)	82 ± 12 (9)	87 ± 5 (27)	t	25	0.8	0.46

[a] t = t-test; F = Watson-Williams F-test.

the DFA procedure using the subset of variables did not improve classification results (21.4%, 6 of 28 nest plots were correctly classified). Female Rock Ptarmigan at Agattu Island may prefer certain habitat features during nest site selection, but these features had little to do with slope, aspect, or general habitat, and were instead influenced by percent cover of vegetation at the nest site, which determined concealment.

Home Range and Movements of Broods

Average number of locations for calculations of home range based on minimum convex polygons was 4.4 (SE = 0.3, range = 3–6, n = 14) and yielded one location per week during the first five weeks of the brood rearing period (30 June–9 August). Our analyses were conducted for established and translocated ptarmigan from a total of 52 and 40 locations, respectively. The home range size of females with broods aged 3–25 days did not differ between established (3.6 ha, SE = 1.6, range = 0.5–10.0 ha, n = 5 broods) and recently translocated females (6.7 ha, SE = 2.4, range = 0.5–16.6 ha, n = 7 broods, t = –0.89, df = 10, P = 0.40). After hatching, translocated females with broods moved greater distances from the nest site to the center of the brood home range (845 m, SE = 243, range = 171–2,185 m, n = 8) than established females attending broods (190 m, SE = 65.2, range = 47–394 m, n = 6, t = –2.27, df = 12, P = 0.04). Females with broods moved to higher elevations above the nest site, but the increase in elevation did not differ between established females (mean = 162 m, SE = 21.8, range =15–157 m, n = 6) and translocated females (mean = +108 m, SE = 25, range = 34–233 m, n = 9, t = –1.40, df = 13, P = 0.19).

DISCUSSION

Our study provides the first data on the behavior and habitat requirements for a subspecies of Rock Ptarmigan endemic to the Aleutian Islands. We had three major findings, which should aid restoration efforts for other island populations impacted by introduced species of mammals. First, translocated females switched from using low-lying coastal areas of Attu Island to alpine habitats at higher elevations of Agattu Island. Second, newly translocated females selected nest sites in areas adjacent to established females and

their offspring. Ptarmigan nest sites were usually located on the lower third of south-facing slopes that provided sufficient cover from predators and inclement weather for nesting females. Last, brood movements and patterns of habitat use were similar between recently translocated and established females. Translocated females moved greater distances with broods than established females, but all females moved to higher elevations and had home ranges of similar size. Overall, our results indicate that translocated and established female ptarmigan had similar behavior and patterns of movement during early stages of establishment of a new island population.

Nest Site Selection

Translocated females were captured at low-lying tundra areas in the coastal area of Massacre Bay at Attu Island but settled and nested on mountain hillsides and rocky alpine areas at Agattu Island. The change in habitat use may imply a preference for upland habitats by Rock Ptarmigan, or could be due to island differences in environmental conditions, including competition for mates, availability of appropriate nesting sites or food resources, or predator communities (Herzog and Boag 1977, Gratson 1988, Martin et al. 1990). Timing of clutch initiation is normally synchronous in Rock Ptarmigan (Wilson and Martin 2010), but breeding attempts by translocated females were delayed one to two weeks due to capture and handling (Holder and Montgomerie 1993, Kaler et al. 2010). Thus, competition between translocated and established females for mates and nest sites may have been reduced at Agattu. Following onset of incubation by established female ptarmigan, pair bonds and territorial boundaries maintained by female–female aggression may have broken down, and males may have been unconstrained in acquiring second mates (Martin et al. 1990). Sex ratios can be male-biased in Rock Ptarmigan populations (Unander and Steen 1985, Cotter 1999, Holder and Montgomerie 1993), and unmated territorial males may have been available during settlement of translocated females at Agattu.

Our finding that translocated and established ptarmigan preferred higher elevations at Agattu is consistent with early reports of habitat associations of Evermann's Rock Ptarmigan at Attu (Bent 1932, Haflinger and Tobish 1977). Habitat preferences may have developed with exposure to high densities of introduced foxes at lower-elevation coastal areas of Attu during the past century, and could explain why ptarmigan persisted at Attu but were extirpated from other islands (Ebbert and Byrd 2002). Unander and Steen (1985) argued that lowland coastal areas were not suited for nesting by ptarmigan at Svalbard Island because of late snowmelt and high predation risk. Snow accumulation at low elevations is limited in the Aleutian Islands because of the maritime climate. However, Glaucous-winged Gulls are potential egg and chick predators, and high densities of gulls at coastal areas could have influenced the preferences of female ptarmigan for alpine nesting sites.

If settlement was constrained by competition, we predicted that translocated and established females might select different nesting habitats, but we found no difference in topographic features or amount of nest cover between the two groups. Furthermore, recently translocated and established females sometimes nested in close proximity (<100 m) and used the same ecological factors for nest site selection. The results of our vegetation and habitat analysis at two scales (5-m and 25-m radius) suggested that after locating a larger area of suitable nest habitat, female ptarmigan appear to select nest sites based on microhabitat features of ground and nest cover. Nesting cover used by incubating females may provide concealment from avian predators and protection from inclement weather (Giesen et al. 1980, Wilson and Martin 2008), which can affect embryo development and thermoregulatory costs of incubating females (Webb 1987; Wiebe and Martin 1997, 2000).

In areas with mammalian predators, ptarmigan may select nest sites with cover that balances the trade-off between increasing nest survival while decreasing the risk of mortality for incubating females (Götmark et al. 1995, Wiebe and Martin 1998). Nests of ptarmigan at Agattu were well concealed and typically placed beneath rocks or among thick vegetation, which provided complete cover from visual predators above the nest but could have restricted a female's ability to detect and elude predators. Nest sites of Rock Ptarmigan at Svalbard Island, where arctic foxes are present, were placed in steep and rocky locations that provided a more open view of the surrounding areas (Pedersen et al. 2005). A lack of native terrestrial mammalian predators in the Aleutian

Islands may explain why female Evermann's Rock Ptarmigan select nest sites that provide concealment from avian predators and greater protection from adverse weather conditions. Wilson and Martin (2008) suggested that lateral cover was an important characteristic for nest sites of Rock Ptarmigan at an alpine site in the Yukon, but we found no evidence of selection for such features, possibly because selection criteria for nest sites vary among populations or because our sample of nests was small.

Nest placement could have been influenced by access to food resources determined by seasonal patterns of snow melt. Female ptarmigan at Agattu were sometimes observed feeding on ericaceous food plants <100 m from their nest sites in moist depressions created by late snowmelt. Gardarsson (1988) reported that Rock Ptarmigan in Iceland nested in sites with greater cover that were close to feeding areas with preferred food plants. If females synchronize clutch initiation with snow melt in areas near nests, they could minimize time off the nest by reducing travel time between foraging areas and nest sites, limiting their exposure to predators and increasing nest attendance during incubation (Wiebe and Martin 2000, Yoder et al. 2004). We lacked information on the quality of vegetation in foraging areas, but proximity of areas with good nest concealment to feeding locations may be important for nest site selection among translocated Rock Ptarmigan.

Brood Movements and Home Range Size

Female ptarmigan attending broods at Agattu departed their nesting territories and moved to higher elevations that were 60–100 m above their nesting site, where they remained in a relatively small home range. Translocated females moved greater distances and tended to have larger brood home ranges than established females. Movement of broods to higher elevations likely addresses the nutritional needs of ptarmigan chicks, which feed on arthropods initially but shift their diet to an increasing proportion of plant material as they mature (Spidsø 1980). Greater brood movements were associated with a 1–2-week delay in clutch initiation among translocated females (Kaler et al. 2010), which could have resulted in a phenological mismatch between the timing of hatching and the availability or nutritional quality of arthropods and plants along an elevational gradient. Alternatively, translocated females may have been forced to select nest sites without regard to locations of brood-rearing habitats, and established females minimized brood movements by familiarity with local areas (Bergerud and Gratson 1988).

Females might be expected to minimize home range size if movements increase the mortality of young by exposing chicks to predators or to inclement weather before the chicks are able to thermoregulate. Our estimates of the size of brood home ranges at Agattu (5.5 ha) are comparable to values for Rock Ptarmigan in the Yukon (≤4 ha; Wilson 2008), but larger brood home ranges have been reported from other field studies of Rock Ptarmigan (24 ha, Favaron et al. 2006; 50 ha, Steen and Unander 1985), Willow Ptarmigan (*L. lagopus*; 14 ha, Erikstad 1985; 25–27 ha, Bergerud and Huxter 1969), and White-tailed Ptarmigan (*L. leucura*; 70 ha, Schmidt 1988). Some variation could be due to estimation technique or number of locations per brood. However, variation could also be due to differences in patterns of space use or ecological conditions. Small home ranges are often associated with a high brood density, and minimal overlap among home ranges of different broods may be evidence of spacing behavior (Erikstad 1985), whereas large home ranges with a high degree of overlap may indicate that food availability is low or patchily distributed (Favaron et al. 2006). Small home ranges of female ptarmigan attending broods at Agattu Island were likely due to a combination of low breeding densities in a newly established population, minimal competition among broods, and high-quality habitats that have not been exploited by ptarmigan since the 1930s. Based on our understanding of ptarmigan distributions at Attu, we predict that low-elevation habitats at Agattu will become occupied in the next decade if the ptarmigan population continues to grow.

Our behavioral data for Evermann's Rock Ptarmigan indicate that translocated birds were able to settle quickly in preferred habitats and had performance similar to established birds. Subsequent population surveys have shown that population numbers have been stable during 2007–2009 after translocations were completed. Translocated birds may survive for several years before the population fails (Woolf et al. 1984), but our evidence for successful settlement, reproduction, and survival are encouraging for persistence of Evermann's Rock Ptarmigan at Agattu Island (Kaler et al. 2010).

A combination of eradication of introduced mammals and translocations should be useful tools for future restoration of island populations of terrestrial birds in the Aleutian Archipelago.

ACKNOWLEDGMENTS

We are indebted to C. E. Braun (Grouse Inc.), W. Taylor (Alaska Department of Fish and Game), and S. E. Ebbert (USFWS Alaska Maritime NWR) for involving us in this project. J. K. Augustine, R. B. Benter, and M. A. Schroeder assisted with capture and translocation of ptarmigan from Attu Island to Agattu Island. L. A. Kenney and G. T. Wann were dedicated research assistants for two field seasons at Agattu Island. The crew of the M/V *Tiglax* included Captain K. D. Bell, D. Erickson, J. Faris, E. Nelson, W. Pepper, and R. Ward, who provided us with safe passage to the outer Aleutians. The LORAN Station of the U.S. Coast Guard at Attu Island provided housing, meals, and other support during our capture effort. K. Martin and S. Wilson offered constructive reviews of our manuscript. Funding for field work for this project was provided by the USFWS Alaska Maritime National Wildlife Refuge and the U.S. Missile Defense Agency. The Division of Biology at Kansas State University provided financial support to R. S. A. Kaler and B. K. Sandercock. Capture and handling of birds was conducted under protocols approved by the Institutional Animal Care and Use Committee at Kansas State University and wildlife permits from the State of Alaska.

LITERATURE CITED

Atkinson, I. A. E. 1985. The spread of commensal species of *Rattus* to oceanic islands and their effects on island avifauna. Pp. 35–81 *in* P. J. Moors (editor), Conservation of Island Birds, Vol. 3. ICBP Technical Publications, Cambridge, UK.

Bailey, E. 1993. Introduction of foxes to Alaskan Islands: history, effects on avifauna, and eradication. U. S. Fish and Wildlife Service Resource Publication 191, Anchorage, AK.

Barg, J. J., J. Jones, and R. L. Robertson. 2005. Describing breeding territories of migratory passerines: suggestions for sampling, choice of estimator, and delineation of core areas. Journal of Animal Ecology 74:139–149.

Bent, A. C. 1932. Life histories of North American gallinaceous birds. U.S. National Museum Bulletin 162.

Bergerud, A. T., and M. W. Gratson. 1988. Survival and breeding strategies of grouse. Pp. 473–576 *in* A. T. Bergerud and M. W. Gratson (editors), Adaptive strategies and population ecology of northern grouse. University of Minnesota Press, Minneapolis, MN.

Bergerud, A. T., and D. S. Huxter. 1969. Breeding season habitat utilization and movement of Newfoundland Willow Ptarmigan. Journal of Wildlife Management 33:967–974.

Braun, C. E., D. H. Nish, and K. M. Giesen. 1978. Release and establishment of White-tailed Ptarmigan in Utah. Southwest Naturalist 23:661–668.

Cotter, R. C. 1999. The reproductive biology of Rock Ptarmigan (*Lagopus mutus*) in the central Canadian Arctic. Arctic 52:23–32.

Courchamp, F., J. L. Chapius, and M. Pascal. 2003. Mammal invaders on island: impact, control, and control impact. Biological Reviews 78:374–383.

Ebbert, S. E., and G. V. Byrd. 2002. Eradications of invasive species to restore natural biological diversity on Alaska Maritime National Wildlife Refuge. Pp. 102–109 in C. R. Veitch and M. N. Clout (editors), Turning the tide: the eradication of invasive species. IUCN SSC invasive species specialist group, IUCN, Cambridge, UK.

Erikstad, K. E. 1985. Growth and survival of Willow Grouse chicks in relation to home range size, brood movements and habitat selection. Ornis Scandinavica 16:181–190.

Favaron, M., G. S. Scherini, D. Preatoni, G. Tosi, and L. A. Wauters. 2006. Spacing behavior and habitat use of Rock Ptarmigan (*Lagopus mutus*) at low density in the Italian Alps. Journal of Ornithology 147:618–628.

Frederick, G. P., and R. J. Gutiérrez. 1992. Habitat use and population characteristics of the White-tailed Ptarmigan in the Sierra Nevada, California. Condor 94:889–902.

Gardarsson, A. 1988. Cyclic population changes and some related events in Rock Ptarmigan in Iceland. Pp. 300–329 in A. T. Bergerud and M. W. Gratson (editors). Adaptive strategies and population ecology of northern grouse. University of Minnesota Press, Minneapolis, MN.

Garrott, R. A., G. C. White, R. M. Bartmann, and D. L. Weybright. 1986. Treatment of reflected signals in biotelemetry triangulation systems. Journal of Wildlife Management 50:747–752.

Gibson, D. D., and G. V. Byrd. 2007. Birds of the Aleutian Islands, Alaska. Nuttall Ornithological Club and The American Ornithologists' Union, Washington, DC.

Giesen, K. M., C. E. Braun, and T. A. May. 1980. Reproduction and nest-site selection by White-tailed Ptarmigan in Colorado. Wilson Bulletin 92:188–199.

Götmark, F., D. Blomqvist, O. C. Johansson, and J. Bergkvist. 1995. Nest-site selection: a trade-off between concealment and view of the surroundings. Journal of Avian Biology 26:305–312.

Gratson, M. W. 1988. Spatial patterns, movements, and cover selection by Sharp-tailed Grouse. Pp. 158–192 in A. T. Bergerud and M. W. Gratson (editors), Adaptive strategies and population ecology of northern grouse University of Minnesota Press, Minneapolis, MN.

Griffith, B., J. M. Scott, J. W. Carpenter, and C. Reed. 1989. Translocation as a species conservation tool: status and strategy. Science 245:477–480.

Haflinger, K., and T. Tobish. 1977. Results of a population survey of the Rock Ptarmigan *Lagopus mutus evermanni* on Attu Island, Alaska May 12–Aug 5, 1977. Unpublished Administrative Report, Alaska Maritime National Wildlife Refuge, Homer, AK.

Hannon, S. J., P. K. Eason, and K. Martin. 1998. Willow Ptarmigan *Lagopus lagopus*. A. Poole and F. Gill. (editors), The birds of North America No. 369. Academy of Natural Sciences, Philadelphia, PA.

Herzog, P. W., and D. A. Boag. 1977. Seasonal changes in aggressive behavior of female Spruce Grouse. Canadian Journal of Zoology 55:1734–1739.

Hirzel, A. H., B. Posse, P. Oggier, Y. Crettendan, C. Glenz, and R. Arlettaz. 2004. Ecological requirements of reintroduced species and the implications for release policy: the case of the Bearded Vulture. Journal of Applied Ecology 41:1103–1116.

Hoffman, R. W., and K. M. Giesen. 1983. Demography of an introduced population of White-tailed Ptarmigan. Canadian Journal of Zoology 61:1758–1764.

Hoffman, R. W., W. D. Snyder, G. C. Miller, and C. E. Braun. 1992. Reintroduction of Greater Prairie-Chickens in northeastern Colorado. Prairie Naturalist 24:197–204.

Holder, K., and R. Montgomerie. 1993. Rock Ptarmigan (*Lagopus mutus*). A. Poole and F. Gill (editors),The birds of North America No. 51. Academy of Natural Sciences, Philadelphia, PA.

Holder, K., R. Montgomerie, and V. L. Friesen. 2000. Glacial vicariance and historical biogeography of Rock Ptarmigan (*Lagopus mutus*) in the Bering region. Molecular Ecology 9:1265–1278.

Holder, K., R. Montgomerie, and V. L. Friesen. 2004. Genetic diversity and management of Nearctic Rock Ptarmigan (*Lagopus mutus*). Canadian Journal of Zoology 82:564–575.

Hooge, P. N., and B. Eichenlaub. 1997. Animal movement extension to ArcView. Version 1.1. Alaska Science Center, Biological Science Office, U.S. Geological Survey, Anchorage, AK.

Johnson, D. E. 1998. Applied multivariate methods for data analysts. Brooks/Cole Publishing Co., Pacific Grove, CA.

Johnson, T. H., and A. J. Stattersfield. 1990. A global review of island endemic birds. Ibis 132:167–180.

Kaler, R. S. A. 2007. Demography, habitat use and movements of a recently reintroduced island population of Evermann's Rock Ptarmigan. M.S. thesis, Kansas State University, Manhattan, KS.

Kaler, R. S. A., S. E. Ebbert, C. E. Braun, and B. K. Sandercock. 2010. Demography of a reintroduced population of Evermann's Rock Ptarmigan in the Aleutian Islands. Wilson Journal of Ornithology 122:1–14.

King, W. B. 1985. Island birds: will the future repeat the past? Pp. 3–15 in P. J. Moors (editor) Conservation of island birds. International Council for Bird Preservation, Cambridge, UK.

Martin, K., S. J. Hannon, and S. Lord. 1990. Female–female aggression in White-tailed Ptarmigan and Willow Ptarmigan during the pre-incubation period. Wilson Bulletin 102:532–536.

Maron, J. L., J. A. Estes, D. A. Croll, E. M. Danner, S. C. Elmendorf, and S. L. Buckelew. 2006. An introduced predator alters Aleutian Island plant communities by thwarting nutrient subsidies. Ecological Monographs 76:3–24.

Martin, K., P. B. Stacey, and C. E. Braun. 2000. Recruitment, dispersal, and demographic rescue in spatially-structured White-tailed Ptarmigan populations. Condor 102:503–516.

Moors, P. J. 1993. Conservation of island birds. International Council for Bird Preservation, Cambridge, UK.

Moors, P. J., and I. A. E. Atkinson. 1984. Predation on seabirds by introduced animals, and factors affecting its severity. Pp. 667–690 in J. P. Croxall, P. G. H. Evans, and R. W. Schreiber (editors), Status and conservation of the world's seabirds, Vol. 2. ICBP Technical Publication, Cambridge, UK.

Murie, O. J. 1959. Fauna of the Aleutian Islands and Alaska Peninsula. North American Fauna 61:1–406.

Musil, D. D., J. W. Connelly, and K. P. Reese. 1993. Movements, survival, and reproduction of sage grouse translocated to central Idaho. Journal of Wildlife Management 57:85–91.

Pedersen, A. Ø., Ø. Overrein, S. Unander, and E. Fuglei. 2005. Svalbard Rock Ptarmigan (*Lagopus mutus hyperboreus*)—a status report. Norwegian Polar Institute Report Series No. 125.

Pruett, C. L., T. N. Turner, C. M. Topp, S. V. Zagrebelny, and K. Winkler. 2010. Divergence in an archipelago and its conservation consequences in Aleutian Island Rock Ptarmigan. Conservation Genetics 11:241–248.

Saltz, D., and D. I. Rubenstein. 1995. Population dynamics of a reintroduced Asiatic wild ass *Equus hemionus*. Ecological Applications 5:237–335.

Sarrazin, F. and R. Barbault. 1996. Reintroduction: challenges and lesson for basic ecology. Trends in Ecology and Evolution 11:474–478.

Savidge, J. A. 1984. Guam: paradise lost for wildlife. Biological Conservation 30:305–317.

Schmidt, R. K. 1988. Behavior of White-tailed Ptarmigan during the breeding season. Pp. 270–299 in A. T. Bergerud and M. W. Gratson (editors), Adaptive strategies and population ecology of northern grouse. University of Minnesota Press, Minneapolis, MN.

Scott, J. M., and J. W. Carpenter. 1987. Release of captive-reared or translocated endangered birds: what do we need to know? Auk 104:544–545.

Spidsø, T. 1980. Food selection by Willow Grouse *Lagopus lagopus* chicks in northern Norway. Ornis Scandinavica 11:99–105.

Steen, J. B., and S. Unander. 1985. Breeding biology of the Svalbard Rock Ptarmigan (*Lagopus mutus hyperboreus*). Ornis Scandinavica 23:366–370.

Swihart, R. K., and N. A. Slade. 1985. Influence of sampling interval on estimates of home range size. Journal of Wildlife Management 49:1019–1025.

Toepfer, J. E., R. L. Eng, and R. K. Anderson. 1990. Translocating prairie grouse: what have we learned? North American Wildlife and Natural Resources Conference 55:569–579.

Turner, L. M. 1886. Contributions to the natural history of Alaska. Arctic series of publications in connection with the Signal Service, No. 2. U.S. Army, Washington, DC.

Unander, S., and J. B. Steen. 1985. Behavior and social structure in Svalbard Rock Ptarmigan (*Lagopus mutus hyperboreus*). Ornis Scandinavica 16:95–98.

Viereck, L. A., C. T. Dyrness, A. R. Batten, and K. J. Wenzlick. 1992. The Alaska vegetation classification. General Technical Report PNW-GTR-286. U.S. Department of Agriculture, Forest Service, Pacific Northwest Research Station, Portland, OR.

Webb, D. R. 1987. Thermal tolerance of avian embryos: a review. Condor 89:874–898.

Western Regional Climate Center. 2009. <http://www.wrcc.dri.edu/cgi-bin/cliMAIN.pl?akshem> (30 November 2009).

Wiebe, K. L., and K. Martin. 1997. Effects of predation, body condition and temperature on incubating rhythms of White-tailed Ptarmigan. Wildlife Biology 3:143–151.

Wiebe, K. L., and K. Martin. 1998. Costs and benefits of nest cover for ptarmigan: changes within and between years. Animal Behavior 56:1137–1144.

Wiebe, K. L., and K. Martin. 2000. The use of incubation behavior to adjust avian reproductive costs after egg laying. Behavioral Ecology and Sociobiology 48:463–470.

Wilson, S. 2008. Influence of environmental variation on habitat selection, life history strategies and population dynamics of sympatric ptarmigan in the southern Yukon Territory, Ph.D. dissertation, University of British Columbia, Vancouver, BC.

Wilson, S., and K. Martin. 2008. Breeding habitat selection of sympatric White-tailed, Rock and Willow Ptarmigan in the southern Yukon Territory, Canada. Journal of Ornithology 149:626–637.

Wilson, S., and K. Martin. 2010. Variable reproductive effort for two sympatric ptarmigan *Lagopus* in response to spring weather conditions in a northern alpine ecosystem. Journal of Avian Biology 41:1–8.

Williams, J. C., G. V. Byrd, and N. B. Konyukhov. 2003. Whiskered Auklets *Aethia pygmaea*, foxes, humans and how to right a wrong. Marine Ornithology 31:175–180.

Woolf, A., R. Norris, and J. Kube. 1984. Evaluation of Ruffed Grouse reintroductions in southern Illinois.

Pp. 59–74 *in* W. L. Robinson (editor). Ruffed Grouse management: state of the art in the early 1980s. North Central Section of The Wildlife Society and the Ruffed Grouse Society. Bookcrafters, Chelsea, MI.

Yoder, J. M., E. A. Marschall, and D. A. Swanson. 2004. The cost of dispersal: predation as a function of movement and site familiarity in Ruffed Grouse. Behavioral Ecology 15:469–476.

Zimmerman, C. E., N. Hillgruber, S. E. Burril, M. A. St. Peters, and J. D. Wetzel. 2005. Offshore marine observation of Willow Ptarmigan, including water landings, Kuskokwim Bay, Alaska. Wilson Bulletin 117:12–14.

CHAPTER TWENTY-THREE

Hunting Lowers Population Size in Greater Sage-Grouse

Robert M. Gibson, Vernon C. Bleich,
Clinton W. McCarthy, and Terry L. Russi

Abstract. How hunting mortality affects population size is an important but understudied problem in the applied ecology of grouse and other upland gamebirds. At issue is whether mortality from recreational hunting is additive and therefore depresses population size, or is compensatory and does not. Empirical analyses of this issue may be inconclusive if harvest levels increase with population size or if statistical analysis fails to control for serial dependence in estimates of population size. We examined the effect of hunting on population size in Greater Sage-Grouse (*Centrocercus urophasianus*) using a lek count time series from an intermittently hunted and relatively isolated population in eastern California. Over a 39-year study period (1960–1998), annual variation in harvest recorded in the field was uncorrelated with the previous spring's lek count. After controlling for a positive correlation between lek counts in successive years, numbers of males on leks in spring decreased significantly as harvest during the previous autumn increased. This pattern is expected if hunting mortality is additive and lowers population size. In light of this and similar results from an independent study in Idaho, we suggest that additive, rather than compensatory, hunting mortality should become the default assumption for wildlife managers when setting hunting regulations for Greater Sage-Grouse.

Key Words: Centrocercus urophasianus, hunting, population dynamics.

How recreational hunting affects population size is particularly relevant to biologists charged with the management of populations of grouse, several species of which are in decline worldwide (Storch 2007). Ecological models of harvesting, such as those used to predict maximum sustainable yield, typically assume that hunting is additive to other sources of mortality and hence reduces population size (Caughley and Sinclair 1994). If so, the short-term recreational benefits of hunting must be weighed against the effects of reduced population size on long-term population viability. In contrast, wildlife managers often assume that hunting is compensatory to

Gibson, R. M., V. C. Bleich, C. W. McCarthy, and T. L. Russi. 2011. Hunting lowers population size in Greater Sage-Grouse. Pp. 307–315 *in* B. K. Sandercock, K. Martin, and G. Segelbacher (editors). Ecology, conservation, and management of grouse. Studies in Avian Biology (no. 39), University of California Press, Berkeley, CA.

other sources of mortality and hence that hunters take only a "doomed surplus" (Errington 1946). This view assumes that the hunting season occurs prior to a period of density-dependent mortality that would reduce population size to the same level as a reduction caused by hunting. Whether hunting mortality is additive or compensatory must therefore depend on the extent of harvest and details of a population's ecology. Not surprisingly, the few studies to have examined this issue have produced variable outcomes. For example, where it has been analyzed in upland gamebirds, hunting mortality is often at least partially additive to natural mortality (Robertson and Rosenberg 1988, Williams et al. 2004). However, Devers et al. (2007) report compensatory mortality in Ruffed Grouse (*Bonasa umbellus*), and the full range of outcomes from compensatory to completely additive hunting mortality has been reported from another species of grouse, the Willow Ptarmigan (*Lagopus lagopus*) (Ellison 1991, Smith and Willebrand 1999, Pedersen et al. 2003). Effects of hunting on mortality rates also vary between different species of waterfowl (Nichols 1991).

Attempts to determine the extent to which hunting depresses population size in gamebirds face at least two methodological difficulties. The first arises because wildlife managers may adopt more liberal hunting regulations as population size increases. Changes in management could confound attempts to demonstrate harvest effects in either of two ways. The first scenario assumes that variation in population size primarily reflects regulation around a stable equilibrium density. If so, population growth rate will decrease with increasing population size. Unfortunately, this is also the pattern expected to result from density-dependent harvest. Hence, under this model, the effect of additive hunting mortality would be difficult or impossible to disentangle from intrinsic density dependence (Sedinger and Rotella 2005). A second possible model assumes that variation in population size primarily represents density-independent variation in growth rate. If so, population growth rate would covary positively with population size, a pattern that could mask the effect of additive hunting mortality. Hence, regardless of which is the more appropriate model, to determine the impact of hunting, harvest levels should be manipulated independently of population size.

A second methodological difficulty concerns measurement of a population's response. Successive values in a population time series are likely to be positively correlated, because a population's size at time t constrains its possible size at time $t + 1$. Consequently, the size of a population following a harvest episode is not by itself a useful response measure. Population biologists sometimes attempt to solve this problem by estimating population growth rate as $r = \ln(N_{t+1}/N_t)$, where N is population size and t and $t + 1$ are consecutive time intervals (Royama 1992). However, this ratio covaries negatively with N_t even for a random time series, and thus exhibits spurious density dependence. Additionally, successive values of r are, by definition, not statistically independent. One solution is to evaluate the effect of harvest on N_{t+1} in a general linear model that includes N_t as a covariate. Use of N_t as a covariate ensures that the autocorrelation between N_t and N_{t+1} is removed prior to estimating the effect of harvest.

The effect of hunting on population size is increasingly relevant to management of the Greater Sage-Grouse (*Centrocercus urophasianus*). This specialized, lek-breeding species was once widely distributed and abundant in sagebrush dominated shrubsteppe habitats throughout western North America. It declined in the early part of the 20th century, staged a recovery in the late 1940s and 1950s, but more recently has declined again across much of its geographic range (Connelly and Braun 1997). Reported population declines are consistent with Schroeder et al.'s (2004) estimate that range-wide habitat for Greater Sage-Grouse has been reduced by 45% from pre-settlement to the present. Population declines have also been associated with habitat fragmentation (Braun 1995) and decreased breeding success (Connelly and Braun 1997). These changes have the potential to reduce population viability, making the effect of recreational hunting a relevant concern.

Over most of its geographic range, the Greater Sage-Grouse is counted on leks in spring and hunted in autumn after the reproductive season. Initial attempts to examine the relationship between hunting and population size correlated lek count time series with temporal variation in harvest levels and concluded that fluctuations in spring lek counts were not attributable to variations in harvest (Crawford 1982, Braun and Beck 1985). However, these studies cannot be considered conclusive, because they did not manipulate harvest levels independently of population size or control for prior population size when analyzing the effect of harvest. More recent studies using radio-tagged birds to estimate variation in mortality rates indicate that hunting mortality rates can be locally high, particularly for females (Connelly et al. 2000), and have associated spatial variation in hunting season mortality rates with variation in harvest (Moynahan et al. 2006, Sedinger et al., this volume, chapter 24). Winter mortality is reported to be low in some areas (Idaho: Connelly et al. 2000), but to exhibit spatial and/or annual variation in others (Montana: Moynahan et al. 2006, Nevada: Sedinger et al., this volume, chapter 24), and two studies have reported elevated mortality rates during severe winters (Moynahan et al. 2006, Anthony and Willis 2009). Low over-winter mortality appears to leave only limited scope for reduced density-dependent mortality to compensate for the effect of autumn hunting mortality on spring population size, which suggests that hunting mortality could often be additive.

The most direct test of the additive hunting mortality hypothesis has been undertaken by Connelly et al. (2003), who compared population growth rates of Greater Sage-Grouse over a seven-year period among 19 study areas in Idaho subject to moderate, limited, or no hunting. They reported that populations in areas closed to hunting grew faster and that depression of population growth by hunting may have been more marked in more xeric habitats. Connelly et al.'s conclusions were challenged by Sedinger and Rotella (2005), who suggested that initial lek sizes (an index of population size) were lower in non-hunted study areas and that under a density-dependent model (see above) these populations may have grown faster because they were initially smaller and not necessarily because additive mortality from hunting was absent. Subsequent clarification by Reese et al. (2005) revealed that Sedinger and Rotella's conclusion about initial lek size was relevant only for a subset of five study areas from mountain valleys. After removing this subset of areas from the sample, the remaining 14 lowland study areas still appeared to exhibit faster growth in areas when not hunted (Connelly et al. 2003). In short, despite Sedinger and Rotella's (2005) critique, Connelly et al.'s (2003) data from xeric, lowland study areas suggest that hunting may slow population growth in Greater Sage-Grouse.

In this paper, we evaluate the effect of hunting on population size using a lek count time series from a relatively isolated and intermittently hunted population of Greater Sage-Grouse in eastern California. Because a correlation between harvest and prior population size could confound analysis of the effect of harvest (see above), we first examined the relationships between spring lek counts and both hunting regulations and harvest the following autumn. Then we use the analytical approach proposed above to examine the effect of harvest on population size in the following spring.

METHODS

We analyzed the dynamics of a population of Greater Sage-Grouse at the southwestern edge of the species' geographic range, in Long Valley, Mono County, California over a 39-year period from 1960 to 1998. The study area is described by Bradbury et al. (1989a). Seasonal movements of Sage-Grouse in Long Valley are confined to the valley floor and surrounding foothills (Bradbury et al. 1989b, Gibson 1996, unpubl. data). The nearest sage grouse populations are in Adobe Valley, 25 km to the north, and the White Mountains 45 km to the east. A larger population occurs in the Bodie Hills, 55 km to the north-northwest. Spring lek counts in the Bodie Hills and Long Valley fluctuated independently during the study period (unpubl. data), indicating that any interchange of birds between these areas was insufficient to mask local population dynamics. The Adobe Valley and White Mountain populations are small, so there is limited potential for immigration from either area to influence population dynamics in Long Valley.

Lek Counts

We used the total count of males on a "core" set of eight leks as a relative measure of spring population size. Leks were counted near sunrise on at least three mornings annually during and immediately after the seasonal mating peak, when male lek attendance is highest. Counts conducted after the onset of mating activity include both territorial adults and non-territorial males of all ages (Dalke et al. 1963, R. Gibson, unpubl. data). From 1960 to 1982, most lek counts were conducted by California Department of Fish and Game (CDFG) biologists using methods described by Bradbury et al. (1989a). From 1983 to 1998, counts were conducted by the authors and their associates. As far as possible, we used teams of observers to count birds attending leks simultaneously during peak male attendance, rather than a single observer visiting leks sequentially, as had occurred before 1983. To avoid disturbing the birds, observers counted leks from a range of approximately 200 m. To minimize the influence of spuriously low counts resulting from bad weather or disturbance of the birds before observers arrived, we used the highest seasonal count at each lek as the annual value. Two steps were taken to avoid double counting of birds that moved among leks. First, where leks were close, and thus more likely to exchange birds on a daily basis, we used the highest daily count for the entire lek complex. Second, to avoid unrepresentative values generated by wintering flocks visiting leks before dispersing to other breeding areas (Bradbury et al. 1989b), we excluded any counts made before 20 March, the earliest copulation date recorded from 1984 to 1998 (R. Gibson, unpubl. data).

All eight core leks were counted annually from 1973 to 1998, except in 1978 and 1982, when only four leks were counted. From 1960 to 1972, coverage was less complete, with subsets of five to seven of the core leks counted annually, except in 1969, when only three sites were surveyed. To ensure that core lek totals were comparable across all years, we used complete core surveys from 1973–1998 ($n = 23$ yr) to develop linear regression models relating the core total to the total for each of the 5–7 lek subsets studied during 1960–1968 and 1970–1972. Counts from 1960–1968 and 1970–1972 were then scaled to the core lek total using the appropriate regression for the subset of leks counted in that year (adjusted r^2 values = 96.7–98.3%). We omitted three years (1969, 1978, and 1982), when four or fewer of the eight core leks were counted. To check the robustness of the core lek total, we compared it to the total for all active leks in the study area for 13 years between 1983 and 1998, when lek surveys were sufficiently extensive to have located all active leks. Within this sample, core leks accounted for 95.2 ± 0.8% of the total male count for Long Valley and accurately predicted the total count (adjusted $r^2 = 0.994$, $Y = 1.453 + 1.044X$, $P < 0.0001$).

Hunting Regulations and Harvest

Over the study period, hunting seasons opened between 1 September and 14 October and were for two days, with the exception of 1970–1972, when the season was three days. Bag limits varied from 0 (season closed) to 2 birds. Seasonal and daily limits were identical. After 1986, permit numbers were restricted (50–250 annually). In most years with a hunting season, check stations were set up on the two main access roads and hunters were interviewed as they left Long Valley. Because of the strategic locations of check stations and consistency in their locations and hours of operation, it is likely that the proportion of hunters sampled was high and relatively constant throughout the study period. We therefore used numbers of birds shot and hunters recorded at check stations as indices of the annual legal harvest and hunter effort. Other data suggest that illegal harvest was slight. Relevant observations include the near absence of harvest citations by game wardens patrolling Long Valley during hunting seasons (unpublished CDFG data) and a low incidence of illegal shooting of radio-tagged birds outside the hunting season (<2%, $n = 215$; R. Gibson, unpubl. data). Check station data were available for open seasons in 1963–1965, 1970–1982, 1987–1988, 1990–1992, and 1997. We assumed a harvest of zero in years when the season was closed.

Data Analysis

Wherever applicable, we used parametric statistical methods to maximize power. We square-root transformed core lek counts, numbers of hunters, and birds checked ("checked harvest") to normalize right-skewed distributions before statistical analysis. However, for clarity, untransformed values are plotted in the figures. We used nonparametric rank correlation (Kendall's) for two analyses where, despite transformations, bivariate relationships remained non-linear. In analyzing correlates of hunting regulations, we present analyses of the effects of bag limit and permit numbers only, because preliminary analyses indicated that date and season length did not predict either hunter participation or checked harvest. Descriptive statistics in the text are given as mean ±SE. P-values are based on two-tailed statistical tests.

RESULTS

Counts at core lek and hunting regulations varied during the study period from 1960 to 1998 (Fig. 23.1). Counts of males at core leks showed no significant linear trend over time (core lek count vs. year: $r = 0.049$, $n = 36$, $P = 0.779$). However,

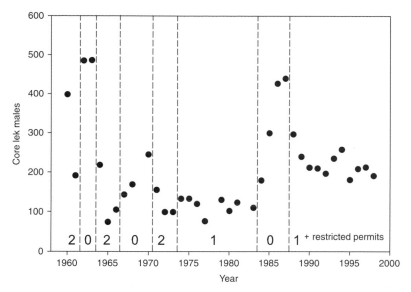

Figure 23.1. Annual variation in male numbers at eight core leks in Long Valley, California, 1960–1998. Vertical dashed lines separate periods with different hunting regulations where the line is drawn after the spring lek count for the year in which regulations changed. Numbers indicate bag limits.

lek counts were positively and significantly auto-correlated from one year to the next ($r = 0.654$, $n = 32$, $P < 0.0001$), though not over longer time lags (2 yr: $r = 0.312$, $n = 31$, $P = 0.088$; 3 yr: $r = 0.152$, $n = 30$, $P = 0.424$; 4 yr: $r = 0.079$, $n = 29$, $P = 0.680$).

Were Hunting Regulations and Harvest Independent of Prior Population Size?

Although managers may increase harvest by liberalizing hunting regulations as population size increases (e.g., Sedinger and Rotella 2005), we found only limited evidence that regulations were adjusted in response to population size and no correlation between checked harvest and lek counts the previous spring.

Hunting Regulations

From 1960 to 1986, three periods when the season was closed (1961–1962, 1966–1969, and 1983–1986) were separated by periods when it was open with bag limits of either one (1973–1982) or two birds (1960, 1963–1965, 1970–1972) (Fig. 23.1), but there was no direct regulation of the numbers of hunters. Over this period there was no statistical association between bag limit and core lek count the preceding spring ($r = 0.034$, $n = 24$,

$P = 0.874$). From 1987 to 1998 the bag limit was one bird; permit numbers were regulated directly and were positively correlated with the previous spring's core lek count (Kendall's $\tau = 0.558$, $n = 12$, $P < 0.02$). Hence, hunting regulations tracked population size only during the last 12 years of the study.

As expected, more liberal regulations were correlated with increased harvest. For example, from 1960 to 1986, significantly more hunters and birds were checked in years with a two- than a one-bird bag limit (hunters: 469.3 ± 66.1 vs. 183.3 ± 14.7, birds: 253.0 ± 63.4 vs. 82.3 ± 7.3, $n = 10$ and 6, unequal variance $t = 5.03$ and 2.80, df = 6.4 and 5.4, $P = 0.002$ and 0.038). Also, during seasons with unregulated numbers of hunters (1960–1986), the mean checked harvest was higher than when permit numbers were restricted (1987–1998) (163 ± 127 vs. 65 ± 15 birds, $n = 18$ and 5, unequal variance $t = 2.92$, df = 19, $P = 0.009$). Permit numbers and harvest were also positively, though not significantly, correlated from 1987 to 1998 (Kendall's $\tau = 0.564$, $n = 5$, $P = 0.112$).

Harvest

The preceding data indicate both that less conservative regulations were correlated with increased harvest and that hunting regulations covaried

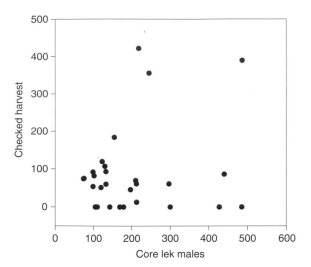

Figure 23.2. The relationship between harvest recorded at check stations and counts at core leks in the preceding spring.

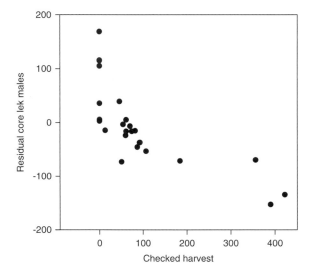

Figure 23.3. The relationship between the count of males at core lek in spring and harvest recorded at check stations in the previous autumn. Core lek counts are scaled as residuals from a regression of the count in year 2 on the count in year 1.

positively with core lek counts for the last 12 years of the study. However, over the entire study period, we found no relationship between spring core lek counts and checked harvest ($r = 0.042$, $n = 28$, $P = 0.831$; Fig. 23.2). The adjusted r^2 value for this relationship was zero, indicating that harvest was effectively independent of prior population size. Thus, retrospective analysis of the effect of harvest on subsequent population size was justified.

Did Harvest Depress Population Size?

Core lek counts increased during three periods when the hunting season was closed, but decreased to lower levels each time it was reopened (Fig. 23.1). As expected if hunting mortality were additive, core

lek counts were negatively and significantly related to the previous fall's harvest in a model that also included the prior year's lek count as a covariate (overall model: $F_{2,22} = 104.321$, $P < 0.0001$; prior year's count: $b' = 0.797$, $P < 0.0001$; checked harvest: $b' = -0.572$, $P < 0.0001$; Fig. 23.3). Together, the prior year's lek count and checked harvest explained 89.6% of the annual variance in core lek counts the following spring (adjusted r^2 based on square-root transformed data).

DISCUSSION

Our analyses indicate that the short-term dynamics of the Long Valley population were dominated by annual variation in harvest, despite a short season (≤ 3 d) and low bag limits (≤ 2 birds). The population increased in each of three periods when the hunting season was closed and declined to lower levels each time hunting was resumed. Fall harvest levels were statistically independent of the previous spring's population size (indexed by lek counts) and, after controlling for a positive correlation between lek counts in successive years, harvest covaried negatively with lek counts the following spring. These results provide evidence that hunting depresses population growth and are not confounded by a correlation between harvest and prior population size (see Introduction).

As with any retrospective analysis, our conclusions are subject to the caveat that other environmental factors affecting population growth might have covaried with harvest rates over time in a way that mimicked the effects of additive hunting mortality. Indeed, it would be surprising if other factors affecting population growth did not exhibit temporal variation. However, our analysis controlled statistically for the most obvious candidate, population size, and it seems unlikely that other factors could have replicated the observed response of the population to the cessation and, particularly, the resumption of hunting during and after each of the three season closures (Fig. 23.1).

In light of the difficulty previous investigators have experienced in demonstrating harvest effects on population size in Greater Sage-Grouse, it may seem surprising that data from a population hunted intermittently for only 2–3 days per year reveal such a clear effect. Four factors may help to explain this. First, our study population is relatively isolated, reducing the opportunity for immigration or emigration to obscure locally determined population dynamics. Second, our statistical analysis accounted for variation in prior population size, a factor overlooked in some earlier studies. Third, due to periodic season closures and the accessibility of the area to large numbers of hunters from population centers in southern and central California when the season was open, the population was exposed to a wide range of hunting pressure. Finally, reproductive success in this population appears to be relatively low, as indicated by juvenile:adult female ratios in wings collected at check stations (CDFG, unpubl. data). Mean juvenile:adult female ratios in Long Valley

for 1960–1984 (2.05 ± 0.50, $n = 15$ yr) and 1985–1995 (1.02 ± 0.15, $n = 7$) fall below mean values for six other western states compiled for the same periods by Connelly and Braun (1997) (2.33 and 1.74, respectively). Low reproductive success must limit population growth and therefore, indirectly, make population size more susceptible to variation in mortality rates.

Our analysis supports the contention of Connelly et al. (2003) that hunting can depress population growth rate in Greater Sage-Grouse. Importantly, our study and Connelly et al.'s derive the same conclusion from complementary lines of evidence (spatial vs. temporal variation in harvest rates) and from widely separated populations. At this point we are not aware of any compelling evidence for compensatory hunting mortality in Greater Sage-Grouse, though scenarios under which this might be the case are conceivable. For example, persistent, deep snow cover can greatly restrict the areas in which sage grouse are able to forage and seek cover during winter (Hupp and Braun 1989) and thus has the potential to cause density-dependent mortality via reduced food availability, increased exposure to predation, or both. If so, density-dependent winter mortality might compensate for the effect of hunting in years or areas subject to these conditions.

Currently, Greater Sage-Grouse are hunted in most states in which they occur and, consequently, each year biologists charged with their management must make decisions on hunting regulations. Both our results and those from low-elevation populations studied by Connelly et al. (2003) indicate that recreational hunting can depress population growth in this species. We therefore suggest that the assumption of additive, rather than compensatory, hunting mortality should be the starting point when setting hunting regulations for Greater Sage-Grouse. A consequence of this conclusion is that hunting should be placed among the factors capable of depressing population size, and hence potentially lowering the long-term viability of small and isolated populations.

ACKNOWLEDGMENTS

Sam Blankenship and Jack Bradbury helped us assemble the data. Many students from UCSD and UCLA and staff from the Bishop BLM office assisted in counting leks. Dan Dawson provided logistical support at the University of California's Sierra Nevada Aquatic Research Laboratory. Financial support was provided by the National Science Foundation and UCLA's Department of Biology. This is Professional Paper 071 from the Eastern Sierra Center for Applied Population Ecology.

LITERATURE CITED

Anthony, R. G., and M. J. Willis. 2009. Survival rates of female Greater Sage-Grouse in autumn and winter in southeastern Oregon. Journal of Wildlife Management 73:538–545.

Bradbury, J. W., S. L. Vehrencamp, and R. M. Gibson. 1989a. Dispersion of displaying male sage grouse. 1. Environmental determinants of temporal variation. Behavioral Ecology and Sociobiology 24:1–14.

Bradbury, J. W., R. M. Gibson, C. E. McCarthy, and S. L. Vehrencamp. 1989b. Dispersion of displaying male sage grouse. 2. The role of female dispersion. Behavioral Ecology and Sociobiology 24:15–24.

Braun, C.E. 1995. Distribution and status of sage grouse in Colorado. Prairie Naturalist 27:1–9.

Braun, C. E., and T. D. I. Beck. 1985. Effects of changes in hunting regulations on sage grouse harvest and populations. Pp. 335–343 in S. L. Beason and S. F. Robertson (editors), Game harvest management. Proceeding of the 3rd International Symposium. Caesar Kleberg Research Institute, Kingsville, TX.

Caughley, G., and A. R. E. Sinclair. 1994. Wildlife ecology and management. Blackwell Science, Cambridge, MA.

Connelly, J. W., and C. E. Braun. 1997. Long-term changes in sage grouse Centrocercus urophasianus populations in western North America. Wildlife Biology 3:229–234.

Connelly, J. W., A. D. Apa, R. B Smith, and K. P. Reese. 2000. Effects of predation and hunting on adult sage grouse Centrocercus urophasianus in Idaho. Wildlife Biology 6:227–232.

Connelly, J. W., K. P. Reese, E. O. Garton, and M. Commons-Kemner. 2003. Response of Greater Sage-Grouse Centrocercus urophasianus populations to different levels of exploitation in Idaho, USA. Wildlife Biology 9:335–340.

Crawford, J. A. 1982. Factors affecting sage grouse harvest in Oregon. Wildlife Society Bulletin 10:374–377.

Dalke, P. D, D. B. Pyrah, D. C. Stanton, J. E. Crawford, and E. F. Schlatterer. 1963. Ecology, productivity and management of sage grouse in Idaho. Journal of Wildlife Management 27:810–841.

Devers, P. K., D. F. Stauffer, G. W. Norman, D. E. Steffen, D. M. Whitaker, J. D. Sole, T. J. Allen, S. L. Bittner, D. A. Buehler, J. W. Edwards, D. E. Figert, S. T. Friedhoff, W. W. Giuliano, C. A. Harper, W. K. Igo, R. L. Kirkpatrick, M. H. Seamster, H. A. Spiker, D. A. Swanson, and B. C. Tefft. 2007. Ruffed Grouse population ecology in the Appalachian region. Wildlife Monographs 168:1–36.

Ellison, L. N. 1991. Shooting and compensatory mortality in tetraonids. Ornis Scandinavica 22:229–240.

Errington, P. 1946. Predation and vertebrate populations. Quarterly Review of Biology 21:144–177.

Gibson, R. M. 1996. A reevaluation of hotspot settlement in lekking sage grouse. Animal Behaviour 52:993–1005.

Hupp, J. W., and C. E. Braun. 1989. Topographic distribution of sage grouse foraging in winter. Journal of Wildlife Management 53:823–829.

Moynahan, B. J., M. S. Lindberg, M. S. Thomas, and J. Ward. 2006. Factors contributing to process variance in annual survival of female Greater Sage-Grouse in Montana. Ecological Applications 16:1529–1538.

Nichols, J. D. 1991. Responses of North American duck populations to exploitation. Pp. 498–525 in C. M. Perrins, J. D. LeBreton, and G. J. M. Hirons (editors), Bird population studies. Oxford University Press, Oxford, UK.

Pedersen, H. C., H. Steen, L. Kastdalen, H. Broseth, R. A. Ims, W. Svendsen, and N. G. H. Yoccoz. 2003. Weak compensation of harvest despite strong density-dependent growth in Willow Ptarmigan. Proceedings of the Royal Society of London Series B 271:281–385.

Reese, K. P., J. W. Connelly, E. O. Garton, and M. Commons-Kemner. 2005. Exploitation and Greater Sage-Grouse Centrocercus urophasianus: a response to Sedinger and Rotella. Wildlife Biology 11:87–91.

Robertson, P. A., and A. A. Rosenberg. 1988. Harvesting gamebirds. Pp. 177–201 in P. J. Hudson and M. R. W. Rands (editors), Ecology and management of gamebirds. BSP Professional Books, Oxford, UK.

Royama, T. 1992. Analytical population dynamics. Chapman and Hall, London, UK.

Schroeder, M. A., C. L. Aldridge, A. D. Apa, J. R. Bohne, C. E. Braun, S. D. Bunnell, J. W. Connelly, P. A. Deibert, S. C. Gardner, M. A. Hilliard, G. D. Kobriger, S. M. McAdam, C. W. McCarthy, J. J. McCarthy, D. L. Mitchell, E. V. Rickerson, and S. J. Stiver. 2004. Distribution of Sage-Grouse in North America. Condor 106:363–376.

Sedinger, J. S., and J. J. Rotella. 2005. Effect of harvest on sage grouse *Centrocercus urophasianus* populations: what can we learn from the current data. Wildlife Biology 11:371–375.

Smith, A., and T. Willebrand. 1999. Mortality causes and survival rates in hunted and unhunted Willow Grouse. Journal of Wildlife Management 63:722–730.

Storch, I. 2007. Conservation status of grouse worldwide: an update. Wildlife Biology 13(Supplement 1): 5–12.

Williams, C. K., R. S. Lutz, and R. Applegate. 2004. Winter survival and additive harvest in Northern Bobwhite coveys in Kansas. Journal of Wildlife Management 68:94–100.

CHAPTER TWENTY-FOUR

Spatial-Temporal Variation in Survival
of Harvested Greater Sage-Grouse

Benjamin S. Sedinger, James S. Sedinger, Shawn Espinosa,
Michael T. Atamian, and Erik J. Blomberg

Abstract. Adult survival is often the demographic parameter to which rate of population change is most sensitive, yet few estimates of survival exist for Greater Sage-Grouse (*Centrocercus urophasianus*). We used relocations of radio-tagged sage grouse to estimate monthly survival throughout their range in Nevada. We also evaluated the relationship between hunter harvest in October and survival during October and November–December, and the relationships between late fall survival and landcover variables in sage grouse home ranges measured from Southwest Regional GAP landcover types. The best performing model of monthly survival allowed survival to vary among seven local sage grouse planning areas in Nevada. This model also allowed survival to differ between October, November–December, and other months and allowed for an additive negative effect of harvest during October. We found no evidence that landscape-level habitat variables in home ranges affected survival during fall. Annual survival varied from 0.16 ± 0.12 in the bistate area along the California–Nevada border to 0.72 ± 0.06 in the south-central area. Total harvest of sage grouse during October was negatively related to survival in October ($\beta = -0.65 \pm 0.17$). Our study demonstrates the potential for substantial spatial variation in survival of sage grouse and the potential for effects of harvest.

Key Words: Centrocercus urophasianus, known fate, Nevada, sage grouse, survival.

reater Sage-Grouse (hereafter sage grouse) are relatively long-lived galliform birds (Schroeder et al. 1999). Sæther and Bakke (2000) suggest that long-lived species that exhibit low reproductive potential should be more sensitive to adult survival than to fecundity. Currently, however, there is a lack of consensus about the relative roles of adult survival and recruitment in the dynamics of sage grouse populations (Aldridge and Brigham 2001, Crawford et al. 2004, Walker 2008), although spread of West Nile virus into sage grouse breeding areas is clearly having population-level consequences (Naugle et al. 2004, Walker 2008). To date, there have been

Sedinger, B. S., J. S. Sedinger, S. Espinosa, M. T. Atamian, and E. J. Blomberg. 2011. Spatial-temporal variation in survival of harvested Greater Sage-Grouse. Pp. 317–328 *in* B. K. Sandercock, K. Martin, and G. Segelbacher (editors). Ecology, conservation, and management of grouse. Studies in Avian Biology (no. 39), University of California Press, Berkeley, CA.

few studies that assess the potential impacts of adult survival on population dynamics of sage grouse. A life table response experiment has shown that the population growth rate (λ) of sage grouse, while greatly impacted by breeding success, is also influenced by adult female survival; adult survival accounted for more than 50% of the variation in λ in Colorado (Sedinger 2007). Walker (2008), using life-stage simulation analysis, showed that adult survival was more weakly related to variation in λ than variables associated with recruitment for sage grouse in the Powder River Basin, Wyoming. In contrast, studies of other large grouse species (e.g., Black Grouse, *Tetrao tetrix*) have shown adult survival to have the greatest influence on population growth (Caizergues and Ellison 1997).

Survival of sage grouse is thought to be influenced by numerous factors. Human harvest has been hypothesized to be additive to other sources of mortality and could negatively influence sage grouse survival (Connelly et al. 2000a, 2003, Reese and Connelly 2011). Survival was lower in an area of Montana that allowed hunting than in another area where hunting was closed (Moynahan et al. 2006). In contrast, Sedinger et al. (2010) did not detect an additive effect of harvest on survival in sage grouse in Colorado. Harvest has been proposed to affect population dynamics of some grouse species (Myrberget 1985, Pedersen et al. 2004), but is thought to be compensatory in others (Ruffed Grouse, *Bonasa umbellus* Devers et al. 2007). Effects of harvest on population growth are often difficult to separate from density-dependent mortality because harvest regulations are often more liberal when abundance is greater, thereby confounding density and harvest effects (Sedinger and Rotella 2005). Finally, heterogeneity in inherent mortality risk among individuals in a sample could produce partially compensatory mortality if individuals with higher inherent risk of mortality are also more vulnerable to harvest. Data to assess this potential are lacking for grouse, but for waterfowl, individuals in poor body condition, who have higher mortality risk (Haramis et al. 1986), are also more vulnerable to harvest (Greenwood et al. 1986, Hepp et al. 1986, Dufour et al. 1993).

Density-dependent mortality is ultimately driven by the availability of habitat. Historically, sage grouse inhabited the majority of the western North American rangelands dominated by sagebrush (Schroeder et al. 1999). Changes in rangeland dynamics at the landscape level and energy development, agriculture, and urban expansion have led to a decrease in sage grouse habitat (Connelly et al. 2004). While it is difficult to envision food limitation in a species that occupies vast tracts of sagebrush, the proportion of individual plants palatable to sage grouse could be limited (Remington and Braun 1985). Deep snow could also reduce access to food in winter (Hupp and Braun 1989), although Zablan et al. (2003) did not detect a negative relationship between annual survival of sage grouse and winter precipitation. Finally, density-dependent mortality processes could involve other mechanisms, such as disease transmission (Naugle et al. 2004) or interference with foraging, not necessarily directly related to per capita food abundance.

In this paper, we assess large-scale spatial variation and seasonal variation in survival across Nevada and we provide estimates of monthly and annual survival for sage grouse in Nevada. We also test for effects of habitat and harvest on adult sage grouse survival in Nevada using known-fate data, harvest estimates, and landcover data. We hypothesized that adult survival would be most affected by landcover types in individuals' local home ranges through effects of landcover type on food and cover. Additionally, we assessed the hypothesis that survival of sage grouse is influenced by harvest and local density.

METHODS

Study Area

All sage grouse in the study were captured in one of seven local area planning units (LAPs) identified by the Nevada Department of Wildlife (NDOW; Fig. 24.1). These units encompass all sage grouse habitat in Nevada and adjacent eastern California and were designated to represent distinct biological units (Nevada sage grouse conservation strategy, Governor Guinn's sage grouse conservation planning team). We excluded samples from the Elko stewardship area because we lacked information about age or gender, but all other LAPs were represented in the analysis. During the study, the state of Nevada managed a two-week hunting season for sage grouse in October with a two-bird daily bag limit and a four-bird possession limit (Nevada Hunt Book,

Figure 24.1. Local area planning units (LAP) for sage-grouse in Nevada and adjacent eastern California, as delineated by the Nevada sage-grouse conservation plan and the Nevada Department of Wildlife.

Nevada Department of Wildlife, Reno, NV). Additionally the Washoe–Modoc–Lassen area encompassed Sheldon National Wildlife Refuge, where there were typically two short (~2 d) sage grouse hunting seasons in September (Nevada Hunt Book, Nevada Department of Wildlife, Reno, NV). The California portion of the Bi-state LAP conducted a permit-only harvest; however, ≤20 sage grouse were harvested annually (California Hunting Regulations: Waterfowl, Upland Game, Hunting and Other Public Uses on State and Federal Areas, California Department of Fish and Game, Sacramento, CA). The Nevada portion of the Bi-state and the Lincoln LAPs were closed to hunting.

Field Techniques

We captured adult sage grouse using night lighting (Giesen et al. 1982) in the vicinity of leks (March–April) and on brood-rearing areas (July–August) from August 2002 through August 2006. Following capture, age and gender were determined (Eng 1955) and all individuals were fitted with a 21.6-g necklace-style radio collar (model A4060 ATS, Isanti, MN). Radios emitted a mortality signal if no movement was detected for eight hours. We

located individual sage grouse approximately monthly by fixed-wing aircraft. Biologists recorded latitude and longitude for each location and status of the individual (mortality or alive). Because individuals were located from the air, frequently in remote locations, it was not possible to determine cause of death for most individuals.

Data Analysis

We restricted the analysis to adults because samples of juveniles were small. We constructed home ranges and estimated 95% kernel density home ranges for each individual sage grouse using kernel density analysis in Hawth's tools (Beyer 2004), an ArcGIS tools package (ESRI 2006). Number of locations per individual was too low for reliable home range estimates (Otis and White 1999), but we produced estimates of home range size to provide an indication of the approximate size of areas in which we estimated landcover. We report mean home range size (±SE).

Sage grouse are sagebrush obligates and rely on sagebrush (*Artemisia* spp.) and other shrubs for food during the non-breeding season (Schroeder et al. 1999). Sage grouse are thought to avoid pinyon–juniper woodlands, which reduce shrub cover as canopies become closed (Miller et al. 2000). Invasive grasses now occupy large areas of the Great Basin, often replacing native shrub communities following fire (Chambers et al. 2007). Invasive grasses do not provide adequate food or cover for sage grouse (Connelly et al. 2000b). While native grasses represent important components of ecosystems used by sage grouse, this landcover type in the Southwestern Regional Gap analysis (SWReGAP), lacked shrubs (Prior-Magee 2007), and we hypothesized that this landcover type would be negatively associated with survival. Similarly, recently burned areas were largely devoid of shrubs and we expected use of these areas to be negatively associated with survival.

We used the SWReGAP for Nevada (USGS National Gap Analysis Program 2004) to estimate the contribution of seven landcover classes to individual home ranges. The SWReGAP combined multi-season satellite imagery (landsat ETM+) with derived landcover data sets to model vegetation structure. Sage grouse home ranges in our study included 74 of the 125 total landcover classifications recognized by the SWReGAP. Of these, we identified 31 landcover classifications that comprised a majority of habitat in home ranges, from which we selected seven landcover types because of their potential positive or negative influence on sage grouse survival. All landcover types we used were combinations of several related landcover classifications (e.g., species of sagebrush). We selected the following classifications for assessment of their relationship to sage grouse survival: (1) sagebrush only (including all sagebrush species in the study area); (2) all shrubs (consisting of all prominent shrub species in the study area, sagebrush included); (3) pinyon–juniper woodland; (4) invasive grasses; (5) native grasses; (6) all grasses (native and invasive); and (7) areas burned before 2002.

We estimated the proportion of each sage grouse home range in each of the seven landcover types and we incorporated these values as individual covariates in candidate models of sage grouse survival. We used general linear models (PROC GLM) in SAS (SAS Institute 2003) to assess variation among LAPs in landcover composition within home ranges. We estimated correlations among habitat types in home ranges using PROC CORR

to assess the extent to which habitat types covaried across home ranges.

We used hunter questionnaire harvest data from NDOW to produce an index of sage grouse harvest for four of the six LAPs (North Central, South Central, Washoe–Modoc–Lassen, and White Pine), which we included in the analysis. We did not include the Lincoln and Bi-state LAPs, which were closed to hunting. The NDOW hunter questionnaire samples ~10% of sage-grouse hunters annually and provides estimates of the number of sage grouse harvested in each county with a hunting season for sage grouse.

We used known-fate analysis in program MARK (White and Burnham 1999) to assess hypotheses about variation in survival of sage grouse in Nevada. We used LAPs as spatial units and months as our minimum time steps for estimating survival. Because of small samples of marked sage grouse in some local planning areas, models allowing for full month by area interactions in monthly survival failed to converge. We therefore considered a series of more constrained models in which we constrained survival to be constant within three-month seasons: January–March (winter), April–June (spring), July–September (summer), and October–December (fall). We also constructed models in which survival was allowed to differ between October and November–December because in areas we considered, hunting occurred in October. We considered more constrained models of temporal variation (e.g., survival in fall differed from that in all other months which were constrained equal). Local area was included only as an additive effect, again because of the small samples available for some LAPs. We included landcover covariates only in fall, where we observed low survival in another analysis of sage grouse survival (E. J. Blomberg and J. S. Sedinger, unpubl. data) or winter because we hypothesized winter was potentially the most nutritionally "stressful" time of year for sage grouse. We included harvest as a covariate for survival in the months of October, when hunting occurred, and November–December, the months immediately following the hunting season. Our rationale for the latter analyses was that the size of the harvest may have been correlated with local population size and harvest could, therefore, have served as an index of local density for assessment of the relationship between density and survival during late fall. All covariates were converted to standard normal variates ($\mu = 0$, $SD = 1$) before analysis. In all models lacking a harvest effect, we combined the Lincoln and White Pine areas because they are adjacent to each other and sample sizes in those two areas were small.

We used information-theoretic approaches to compare models (hypotheses) about variation in survival of sage grouse in Nevada (Burnham and Anderson 2002). We report model-averaged estimates of monthly survival and resulting estimates of annual survival for each of the LAPs. We also report estimates of slope coefficients (β's) linking covariates to monthly survival from the best models containing these covariates as well as averaged across models. We used the logit link to model covariate effects and the delta method to calculate SEs for annual survival estimates (Powell 2007).

RESULTS

We released 174 adult sage grouse with radios (36 males and 138 females) between August 2002 and December 2006. Individuals were located between 3 and 33 times ($\bar{x} = 5.9$). Ninety-five percent kernel home ranges averaged 750.4 ± 32.5 ha. Proportion of home ranges composed of pinyon–juniper, sagebrush, total shrubs, and grasses all varied among LAPs ($F > 6.62$, $df = 5,168$, $P < 0.001$; Table 24.1). Sagebrush covered the largest proportion of home ranges in all LAPs; pinyon–juniper represented 36% of home ranges in the Washoe–Lassen–Modoc LAP. Grasses represented <3% of home ranges, except in the North Central LAP, where grasses averaged 11% of home ranges. Sage grouse home ranges contained an average of 75.9% sagebrush, 80.8% shrubs, 8.0% pinyon–juniper, 2.0% invasive grasses, 4.5% native grasses, 6.5% all grasses, and 0.5% recently burned areas. Correlations between landcover types ranged from −0.56 to 0.83 (Table 24.2). As expected, sagebrush and shrub landcover types were negatively correlated with grass landcover types ($-0.40 < r < -0.21$), and the pinyon–juniper landcover type was negatively correlated with shrub landcover types ($-0.56 < r < -0.38$). The recently burned landcover type was negatively correlated with both shrub landcover types ($-0.41 < r < -0.34$). Harvest varied annually and among LAPs, ranging between

39 and 85 sage grouse for White Pine, 120 and 375 sage-grouse for Washoe–Modoc–Lassen, 129 and 418 sage grouse for South Central, and 694 and 1538 sage grouse for North Central.

Four models received some support ($w_i > 0.1$; Table 24.3). All competitive models contained an additive effect of LAP and a difference in survival between October and other months. The top three models (Akaike weights, $w_i = 0.47$, 0.23, and 0.17) all contained an additive effect of harvest during October. The first- and third-ranked models also contained a difference between survival in November–December and other months. The only other model receiving support ($w_i = 0.11$) contained a difference between October and November–December survival in the North Central LAP and other areas.

Models lacking an area effect were not competitive ($\Delta AIC > 8$), nor were models with seasonal structure other than a differentiation between fall (October–December) and other months ($\Delta AIC > 5$). Models with gender effects performed poorly ($\Delta AIC > 13$). The best model containing a difference between males and females produced only slight differences between the sexes in mean monthly survival (males = 0.961 ± 0.0105, females = 0.958 ± 0.0105). Percent native grass was the only landcover covariate that received any support (sum of AIC weights = 0.18), but the β for the grass effect was small and not different from 0 ($\beta = 0.04 \pm 0.26$). No models containing the variables all grass or invasive grass were competitive. Models containing an effect of harvest during October received 0.87 of Akaike weights, while models including an effect of harvest during October on survival in November–December were not competitive. Harvest in October was negatively related to October survival (best model: $\beta = -0.72 \pm 0.16$; model averaged: $\beta = -0.65 \pm 0.17$; Fig. 24.2).

The model-averaged estimate (models with $\Delta AIC < 4$) of monthly survival was lowest in the Bi-state LAP during November–December (0.71 ± 0.14), followed by October in North Central (0.86 ± 0.02) (Table 24.4). Annual survival was highest in the South Central LAP (0.72 ± 0.06) and lowest in the Bi-state LAP (0.16 ± 0.12) based on model-averaged estimates of monthly survival. Survival estimates for other local areas were intermediate between those for the South Central and Bi-state LAPs.

TABLE 24.1

Mean proportion of seven landcover classifications comprising sage-grouse home ranges for six local area planning units (LAP) in Nevada, 2002–2006.

LAP	n	Pinyon-juniper	Sagebrush	All shrubs	Native grass	Invasive grass	All grass	Recently burned
Bi State	6	0.16 ± 0.05	0.76 ± 0.07	0.79 ± 0.07	0.01 ± <0.01	<0.01 ± <0.01	0.01 ± <0.01	0
Lincoln	14	0.11 ± 0.03	0.84 ± 0.03	0.89 ± 0.03	0	0	0	0
North Central	88	0.05 ± 0.01	0.72 ± 0.01	0.78 ± 0.01	0.08 ± 0.01	0.03 ± 0.01	0.11 ± 0.01	0.01 ± 0.01
South Central	49	0.10 ± 0.02	0.83 ± 0.02	0.85 ± 0.02	0.01 ± <0.01	0.01 ± <0.01	0.02 ± 0.01	0
Washoe–Modoc–Lassen	5	0.36 ± 0.06	0.57 ± 0.03	0.59 ± 0.03	<0.01 ± <0.01	0.03 ± 0.03	0.03 ± 0.03	0
White Pine	12	0.03 ± 0.01	0.73 ± 0.07	0.91 ± 0.02	0.02 ± 0.01	0.03 ± 0.01	0.02 ± 0.01	0

NOTE: Habitat classes include the following SWReGAP landcover classifications: Pinyon-Juniper = Great Basin Pinyon-Juniper woodland; Sagebrush = Inter-Mountain Basins Big Sagebrush Shrubland, Great Basin Xeric Mixed Sagebrush Shrubland, Inter-Mountain Basins Big Sagebrush Steppe, and Inter-Mountain Basins Montane Sagebrush Steppe; All Shrubs = Inter-Mountain Basins Big Sagebrush Shrubland, Great Basin Xeric Mixed Sagebrush Shrubland, Inter-Mountain Basins Big Sagebrush Steppe, Inter-Mountain Basins Mountain Mahogany Woodland and Shrubland, Inter-Mountain Basins Semi-Desert Shrub Steppe, Inter-Mountain Basins Montane Sagebrush Steppe, Inter-Mountain Basins Mixed Salt Desert Scrub, and Inter-Mountain Basins Greasewood Flat; Native Grass = North Pacific Montane Grassland and Inter-Mountain Basins Semi-Desert Grassland; Invasive Grass = Invasive Perennial Grassland and Invasive Annual Grassland; All Grasslands = North Pacific Montane Grassland, Inter-Mountain Basins Semi-Desert Grassland, Invasive Perennial Grassland, and Invasive Annual Grassland; Recently Burned = Burned before 2002.

TABLE 24.2

Pearson correlation coefficients among landcover types in sage-grouse home ranges (n = 174) in Nevada, 2002–2006.

Landcover Type	Pinyon-juniper	Sagebrush	All shrubs	Invasive grasses	Native grasses	All grasses	Recently burned
Pinyon-juniper	1.00						
Sagebrush	−0.38	1.00					
All shrubs	−0.56	0.83	1.00				
Invasive grasses	−0.15	−0.30	−0.21	1.00			
Native grasses	−0.26	−0.23	−0.23	−0.14	1.00		
All grasses	−0.31	−0.40	−0.33	0.68	0.63	1.00	
Recently burned	−0.06	−0.34	−0.41	−0.02	−0.05	−0.06	1.00

TABLE 24.3
Model selection for known-fate models of monthly survival for sage-grouse in Nevada, 2002–2006.

Model	AIC$_c$	Model weight	No. parameters	Deviance
Area+(Oct + Nov-Dec)+H	0	0.47	9	504.35
Area+(Oct)+H	1.43	0.23	7	509.82
Area+(Oct+Nov-Dec)+H+G	2.00	0.17	10	504.32
Area+Oct*NC+(Nov-Dec)*NC	2.92	0.11	10	505.24
Area+(Oct+Nov-Dec)*H	6.99	0.01	10	509.31
Area+(Oct+Nov-Dec)*H*G	8.77	0.01	11	509.06
Area+(Oct+Nov-Dec)	10.48	0.0	8	516.85
Area+(Oct,Nov-Dec)*Annual	12.68	0.0	15	504.83
Constant	13.65	0.0	1	534.11
Area+(Nov-Dec)*H	13.78	0.0	8	520.15
Area+(Oct,Nov-Dec)*G	15.27	0.0	7	523.66
Sex	15.61	0.0	2	534.07
Area	16.31	0.0	6	526.72

NOTE: Notation follows Lebreton et al. (1992); + = main effect or additive model, * = indicates factorial model with interactions. Area = birds grouped by local area planning unit (Lincoln, Bi-state, North Central, South Central, Washoe–Modoc–Lassen, White Pine); Oct − October (month of legal harvest in Nevada) survival independent of other months; Nov-Dec = Survival combined for November and December and independent of other months; G = covariate examining the effect of native grasses; Annual = annual variation; H = covariate examining the effect of harvest during October, Constant = model with constant monthly survival.

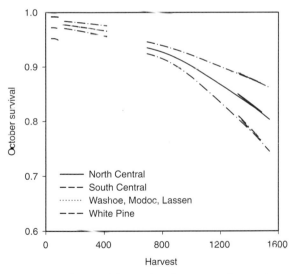

Figure 24.2. Influence of the number of sage-grouse harvested during the October hunting season on October survival by local area planning unit. Estimates and SEs from the best performing model, which included fixed effects of area and an additive effect of harvest on survival. Harvest in White Pine County was <100 sage-grouse per year. Survival and harvest in the South Central and Washoe–Modoc–Lassen LAPs were similar, and estimates for these areas overlap each other. Gaps in estimates of the relationship between number harvested and October survival represent gaps in number harvested among local area planning units.

TABLE 24.4

Seasonal and spatial variation in monthly survival (± SE) for sage-grouse in Nevada, 2002–2006.

Area	n	October	November–December	January–September	Annual
Bi-State	6	0.93 ± 0.05	0.71 ± 0.14	0.89 ± 0.04	0.16 ± 0.12
Lincoln	14	0.96 ± 0.02	0.99 ± 0.01	0.93 ± 0.01	0.49 ± 0.11
North Central	88	0.86 ± 0.02	0.96 ± 0.01	0.97 ± 0.00	0.59 ± 0.08
South Central	49	0.97 ± 0.01	0.97 ± 0.01	0.97 ± 0.01	0.72 ± 0.06
Wahoe–Modoc–Lassen	5	0.97 ± 0.01	0.97 ± 0.01	0.97 ± 0.01	0.71 ± 0.13
White Pine	12	0.97 ± 0.01	0.94 ± 0.03	0.96 ± 0.02	0.57 ± 0.19

NOTE: Parameter estimates based on model-averaging of models with ΔAIC < 4 (Table 24.3).

DISCUSSION

Our estimates of annual survival for sage grouse were comparable to those for Montana, where annual survival ranged from 0.32 ± 0.06 to 0.96 ± 0.02, depending on year, location, and reproductive status (Moynahan et al. 2006). Similar to Walker (2008), we detected substantial spatial variation in monthly and annual survival of sage grouse in Nevada. Our estimate of annual survival (0.16 ± 0.12) for the Bi-state area was unusually low and certainly would not be consistent with a stable population, but was based on a small sample of radio-tagged sage grouse, consistent with its low precision, and should be viewed as preliminary. Our estimate, however, was similar to an independent estimate from an adjacent area in Mono County, California (M. Farinha, unpubl. data), suggesting low annual survival in this part of the range. Walker (2008) also reported annual survival <0.2 for one area of the Powder River Basin during one year, although some mortality was attributable to West Nile virus. Sage grouse in the Bi-state area are genetically distinct from all other sage grouse (Oyler-McCance et al. 2005) and are of substantial conservation concern, so additional research and conservation effort for this population is warranted.

We are the second study to report annual survival for both male and female sage grouse. Zablan et al. (2003) reported that adult females and males had annual survival probabilities of 0.59 and 0.37, respectively. In contrast, our estimates of annual survival were nearly identical for the two genders (males = 0.62 ± 0.01, females = 0.60 ± 0.01)

based on the best model allowing for differences between genders (Table 24.3). We lack detailed histories of reproduction and other activities, but the similarity of annual survival estimates for the two genders suggests that mortality risks were similar between the genders in this study.

Sage grouse in the South Central LAP had among the highest estimated annual survival (0.72 ± 0.08) yet estimated for sage grouse. Populations in that local area were stable during this study (NDOW, unpubl. data). There are few other estimates of annual survival for sage grouse (Zablan et al. 2003, Moynahan et al. 2006, Walker 2008, Sedinger et al. 2010), but these other studies reported variation in estimates of annual survival similar to those we report here. Clearly, such variation indicates that factors affecting adult survival are likely to influence local population dynamics (Sedinger 2007), although we were unable to determine cause of death in this study.

Lowest monthly survival occurred during October (North Central area) or November–December (Bi-state, White Pine). Only the Lincoln area experienced lowest monthly survival during other months of the year (January–September), which include the breeding season. Moynahan et al. (2006) also estimated low monthly survival over winter (November–April) in a year associated with severe winter weather, particularly in December. Anthony and Willis (2009) attributed low monthly survival in winter to severe weather in a high-altitude population in southeastern Oregon. These studies contrast with the hypothesis that mortality outside the breeding or hunting seasons is low (Connelly et al. 2011), indicating that mortality risk varies substantially both temporally and spatially, and mortality outside of the breeding and hunting seasons is sufficient to affect population dynamics.

The magnitude of harvest in October was negatively correlated with survival during October, but not November–December. One hypothesis to explain this result is that harvest was an additive source of mortality (*sensu* Anderson and Burnham 1976). Moynahan et al. (2006) also found that survival during the hunting season was lower in an area that allowed harvest versus an area where harvest was closed. An estimate of the process correlation between harvest rate and survival for sage grouse in Colorado was >0, however, indicating no negative correlation between harvest rate and survival (Sedinger et al. 2010). Powell et al. (this volume, chapter 25) also concluded that harvest was primarily compensatory in Greater Prairie-Chickens (*Tympanuchus cupido*) in Nebraska.

Three LAPs (Bi-state, Lincoln, and White Pine) with substantially lower harvest (harvest was closed in Bi-state and Lincoln) all had lower annual survival than the North Central LAP, which had the highest harvest. We cannot estimate per capita harvest rates for LAPs because we lack estimates of population size or samples of marked individuals that could be used to estimate harvest rate from band recoveries. Nevertheless, substantial variation among LAPs in annual survival, independent of harvest, suggests that even if harvest was an additive source of mortality, other sources of mortality were more important in determining annual survival in Nevada.

Harvest of gallinaceous birds is frequently positively correlated with the number available (Errington 1945, Hudson and Dobson 2001, Cattadori et al. 2003, Guthery et al. 2004), associated with both functional and numerical relationships between harvest and abundance. Also, harvest regulations may reflect local population density (Connelly et al. 2003, Sedinger and Rotella 2005). Consequently, harvest could serve as a surrogate indicator of local population size and apparent effects of harvest could reflect local density-dependent effects on survival. We failed, however, to detect a negative relationship between harvest in October and survival during November–December, which would have been consistent with density-dependent mortality during these months. Our results do not rule

out the potential for density-dependent effects during other times of year. Nevertheless, we caution that the potential association between local density and harvest could render the apparent correlation between harvest and survival spurious, and results we present here are preliminary. Gibson et al. (this volume, chapter 23) found that number of male sage grouse displaying on leks was negatively associated with harvest the previous fall, suggesting an additive effect of harvest on survival. Gibson et al. (this volume, chapter 23) were able to exclude the potential that harvest was correlated with population size, excluding the potential that apparent harvest effects represented spurious effects of local population density. Gibson et al. (this volume, chapter 23) were forced to exclude years when no harvest data were collected, and in several of these years ($n = 6$) lek counts increased following harvest, having an unknown effect on their analysis. Overall, lek counts increased in 11 years and declined in 13 years following open seasons (Gibson et al. this volume, chapter 23; Fig. 23.1). Also, examining harvest data and lek counts in Gibson et al. (this volume, chapter 23) suggests that in several years harvest rate substantially exceeded the 8–10% of the population that has been thought to be compensatory in sage grouse (Sedinger et al. 2010), but the effect of these heavy harvest years on the overall analysis is unknown. Improving understanding of the role of harvest on survival during the hunting season itself will require banding studies where harvest rate and survival can be simultaneously estimated for a given population (e.g., Zablan et al. 2003).

Monthly survival of sage grouse in November–December was not associated with any of the measures of vegetation within home ranges that we assessed. Moynahan et al. (2006) also failed to detect relationships between vegetation features and survival of sage grouse. There are two explanations for the failure of our study and that of Moynahan et al. (2006) to detect effects of vegetation on survival. First, it is possible that vegetation variables used in analyses were unrelated to survival. Alternatively, sage grouse may only use areas containing vegetation that best supports survival. Under this second hypothesis, we would not expect to detect a relationship between survival and vegetation in individual home ranges because individuals have already selected home ranges that best support their survival. Sagebrush or all shrubs dominated landcover types in sage grouse home ranges, as expected. Invasive grasses, recently burned area, and pinyon–juniper comprised an average of 2.0%, 0.5%, and 7.6%, respectively, of sage grouse home ranges in Nevada during the study. Only 3 of 174 sage grouse had recently burned areas in their home ranges. Our results and those of Moynahan et al. (2006), suggest that sage grouse may often occupy home ranges meeting their minimum habitat requirements for survival, making it difficult to detect a relationship between survival and vegetation at the scale of individual home ranges.

We believe the low representation of recently burned areas, pinyon–juniper, and invasive grasses in sage grouse home ranges indicates avoidance because all these habitats were common in Nevada during the study (USGS National Gap Analysis Program 2004). The low use of areas covered by invasive grasses, recently burned area, and pinyon–juniper likely explains the poor performance of models of survival containing these variables. Low use of these landcover types also suggests that their presence reduces land area that will support sage grouse. Substantial increases in recently burned areas and invasive grasses (Chambers et al. 2007, Eiswerth et al. 2008) thus requires that either density of sage grouse must have increased on remaining habitats or numbers of sage grouse have declined.

Overall, we found substantial spatial-temporal variation in survival within Nevada, similar to the findings of Moynahan et al. (2006). Survival of sage grouse tended to be lower in late fall. We did not detect lower survival during the breeding season (March–June) which has been hypothesized to result from predation during that period (Connelly et al. 2000a). We also found that survival was lower during October, when hunting occurs. Whether this finding reflects an additive effect of harvest on survival, an effect of local density on survival, or some other unknown ecological factor (e.g., disease, predation, poor nutrition) remains uncertain. We found no evidence that males differed from females in annual survival and we found no effect of vegetation within home ranges on survival. Further elucidation of the roles of harvest and habitat in annual mortality processes will require carefully designed studies with adequate samples of marked individuals.

ACKNOWLEDGMENTS

Numerous biologists from the Nevada Department of Wildlife captured and radio-tagged sage grouse. We thank B. K. Sandercock for helpful comments on an earlier draft of the manuscript.

LITERATURE CITED

Aldridge, C. L., and R. M. Brigham. 2001. Nesting and reproductive activities of Greater Sage-Grouse in a declining northern fringe population. Condor 103:537–543.

Anderson, D. R., and K. P. Burnham. 1976. Population ecology of the Mallard. VI. The effect of exploitation on survival. U.S. Fish and Wildlife Service Resource Publication 128:1–66.

Anthony, R. G., and M. J. Willis. 2009. Survival rates of female Greater Sage-Grouse in autumn and winter in southeastern Oregon. Journal of Wildlife Management 73:538–545.

Beyer, H. L. 2004. Hawth's analysis tools for ArcGIS. <http://www.spatialecology.com/htools>.

Burnham, K. P., and D. R. Anderson. 2002. Model selection and multimodel inference a practical information theoretic approach. 2nd ed. Springer, New York, NY.

Caizergues, A., and L. N. Ellison. 1997. Survival of Black Grouse Tetrao tetrix in the French Alps. Wildlife Biology 3:177–186.

Cattadori, I. M., D. T. Haydon, S. J. Thirgood, and P. J. Hudson. 2003. Are indirect measures of abundance a useful index of population density? The case of Red Grouse harvesting. Oikos 100:439–446.

Chambers, J. C., B. A. Roundy, R. R. Blank, S. E. Meyer, and A. Whittaker. 2007. What makes great basin sagebrush ecosystems invasible by Bromus tectorum? Ecological Monographs 77:117–145.

Connelly, J. W., A. D. Apa, R. B. Smith, and K. P. Reese. 2000a. Effects of predation and hunting on adult sage-grouse Centrocercus urophasianus in Idaho. Wildlife Biology 6:227–232.

Connelly, J. W., C. A. Hagen, and M. A. Schroeder. 2011. Characteristics and dynamics of Greater Sage-Grouse populations. Studies in Avian Biology 38:53–68.

Connelly, J. W., S. T. Knick, M. A. Schroeder, and S. J. Stiver. 2004. Conservation assessment of Greater Sage-Grouse and sagebrush habitats. Western Association of Fish and Wildlife Agencies, Cheyenne, WY.

Connelly, J. W., K. P. Reese, E. O. Garton, and M. L. Commons-Kemner. 2003. Response of Greater Sage-Grouse Centrocercus urophasianus populations to different levels of exploitation in Idaho, USA. Wildlife Biology 9:335–340.

Connelly, J. W., M. A. Schroeder, A. R. Sands, and C. E. Braun. 2000b. Guidelines to manage sage-grouse populations and their habitats. Wildlife Society Bulletin 28:967–985.

Crawford, J. A., R. A. Olson, N. E. West, J. C. Mosley, M. A. Schroeder, T. D. Whitson, R. F. Miller, M. A. Gregg, and C. S. Boyd. 2004. Synthesis paper: Ecology and management of sage-grouse and sage-grouse habitat. Journal of Range Management 57:2–19.

Devers, P. K., D. F. Stauffer, G. W. Norman, D. E. Steffan, D. M. Whitaker, J. D. Sole, T. J. Allen, S. L. Bittner, D. A. Buehler, J. W. Edwards, D. E. Figert, S. T. Friedhoff, W. W. Giuliano, C. A. Harper, W. K. Igo, R. L. Kirkpatrick, M. H. Seamster, H. A. Spiker, Jr., D. A. Swanson, and B. C. Tefft. 2007. Ruffed Grouse population ecology in the Appalachian Region. Wildlife Monographs 168:1–36.

Dufour, K. W., C. D. Ankney, and P. J. Weatherhead. 1993. Condition and vulnerability to hunting among Mallards staging at Lake St. Clair, Ontario. Journal of Wildlife Management 57:209–215.

Eiswerth, M. E., K. Krauter, S. R. Swanson, and M. Zielinski. 2008. Post-fire seeding on Wyoming big sagebrush ecological sites: regression analyses of seeded nonnative and native species densities. Journal of Environmental Management 90:1320–1325.

Eng, R. L. 1955. A method for obtaining sage-grouse age and sex ratios from wings. Journal of Wildlife Management 19:141–146.

Errington, P. L. 1945. Some contributions of a fifteen-year study of Northern Bobwhite to a knowledge of population phenomena. Ecological Monographs 15:1–34.

Environmental Systems Research Institute (ESRI). 2006. ArcGIS, release 9.2. ESRI, Redlands, CA.

Giesen, K. M., T. J. Schoenberg, and C. E. Braun. 1982. Methods for trapping sage-grouse in Colorado. Wildlife Society Bulletin 10:224–231.

Greenwood, H., R. G. Clark, and P. J. Weatherhead. 1986. Condition bias of hunter-shot Mallards. Canadian Journal of Zoology 64:599–601.

Guthery, F. S., M. J. Peterson, J. J. Lusk, M. J. Rabe, S. J. DeMaso, M. Sams, R. D. Applegate, and T. V. Dailey. 2004. Multistate analysis of fixed, liberal regulations in quail harvest management. Journal of Wildlife Management 68:1104–1113.

Haramis, G. M., J. D. Nichols, K. H. Pollock, and J. E. Hines. 1986. The relationship between body mass and survival of wintering Canvasbacks. Auk 103:506–514.

Hepp, G. R., R. J. Blohm, R. E. Reynolds, J. E. Hines, and J. D. Nichols. 1986. Physiological condition of autumn banded Mallards and its relationship to hunting vulnerability. Journal of Wildlife Management 50:177–183.

Hudson, P. J., and A. P. Dobson. 2001. Harvesting unstable populations: Red Grouse Lagopus lagopus scoticus (Lath.) in the United Kingdom. Wildlife Biology 7:189–195.

Hupp, J. W., and C. E. Braun. 1989. Topographic distribution of sage-grouse foraging in winter. Journal of Wildlife Management 53:823–829.

Miller, R. F., T. Svejcar, and J. A. Rose. 2000. Western juniper succession in shrub steppe: impacts on community composition and structure. Journal of Range Management 53:574–585.

Moynahan, B. J., M. S. Lindberg, and J. W. Thomas. 2006. Factors contributing to process variance in annual survival of female Greater Sage-Grouse in Montana. Ecological Applications 16:1529–1538.

Myrberget, S. 1985. Is hunting mortality compensated for in grouse populations with special reference to Willow Grouse. Proceedings of the International Union of Game Biologists 12:329–336.

Naugle, D. E., C. L. Aldridge, B. L. Walker, T. E. Cornish, B. J. Moynahan, M. J. Holloran, K. Brown, G. D. Johnson, E. T. Schmidtmann, R. T. Mayer, C. Y. Kato, M. R. Matchett, T. J. Christiansen, W. E. Cook, T. Creekmore, R. D. Falise, E. T. Rinkes, and M. S. Boyce. 2004. West Nile virus: pending crisis for Greater Sage-Grouse. Ecology Letters 7:704–713.

Otis, D. L., and G. C. White. 1999. Autocorrelation of location estimates and the analysis of radiotracking data. Journal of Wildlife Management 63:1039–1044.

Oyler-McCance, S. J., S. E. Taylor, and T. W. Quinn. 2005. A multilocus population genetic survey of the Greater Sage-Grouse across their range. Molecular Ecology 14:1293–1310.

Pedersen, H. C., H. Steen, L. Kastdalen, H. Brøseth, R. A. Ims, W. Svendsen, and N. G. Yoccoz. 2004. Weak compensation of harvest despite strong density-dependent growth in Willow Ptarmigan. Proceedings of the Royal Society, London B 271:381–385.

Powell, L. A. 2007. Approximating variance of demographic parameters using the delta method: a reference for avian biologists. Condor 109:949–954.

Prior-Magee, J. S. 2007. Product use and availability. Pp. 186–191 in J. S. Prior-Magee et al. (editors), Southwest Regional Gap Analysis final report. U.S. Geological Survey, Gap Analysis Program, Moscow, ID.

Reese, K. P., and J. W. Connelly. 2011. Harvest management of Greater Sage-Grouse: a changing paradigm for game bird management. Studies in Avian Biology 38:101–112.

Remington, T. E., and C. E. Braun. 1985. Sage-grouse food selection in winter, North Park, Colorado. Journal of Wildlife Management 49:1055–1061.

Sæther, B. E., and Ø. Bakke. 2000. Avian life history variation and contribution of demographic traits to the population growth rate. Ecology 81:642–653.

SAS Institute. 2003. SAS/STAT user's guide, version 9.1. SAS Institute, Inc., Cary, NC.

Schroeder M. A., J. R. Young, and C. E. Braun. 1999. Sage-grouse (*Centrocercus urophasianus*). A. Poole and F. Gill (editors), The birds of North America No. 425. The Birds of North America, Inc., Philadelphia, PA.

Sedinger, J. S. 2007. Improving understanding and assessment of sage-grouse populations. Bulletin, University of Idaho College of Natural Resources Experiment Station 88:43–56.

Sedinger J. S., and J. J. Rotella. 2005. Effect of harvest on sage-grouse *Centrocercus urophasianus* populations: what can we learn from the current data? Wildlife Biology 11:371–375.

Sedinger, J. S., G. C. White, S. Espinosa, E. T. Partee, and C. E. Braun. 2010. An approach to assessing compensatory versus additive harvest mortality: an example using Greater Sage-Grouse *Centrocercus urophasianus*. Journal of Wildlife Management 74:326–332.

USGS National Gap Analysis Program. 2004. Provisional digital land cover map for the southwestern United States, version 1.0. RS/GIS Laboratory, College of Natural Resources, Utah State University, Logan, UT.

Walker, B. L. 2008. Greater Sage-Grouse response to coal-bed natural gas development and West Nile virus in the Powder River Basin, Montana and Wyoming, USA. Ph.D. dissertation, University of Montana, Missoula, MT.

White, G. C., and K. P. Burnham. 1999. Program MARK: survival estimation from populations of marked animals. Bird Study 46(Supplement):120–138.

Wisdom, M. J., M. M. Rowland, B. C. Wales, M. A. Hemstrom, W. J. Hann, M. G. Raphael, R. S. Holtausen, R. A. Gravenmier, and T. D. Rich. 2002. Modeled effects of sagebrush-steppe restoration on Greater Sage-Grouse in the interior Columbia Basin, U.S.A. Conservation Biology 16:1123–1231.

Zablan M. A., C. E. Braun, and G. C. White. 2003. Estimation of Greater Sage-Grouse survival in North Park, Colorado. Journal of Wildlife Management 67:144–154.

CHAPTER TWENTY-FIVE

Adaptive Harvest Management and Harvest Mortality of Greater Prairie-Chickens

Larkin A. Powell, J. Scott Taylor, Jeffrey J. Lusk,
and Ty W. Matthews

Abstract. Adaptive harvest management (AHM) can assist biologists with decisions made under uncertainty. There have been few applications of AHM to manage wildlife at the state level, and we provide a theoretical exercise using AHM in the context of Greater Prairie-Chicken harvest in southeast Nebraska. Our goals were to develop and evaluate an AHM framework for a state-specific harvest decision, and to use the AHM process to evaluate uncertainties associated with harvest mortality for Greater Prairie-Chickens in Nebraska. Harvest of prairie chickens in southeast Nebraska was restarted 2000, using a special limited permit system, and was controversial with respect to the potential impacts of harvest on a recovering population. We followed standard steps to develop our AHM framework and created a formal utility function to reward harvest regulations that would meet management objectives. We used observed spring counts of males at leks and predicted counts from two competing alternative models based on additive and compensatory harvest mortality to weight our confidence in each model. Our AHM framework provided a framework to select the optimal harvest regulation package. Harvest rates averaged 0.057 as a proportion of the fall population during 2000–2007, and count data suggested that the population was relatively stable. The compensatory harvest mortality model had achieved >99% confidence by 2004, which suggests that harvest mortality in this population may be compensatory for harvest rates <0.06. Our exercise shows that AHM can be effectively applied to harvest decisions at a small geographic scale, and we encourage biologists to consider using data on harvest to formally gain information that will enhance harvest management.

Key Words: adaptive harvest management, Greater Prairie-Chicken, monitoring data, *Tympanuchus cupido.*

Powell, L. A., J. S. Taylor, J. J. Lusk, and T. W. Matthews. 2011. Adaptive harvest management and harvest mortality of Greater Prairie-Chickens. Pp. 329–339 *in* B. K. Sandercock, K. Martin, and G. Segelbacher (editors). Ecology, conservation, and management of grouse. Studies in Avian Biology (no. 39), University of California Press, Berkeley, CA.

nformed harvest management decisions are critical to sustain game populations. Harvest decisions can be controversial, especially for species that are not abundant; thus, decisions must be defensible. Monitoring data can provide guidance for decisions if gathered and interpreted correctly (Lyons et al. 2008). However, population fluctuations can be complicated by environmental factors other than harvest. Moreover, it is common for harvest decisions to have complex sets of multiple decisions, subjective values of stakeholders, and uncertainties about the dynamics of the game population's response to harvest mortality. Adaptive management (AM) is an iterative, learning-based framework for making decisions in wildlife management and conservation biology (Williams et al. 2007). AM has emerged as an effective process to manage natural resources in complex situations in which key components needed to make optimal decisions are unknown or uncertain. Under an AM framework, data from population monitoring is formally incorporated into the decision-making process.

A process for adaptive harvest management (AHM) of waterfowl has been incorporated into the North American Waterfowl Management Plan (NAWMP; Johnson and Williams 1999, NAWMP Committee 2004). Annual harvest regulations are determined through an international decision-making process, and AHM has been used to incorporate uncertainties in system structure, stochastic environmental effects, and incomplete management control of harvest rates. However, harvest regulations for non-migratory species of grouse and other upland gamebirds are made at a state or provincial level, and there have been no local applications of AHM at a state or provincial level.

Greater Prairie-Chickens (*Tympanuchus cupido*) in southeastern Nebraska are thought to be part of the northern extent of the Flint Hills population. Anecdotal evidence suggested low populations until 1990, and grouse hunting was not permitted in southeastern Nebraska during 1930–1999 (S. Taylor, pers. comm.). Harvest of prairie chickens in southeastern Nebraska began in 2000, using a special limited permit system (300 seasonal permits, limit of two birds per permit). This harvest was controversial because of uncertainty with regard to the additive or compensatory response to harvest mortality (Ellison 1991).

Nebraska's southeastern population of Greater Prairie-Chickens is well suited for use as a case study of AHM at the state level. One regulatory agency sets the harvest regulations, and prairie chickens inhabit five counties (approximately 740,000 ha), which is a manageable spatial scale for monitoring. The Nebraska Game and Parks Commission (NGPC) has three monitoring programs in place: spring surveys of booming grounds ("lek counts"), wing surveys from hunter-bagged birds, and a hunter success survey.

At present, AHM is not formally used to make decisions regarding regulations for prairie chicken harvest in Nebraska; rather, NGPC uses a "monitor-and-modify" decision-making process. Annual decisions are made using the best available data, and monitoring information is used to provide annual evaluations of the allowable harvest (Johnson 1999). Our goals were to develop and evaluate an AHM framework for a state-specific harvest management decision, and to use the AHM process to gain information that would decrease the uncertainties associated with harvest mortality of Greater Prairie-Chickens in Nebraska. Our general approach will be suitable for management of grouse populations in other jurisdictions.

METHODS

We followed the example of the North American Waterfowl Management Plan to establish our AHM framework (Williams and Johnson 1995; Williams et al. 2002, 2007). The five steps included: (1) Determine objectives, (2) define the sets of regulatory options, (3) define a set of competing models to represent uncertainties, (4) design an annual monitoring program, and (5) define a method to measure model credibility.

Objectives

Taylor (2000) set two objectives for prairie chicken harvest in southeastern Nebraska: (1) Maintain a spring population of approximately 1,500 males, and (2) maximize recreational opportunities associated with harvest of prairie chickens. We translated these objectives into a utility function (Williams et al. 2002). NGPC biologists determined that roadside surveys in 1996 detected approximately 40% of the population, as estimated from a county-wide survey conducted in 1995 (considered a near-complete count; S. Taylor, unpubl. data). The design of the current roadside survey has not changed since 1996; hence we assumed that counts of 650 males on the survey (C_m) approximated 1,500 males in the population. Thus, biologists stated that a population of 1,000 males (400 males detected on surveys) would be the point at which harvests would be suspended. Our utility function (R) was equal to 1.0 if $C_m \geq 650$; R = 0.0 if $C_m \leq 400$. If $400 < C_m < 650$:

$$R = \frac{C_m - 400}{650 - 400}$$

Regulation Options

The NGPC established four regulation sets for prairie chickens in southeastern Nebraska prior to the opening of the first harvest (Taylor 2000). The most restrictive set was no harvest, which was implemented from 1930 to 1999. The regulation package selected during 2000–2002 was a restrictive set, with 300 permits allowing a harvest of up to two birds (either sex) per permit. A moderate regulation package was selected during 2003, which provided for 400 permits allowing a harvest of up to three birds each. A liberal regulation package, yet to be selected in southeastern Nebraska, would be to use the same regulation set as the set currently used for prairie grouse in western Nebraska, which has a daily bag limit of up to three birds and a possession limit of 12. Seasons have opened on the Saturday closest to 15 September and closed on 31 December since harvest began in 2000; opening and closing dates do not differ among the sets of regulations.

Competing Models

We selected a simple set of two competing models for our exercise, which differed by the potential effects of harvest mortality on the population. Effects of harvest on grouse species are not well known (Ellison 1991) and were of interest to biologists in Nebraska. Grassland habitat in southeastern Nebraska was stable during the years of our study; thus, we did not include carrying capacity in our population model. Because the booming ground counts were of males, our models predicted the number of males in subsequent springs. Under a system of compensatory mortality, harvest mortality leads to density-dependent improvements in survival or reproduction to compensate for losses to harvest (Nichols et al.

1984), and the finite rate of population growth (λ_t) can be modeled as:

$$\lambda_t = \frac{N_{t+1}}{N_t} = \left(S_A + \beta_t \cdot S_J\right)$$

where N is the male population size in years t and $t + 1$, S_A is the annual survival rate of adults, β_t is the number of juveniles produced per adult that survive until harvest (determined by harvest wing ratios) in year t, and S_J is the 7-month survival rate of juveniles from harvest to the spring mating season. Under a completely additive system, harvest mortality directly lowers survival rates, and the population can be modeled as:

$$\lambda_t = \left(S_A - \left(\frac{H_t}{1-c}\right)\right) + \beta_t\left(S_J - \left(\frac{H_t}{1-c}\right)\right)$$

where H is the harvest rate (proportion of population harvested) and c is the crippling loss rate (proportion of shot birds that die without reaching the hunter bag).

We incorporated annual survival rates in our model from Hamerstrom and Hamerstrom (1973; sensu Wisdom and Mills 1997; $S_A = 0.47$). To be conservative, we used 0.20 for c, after Anderson and Burnham (1976; continental average for Mallards, Anas platyrhynchos). Elsewhere, DeStefano and Rusch (1986) reported a hunter-reported crippling rate of 0.13 for Ruffed Grouse (Bonasa umbellus) in Wisconsin, and Durbian et al. (1999) found a hunter-reported crippling rate of 0.06 for Greater Prairie-Chickens in Kansas. We assumed harvest rates did not vary by age class.

Population Monitoring

NGPC conducted a county-wide survey of booming grounds in southeastern Nebraska during 1995, which estimated 4,400 prairie chickens. Since 1996, NGPC has conducted 32.2-km (20.0 mi) roadside surveys through each county using the same routes each year. The goal of the monitoring program is to detect changes in size of the localized population of prairie chickens; thus, routes were designed as road transects through the primary prairie chicken range in each county with known breeding populations. The surveys are conducted over two days; first, biologists stop approximately every 1.6 km (1 mi) at intersections of roads and listen for booming grounds. On the second day, biologists visit the booming grounds recorded the previous day to count the number of displaying males.

Assessing Model Credibility

We made annual decisions regarding harvest regulations. We had a 9-year data set (2000–2008) available for our analyses, but used an iterative, annual approach that would illustrate a real situation, as if NGPC had formally used AHM since harvest began in 2000. We followed seven steps in the AHM decision-making process (Johnson and Williams 1999, Williams et al. 2002): (1) Specify initial certainty in competing models; (2) apply harvest decision; (3) predict next year's survey results, following harvest, under competing models; (4) determine population trend by monitoring; (5) assess probability that competing models predict the current survey results; (6) adjust cumulative model weights; and (7) use utility values (R) to make harvest decisions.

The adaptive process of incorporating prior knowledge with newly acquired information is a fundamental Bayesian modeling approach (Sit and Taylor 1998). For most resource decisions, some uncertainty exists about the status of the current system; in our case, the uncertainty was the potential effects of harvest mortality. After each management cycle, new data are available, which can be used to assign probabilities to alternate possible states of the system: compensatory and additive mortality. Probabilities for system states are useful for guiding decision making, and we followed the Bayesian approach described for waterfowl in North America (Williams et al. 2002).

We began by assigning prior probabilities, or certainties, in each competing model at the point before harvests had begun. Taylor (2000) stated that NGPC felt that prairie chicken harvest mortality might be partially additive. We had no data to suggest otherwise, so we set our initial certainty [w_i^j, where i = model (A: additive; C: compensatory) and j = year] in each competing model to slightly favor the additive model (w_A^0 = 0.55, w_C^0 = 0.45). Harvest began in 2000 with a restrictive set of regulations, which were relaxed to a moderate set of regulations in 2003. We used additive and compensatory population models to predict the spring population of males, and we used year-specific productivity information (hunter wing surveys) and harvest rates (hunter success surveys) to account for annual variation

in productivity and effort. Wing and success surveys are conducted in a single mailing; all prairie chicken hunters are provided return envelopes and cards, and hunters are reminded at the end of the season if they have not returned their survey form and wings.

A second step in the Bayesian approach was to compare predictions from our competing models with the spring counts of males from booming grounds. First, we established a probability density function with a normal distribution (μ = 0, SD = 90; Fig. 25.1), which provided conditional probabilities, P_A^j and (P_C^j, for each model's predictions of spring male counts during year j. We used a standard deviation of 90 to represent approximately 15% of the maximum spring count (ca. 600) during our exercise (Fig. 25.1). For example, if the additive model's prediction matched the spring survey counts (difference of zero), it would receive the highest conditional probability (P_A^j = 0.004). If a model's prediction differed from the spring survey counts, it received a lower score based on the value of the density function. For example, if the compensatory model's prediction of spring male counts was 150 males lower than the actual count, it would receive a conditional probability, P_C^j, of 0.001 (Fig. 25.1). The magnitude of changes in model weights through time has the potential to be dependent on the variance of the density function used; a density function with a small SD will penalize the conditional probability more severely than a density function with a larger SD (Fig. 25.1). We used SD = 90 to create a conservative density function.

The last step in the Bayesian approach was to update our prior beliefs (w_i^{j-1}) with information from our model comparisons (P_i^j) during the last time step (Williams et al. 2002). Under each scenario, we used the year-specific conditional probabilities (P_i^j) to update the cumulative model weights for the additive and compensatory model, w_A^j and w_C^j, annually according to:

$$w_A^j = \frac{w_A^{j-1} \cdot P_A^j}{(w_A^{j-1} \cdot P_A^j) + (w_C^{j-1} \cdot P_C^j)} \text{ and}$$

$$w_C^j = \frac{w_C^{j-1} \cdot P_C^j}{(w_C^{j-1} \cdot P_C^j) + (w_A^j - 1 \cdot P_A^j)}$$

We plotted cumulative model weights to gain information about the relative confidence in the competing models of harvest mortality.

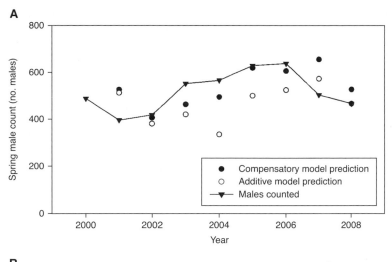

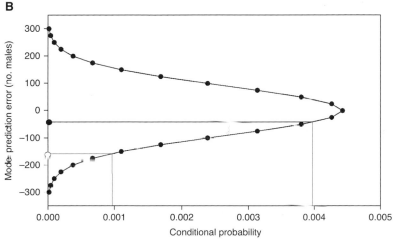

Figure 25.1. (A) Spring counts of male Greater Prairie-Chickens in southeast Nebraska during 2000–2008, and model predictions of spring counts from two alternative models. (B) Derivation of conditional probabilities from a probability density function (μ = 0, SD = 90) as a function of the difference between spring counts of males and model predictions of counts.

Once we had cumulative model weights, we calculated the year-specific (*i*) utility value (R^i_A, R^i_C) for each model's prediction of counts of males for the following spring. We calculated R for each regulation package; that is, given the current population count, would a given regulation package be predicted to achieve the population objective? We used the predictive population models to predict the next spring's count numbers using harvest rates specific to each of the four regulation sets (closed: 0.0, restrictive: 0.03, moderate: 0.08, liberal: 0.12). We used the following formula to calculate a weighted utility value, $\overline{R}^i$, for each set of regulations given the current model weights for our competing models:

$$\overline{R}^i_j = \left(w^i_A + R^i_A\right) + \left(w^i_C + R^i_C\right)$$

We selected the best set of regulations as the set with the highest $\overline{R}^i$. If two or more regulation sets had equal or similar utility values, we selected the most liberal set. The decision-making framework used two objectives for the harvest. Thus, we first selected the regulation that would be most likely to meet the population objectives, given the current knowledge of the effects of harvest mortality. Second, in cases of similar utility values, we always selected the most liberal choice to maximize recreational opportunities for hunters, given that the population that would be sustained. Last, we evaluated the ability of our adaptive harvest management framework to provide proper decisions during population declines. We conducted a hypothetical exercise in which counts declined by 15% per year. All analyses were performed using a spreadsheet designed in Microsoft Excel.

RESULTS

Spring counts of male prairie chickens have remained stable (mean count: 517.6, CV: 0.17) since harvest was initiated in southeastern Nebraska in 2000 (Fig. 25.1). Moderate declines in counts occurred during 2001, 2007, and 2008. Mean response of hunters to harvest surveys was 64% (range: 58–74%) during 2000–2007. The mean harvest rate during 2000–2007 was 5.7% of the population, but harvest rates varied under constant regulations (Fig. 25.2). Harvest rates averaged 0.026 under restrictive regulations (range: 0.01–0.04) and 0.076 during moderate regulations (range: 0.05–0.136).

Predictions of the competing models were similar during 2000–2002, but the predictions began to differ more substantially when harvest rates increased in 2003. Model weights, as of 2004, shifted to almost complete confidence in the compensatory model of harvest mortality (Fig. 25.3). The shape of the probability density function had relatively small effects on the cumulative model weights.

During 2000–2002, the "no harvest" and "restrictive harvest" regulations had similar weighted utility values, $\bar{R}^i$, and their utility values were higher than the utility values for the moderate and liberal regulation packages (Fig. 25.4). As the prairie chicken counts drew closer to the population objective of 1,500 males in 2003, the weighted utility values for "moderate" harvests became similar to utility values for the "no harvest" and "restrictive" regulations. Since 2004, the utility values for the "liberal" regulation package were similar to the utility values for the more restrictive regulations.

Our framework for selecting harvest regulations would indicate that restrictive regulations should be chosen during 2000–2002, moderate regulations in 2003, and liberal regulations since 2004. NGPC changed harvest regulations from restrictive to moderate in 2003.

Under the hypothetical 15% decline scenario, the additive mortality model accumulated over 80% of the total confidence within three years (Fig. 25.5A). Weighted utility values dropped below 0.4 in two years (Fig. 25.5B), which would indicate that harvest should be suspended; utility values dropped to 0 by 2003, indicating no value of the harvest to achieve objectives.

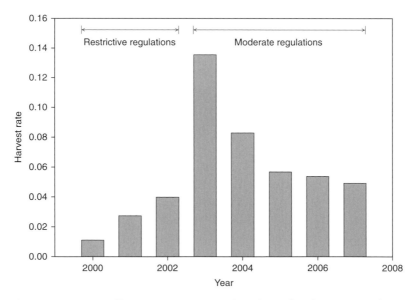

Figure 25.2. Estimates of harvest rates (proportion of population) from hunter surveys for Greater Prairie-Chickens in southeast Nebraska. Regulations for harvest were restrictive during 2000–2002 and moderate during 2003–2007.

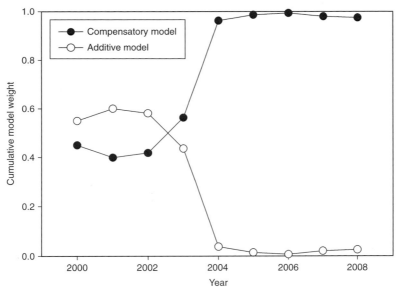

Figure 25.3. Cumulative model weights of two competing models (compensatory vs. additive) for harvest mortality of Greater Prairie-Chickens in southeast Nebraska.

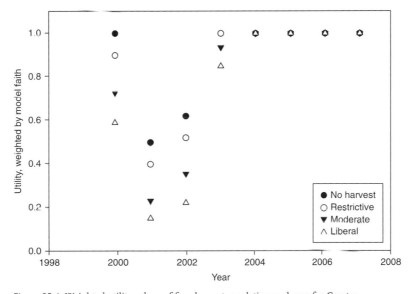

Figure 25.4. Weighted utility values of four harvest regulation packages for Greater Prairie-Chickens in southeast Nebraska.

DISCUSSION

Our exercise effectively shows how an AHM framework can be applied to a harvest management decision at a local, state, or provincial level. Although adaptive management is especially suited for complex problems at continental scales, it can be a useful exercise for state or provincial agencies at regional scales as well (Johnson 1999).

All calculations were easily accomplished in a spreadsheet, and our template could be modified to fit similar harvest scenarios in other states and for other species.

Harvest rates responded to liberalizing of harvest regulations, especially in the first year after change. Harvest rates were variable under constant regulations. Taylor (2000) anticipated harvest rates of up to 0.15. But we documented

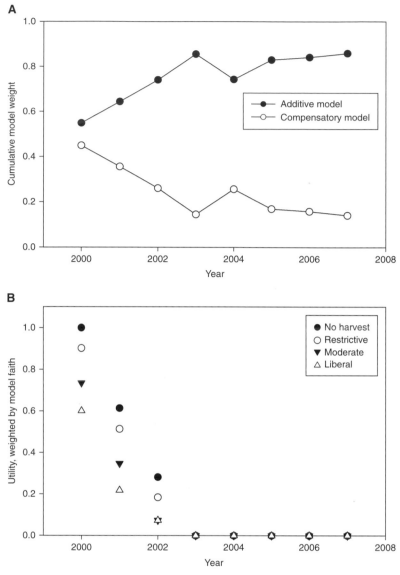

Figure 25.5. (A) Cumulative model weights and (B) weighted utility values under a hypothetical annual 15% decline in spring counts of Greater Prairie-Chickens in southeast Nebraska.

that harvest rates only approached that level during 2003, the year that harvest regulations were changed and the previous lottery system was replaced with a first-come, first-served system of permit allocation.

Currently, the Greater Prairie-Chicken population in southeastern Nebraska appears to be stable. Harvest mortality appears to be compensatory for this population at the low harvest levels we documented. Our competing models were completely additive and completely compensatory; it is possible that a threshold density exists where

a harvest rate of 0.05 becomes additive, and this possibility could be added as a competing model in the future if regulations are liberalized and harvest rates exceed current levels.

Ellison (1991) reported that few studies had shown evidence of compensatory mortality for tetraonid species. High-density grouse and partridge populations have capacity for compensation to harvest because territorial and/or lekking behavior often results in a significant portion of non-breeding males. Harvest of non-breeding males will not affect population productivity, and the non-breeders are available to replace breeding males removed by harvest. Gibson et al. (this volume, chapter 23) reported additive harvest mortality effects for a smaller, lower-density (approximately half the density of prairie chickens at our study site; Bradbury et al. 1989), isolated population of Greater Sage-Grouse (*Centrocercus urophasianus*) in eastern California. Small et al. (1991) suggested that Ruffed Grouse (*Bonasa umbellus*) experience additive harvest mortality at high (>0.50) harvest rates on public lands in Wisconsin. Pedersen et al. (2004) confirmed low rates of compensatory harvest mortality (~0.30) for Willow Ptarmigan (*Lagopus lagopus*) on Norwegian estates (size range: 20–54 km²). The population of prairie chickens in southeastern Nebraska could be considered, locally, high density; some leks have >50 males, and it is likely that there are many non-breeding males. At the low harvest rates we observed, it is likely that the compensatory response to harvest is a function of the non-breeding males. The hunter wing survey does not provide a sex ratio for the harvest; harvest of hens also occurs, which directly impacts production. The low productivity we observed may mask the impact of harvesting hens, and we would expect higher harvest rates to produce an additive impact on the population. It is also possible that density-dependent productivity or survival account for the compensation to harvest. During the time period of our exercise, prairie chicken productivity, as measured by hunter wing ratios, exhibited a negative, but nonsignificant, relationship with abundance (spring male counts; J. Lusk, unpubl. data). The characteristics of this population and availability of monitoring data provide opportunity for further field research and simulation modeling to gain insights into harvest dynamics.

Empirical studies that address effects of harvest mortality are essential to provide information for management. However, annual variation in population dynamics makes it difficult to directly estimate harvest effects (Cox et al. 2004), and we believe the use of monitoring data can guide management effectively. Our AHM exercise relied on estimates of demographic parameters for survival, productivity, and harvest rate. We used year-specific rates of productivity and harvest that were available for our population; however, we did not have annual survival estimates for our population. The rate of 0.47 that we used from Hamerstrom and Hamerstrom (1973) is similar to the annual

survival rate estimate of 0.45 (95% CI: 0.33–0.56) for Lesser Prairie-Chickens (*T. pallidicinctus*) in southwestern Kansas reported by Hagen et al. (2005), but lower than the rate of 0.55 (95% CI: 0.46–0.66) reported for Greater Prairie-Chickens in northeastern Kansas by Nooker and Sandercock (2008). The AHM framework provides the opportunity to identify research needs, and we initiated a field research project on our study site in 2006 which will soon provide site-specific estimates of annual survival. To incorporate uncertainty of parameter estimates into the AHM exercise, the annual model weights could be produced as the mean of repeated simulations with demographic parameters selected randomly from distributions. Such an approach could be especially important when significant uncertainty in parameter estimates exists.

Theoretically, long-distance immigration to our study area could keep populations higher than expected following harvest (Smith and Willebrand 1999). We did not have immigration data for our study population to include in a competing model, but it is possible to gather such data and assess this hypothesis in future model comparisons under the adaptive framework we describe. Nooker and Sandercock (2008) found high rates of between year lek specificity of breeding males in northeastern Kansas, but dispersal information for juveniles is lacking. However, our monitoring surveys were conducted over a multi-county area; our survey data provides no evidence that counties closer to Kansas have different trends than counties to the north. Juvenile immigration is unlikely to sustain the population of prairie chickens in our study area, as dispersal would have to occur at distances several orders of magnitude greater than the ca. 1-km mean dispersal distances of juveniles reported by Bowman and Robel (1977).

The AHM process allows a context for management discussion, and additional models could be added to our simple set of two competing models if other hypotheses were proposed by a stakeholder. For example, we would also consider adding density-independent models of precipitation effects on production, population limitation by carrying capacity if grassland habitat changed substantially in our study site, and density-dependent reproduction, as considered by NAWMP (NAWMP Committee 2004).

Harvest decisions can be complex and controversial. Connelly et al. (2003), Sedinger and

Rotella (2005), and Sedinger et al. (this volume, chapter 24) describe the uncertainties surrounding harvest of Greater Sage-Grouse in Idaho and Nevada; their problem is similar in scope, uncertainty, and landscape scale to our exercise. AHM is a formal mechanism to provide defendable criteria for decisions made under some level of uncertainty. The AHM framework is unique in its synthesis of survey and harvest data. Agencies using "monitor-and-modify" decision-making processes (Johnson 1999) usually have the type of data needed to implement AHM. We encourage wildlife managers to consider AHM as a process that can provide information about harvested populations of grouse.

ACKNOWLEDGMENTS

We acknowledge NGPC biologists M. Remund, B. Goracke, B. Seitz, J. Hoffman, J. Bruner, and P. Molini, who collected survey data used in our analyses. LAP was supported by Polytechnic of Namibia and a Fulbright award through the U.S. State Department during preparation of this manuscript. D. Tyre and C. Moore provided comments on early versions of our model structure. This research was supported by Hatch Act funds through the University of Nebraska Agricultural Research Division, Lincoln, Nebraska. The spreadsheet used for this exercise is available from LAP.

LITERATURE CITED

Anderson, D. R., and K. P. Burnham. 1976. Population ecology of the Mallard. VI. The effect of exploitation on survival. U.S. Fish and Wildlife Service, Resource Publication 128.

Bowman, T. J., and R. J. Robel. 1977. Brood break-up, dispersal, mobility, and mortality of juvenile prairie chickens. Journal of Wildlife Management 41:27–34.

Bradbury, J. W., S. L. Vehrencamp, and R. M. Gibson. 1989. Dispersion of displaying male sage grouse. 1. Environmental determinants of temporal variation. Behavioral Ecology and Sociobiology 24:1–14.

Connelly, J. W., K. P. Reese, E. O. Garton, and M. L. Commons-Kemner. 2003. Response of Greater Sage-Grouse Centrocercus urophasianus populations to different levels of exploitation in Idaho, USA. Wildlife Biology 9:335–340.

Cox, S. A., A. D. Peoples, S. J. DeMaso, J. J. Lusk, and F. S. Guthery. 2004. Survival and cause-specific mortality of Northern Bobwhites in western Oklahoma. Journal of Wildlife Management 68:663–671.

DeStefano, S., and D. H. Rusch. 1986. Harvest rates of Ruffed Grouse in northeastern Wisconsin. Journal of Wildlife Management 50:361–367.

Durbian, F. E., III, E. J. Finck, and R. D. Applegate. 1999. Greater Prairie-Chicken harvest in Kansas: early vs. regular seasons. Great Plains Research 9:87–94.

Ellison, L. N. 1991. Shooting and compensatory mortality in Tetraonids. Ornis Scandinavica 22:229–240.

Hagen, C. A., J. C. Pitman, B. K. Sandercock, R. J. Robel, and R. D. Applegate. 2005. Age-specific variation in apparent survival rates of male Lesser Prairie-Chickens. Condor 107:78–86.

Hamerstrom, F. N., Jr., and F. Hamerstrom. 1973. The prairie chicken in Wisconsin. Technical Bulletin No. 64. Department of Natural Resources, Madison, WI.

Johnson, B. L. 1999. The role of adaptive management as an operational approach for resource management agencies. Conservation Ecology 3(2):8. <http://www.consecol.org/vol3/iss2/art8/> (15 February 2009).

Johnson, F., and K. Williams. 1999. Protocol and practice in the adaptive management of waterfowl harvests. Conservation Ecology 3(1):8. <http://www.consecol.org/vol3/iss1/art8/> (15 February 2009).

Lyons, J. E., M. C. Runge, H. P. Laskowski, and W. L. Kendall. 2008. Monitoring in the context of structured decision-making and adaptive management. Journal of Wildlife Management 72:1683–1692.

Nichols, J. D., M. J. Conroy, D. R. Anderson, and K. P. Burnham. 1984. Compensatory mortality in waterfowl populations: a review of the evidence and implications for research and management. Transactions of the North American Wildlife and Natural Resources Conference 49:535–554.

Nooker, J. K., and B. K. Sandercock. 2008. Correlates and consequences of male mating success in lek-mating Greater Prairie-chickens (Tympanuchus cupido). Behavioral Ecology and Sociobiology 62:1377–1388.

North American Waterfowl Management Plan Committee. 2004. North American Waterfowl Management Plan 2004. Implementation framework: strengthening the biological foundation. Canadian Wildlife Service, U.S. Fish and Wildlife Service, and Secretaria de Medio Ambiente y Recursos Naturales.

Pedersen, H. C., H. Steen, L. Kastdalen, H. Brøseth, R. A. Ims, W. Svendsen, and N. G. Yoccoz. 2004. Weak compensation of harvest despite strong density-dependent growth in Willow Ptarmigan. Proceedings of the Royal Society of London, Series B 271:381–385.

Sedinger, B. S., J. S. Sedinger, S. Espinosa, M. T. Atamian, and E. J. Blomberg. Spatial-temporal variation in Greater Sage-Grouse survival in Nevada. Studies in Avian Biology 39:xxx–xxx.

Sedinger, J. S., and J. J. Rotella. 2005. Effect of harvest on sage-grouse *Centrocercus urophasianus* populations: what can we learn from the current data? Wildlife Biology 11:371–375.

Sit, V., and B. Taylor. 1998. Statistical methods for adaptive management studies. Land Management Handbook No. 42. Research Branch, B.C. Ministry of Forests, Victoria, BC.

Small, R. J., J. C. Holzwart, and D. H. Rusch. 1991. Predation and hunting mortality of Ruffed Grouse in central Wisconsin. Journal of Wildlife Management 55:512–520.

Smith, A., and T. Willebrand. 1999. Mortality causes and survival rates of hunted and unhunted willow grouse. Journal of Wildlife Management 63:722–730.

Taylor, J. S. 2000. Greater Prairie-Chicken in southeast Nebraska: an overview of population status and management considerations. Unpublished Report. Nebraska Game and Parks Commission, Lincoln, NE.

Williams, B. K., and F. A. Johnson. 1995. Adaptive management and the regulation of waterfowl harvests. Wildlife Society Bulletin 23:430–436.

Williams, B. K., J. D. Nichols, and M. J. Conroy. 2002. Analysis and management of animal populations. Academic Press, San Diego, CA.

Williams, B. K., R. C. Szaro, and C. D. Shapiro. 2007. Adaptive management: the U.S. Department of the Interior technical guide. U.S. Department of the Interior.

Wisdom, M. J., and L. S. Mills. 1997. Sensitivity analysis to guide population recovery: prairie-chickens as an example. Journal of Wildlife Management 61:302–312.

INDEX

greens, beach (*Honckenya peploides*), 298
ground net, 92, 269, 298
Grouse, Black (*Tetrao tetrix*)
    adult survival and population growth, 318
    natal dispersal distance, 98
Grouse, Blue [Dusky] (*Dendragapus obscurus*), 132
    natal dispersal distance, 97–98
Grouse, Caucasian (*Tetrao mlokosiewiczi*), 98
Grouse, Chinese (*Bonasa sewersowi*), 98
Grouse, Hazel (*Bonasa bonasia*)
    behavior pattern of juvenile dispersers, 93–98
    colonization rate, 90
    forest habitat characteristics, 91–92
    landscape structure and movement pattern, 93–94,
      99, 100
    natal dispersal and population dynamics in
      fragmented landscape, 89–103
    patch isolation, 90
    range expansion, 90
    sex ratio, 98
grouse, prairie (*Tympanuchus* spp.), 64
    cause of population declines, 4
    spatially explicit models, 3–19
Grouse, Red (*Lagopus lagopus scoticus*), 97, 162, 196, 198,
    205, 277, 286
    chick growth rate and survival, 162
    experimental testosterone implant and male
      behavior, 196
    monogamous mating system, 196
Grouse, Ruffed (*Bonasa umbellus*), 90, 308
    additive harvest mortality, 336
    compensatory harvest mortality, 318
    crippling loss rate, 331
    natal dispersal distance, 98
Grouse, Sharp-tailed (*Tympanuchus phasianellus*), 12, 14,
    60, 146
    distribution in Alberta, 34
    estimating lek occurrence and density, 33–49
    experimental testosterone implant and male
      behavior, 196
    historical distribution, 4
    lek-mating system, 196
    nest parasitism, 218
    population decline, 224
    weather and productivity, 183
Grouse, Siberian (*Falcipennis falcipennis*), 98
Grouse, Sooty (*Dendragapus fuliginosus*), 242
Grouse, Spruce (*Falcipennis canadensis*)
    natal dispersal distance, 98
Gull, Glaucous-winged (*Larus glaucescens*), 298, 302
guppy (*Poecilia reticulata*), 263

habitat
    edge density and population declines, 73
    factors selected by brood-rearing Greater Sage-Grouse
      and spatial scale, 152
    grass structure and Greater Sage-Grouse, 115
    litter accumulation and chick movement, 188
    litter accumulation and small mammal
      abundance, 188
    post-settlement conversion of grasslands, 180

shinnery oak and Lesser Prairie-Chicken, 229
    spatial configuration, 8
habitat fragmentation. *See* fragmentation
habitat models
    discrete choice analysis of brood macrohabitat, 182
    Greater Sage-Grouse, 155–64, 171–75
    index of complexity, 24
    index of continuity, 54
    landscape metrics for Greater Sage-Grouse brood
      success, 145 (table)
    landscape metrics for Greater Sage-Grouse nest
      success, 143 (table)
    lek habitats of Greater Prairie-Chicken, 21–32
    macrohabitat selection by Greater Prairie-Chicken
      broods, 182 (table)
    microhabitat selection by Greater Prairie-Chicken
      broods, 183–85
    nest and brood habitats for Greater Sage-Grouse,
      137–50
    nest and brood success for Greater Sage-Grouse, 142
      (table), 144 (table)
    resource selection analysis, 171–73
    spatially explicit for prairie grouse, 3–20
    suitability index for Greater Prsairie-Chicken,
      23–25
habitat monitoring and evaluation, 121–23, 132–34,
    154–55, 171
    effective height of vegetation, 129–31, 133
    estimation of canopy coverage, 171, 181
    horizontal cover, 120, 133
    line-intercept method, 154
    point-centered quarter method, 109, 171
    remote sensing, 154–55
    Robel pole, 109, 122, 154, 171
    vegetation droop height, 120–22
    visual obstruction reading (VOR), 181, 225
habitat selection. *See* habitat use
habitat suitability
    ecological niche modeling, 25
    limiting factor in Greater Prairie-Chicken population
      decline, 30
    modeling using leks, 21–30
habitat treatment. *See* management treatments
habitat use
    agricultural lands and Greater Prairie-Chicken, 72,
      186, 217, 262
    avoidance of burned areas by Greater
      Sage-Grouse, 326
    brood success of Greater Sage-Grouse, 151–67
    cover and Greater Sage-Grouse brood use, 174 (fig.)
    dwarf sagebrush and nesting of Greater Sage-Grouse,
      119–36
    habitat selection and Greater Prairie-Chicken brood
      survival, 179–91
    interspersion of cover types and use by Greater
      Sage-Grouse, 77–88
    sagebrush cover and winter use by Greater
      Sage-Grouse, 138
    seasonal for Greater Sage-Grouse, 86
    Sharp-tailed Grouse winter, 34
    thermoregulation and Greater Prairie-Chicken, 61

mating success
  comb size in Greater Prairie-Chicken, 205
  copulation frequency and distance to lek center for
      Greater-Prairie Chicken, 197
  male aggressive behavior in Greater Prairie-Chicken, 205
mating systems, 196
migration
  Greater Sage-Grouse, 80
  historical, 6
  partial, 6, 59
  reciprocal dispersal, 59
model
  daily nest survival for Greater Sage-Grouse, 113 (table)
  development, 5
  estimating lek occurrence and density for Sharp-tailed
      Grouse, 33–49
  estimation of renesting probabilities for Greater
      Prairie-Chicken, 216 (table)
  habitat selection and brood success for Greater
      Sage-Grouse, 157–58
  habitat selection and design II approach, 155
  harvest mortality and population size, 331–38
  hierarchical modeling of lek habitats for Greater
      Prairie-Chicken, 21–32
  metapopulation, 4
  predicting Greater Sage-Grouse nest sites, 113 (table)
  spatially explicit, 4
  species distribution, 4
  variables for Greater Sage-Grouse brood habitat use
      and survival, 156 (table)
molt
  movement by Greater Prairie-Chicken, 58
  White-tailed Ptarmigan, 286
monitoring techniques. See also marking techniques;
      radiotelemetry
Monte Carlo simulation
  anthropogenic features, 66
  distance to anthropogenic features, 69–70
  extinction probability for Rock Ptarmigan, 271, 274
  nesting habitat for Greater Sage-Grouse, 123
  survival of Greater Prairie-Chicken, 250
  White-tailed Ptarmigan, 288
morphometrics
  body mass of female Greater Prairie-Chicken by
      age-class, 260
  body mass of female Greater Sage-Grouse by
      age-class, 112
  index of body size, 258
mortality. See also harvest mortality; predation
  density dependent, 318
  Greater Prairie-Chicken chicks, 250–52
  risk and movement behavior, 60
mortality causes. See also harvest mortality; predation
  collision with ski-lift cable, 273, 277
  death in snow burrow, 273
  fence collision, 6, 257
  hypothermia, 251
  power lines, 6
  predation, 96, 251, 263, 273
  recreational hunting, 307–15
  vehicle collision, 70

movement. See also brood movement; dispersal;
      migration
  costs and benefits, 60–61
  daily and seasonal pattern for Greater Prairie-Chicken,
      55–59
  daily for Hazel Grouse, 93
  distance for brood-rearing Greater Sage-Grouse, 155
  distance for Greater Prairie-Chicken, 6, 184
  distance for Hazel Grouse broods, 97
  effect of habitat fragmentation on Hazel Grouse, 99
  elevation and Evermann's Rock Ptarmigan
      broods, 302
  endogenous and exogenous factors, 52
  estimation biases, 53
  female behavior after disappearance of chicks, 251
  female Lesser Prairie-Chicken during breeding
      season, 71, 228
  historical migration of Greater Prairie-Chicken, 6
  methodology for analyses, 53–54
  pattern and ambient temperature, 56
  patterns and phases of dispersal for Hazel Grouse, 93
  photoperiod, 52
  post-release for translocated Evermann's Rock
      Ptarmigan, 300
  ptarmigan in the Aleutian Islands, 297
  seasonal for Greater Sage-Grouse, 80
  thermoregulation, 58
multinomial discrete choice model, 200, 202
multi-response permutation procedures, 109

natal dispersal. See dispersal
needle-and-thread (Stipa comata), 120, 129, 131, 133
needlegrass (Nassella spp.), 170
needlegrass, green (Nassella viridula), 108
nest abandonment, 121, 213, 225–28, 271
nest initiation. See also renesting
  drought and Lesser Prairie-Chicken, 228
  follicular development in Greater Sage-Grouse, 114
  latitudinal variation for Greater Prairie-Chicken, 217
  rate for Greater Sage-Grouse, 112, 114
  rate for Lesser Prairie-Chicken, 226
  timing and weather, 217
  timing for Attwater's Prairie-Chicken, 217
  timing for Greater Prairie-Chicken, 214, 217
  timing for Greater Sage-Grouse by nest order, 112
nest parasitism, 6, 216, 218
nest site characteristics
  concealment cover, 132, 217
  cover type and nest success for Greater
      Sage-Grouse, 141
  distance to lek for Greater Sage-Grouse, 112
  dwarf sagebrush habitat and Greater Sage-Grouse,
      130 (table)
  Evermann's Rock Ptarmigan, 300–2
  Greater Prairie-Chicken, 22
  Greater Sage-Grouse, 111 (table), 113–15, 124–34,
      138, 141
  influence of anthropogenic disturbance on Lesser
      Prairie-Chicken, 71–72
  Lesser Prairie-Chicken, 226, 228–29
  nest fate for Lesser Prairie-Chicken, 227

26. Sogge, M. K., B. E. Kus, S. J. Sferra, and M. J. Whitfield, editors. 2003. *Ecology and Conservation of the Willow Flycatcher.*

27. Shuford, W. D. and K. C. Molina, editors. 2004. *Ecology and Conservation of Birds of the Salton Sink: An Endangered Ecosystem.*

28. Carmen, W. J. 2004. *Noncooperative Breeding in the California Scrub-Jay.*

29. Ralph, C. J. and E. H. Dunn, editors. 2004. *Monitoring Bird Populations Using Mist Nets.*

30. Saab, V. A. and H. D. W. Powell, editors. 2005. *Fire and Avian Ecology in North America.*

31. Morrison, M. L. editor. 2006. *The Northern Goshawk: A Technical Assessment of Its Status, Ecology, and Management.*

32. Greenberg, R., J. E. Maldonado, S. Droege, and M.V. McDonald, editors. 2006. *Terrestrial Vertebrates of Tidal Marshes: Evolution, Ecology, and Conservation.*

33. Mason, J. W., G. J. McChesney, W. R. McIver, H. R. Carter, J. Y. Takekawa, R. T. Golightly, J. T. Ackerman, D. L. Orthmeyer, W. M. Perry, J. L. Yee, M. O. Pierson, and M. D. McCrary. 2007. *At-Sea Distribution and Abundance of Seabirds off Southern California: A 20-Year Comparison.*

34. Jones, S. L. and G. R. Geupel, editors. 2007. *Beyond Mayfield: Measurements of Nest-Survival Data.*

35. Spear, L. B., D. G. Ainley, and W. A. Walker. 2007. *Foraging Dynamics of Seabirds in the Eastern Tropical Pacific Ocean.*

36. Niles, L. J., H. P. Sitters, A. D. Dey, P. W. Atkinson, A. J. Baker, K. A. Bennett, R. Carmona, K. E. Clark, N. A. Clark, C. Espoz, P. M. González, B. A. Harrington, D. E. Hernández, K. S. Kalasz, R. G. Lathrop, R. N. Matus, C. D. T. Minton, R. I. G. Morrison, M. K. Peck, W. Pitts, R. A. Robinson, and I. L. Serrano. 2008. *Status of the Red Knot* (Calidris canutus rufa) *in the Western Hemisphere.*

37. Ruth, J. M., T. Brush, and D. J. Krueper, editors. 2008. *Birds of the US-Mexico Borderland: Distribution, Ecology, and Conservation.*

38. Knick, S. T. and J. W. Connelly, editors. 2011. *Ecology and Conservation of Greater Sage-Grouse: A Landscape Species and its Habitats.*

39. Sandercock, B. K., K. Martin, and G. Segelbacher, editors. 2011. *Ecology, Conservation, and Management of Grouse.*

Indexer:	Leslie A. Robb
Composition:	MPS Limited, a Macmillan Company
Text:	Scala
Display:	Scala Sans
Printer and Binder:	Thomson-Shore Printing

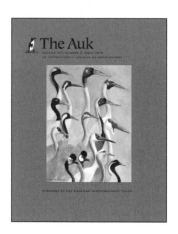